Mathematical and Physical Fundamentals of Climate Change

Mathematical and Physical Fundamentals of Climate Change

Zhihua Zhang
Beijing Normal University, China

John C. Moore
University of Lapland, Finland & Beijing Normal University, China

ELSEVIER AMSTERDAM • BOSTON • HEIDELBERG • LONDON • NEW YORK • OXFORD
PARIS • SAN DIEGO • SAN FRANCISCO • SINGAPORE • SYDNEY • TOKYO

Elsevier
Radarweg 29, PO Box 211, 1000 AE Amsterdam, Netherlands
The Boulevard, Langford Lane, Kidlington, Oxford OX5 1GB, UK
225 Wyman Street, Waltham, MA 02451, USA

Notices
Knowledge and best practice in this field are constantly changing. As new research and experience
broaden our understanding, changes in research methods, professional practices, or medical
treatment may become necessary.

Practitioners and researchers must always rely on their own experience and knowledge in
evaluating and using any information, methods, compounds, or experiments described herein. In
using such information or methods they should be mindful of their own safety and the safety of
others, including parties for whom they have a professional responsibility.

To the fullest extent of the law, neither the Publisher nor the authors, contributors, or editors,
assume any liability for any injury and/or damage to persons or property as a matter of products
liability, negligence or otherwise, or from any use or operation of any methods,
products,instructions, or ideas contained in the material herein.

ISBN: 978-0-12-800066-3

British Library Cataloguing-in-Publication Data
A catalogue record for this book is available from the British Library

Library of Congress Cataloging-in-Publication Data
A catalog record for this book is available from the Library of Congress

For information on all Elsevier publications
visit our web site at http://store.elsevier.com/

Working together
to grow libraries in
developing countries

www.elsevier.com • www.bookaid.org

Contents

6. Empirical Orthogonal Functions

7. Random Processes and Power Spectra

8. Autoregressive Moving Average Models

9. Data Assimilation

10. Fluid Dynamics

Preface: Interdisciplinary Approaches to Climate Change Research

Climate change is now widely recognized as the major environmental problem facing human societies. Its impacts and costs will be large, serious, and unevenly spread. Owing to the observed increases in temperature, decreases in snow and ice extent, and increases in sea level, global warming is unequivocal.

The main factor causing climate change and global warming is the increase of global carbon dioxide emissions. The Fourth Assessment Report (2007) of the Intergovernmental Panel on Climate Change of the United Nations indicated that most of the observed warming over the last 50 years is likely to have been due to the increasing concentrations of greenhouse gases produced by human activities such as deforestation and burning fossil fuels. This conclusion was made even stronger by the Fifth Assessment Report released in 2013. The concentration of carbon dioxide in the atmosphere increased from a preindustrial value of about 280 to 391 ppm in 2011. Continued increases in carbon dioxide emissions will cause further warming and induce many changes in the global climate system. It is likely that global warming will exceed 2 °C this century unless global carbon dioxide emissions are cut by at least 50% of the 1990 levels by 2050, and by much more thereafter.

In current climate change research, scientists exploit various complicated techniques in order to squeeze useful information out of the available observation data, unravel the causes of climate change, identify significant changes in the climate, interpret the properties of the associated variability, deal with extreme climate events, and make predictions about the future climate.

This book covers the comprehensive range of mathematical and physical techniques used widely in climate change research. The main topics include signal processing, time-frequency analysis, data analysis, statistical diagnosis, power spectra, autoregressive moving average models, data assimilation, atmospheric dynamics, oceanic dynamics, glaciers and sea level rise, and Earth system modeling. This book is self-contained, assuming only a basic knowledge of calculus. Much of the latest research is also included. Various theories and algorithms in this book are used widely not only in climate change research, but also in geoscience and applied science. This book will be of great value to researchers and advanced students in a wide range of disciplines. Researchers in and students of meteorology, climatology, oceanography, and environmental science can grasp advanced mathematical and physical methods used in climate change research and geoscience, and researchers in and students of applied mathematics, statistics, physics, computer science, and electrical engineering can learn how to use advanced mathematical and physical methods in climate change research, geoscience, and applied science.

Please find the companion website at http://booksite.elsevier.com/9780128000663

Chapter 1

Fourier Analysis

Motivated by the study of heat diffusion, Joseph Fourier claimed that any periodic signals can be represented as a series of harmonically related sinusoids. Fourier's idea has a profound impact in geoscience. It took one and a half centuries to complete the theory of Fourier analysis. The richness of the theory makes it suitable for a wide range of applications such as climatic time series analysis, numerical atmospheric and ocean modeling, and climatic data mining.

1.1 FOURIER SERIES AND FOURIER TRANSFORM

Assume that a system of functions $\{\varphi_n(t)\}_{n \in \mathbb{Z}_+}$ in a closed interval $[a, b]$ satisfies $\int_a^b |\varphi_n(t)|^2 \, dt < \infty$. If

$$\int_a^b \varphi_n(t) \overline{\varphi}_m(t) \, dt = \begin{cases} 0 & (n \neq m), \\ 1 & (n = m), \end{cases}$$

and there does not exist a nonzero function f such that

$$\int_a^b |f(t)|^2 \, dt < \infty, \quad \int_a^b f(t) \overline{\varphi}_n(t) \, dt = 0 \quad (n \in \mathbb{Z}_+),$$

then this system is said to be an *orthonormal basis* in the interval $[a, b]$.

For example, the trigonometric system $\{\frac{1}{\sqrt{2\pi}}, \frac{1}{\sqrt{\pi}} \cos(nt), \frac{1}{\sqrt{\pi}} \sin(nt)\}_{n \in \mathbb{Z}_+}$ and the exponential system $\{\frac{1}{\sqrt{2\pi}} e^{int}\}_{n \in \mathbb{Z}}$ are both orthonormal bases in $[-\pi, \pi]$.

Let $f(t)$ be a periodic signal with period 2π and be integrable over $[-\pi, \pi]$, write $f \in L_{2\pi}$. In terms of the above orthogonal basis, let $a_0(f) = \frac{1}{\pi} \int_{-\pi}^{\pi} f(t) \, dt$ and

$$a_n(f) = \frac{1}{\pi} \int_{-\pi}^{\pi} f(t) \cos(nt) \, dt \quad (n \in \mathbb{Z}_+),$$

$$b_n(f) = \frac{1}{\pi} \int_{-\pi}^{\pi} f(t) \sin(nt) \, dt \quad (n \in \mathbb{Z}_+).$$

Then $a_0(f), a_n(f), b_n(f) (n \in \mathbb{Z}_+)$ are said to be *Fourier coefficients* of f. The series

$$\frac{a_0(f)}{2} + \sum_{1}^{\infty} (a_n(f) \cos(nt) + b_n(f) \sin(nt))$$

Mathematical and Physical Fundamentals of Climate Change

is said to be the *Fourier series* of f. The sum

$$S_n(f;t) := \frac{a_0(f)}{2} + \sum_{1}^{n}(a_k(f)\cos(kt) + b_k(f)\sin(kt))$$

is said to be the *partial sum* of the Fourier series of f. It can be rewritten in the form

$$S_n(f;t) = \sum_{-n}^{n} c_k(f)e^{ikt},$$

where

$$c_k(f) = \frac{1}{2\pi} \int_{-\pi}^{\pi} f(t)e^{-ikt}\,dt \quad (k \in \mathbb{Z})$$

are also called the Fourier coefficients of f.

It is clear that these Fourier coefficients satisfy

$$a_0(f) = 2c_0(f), \quad a_n(f) = c_{-n}(f) + c_n(f), \quad b_n(f) = i(c_{-n}(f) - c_n(f)).$$

Let $f \in L_{2\pi}$. If f is a real signal, then its Fourier coefficients $a_n(f)$ and $b_n(f)$ must be real. The identity

$$a_n(f)\cos(nt) + b_n(f)\sin(nt) = A_n(f)\sin(nt + \theta_n(f))$$

shows that the general term in the Fourier series of f is a sine wave with circle frequency n, amplitude A_n, and initial phase θ_n. Therefore, the Fourier series of a real periodic signal is composed of sine waves with different frequencies and different phases.

Fourier coefficients have the following well-known properties.

Property. Let $f, g \in L_{2\pi}$ and α, β be complex numbers.

(i) (Linearity). $c_n(\alpha f + \beta g) = \alpha c_n(f) + \beta c_n(g)$.

(ii) (Translation). Let $F(t) = f(t + \alpha)$. Then $c_n(F) = e^{in\alpha}c_n(f)$.

(iii) (Integration). Let $F(t) = \int_0^t f(u)\,du$. If $\int_{-\pi}^{\pi} f(t)\,dt = 0$, then $c_n(F) = \frac{c_n(f)}{in}$ $(n \neq 0)$.

(iv) (Derivative). If $f(t)$ is continuously differentiable, then $c_n(f') = inc_n(f)$ $(n \neq 0)$.

(v) (Convolution). Let the convolution $(f * g)(t) = \int_{-\pi}^{\pi} f(t - x)g(x)\,dx$. Then $c_n(f * g) = 2\pi c_n(f)c_n(g)$.

Proof. Here we prove only (v). It is clear that $f * g \in L_{2\pi}$ and

$$c_n(f * g) = \frac{1}{2\pi} \int_{-\pi}^{\pi} (f * g)(t)e^{-int}\,dt = \frac{1}{2\pi} \int_{-\pi}^{\pi} \left(\int_{-\pi}^{\pi} f(t - u)g(u)\,du \right) e^{-int}\,dt.$$

Interchanging the order of integrals, we get

$$c_n(f * g) = \frac{1}{2\pi} \int_{-\pi}^{\pi} \left(\int_{-\pi}^{\pi} f(t - u)e^{-int}\,dt \right) g(u)\,du.$$

Let $v = t - u$. Since $f(v)e^{-inv}$ is a periodic function with period 2π, the integral in brackets is

$$\int_{-\pi}^{\pi} f(t-u)e^{-int}\, dt = e^{-inu} \int_{-\pi-u}^{\pi-u} f(v)e^{-inv}\, dv$$

$$= e^{-inu} \int_{-\pi}^{\pi} f(v)e^{-inv}\, dv = 2\pi c_n(f)e^{-inu}.$$

Therefore,

$$c_n(f * g) = c_n(f) \int_{-\pi}^{\pi} g(u)e^{-inu}\, du = 2\pi c_n(f)c_n(g).$$

Throughout this book, the notation $f \in L(\mathbb{R})$ means that f is integrable over $\mathbb{R}$ and the notation $f \in L[a, b]$ means that $f(t)$ is integrable over a closed interval $[a, b]$, and the integral $\int_{\mathbb{R}} = \int_{-\infty}^{\infty}$. ☐

Riemann-Lebesgue Lemma. *If* $f \in L(\mathbb{R})$, *then* $\int_{\mathbb{R}} f(t)e^{-i\omega t}\, dt \to 0$ *as* $|\omega| \to \infty$. *Especially*,

(i) *if* $f \in L[a, b]$, *then* $\int_a^b f(t)e^{-i\omega t}\, dt \to 0(|\omega| \to \infty)$;
(ii) *if* $f \in L_{2\pi}$, *then* $c_n(f) \to 0(|n| \to \infty)$ *and* $a_n(f) \to 0$, $b_n(f) \to 0(n \to \infty)$.

The Riemann-Lebesgue lemma (ii) states that Fourier coefficients of $f \in L_{2\pi}$ *tend to zero as* $n \to \infty$.

Proof. If f is a simple step function and

$$f(t) = \begin{cases} c, & a \le t \le b, \\ 0, & \text{otherwise,} \end{cases}$$

where c is a constant, then

$$\left| \int_{\mathbb{R}} f(t)e^{-i\omega t}\, dt \right| = \left| \int_a^b ce^{-i\omega t}\, dt \right| = \left| \frac{c}{i\omega}(e^{-ib\omega} - e^{-ia\omega}) \right| \le 2\left| \frac{c}{\omega} \right| \quad (\omega \ne 0),$$

and so $\int_{\mathbb{R}} f(t)e^{-i\omega t}\, dt \to 0(|\omega| \to \infty)$. Similarly, it is easy to prove that for any step function $s(t)$,

$$\int_{\mathbb{R}} s(t)e^{-i\omega t}\, dt \to 0 \quad (|\omega| \to \infty).$$

If f is integrable over $\mathbb{R}$, then, for $\epsilon > 0$, there exists a step function $s(t)$ such that

$$\int_{\mathbb{R}} |f(t) - s(t)|\, dt < \epsilon.$$

Since $s(t)$ is a step function, for the above ϵ, there exists an N such that

$$\left| \int_{\mathbb{R}} s(t)e^{-i\omega t}\, dt \right| < \epsilon \quad (|\omega| > N).$$

From this and $|e^{-i\omega t}| \leq 1$, it follows that

$$\left| \int_{\mathbb{R}} f(t) e^{-i\omega t} \, dt \right| \leq \int_{\mathbb{R}} |f(t) - s(t)| \, dt + \left| \int_{\mathbb{R}} s(t) e^{-i\omega t} \, dt \right| < 2\epsilon \quad (|\omega| > N),$$

i.e., $\int_{\mathbb{R}} f(t) e^{-i\omega t} \, dt \to 0 (|\omega| \to \infty)$.

Especially, if $f \in L[a, b]$, take

$$F(t) = \begin{cases} f(t), & a \leq t \leq b, \\ 0, & \text{otherwise.} \end{cases}$$

Then $F \in L(\mathbb{R})$, and so $\int_{\mathbb{R}} F(t) e^{-i\omega t} \, dt \to 0 (|\omega| \to \infty)$. From

$$\int_{\mathbb{R}} F(t) e^{-i\omega t} \, dt = \int_a^b f(t) e^{-i\omega t} \, dt,$$

it follows that $\int_a^b f(t) e^{-i\omega t} \, dt \to 0 (|\omega| \to \infty)$.

Take $a = -\pi$, $b = \pi$, and $\omega = n$. Then $\int_{-\pi}^{\pi} f(t) e^{-int} \, dt \to 0$ as $|n| \to \infty$, i.e.,

$$c_n(f) \to 0 \quad (|n| \to \infty).$$

Combining this with $a_n(f) = c_{-n}(f) + c_n(f)$ and $b_n(f) = i(c_{-n}(f) - c_n(f))$, we get

$$a_n(f) \to 0, \quad b_n(f) \to 0 \quad (n \to \infty).$$

$\square$

The partial sums of Fourier series can be written in an integral form as follows.

By the definition of Fourier coefficients,

$$S_n(f; t) = \sum_{-n}^{n} c_k(f) e^{ikt} = \sum_{-n}^{n} \left(\frac{1}{2\pi} \int_{-\pi}^{\pi} f(u) e^{-iku} \, du \right) e^{ikt}$$

$$= \int_{-\pi}^{\pi} f(u) \left(\frac{1}{2\pi} \sum_{-n}^{n} e^{ik(t-u)} \right) du.$$

Let $v = t - u$. Then

$$S_n(f; t) = \int_{-\pi}^{\pi} f(t - v) D_n(v) \, dv, \tag{1.1}$$

where $D_n(v) = \frac{1}{2\pi} \sum_{-n}^{n} e^{ikv}$ and is called the *Dirichlet kernel*.

The Dirichlet kernel possesses the following properties:

(i) $D_n(-v) = D_n(v)$, i.e., the Dirichlet kernel is an even function.
(ii) $D_n(v + 2\pi) = D_n(v)$, i.e., the Dirichlet kernel is a periodic function with period 2π.

(iii) $D_n(v) = \dfrac{\sin\left(n+\frac{1}{2}\right)v}{2\pi \sin \frac{v}{2}}$. This is because

$$D_n(v) = \frac{1}{2\pi} \sum_{-n}^{n} e^{ikv} = \frac{e^{-inv} - e^{i(n+1)v}}{2\pi(1 - e^{iv})} = \frac{\sin\left(n + \frac{1}{2}\right)v}{2\pi \sin \frac{v}{2}}.$$

(iv) $\int_{-\pi}^{\pi} D_n(v)\,dv = 1$. This is because

$$\int_{-\pi}^{\pi} D_n(v)\,dv = \int_{-\pi}^{\pi} \left(\frac{1}{2\pi} \sum_{-n}^{n} e^{ikv}\right) dv = \frac{1}{2\pi} \sum_{-n}^{n} \left(\int_{-\pi}^{\pi} e^{ikv}\,dv\right) = 1.$$

We will give the Jordan criterion for Fourier series. Its proof needs the following proposition.

Proposition 1.1. *For any real numbers a and b, the following inequality holds:*

$$\left| \int_{a}^{b} \frac{\sin u}{u}\,du \right| \leq 6.$$

Proof. When $1 \leq a \leq b$, by the second mean-value theorem for integrals, there exists a $\xi (a \leq \xi \leq b)$ such that

$$\left| \int_{a}^{b} \frac{\sin u}{u}\,du \right| = \frac{1}{a} \left| \int_{a}^{\xi} \sin u\,du \right| \leq 2.$$

When $0 \leq a \leq b \leq 1$, with use of the inequality $|\sin u| \leq |u|$, it follows that

$$\left| \int_{a}^{b} \frac{\sin u}{u}\,du \right| \leq \int_{a}^{b} \left| \frac{\sin u}{u} \right| du \leq 1.$$

When $0 \leq a \leq 1 \leq b$,

$$\left| \int_{a}^{b} \frac{\sin u}{u}\,du \right| \leq \left| \int_{a}^{1} \frac{\sin u}{u}\,du \right| + \left| \int_{1}^{b} \frac{\sin u}{u}\,du \right| \leq 3.$$

Noticing that $\frac{\sin u}{u}$ is a even function, it can easily prove that for all cases of real numbers a and b,

$$\left| \int_{a}^{b} \frac{\sin u}{u}\,du \right| \leq 6.$$

$\square$

If a signal is the difference of two monotone increasing signals in an interval, then this signal is called a signal of *bounded variation* in this interval. Almost all geophysical signals are signals of bounded variation.

Jordan Criterion. *Suppose that a signal $f \in L_{2\pi}$ is of bounded variation in $(t - \eta, t + \eta), \eta > 0$. Then the partial sums of the Fourier series of f*

$$S_n(f;t) \to \frac{1}{2}(f(t+0) + f(t-0)) \quad (n \to \infty) \quad \text{at} \, t.$$

Proof. The assumption that $f(t)$ is of bounded variation in $(t-\eta, t+\eta)$ shows that $f(t+0)$ and $f(t-0)$ exist. By (1.1) and the properties of Dirichlet kernel, it follows that

$$S_n(f;t) - \frac{1}{2}(f(t+0) + f(t-0)) = \int_{-\pi}^{\pi} \left\{ f(t-v) - \frac{1}{2}(f(t+0) + f(t-0)) \right\}$$

$$D_n(v)\, dv = \frac{1}{\pi} \int_0^{\pi} \psi_t(v) \frac{\sin\left(n + \frac{1}{2}\right)v}{2\sin\frac{v}{2}}\, dv,$$

where $\psi_t(v) = f(t+v) + f(t-v) - f(t+0) - f(t-0)$. It is clear that

$$\frac{\sin\left(n + \frac{1}{2}\right)v}{2\sin\frac{v}{2}} = \frac{1}{v}\sin(nv) + \left(\frac{1}{2}\coth\frac{v}{2} - \frac{1}{v}\right)\sin(nv) + \frac{1}{2}\cos(nv).$$

Therefore,

$$S_n(f;t) - \frac{1}{2}(f(t+0) + f(t-0)) = \frac{1}{\pi}\int_0^{\pi} \psi_t(v)\frac{1}{v}\sin(nv)\, dv$$

$$+ \frac{1}{\pi}\int_0^{\pi} \psi_t(v)\left(\frac{1}{2}\coth\frac{v}{2} - \frac{1}{v}\right)\sin(nv)\, dv$$

$$+ \frac{1}{\pi}\int_0^{\pi} \psi_t(v)\frac{1}{2}\cos(nv)\, dv. \tag{1.2}$$

Note that $\frac{\psi_t(v)}{v} \in L[\delta, \pi]$. Here δ will be determined, $\psi_t(v)\left(\frac{1}{2}\coth\frac{v}{2} - \frac{1}{v}\right) \in L[0, \pi]$, and $\psi_t(v) \in L[0, \pi]$. By Riemann-Lebesgue Lemma, it follows that

$$\int_{\delta}^{\pi} \frac{\psi_t(v)}{v}\sin(nv)\, dv \to 0 \quad (n \to \infty),$$

$$\int_0^{\pi} \psi_t(v)\left(\frac{1}{2}\coth\frac{v}{2} - \frac{1}{v}\right)\sin(nv)\, dv \to 0 \quad (n \to \infty),$$

$$\int_0^{\pi} \psi_t(v)\cos(nv)\, dv \to 0 \quad (n \to \infty).$$

Combining this with (1.2), we get

$$S_n(f;t) - \frac{1}{2}(f(t+0) + f(t-0)) - \frac{1}{\pi}\int_0^{\delta} \psi_t(v)\frac{1}{v}\sin(nv)\, dv \to 0 \quad (n \to \infty), \tag{1.3}$$

where $\psi_t(v) = f(t+v) + f(t-v) - f(t+0) - f(t-0)$.

Since $\psi_t(v)$ is of bounded variation in $(-\eta, \eta)$ and $\psi_t(0+0) = 0$, there exist two monotone increasing functions $h_1(v)$ and $h_2(v)$ satisfying $h_1(0+0) = h_2(0+0) = 0$ such that

$$\psi_t(v) = h_1(v) - h_2(v).$$

Since $h_1(0 + 0) = h_2(0 + 0) = 0$, for any given $\epsilon > 0$, there is a $\delta(0 < \delta < \pi)$ such that

$$0 \le h_1(v) \le \epsilon, \quad 0 \le h_2(v) \le \epsilon \quad (0 < v \le \delta).$$

For the fixed δ, by (1.3), there exists an N such that

$$\left| S_n(f; t) - \frac{1}{2}(f(t + 0) + f(t - 0)) - \frac{1}{\pi} \int_0^\delta h_1(v) \frac{\sin(nv)}{v} dv \right.$$
$$\left. + \frac{1}{\pi} \int_0^\delta h_2(v) \frac{\sin(nv)}{v} dv \right| < \epsilon \quad (n \ge N),$$

and so

$$\left| S_n(f; t) - \frac{1}{2}(f(t + 0) + f(t - 0)) \right| \le \left| \frac{1}{\pi} \int_0^\delta h_1(v) \frac{\sin(nv)}{v} dv \right|$$
$$+ \left| \frac{1}{\pi} \int_0^\delta h_2(v) \frac{\sin(nv)}{v} dv \right| + \epsilon \quad (n \ge N).$$

However, using the second mean-value theorem, there exist $\zeta_i(0 < \zeta_i < \delta)$ such that

$$\frac{1}{\pi} \int_0^\delta h_i(v) \frac{\sin(nv)}{v} dv = \frac{1}{\pi} h_i(\delta) \int_{\zeta_i}^\delta \frac{\sin(nv)}{v} dv \quad (i = 1, 2),$$

and by Proposition 1.1,

$$\left| \frac{1}{\pi} \int_0^\delta h_i(v) \frac{\sin(nv)}{v} dv \right| = \left| \frac{1}{\pi} h_i(\delta) \int_{\zeta_i}^\delta \frac{\sin(nv)}{v} dv \right|$$
$$\le \frac{\epsilon}{\pi} \left| \int_{n\zeta_i}^{n\delta} \frac{\sin v}{v} dv \right| \le \frac{6\epsilon}{\pi} \quad (i = 1, 2).$$

Therefore,

$$\left| S_n(f; t) - \frac{1}{2}(f(t + 0) + f(t - 0)) \right| \le \left(\frac{12}{\pi} + 1 \right) \epsilon \quad (n \ge N),$$

i.e., $S_n(f; t) \to \frac{1}{2}(f(t + 0) + f(t - 0))(n \to \infty)$ at t. $\qquad\qquad \square$

In general, let $f(t) \in L[-\frac{T}{2}, \frac{T}{2}]$ be a periodic function with period T. Then its Fourier series is

$$\frac{a_0(f)}{2} + \sum_1^\infty \left(a_n(f) \cos \frac{2n\pi t}{T} + b_n(f) \sin \frac{2n\pi t}{T} \right),$$

where the Fourier coefficients are

$$a_0(f) = \frac{2}{T} \int_{-T/2}^{T/2} f(t)\, dt,$$

$$a_n(f) = \frac{2}{T} \int_{-T/2}^{T/2} f(t) \cos \frac{2n\pi t}{T}\, dt \quad (n \in \mathbb{Z}_+),$$

and

$$b_n(f) = \frac{2}{T} \int_{-T/2}^{T/2} f(t) \sin \frac{2n\pi t}{T}\, dt \quad (n \in \mathbb{Z}_+).$$

An orthogonal basis and an orthogonal series on $[-1, 1]$ used often are stated as follows.

Denote Legendre polynomials by $X_n(t)$ $(n = 0, 1, \ldots)$:

$$X_n(t) = \frac{1}{2^n n!} \frac{d^n (t^2 - 1)^n}{dt^n} \quad (n = 0, 1, \ldots).$$

Especially, $X_0(t) = 1$, $X_1(t) = t$, and $X_2(t) = \frac{3}{2}t^2 - \frac{1}{2}$.

By use of Leibnitz's formula, the Legendre polynomials are

$$X_n(t) = \frac{1}{2^n n!} \left\{ (t-1)^n \frac{d^n (t+1)^n}{dt^n} + C_n^1 n(t-1)^{n-1} \frac{d^{n-1}(t+1)^n}{dt^{n-1}} \right.$$

$$\left. + \cdots + C_n^n n! (t+1)^n \right\},$$

where $C_n^k = \frac{n!}{k!(n-k)!}$. Let $t = 1$ and $t = -1$. Then

$$X_n(1) = 1, \quad X_n(-1) = (-1)^n \quad (n = 0, 1, 2, \ldots).$$

Legendre polynomials possess the property:

$$\int_{-1}^{1} X_n(t) X_m(t)\, dt = \begin{cases} 0, & n \neq m, \\ \frac{2}{2n+1}, & n = m. \end{cases}$$

So Legendre polynomials conform to an orthogonal basis on the interval $[-1, 1]$. In terms of this orthogonal basis, any signal f of finite energy on $[-1, 1]$ can be expanded into a Legendre series $\sum_0^\infty l_n X_n(t)$, where

$$l_n = \frac{2n+1}{2} \int_{-1}^{1} f(t) X_n(t)\, dt.$$

The coefficients l_n are called *Legendre coefficients*.

Now we turn to introduce the concept of the Fourier transform. Suppose that $f \in L(\mathbb{R})$. The integral

$$\widehat{f}(\omega) := \int_{\mathbb{R}} f(t) e^{-it\omega}\, dt \quad (\omega \in \mathbb{R})$$

is called the *Fourier transform* of f. Suppose that $\widehat{f} \in L(\mathbb{R})$. The integral

$$\frac{1}{2\pi} \int_{\mathbb{R}} \widehat{f}(\omega) e^{it\omega} \, d\omega \quad (t \in \mathbb{R})$$

is called the *inverse Fourier transform*. Suppose that $f \in L(\mathbb{R})$ and $\widehat{f} \in L(\mathbb{R})$. It can be proved easily that

$$\frac{1}{2\pi} \int_{\mathbb{R}} \widehat{f}(\omega) e^{i\omega t} \, d\omega = f(t).$$

Theorem 1.1. *Let* $f \in L(\mathbb{R})$. *Then*

(i) $\lim_{|\omega| \to \infty} \widehat{f}(\omega) = 0$,
(ii) $|\widehat{f}(\omega)| \leq \int_{\mathbb{R}} |f(t)| \, dt =: \| f \|_1$,
(iii) $\widehat{f}(\omega)$ *is continuous uniformly on* $\mathbb{R}$.

Proof. The first conclusion is just the Riemann-Lebesgue lemma. It follows from the definition that

$$|\widehat{f}(\omega)| = \left| \int_{\mathbb{R}} f(t) e^{-i\omega t} \, dt \right| \leq \int_{\mathbb{R}} |f(t)| \, dt = \| f \|_1 \, .$$

Since

$$|\widehat{f}(\omega + h) - \widehat{f}(\omega)| \leq \int_{\mathbb{R}} |f(t)| |e^{-iht} - 1| \, dt,$$

with use of the dominated convergence theorem, it follows that for any $\omega \in \mathbb{R}$,

$$\lim_{h \to 0} |\widehat{f}(\omega + h) - \widehat{f}(\omega)| \leq \int_{\mathbb{R}} |f(t)| \left(\lim_{h \to 0} |e^{-iht} - 1| \right) \, dt = 0,$$

i.e., $\widehat{f}(\omega)$ is continuous uniformly on $\mathbb{R}$. $\qquad\square$

Fourier transforms have the following properties.
Property. Let $f, g \in L(\mathbb{R})$. Then

(i) (Linearity). $(\alpha f + \beta g)^{\wedge}(\omega) = \alpha \widehat{f}(\omega) + \beta \widehat{g}(\omega)$, where α, β be constants.
(ii) (Dilation). $(D_a f)^{\wedge}(\omega) = \frac{1}{|a|} \widehat{f}\left(\frac{\omega}{a}\right)$ $(a \neq 0)$, where $D_a f = f(at)$ is the *dilation operator*.
(iii) (Translation). $(T_\alpha f)^{\wedge}(\omega) = \widehat{f}(\omega) e^{-i\omega\alpha}$, where $T_\alpha f = f(t - \alpha)$ is the *translation operator*.
(iv) (Modulation and conjugate). $\left(f(t) e^{i\alpha t}\right)^{\wedge}(\omega) = \widehat{f}(\omega - \alpha)$, $\widehat{\overline{f}}(\omega) = \overline{\widehat{f}(-\omega)}$.
(v) (Symmetry). If $\widehat{f} \in L(\mathbb{R})$, then $\widehat{\widehat{f}}(t) = 2\pi f(-t)$.
(vi) (Time derivative). If $f^{(j)} \in L(\mathbb{R})(j = 1, \ldots, n)$, then $\widehat{f^{(n)}}(\omega) = (i\omega)^n \widehat{f}(\omega)$.
(vii) (Convolution in time). Let the convolution $(f * g)(t) = \int_{\mathbb{R}} f(t - u) g(u) \, du$. Then

$$(f * g)^{\wedge}(\omega) = \widehat{f}(\omega) \cdot \widehat{g}(\omega),$$

i.e., the Fourier transform of the convolution of two signals equals the product of their Fourier transforms.

Proof. These seven properties are derived easily by the definition. We prove only (ii), (iii), and (vii).

The Fourier transform of $D_a(f)$ is

$$(D_af)^\wedge(\omega) = \int_{\mathbb{R}} f(at)e^{-i\omega t}\,dt.$$

If $a > 0$, then $|a| = a$ and

$$\int_{\mathbb{R}} f(at)e^{-i\omega t}\,dt = \int_{\mathbb{R}} f(u)e^{-i(\frac{\omega}{a})u}\frac{du}{a} = \frac{1}{|a|}\widehat{f}\left(\frac{\omega}{a}\right).$$

If $a < 0$, then $|a| = -a$ and

$$\int_{\mathbb{R}} f(at)e^{-i\omega t}\,dt = -\int_{\mathbb{R}} f(u)e^{-i(\frac{\omega}{a})u}\frac{du}{a} = -\frac{1}{a}\widehat{f}\left(\frac{\omega}{a}\right) = \frac{1}{|a|}\widehat{f}\left(\frac{\omega}{a}\right).$$

We get (ii).

The Fourier transform of $T_\alpha f$ is

$$(T_\alpha f)^\wedge(\omega) = \int_{\mathbb{R}} f(t - \alpha)e^{-i\omega t}\,dt.$$

Let $u = t - \alpha$. Then

$$(T_\alpha f)^\wedge(\omega) = \int_{\mathbb{R}} f(u)e^{-i\omega(u+\alpha)}\,du = e^{-i\omega\alpha}\int_{\mathbb{R}} f(u)e^{-i\omega u}\,du = \widehat{f}(\omega)e^{-i\omega\alpha}.$$

We get (iii).

By the definition of the Fourier transform,

$$(f * g)^\wedge(\omega) = \int_{\mathbb{R}} (f * g)(t)e^{-it\omega}\,dt = \int_{\mathbb{R}} \left(\int_{\mathbb{R}} F(t - u)g(u)\,du\right)e^{-it\omega}\,dt.$$

Interchanging the order of integrals, and then letting $v = t - u$, we get

$$(f * g)^\wedge(\omega) = \int_{\mathbb{R}} \left(\int_{\mathbb{R}} f(t - u)e^{-it\omega}\,dt\right)g(u)\,du$$

$$= \int_{\mathbb{R}} \left(\int_{\mathbb{R}} f(v)e^{-i(v+u)\omega}\,dv\right)g(u)\,du$$

$$= \int_{\mathbb{R}} f(v)e^{-iv\omega}\,dv \cdot \int_{\mathbb{R}} g(u)e^{-iu\omega}\,du = \widehat{f}(\omega) \cdot \widehat{g}(\omega).$$

So we get (vii). □

The notation $f \in L^2(\mathbb{R})$ means that f is a signal of finite energy on $\mathbb{R}$, i.e., $\int_{\mathbb{R}} |f(t)|^2\,dt < \infty$. The definition of the Fourier transform of $f \in L^2(\mathbb{R})$ is based on the Schwartz space.

A space consists of the signals f satisfying the following two conditions:

(i) f is infinite-time differentiable on $\mathbb{R}$;

(ii) for any non-negative integers p, q,

$$t^p f^{(q)}(t) \to 0 \quad (|t| \to \infty).$$

This space is called the *Schwartz space*. Denote it by $f \in S$.

From the definition of the Schwartz space, it follows that if $f \in S$, then $f \in L(\mathbb{R})$ and $f \in L^2(\mathbb{R})$. It can be proved easily that if $f \in S$, then $\widehat{f} \in S$.

On the basis of the Schwartz space, the Fourier transform of $f \in L^2(\mathbb{R})$ is defined as follows.

Definition 1.1. Let $f \in L^2(\mathbb{R})$. Take arbitrarily $f_n(t) \in S$ such that $f_n(t) \to f(t) (L^2)$. The limit of $\{\widehat{f_n}(\omega)\}$ in $L^2(\mathbb{R})$ is said to be the Fourier transform of $f(t)$, denoted by $\widehat{f}(\omega)$, i.e., $\widehat{f_n}(\omega) \to \widehat{f}(\omega)(L^2)$.

Remark. $f_n(t) \to f(t)(L^2)$ means that $\int_{\mathbb{R}} (f_n(t) - f(t))^2 \, dt \to 0 (n \to \infty)$.

Similarly, on the basis of Definition 1.1, Fourier transforms for $L^2(\mathbb{R})$ have the following properties.

Property. Let $f, g \in L^2(\mathbb{R})$ and α, β be constants. Then

(i) (Linearity). $(\alpha f + \beta g)^{\wedge}(\omega) = \alpha \widehat{f}(\omega) + \beta \widehat{g}(\omega)$.

(ii) (Dilation). $(D_a f)^{\wedge}(\omega) = \frac{1}{|a|} \widehat{f}\left(\frac{\omega}{a}\right)$, where $D_a f = f(at)$ and $a \neq 0$ is a constant.

(iii) (Translation). $(T_\alpha f)^{\wedge}(\omega) = \widehat{f}(\omega) e^{-i\omega\alpha}$, where $T_\alpha f = f(t - \alpha)$.

(iv) (Modulation). $(f(t) e^{i\alpha t})^{\wedge}(\omega) = \widehat{f}(\omega - \alpha)$.

(v) $\widehat{f'}(\omega) = (i\omega)\widehat{f}(\omega), \widehat{\widehat{f}}(t) = 2\pi f(-t)$, and $\widehat{\overline{f}}(\omega) = \overline{\widehat{f}(-\omega)}$.

A linear continuous functional F, which is defined as a linear map from the Schwartz space to the real axis, is called a *generalized distribution* on the Schwartz space. Denote it by $F \in S'$. For any $g \in S$, denote $F(g)$ by $\langle F, g \rangle$. For each $f \in L^2(\mathbb{R})$, we can define a linear continuous functional on the Schwartz space as follows:

$$\langle f, g \rangle := \int_{\mathbb{R}} f(t) g(t) dt \quad \text{for any } g \in S,$$

which implies that $L^2(\mathbb{R}) \subset S'$.

The operation rules for generalized distributions on the Schwartz space are as follows:

(i) (Limit). Let $F_n \in S'(n = 1, 2, \ldots)$ and $F \in S'$. For any $g \in S$, define $F_n \to F(S')(n \to \infty)$ as

$$\langle F_n, g \rangle \to \langle F, g \rangle.$$

(ii) (Multiplier). Let $F \in S'$ and α be a constant. For any $g \in S$, define αF as

$$\langle \alpha F, g \rangle = \langle F, \alpha g \rangle.$$

(iii) (Derivative). Let $F \in S'$. For any $g \in S$, define the derivative $F' \in S'$ as

$$\langle F', g \rangle = -\langle F, g' \rangle.$$

(iv) (Dilation). Let $F \in S'$. For any $g \in S$, define $D_a F = F(at)$ as

$$\langle D_a F, g \rangle = \left\langle F, \frac{1}{|a|} g \left(\frac{t}{a} \right) \right\rangle,$$

where $a \neq 0$ is a constant.

(v) (Translation). Let $F \in S'$. For any $g \in S$, define $T_a F = F(t - a)$ as

$$\langle T_a F, g \rangle = \langle F, g(t + a) \rangle,$$

where a is a constant.

(vi) (Antiderivative). Let $F \in S'$. For any $g \in S$, define the antiderivative F^{-1} as

$$\langle F^{-1}, g \rangle = -\left\langle F, \int_{-\infty}^{t} \Phi_g(u) \, du \right\rangle,$$

where $\Phi_g(u) = g(u) - \frac{1}{\sqrt{\pi}} e^{-u^2} \int_{\mathbb{R}} g(t) \, dt$.

Definition 1.2. Let $F \in S'$.

(i) The Fourier series of F is defined as $\sum_n C_n e^{int}$, where the Fourier coefficients are

$$C_n = -\frac{1}{2\pi} \left\{ T_{2\pi} (Fe^{-int})^{-1} - (Fe^{-int})^{-1} \right\},$$

where $T_{2\pi}$ is the translation operator and $(Fe^{-int})^{-1}$ is the antiderivative of Fe^{-int}.

(ii) The Fourier transform of F is defined as $\langle \widehat{F}, g \rangle = \langle F, \widehat{g} \rangle$ for any $g \in S$.

Fourier transforms of generalized distributions on the Schwartz space have the following properties.

Property. Let $F \in S'$. Then

(i) (Derivative). $\widehat{F'}(\omega) = i\omega \widehat{F}(\omega)$.

(ii) (Translation). $(T_a F)^{\wedge}(\omega) = e^{-ia\omega} \widehat{F}(\omega)$, where a is a constant and $T_a F = F(t - a)$.

(iii) (Delation). $(D_a F)^{\wedge}(\omega) = \frac{1}{|a|} \widehat{F}(\frac{\omega}{a})$, where $a \neq 0$ and $D_a F = F(at)$.

The Dirac function and the Dirac comb are both important tools in geophysical signal processing. Define the Dirac function δ as a generalized distribution on the Schwartz space which satisfies for any $g \in S$,

$$\langle \delta, g \rangle = g(0).$$

In general, define δ_{t_0} as a generalized distribution on the Schwartz space which satisfies for any $g \in S$,

$$\langle \delta_{t_0}, g \rangle = g(t_0) \quad (t_0 \in \mathbb{R}).$$

Clearly, $\delta_0 = \delta$. Therefore, δ_{t_0} is the generalization of the Dirac function δ.

By operation rule (iv) of generalized distributions on a Schwartz space, it is easy to prove that for any $g \in S$, the first-order generalized derivative of the Dirac function is

$$\langle \delta', g \rangle = -\langle \delta, g' \rangle = -g'(0);$$

and the second-order generalized derivative of the Dirac function is

$$\langle \delta'', g \rangle = -\langle \delta', g' \rangle = \langle \delta, g'' \rangle = g''(0).$$

In general, the n-order generalized derivative of the Dirac function is

$$\langle \delta^{(n)}, g \rangle = (-1)^n g^{(n)}(0).$$

Denote the Fourier transform of δ_{t_0} by $\widehat{\delta_{t_0}}$. By Definition 1.2(ii), the Fourier transform of δ_{t_0} satisfies

$$\langle \widehat{\delta_{t_0}}, g \rangle = \langle \delta_{t_0}, \widehat{g} \rangle = \widehat{g}(t_0) \quad \text{for any } g \in S.$$

Since $g \in S \subset L(\mathbb{R})$, by the definition of the Fourier transform, we have

$$\widehat{g}(t_0) = \int_{\mathbb{R}} g(\omega) e^{-it_0\omega} \, d\omega = \langle e^{-it_0\omega}, g \rangle.$$

Therefore, $\langle \widehat{\delta_{t_0}}, g \rangle = \langle e^{-it_0\omega}, g \rangle$. This means $\widehat{\delta_{t_0}} = e^{-it_0\omega}$. Especially, noticing that $\delta_0 = \delta$, we find that the Fourier transform of the Dirac function is equal to 1.

On the other hand, by Definition 1.2(ii), for any $g \in S$,

$$\left\langle \left(e^{-it_0\omega} \right)^\wedge, g \right\rangle = \left\langle e^{-it_0\omega}, \widehat{g} \right\rangle = \int_{\mathbb{R}} \widehat{g}(\omega) e^{-it_0\omega} \, d\omega.$$

Since $g \in L(\mathbb{R})$ and $\widehat{g} \in L(\mathbb{R})$, the identity $\frac{1}{2\pi} \int_{\mathbb{R}} \widehat{g}(\omega) e^{-it_0\omega} \, d\omega = g(-t_0)$ holds. So

$$\left\langle \left(e^{-it_0\omega} \right)^\wedge, g \right\rangle = 2\pi g(-t_0).$$

From this and the definition $\langle \delta_{-t_0}, g \rangle = g(-t_0)$, it follows that

$$\left\langle \left(e^{-it_0\omega} \right)^\wedge, g \right\rangle = 2\pi \langle \delta_{-t_0}, g \rangle.$$

This means that $\left(e^{-it_0\omega} \right)^\wedge = 2\pi \delta_{-t_0}$. Noticing that $\delta_0 = \delta$, we obtain that the Fourier transform of 1 is equal to $2\pi\delta$.

Summarizing all the results, we have the following.

Formula 1.1.

(i) $\widehat{\delta_{t_0}} = e^{-it_0\omega}$ and $\left(e^{-it_0\omega} \right)^\wedge = 2\pi \delta_{-t_0}$,

(ii) $\widehat{\delta} = 1$ and $\widehat{1} = 2\pi\delta$.

Remark. In engineering and geoscience, instead of the rigid definition, one often uses the following alternative definition for the Dirac function δ:

(i) $\delta(t) = \begin{cases} \infty, & t = 0, \\ 0, & t \neq 0, \end{cases}$

(ii) $\int_{\mathbb{R}} \delta(t)\, dt = 1,$

(iii) $\int_{\mathbb{R}} \delta(t) g(t)\, dt = g(0)$ for any $g(t)$.

The series $\sum_n \delta_{2n\pi}$ is called the *Dirac comb* which is closely related to sampling theory. In order to show that it is well defined, we need to prove that the series $\sum_n \delta_{2n\pi}$ is convergent.

Let S_n be its partial sums and $S_n = \sum_{-n}^{n} \delta_{2k\pi}$. Clearly, S_n are generalized distributions on the Schwartz space, i.e., $S_n \in S'$ and for any $g \in S$,

$$\langle S_n, g \rangle = \left\langle \sum_{-n}^{n} \delta_{2k\pi}, g \right\rangle = \sum_{-n}^{n} \langle \delta_{2k\pi}, g \rangle.$$

Combining this with the definition $\langle \delta_{2k\pi}, g \rangle = g(2k\pi)$, we get

$$\langle S_n, g \rangle = \sum_{-n}^{n} g(2k\pi).$$

Since $g \in S$, the series $\sum_n g(2n\pi)$ converges. So there exists a $\delta^* \in S'$ such that

$$\langle S_n, g \rangle \to \langle \delta^*, g \rangle \quad \text{or} \quad S_n \to \delta^*(S') \quad (n \to \infty),$$

i.e., the series $\sum_n \delta_{2n\pi}$ converges to δ^*, and $\langle \delta^*, g \rangle = \sum_n g(2n\pi)$ for any $g \in S$.

Secondly, we prove that δ^* is a 2π-periodic generalized distribution.

By operation rule (v) of generalized distributions on a Schwartz space, for any $g \in S$,

$$\langle T_{2\pi} \delta^*, g \rangle = \langle \delta^*, g(t + 2\pi) \rangle = \sum_n g(2(n+1)\pi) = \sum_n g(2n\pi) = \langle \delta^*, g \rangle.$$

This means that δ^* is a periodic generalized distribution with period 2π.

Third, by Definition 1.2(i), we will find the Fourier series of δ^*. We only need to find its Fourier coefficients.

Denote the Fourier coefficients of δ^* by C_n. Since $\delta^* \in S'$, by Definition 1.2(i), for any $g \in S$,

$$\langle C_n, g \rangle = -\frac{1}{2\pi} \langle T_{2\pi} (\delta^* e^{-int})^{-1} - (\delta^* e^{-int})^{-1}, g \rangle.$$

Using operation rule (v) of generalized distributions on a Schwartz space, we get

$$\langle T_{2\pi} (\delta^* e^{-int})^{-1} - (\delta^* e^{-int})^{-1}, g \rangle = \langle (\delta^* e^{-int})^{-1}, \widetilde{g}(t) \rangle,$$

where $\widetilde{g}(t) = g(t + 2\pi) - g(t)$. Therefore

$$\langle C_n, g \rangle = -\frac{1}{2\pi} \langle (\delta^* e^{-int})^{-1}, \widetilde{g}(t) \rangle.$$

Using operation rule (vi) of generalized distributions on a Schwartz space, we get

$$\langle C_n, g \rangle = \frac{1}{2\pi} \left\langle \delta^* e^{-int}, \int_{-\infty}^{t} \Phi_{\widetilde{g}}(u) \, du \right\rangle,$$

where

$$\Phi_{\widetilde{g}}(u) = \widetilde{g}(u) - \frac{1}{\sqrt{\pi}} e^{-u^2} \int_{\mathbb{R}} \widetilde{g}(t) \, dt.$$

Since $\int_{\mathbb{R}} \widetilde{g}(t) dt = \int_{\mathbb{R}} g(t + 2\pi) \, dt - \int_{\mathbb{R}} g(t) \, dt = 0$, we get

$$\int_{-\infty}^{t} \Phi_{\widetilde{g}}(u) \, du = \int_{-\infty}^{t} \widetilde{g}(u) \, du = \int_{-\infty}^{t} (g(u + 2\pi) - g(u)) \, du = \int_{t}^{t+2\pi} g(u) \, du,$$

and so

$$\langle C_n, g \rangle = \frac{1}{2\pi} \left\langle \delta^* e^{-int}, \int_{t}^{t+2\pi} g(u) \, du \right\rangle.$$

Using operation rule (ii) of generalized distributions on a Schwartz space, we get

$$\left\langle \delta^* e^{-int}, \int_{t}^{t+2\pi} g(u) \, du \right\rangle = \left\langle \delta^*, \quad e^{-int} \int_{t}^{t+2\pi} g(u) \, du \right\rangle,$$

and so

$$\langle C_n, g \rangle = \frac{1}{2\pi} \left\langle \delta^*, \quad e^{-int} \int_{t}^{t+2\pi} g(u) \, du \right\rangle.$$

We have proved $\langle \delta^*, g \rangle = \sum_k g(2k\pi)$ for any $g \in S$. Noticing that $e^{-in2k\pi} = 1$, we find the right-hand side is

$$\frac{1}{2\pi} \left\langle \delta^*, e^{-int} \int_{t}^{t+2\pi} g(u) \, du \right\rangle = \frac{1}{2\pi} \sum_k e^{-in2k\pi} \int_{2k\pi}^{2k\pi+2\pi} g(u) du$$

$$= \frac{1}{2\pi} \sum_k \int_{2k\pi}^{2(k+1)\pi} g(u) \, du,$$

and so

$$\langle C_n, g \rangle = \frac{1}{2\pi} \sum_k \int_{2k\pi}^{2(k+1)\pi} g(u) \, du = \frac{1}{2\pi} \int_{\mathbb{R}} g(u) \, du = \left\langle \frac{1}{2\pi}, g \right\rangle,$$

i.e., $C_n = \frac{1}{2\pi} (n \in \mathbb{Z})$. By Definition 1.2(i), the Fourier series of δ^* is $\frac{1}{2\pi} \sum_n e^{int}$.

Finally, we prove the Fourier series $\frac{1}{2\pi} \sum_n e^{int}$ converges to δ^*, i.e., $\frac{1}{2\pi} \sum_n e^{int} = \delta^*(t)(S')$.

Its partial sum is $S_n(t) = \frac{1}{2\pi} \sum_{-n}^{n} e^{ikt}$. This is the Dirichlet kernel $D_n(t)$. Using property (ii) of the Dirichlet kernel, we get

$$\langle S_n, g \rangle = \langle D_n, g \rangle = \int_{\mathbb{R}} D_n(t)g(t)\,dt = \sum_k \int_{(2k-1)\pi}^{(2k+1)\pi} D_n(t)g(t)\,dt$$

$$= \sum_k \int_{-\pi}^{\pi} D_n(t)g(t + 2k\pi)\,dt = \int_{-\pi}^{\pi} D_n(t) \sum_k g(t + 2k\pi)\,dt.$$

By the Jordan criterion for Fourier series, we have

$$\int_{-\pi}^{\pi} D_n(t) \sum_k g(t + 2k\pi)\,dt \rightarrow \sum_k g(2k\pi) \quad (n \rightarrow \infty),$$

and so $\langle S_n, g \rangle \rightarrow \sum_k g(2k\pi)(n \rightarrow \infty)$. From this and $\langle \delta^*, g \rangle = \sum_k g(2k\pi)$, it follows that

$$\langle S_n, g \rangle \rightarrow \langle \delta^*, g \rangle \quad (n \rightarrow \infty).$$

This means that $S_n \rightarrow \delta^*(S')(n \rightarrow \infty)$. From this and $\delta^* = \sum_n \delta_{2n\pi}$, we get

$$\sum_n \delta_{2n\pi} = \frac{1}{2\pi} \sum_n e^{int} \quad (S').$$

Taking the Fourier transform on both sides and using Formula 1.1, we get

$$\left(\sum_n \delta_{2n\pi} \right)^{\wedge} = \frac{1}{2\pi} \sum_n \left(e^{int} \right)^{\wedge} = \frac{1}{2\pi} \sum_n \delta_n.$$

Formula 1.2. The Fourier transform of a Dirac comb is still a Dirac comb, i.e.,

$$\left(\sum_n \delta_{2n\pi} \right)^{\wedge} = \frac{1}{2\pi} \sum_n \delta_n.$$

The Laplace transform is a generalization of the Fourier transform. Since it can convert differential or integral equations into algebraic equations, the Laplace transform can be used to solve differential/integral equations with initial conditions.

Let $f \in L[0, \infty]$. The *Laplace transform* of a signal $f(t)$ is defined as

$$L[f(t)] := \int_0^{\infty} f(t)e^{-st}\,dt \quad (\text{Re}\,s \geq 0).$$

It is sometimes called the *one-sided Laplace transform*.

Laplace transforms possess the following properties:

(i) Let $f, g \in L[0, \infty]$ and c, d be constants. Then $L[cf(t) + dg(t)] = cL[f(t)] + dL[g(t)]$.

(ii) Let $f^{(j)} \in L[0, \infty] (j = 1, \ldots, N)$. Then

$$L[f^{(N)}(t)] = -f^{(N-1)}(0) - \cdots - s^{N-3}f''(0) - s^{N-2}f'(0)$$
$$- s^{N-1}f(0) + s^N L[f(t)].$$

(iii) Let $f \in L[0, \infty]$. Then $L\left[\int_0^t f(u)\, du\right] = \frac{1}{s}L[f(t)]$.

By the definition and properties of Laplace transforms, it follows further that

$$L[1] = \int_0^\infty e^{-st}\, dt = \frac{1}{s},$$

$$L[e^{-at}] = \int_0^\infty e^{-(a+s)t}\, dt = \frac{1}{s+a},$$

$$L\left[\frac{e^{-at} - e^{-bt}}{a-b}\right] = \frac{1}{a-b}\{L\{e^{-at}\} - L\{e^{-bt}\}\}$$

$$= \frac{1}{a-b}\left\{\frac{1}{s+a} - \frac{1}{s+b}\right\} = -\frac{1}{(s+a)(s+b)},$$

$$L\left[\frac{ae^{-at} - be^{-bt}}{a-b}\right] = \frac{1}{a-b}\{aL\{e^{-at}\} - bL\{e^{-bt}\}\}$$

$$= \frac{1}{a-b}\left\{\frac{a}{s+a} - \frac{b}{s+b}\right\} = \frac{s}{(s+a)(s+b)},$$

$$L[t^N] = \int_0^\infty t^N e^{-st}\, dt = \frac{N!}{s^{N+1}}.$$

Finally, we consider the two-dimensional case. If $f(t_1, t_2) \in L(\mathbb{R}^2)$, the *two-dimensional Fourier transform* is defined as

$$\widehat{f}(\omega_1, \omega_2) := \int\int_{\mathbb{R}^2} f(t_1\, t_2)e^{-i(\omega_1 t_1 + \omega_2 t_2)}\, dt_1\, dt_2.$$

The *two-dimensional inverse Fourier transform* is defined as

$$\frac{1}{(2\pi)^2}\int\int_{\mathbb{R}^2}\widehat{f}(\omega_1, \omega_2)e^{i(\omega_1 t_1 + \omega_2 t_2)}\, d\omega_1\, d\omega_2.$$

It can be proved that if $f \in L(\mathbb{R}^2)$ and $\widehat{f} \in L(\mathbb{R}^2)$, then

$$f(t_1, t_2) = \frac{1}{(2\pi)^2}\int\int_{\mathbb{R}^2}\widehat{f}(\omega_1, \omega_2)e^{i(\omega_1 t_1 + \omega_2 t_2)}\, d\omega_1\, d\omega_2.$$

Two-dimensional Fourier transforms have the following similar properties:

(i) (Translation). Let $f \in L(\mathbb{R}^2)$ and $a = (a_1, a_2) \in \mathbb{R}^2$. Then

$$(f(t_1 + a_1, t_2 + a_2))^\wedge(\omega_1, \omega_2) = e^{i(\omega_1 a_1 + \omega_2 a_2)} \widehat{f}(\omega_1, \omega_2).$$

(ii) (Delation). Let $f \in L(\mathbb{R}^2)$ and λ be a real constant. Then

$$(f(\lambda t_1, \lambda t_2))^\wedge(\omega_1, \omega_2) = \frac{1}{|\lambda|^2} \widehat{f}\left(\frac{\omega_1}{\lambda}, \frac{\omega_2}{\lambda}\right).$$

(iii) (Convolution). Let $f, g \in L(\mathbb{R}^2)$ and the convolution

$$(f * g)(t_1, t_2) = \int \int_{\mathbb{R}^2} f(t_1 - u_1, t_2 - u_2) g(u_1, u_2) \, du_1 du_2.$$

Then

$$(f * g)^\wedge(\omega_1, \omega_2) = \widehat{f}(\omega_1, \omega_2) \widehat{g}(\omega_1, \omega_2).$$

1.2 BESSEL'S INEQUALITY AND PARSEVAL'S IDENTITY

Bessel's inequality and Parseval's identity are fundamental results of Fourier series and Fourier transform. Bessel's inequality is a stepping stone to the more powerful Parseval's identity.

Bessel's Inequality for Fourier Series. *Let $f \in L_{2\pi}$ and a_n, b_n, c_n be its Fourier coefficients. Then*

$$\left(\frac{a_0}{2} + \sum_{1}^{n}(a_k^2 + b_k^2)\right) \leq \frac{1}{\pi} \int_{-\pi}^{\pi} f^2(t) \, dt$$

or

$$\sum_{-n}^{n} |c_k|^2 \leq \frac{1}{2\pi} \int_{-\pi}^{\pi} f^2(t) \, dt.$$

Proof. Denote partial sums of the Fourier series of f by $S_n(f; t)$. Since

$$(S_n(f; t) - f(t))^2 = S_n^2(f; t) - 2f(t)S_n(f; t) + f^2(t),$$

integrating over the interval $[-\pi, \pi]$, we get

$$\int_{-\pi}^{\pi} (S_n(f; t) - f(t))^2 \, dt = \int_{-\pi}^{\pi} S_n^2(f; t) \, dt - 2\int_{-\pi}^{\pi} f(t)S_n(f; t) \, dt + \int_{-\pi}^{\pi} f^2(t) \, dt$$

$$= I_1 - I_2 + \int_{-\pi}^{\pi} f^2(t) \, dt.$$

We compute I_1. The partial sums of the Fourier series of f are

$$S_n(f; t) = \frac{a_0}{2} + \sum_{1}^{n}(a_k \cos(kt) + b_k \sin(kt)).$$

So

$$I_1 = \int_{-\pi}^{\pi} S_n^2(f; t)\, dt = \int_{-\pi}^{\pi} \left(\frac{a_0}{2} + \sum_1^n (a_k \cos(kt) + b_k \sin(kt)) \right)^2 dt$$

$$= \int_{-\pi}^{\pi} \frac{a_0^2}{4}\, dt + \int_{-\pi}^{\pi} a_0 \left(\sum_1^n (a_k \cos(kt) + b_k \sin(kt)) \right) dt$$

$$+ \int_{-\pi}^{\pi} \left(\sum_1^n (a_k \cos(kt) + b_k \sin(kt)) \right)^2 dt.$$

By the orthogonality of trigonometric system $\{1, \cos(nt), \sin(nt)\}_{n \in \mathbb{Z}_+}$, we obtain that

$$I_1 = \pi \left(\frac{a_0^2}{2} + \sum_1^n (a_k^2 + b_k^2) \right).$$

We compute I_2. Since

$$I_2 = 2 \int_{-\pi}^{\pi} f(t) S_n(f; t)\, dt = 2 \int_{-\pi}^{\pi} f(t) \left(\frac{a_0}{2} + \sum_1^n (a_k \cos(kt) + b_k \sin(kt)) \right) dt$$

$$= a_0 \int_{-\pi}^{\pi} f(t)\, dt + 2 \sum_1^n \left(a_k \int_{-\pi}^{\pi} f(t) \cos(kt) dt + b_k \int_{-\pi}^{\pi} f(t) \sin(kt)\, dt \right),$$

by the definition of the Fourier coefficients, we get

$$I_2 = 2\pi \left(\frac{a_0^2}{2} + \sum_1^n (a_k^2 + b_k^2) \right).$$

Therefore,

$$\int_{-\pi}^{\pi} (S_n(f; t) - f(t))^2\, dt = -\pi \left(\frac{a_0^2}{2} + \sum_1^n (a_k^2 + b_k^2) \right) + \int_{-\pi}^{\pi} f^2(t)\, dt. \quad (1.4)$$

Noticing that $a_0 = 2c_0$, $a_k = c_k + c_{-k}$, $b_k = i(c_k - c_{-k})$, and

$$a_k^2 + b_k^2 = |c_{-k} + c_k|^2 + |i(c_{-k} - c_k)|^2$$

$$= (c_{-k} + c_k)(\overline{c}_{-k} + \overline{c}_k) + (c_{-k} - c_k)(\overline{c}_{-k} - \overline{c}_k)$$

$$= 2(c_{-k}\overline{c}_{-k} + c_k\overline{c}_k) = 2\left(|c_{-k}|^2 + |c_k|^2 \right),$$

the first term on the right-hand side of (1.4):

$$-\pi \left(\frac{a_0^2}{2} + \sum_1^n (a_k^2 + b_k^2) \right) = -\pi \left(2|c_0|^2 + \sum_1^n 2(|c_{-k}|^2 + |c_k|^2) \right)$$

$$= -2\pi \sum_{-n}^n |c_k|^2.$$

From this and (1.4), it follows that

$$\int_{-\pi}^{\pi} (S_n(f;t) - f(t))^2 \, dt = -2\pi \sum_{-n}^{n} |c_k|^2 + \int_{-\pi}^{\pi} f^2(t) \, dt. \qquad (1.5)$$

Noticing that $\int_{-\pi}^{\pi} (S_n(f;t) - f(t))^2 \, dt \geq 0$, we find from (1.4) and (1.5) that

$$\left(\frac{a_0}{2} + \sum_{1}^{n} (a_k^2 + b_k^2) \right) \leq \frac{1}{\pi} \int_{-\pi}^{\pi} f^2(t) \, dt$$

and

$$\sum_{-n}^{n} |c_k|^2 \leq \frac{1}{2\pi} \int_{-\pi}^{\pi} f^2(t) \, dt.$$

$\square$

Parseval's Identity for Fourier Series. *Let $f \in L_{2\pi}$ and a_n, b_n, c_n be its Fourier coefficients. If the partial sums of its Fourier series $S_n(f;t)$ tend to $f(t)$ as $n \to \infty$, then*

$$\int_{-\pi}^{\pi} f^2(t) \, dt = \pi \left(\frac{a_0^2}{2} + \sum_{1}^{\infty} (a_n^2 + b_n^2) \right)$$

and

$$\int_{-\pi}^{\pi} f^2(t) \, dt = 2\pi \sum_{n} |c_n|^2.$$

Parseval's identity is sometimes called the law of conservation of energy.

Proof. In the proof of Bessel's inequality, we have obtained (1.4) and (1.5). Letting $n \to \infty$ in (1.4) and (1.5), and using the assumption $S_n(f;t) \to f(t)$ ($n \to \infty$), we obtain immediately the desired results:

$$\int_{-\pi}^{\pi} f^2(t) \, dt = \pi \left(\frac{a_0^2}{2} + \sum_{1}^{\infty} (a_k^2 + b_k^2) \right)$$

and

$$\int_{-\pi}^{\pi} f^2(t) \, dt = 2\pi \sum_{k} |c_k|^2.$$

$\square$

For a Schwartz space, the original signals and their Fourier transforms have the following relation.

Theorem 1.2. *If $f, g \in S$, then*

$$\int_{\mathbb{R}} f(t) \overline{g}(t) \, dt = \frac{1}{2\pi} \int_{\mathbb{R}} \widehat{f}(\omega) \overline{\widehat{g}}(\omega) \, d\omega.$$

Proof. It follows from $g \in S$ that $g \in L(\mathbb{R})$ and $\widehat{g} \in L(\mathbb{R})$). Thus,

$$g(t) = \frac{1}{2\pi} \int_{\mathbb{R}} \widehat{g}(\omega) e^{i\omega t} \, d\omega.$$

Taking the conjugate on both sides, we get

$$\overline{g}(t) = \frac{1}{2\pi} \int_{\mathbb{R}} \overline{\widehat{g}}(\omega) e^{-i\omega t} \, d\omega,$$

and so

$$\int_{\mathbb{R}} f(t) \overline{g}(t) \, dt = \frac{1}{2\pi} \int_{\mathbb{R}} f(t) \left(\int_{\mathbb{R}} \overline{\widehat{g}}(\omega) e^{-i\omega t} \, d\omega \right) dt.$$

Interchanging the order of integrals and using the definition of the Fourier transform, the right-hand side is

$$\frac{1}{2\pi} \int_{\mathbb{R}} f(t) \left(\int_{\mathbb{R}} \overline{\widehat{g}}(\omega) e^{-i\omega t} \, d\omega \right) dt = \frac{1}{2\pi} \int_{\mathbb{R}} \left(\int_{\mathbb{R}} f(t) e^{-i\omega t} \, dt \right) \overline{\widehat{g}}(\omega) d\omega$$

$$= \frac{1}{2\pi} \int_{\mathbb{R}} \widehat{f}(\omega) \overline{\widehat{g}}(\omega) \, d\omega.$$

Therefore,

$$\int_{\mathbb{R}} f(t) \overline{g}(t) \, dt = \frac{1}{2\pi} \int_{\mathbb{R}} \widehat{f}(\omega) \overline{\widehat{g}}(\omega) \, d\omega.$$

$\square$

Let $f(t) = g(t)$ in Theorem 1.2. Then the following identity holds.
Parseval's Identity for a Schwartz Space. *If* $f \in S$, *then*

$$\int_{\mathbb{R}} |f(t)|^2 \, dt = \frac{1}{2\pi} \int_{\mathbb{R}} |\widehat{f}(\omega)|^2 \, d\omega.$$

Theorem 1.2 can be extended from S to $L^2(\mathbb{R})$ as follows.
Theorem 1.3. *If* $f, g \in L^2(\mathbb{R})$, *then*

$$\int_{\mathbb{R}} f(t) \overline{g}(t) \, dt = \frac{1}{2\pi} \int_{\mathbb{R}} \widehat{f}(\omega) \overline{\widehat{g}}(\omega) \, d\omega.$$

Proof. Take arbitrarily $f_n \in S$, $g_n \in S$ such that $f_n \to f(L^2)$, $g_n \to g(L^2)$ as $n \to \infty$. By Definition 1.1,

$$\widehat{f_n}(\omega) \to \widehat{f}(\omega)(L^2),$$

$$\widehat{g_n}(\omega) \to \widehat{g}(\omega)(L^2),$$

and so

$$\frac{1}{2\pi} \int_{\mathbb{R}} \widehat{f_n}(\omega) \overline{\widehat{g_n}}(\omega) \, d\omega \to \frac{1}{2\pi} \int_{\mathbb{R}} \widehat{f}(\omega) \overline{\widehat{g}}(\omega) \, d\omega.$$

On the other hand, since $f_n \in S$ and $g_n \in S$, Theorem 1.2 shows that

$$\frac{1}{2\pi} \int_{\mathbb{R}} \widehat{f}_n(\omega) \overline{\widehat{g}_n}(\omega) \, d\omega = \int_{\mathbb{R}} f_n(t) \overline{g}_n(t) \, dt.$$

Since $f_n \to f$ and $g_n \to g$, the integral on the right-hand side has a limit, i.e., as $n \to \infty$

$$\int_{\mathbb{R}} f_n(t) \overline{g}_n(t) \, dt \to \int_{\mathbb{R}} f(t) \overline{g}(t) \, dt,$$

and so

$$\frac{1}{2\pi} \int_{\mathbb{R}} \widehat{f}_n(\omega) \overline{\widehat{g}_n}(\omega) \, d\omega \to \int_{\mathbb{R}} f(t) \overline{g}(t) \, dt.$$

Since the limit is unique, we get

$$\int_{\mathbb{R}} f(t) \overline{g}(t) \, dt = \frac{1}{2\pi} \int_{\mathbb{R}} \widehat{f}(\omega) \overline{\widehat{g}}(\omega) \, d\omega.$$

$\square$

Let $g(t) = f(t)$ in Theorem 1.3. Then the following identity holds.
Parseval's Identity of the Fourier Transform. *If $f \in L^2(\mathbb{R})$, then*

$$\int_{\mathbb{R}} |f(t)|^2 \, dt = \frac{1}{2\pi} \int_{\mathbb{R}} |\widehat{f}(\omega)|^2 \, d\omega.$$

In a similar way, for the two-dimensional signal, the following theorem can be derived.
Theorem 1.4. *If $f, g \in L^2(\mathbb{R}^2)$, then*

$$\int \int_{\mathbb{R}^2} f(t_1, t_2) \overline{g}(t_1, t_2) \, dt_1 dt_2 = \frac{1}{(2\pi)^2} \int \int_{\mathbb{R}^2} \widehat{f}(\omega_1, \omega_2) \overline{\widehat{g}}(\omega_1, \omega_2) \, d\omega_1 d\omega_2.$$

Let $f = g$ in Theorem 1.4. Then the following identity holds.
Parseval's Identity. *Let $f(t_1, t_2) \in L^2(\mathbb{R}^2)$. Then*

$$\int \int_{\mathbb{R}^2} |f(t_1, t_2)|^2 \, dt_1 dt_2 = \frac{1}{(2\pi)^2} \int \int_{\mathbb{R}^2} |\widehat{f}(\omega_1, \omega_2)|^2 \, d\omega_1 d\omega_2.$$

1.3 GIBBS PHENOMENON

If a function $f(t)$ is defined in a neighborhood of t_0 and $f(t_0 + 0), f(t_0 - 0)$ exist but $f(t_0 + 0) \neq f(t_0 - 0)$, then t_0 is called the *first kind of discontinuity* of $f(t)$.

Suppose that functions $\{f_n(t)\}_{n \in \mathbb{Z}_+}$ and $f(t)$ are defined in a neighborhood of t_0 and $f_n(t) \to f(t)$ as $n \to \infty$ in the neighborhood, and t_0 is the first kind of discontinuity of $f(t)$. Without loss of generality, we may assume $f(t_0 - 0) < f(t_0 + 0)$. If $\{f_n(t)\}$ has a double sublimit lying outside the closed interval $[f(t_0 - 0), f(t_0 + 0)]$ as $t \to t_0, n \to \infty$, then we say that for the sequence of functions $\{f_n(t)\}$ the *Gibbs phenomenon* occurs at t_0.

Example 1.1. Consider a function

$$\varphi(t) = \begin{cases} \frac{\pi - t}{2}, & 0 < t < 2\pi, \\ 0, & t = 0, \end{cases} \quad \text{and} \quad \varphi(t + 2\pi) = \varphi(t), \quad \text{and} \quad t_0 = 0.$$

Clearly, $\varphi(t)$ is continuous in $0 < |t| < \pi$ and $\varphi(0 + 0) = \frac{\pi}{2}$, $\varphi(0 - 0) = -\frac{\pi}{2}$, and the point $t_0 = 0$ is the first kind of discontinuity of $\varphi(t)$. It is well known that the Fourier series of $\varphi(t)$ is

$$\sum_1^\infty \frac{\sin(kt)}{k} \quad (t \in \mathbb{R}).$$

Consider the sequence of partial sums of the Fourier series of $\varphi(t)$:

$$S_n(\varphi; t) = \sum_1^n \frac{\sin(kt)}{k} \quad (t \in \mathbb{R}).$$

Since $\varphi(t) \in L_{2\pi}$ and is of bounded variation in $0 < |t| < \pi$, the Jordan criterion shows that the sequence of partial sums of its Fourier series converges at $t_0 = 0$ and

$$S_n(\varphi; 0) \to \frac{1}{2}(\varphi(0 + 0) + \varphi(0 - 0)) \quad (n \to \infty).$$

Since $\varphi(0 + 0) = \frac{\pi}{2}$ and $\varphi(0 - 0) = -\frac{\pi}{2}$, we get $S_n(\varphi; 0) \to 0(n \to \infty)$.

Now we prove $S_n(\varphi; t)$ has a double sublimit lying outside the closed interval $[-\frac{\pi}{2}, \frac{\pi}{2}]$ as $n \to \infty, t \to 0$.

Note that

$$\sum_1^n \cos(kv) = \sum_1^n \frac{e^{-ikv} + e^{ikv}}{2} = \frac{1}{2}\left(\sum_{-n}^n e^{ikv} - 1\right) = \pi D_n(v) - \frac{1}{2},$$

where $D_n(v)$ is the Dirichlet kernel. Using property (iii) of the Dirichlet kernel, the partial sums of the Fourier series of $\varphi(t)$ can be rewritten as follows:

$$S_n(\varphi; t) = \sum_1^n \frac{\sin(kt)}{k} = \sum_1^n \int_0^t \cos(kv)\, dv = \int_0^t \sum_1^n \cos(kv)\, dv$$

$$= \int_0^t \frac{\sin\left(n + \frac{1}{2}\right)v}{2\sin\frac{v}{2}}\, dv - \frac{t}{2}$$

$$= \int_0^t \frac{\sin\left(n + \frac{1}{2}\right)v}{v}\, dv + \int_0^t \left(\frac{\sin\left(n + \frac{1}{2}\right)v}{2\sin\frac{v}{2}} - \frac{\sin\left(n + \frac{1}{2}\right)v}{v}\right)$$

$$dv - \frac{t}{2}. \tag{1.6}$$

Let $u = (n + \frac{1}{2})v$. Then the first integral on the right-hand side of (1.6) is

$$\int_0^t \frac{\sin\left(n + \frac{1}{2}\right)v}{v}\,dv = \int_0^{(n+\frac{1}{2})t} \frac{\sin u}{u}\,du.$$

Take $t = t_n = \frac{a}{n}$, where a is any real number. Then, as $n \to \infty$ and $t \to 0$,

$$\int_0^{t_n} \frac{\sin\left(n + \frac{1}{2}\right)v}{v}\,dv = \int_0^{(n+\frac{1}{2})\frac{a}{n}} \frac{\sin u}{u}\,du \to \int_0^a \frac{\sin u}{u}\,du.$$

By inequalities $\left|\sin\left(n + \frac{1}{2}\right)v\right| \leq 1$ and $\left|v - 2\sin\frac{v}{2}\right| \leq \left|\frac{v^3}{24}\right|$, and $\sin v \geq \frac{2}{\pi}v \left(0 < v \leq \frac{\pi}{2}\right)$, it follows that

$$\left|\sin\left(n + \frac{1}{2}\right)v\left(\frac{1}{2\sin\frac{v}{2}} - \frac{1}{v}\right)\right| = \left|\sin(n + \frac{1}{2})v\frac{v - 2\sin\frac{v}{2}}{2v\sin\frac{v}{2}}\right|$$

$$\leq \left|\frac{\frac{v^3}{24}}{\frac{2}{\pi}v^2}\right| = \frac{\pi}{12}|v|,$$

and so the second integral on the right-hand side of (1.6) is

$$\left|\int_0^t \left(\frac{\sin\left(n + \frac{1}{2}\right)v}{2\sin\frac{v}{2}} - \frac{\sin\left(n + \frac{1}{2}\right)v}{v}\right)\,dv\right| \leq \frac{\pi}{24}t^2.$$

Take $t = t_n = \frac{a}{n}$. Then

$$\left|\int_0^{t_n} \left(\frac{\sin\left(n + \frac{1}{2}\right)v}{2\sin\frac{v}{2}} - \frac{\sin\left(n + \frac{1}{2}\right)v}{v}\right)\,dv\right| \leq \frac{\pi a^2}{24n^2}.$$

As $n \to \infty$ and $t \to 0$,

$$\int_0^{t_n} \left(\frac{\sin\left(n + \frac{1}{2}\right)v}{2\sin\frac{v}{2}} - \frac{\sin\left(n + \frac{1}{2}\right)v}{v}\right)\,dv \to 0$$

It is clear that the last term on the right-hand side of (1.6) $\frac{t_n}{2} \to 0$ as $n \to \infty$ and $t \to 0$.

Therefore, take $t = t_n = \frac{a}{n}$, where a is any real number. By (1.6), we have

$$S_n(\varphi; t_n) \to \int_0^a \frac{\sin u}{u}\,du =: I(a) \quad (n \to \infty, \quad t \to 0),$$

i.e., $S_n(\varphi; t)$ has double sublimits $I(a)$ as $n \to \infty, t \to 0$. Since a is any real number, all values of $I(a)$ consist of a closed interval $[I(-\pi), I(\pi)]$, and

$$I(\pi) = \int_0^\pi \frac{\sin u}{u}\,du > \frac{\pi}{2}, \quad I(-\pi) = \int_0^{-\pi} \frac{\sin u}{u}\,du < -\frac{\pi}{2},$$

and so $[I(-\pi), I(\pi)] \supset [-\frac{\pi}{2}, \frac{\pi}{2}]$.

Therefore, for the sequence of partial sums $\{S_n(\varphi; t)\}$ the Gibbs phenomenon occurs at $t_0 = 0$.

Theorem 1.5. *Suppose that $f(t)$ is a 2π-periodic function of bounded variation and continuous in a neighborhood of t_0, and t_0 is the first kind of discontinuity of $f(t)$. Then for the sequence of partial sums of the Fourier series of $f(t)$ the Gibbs phenomenon occurs at t_0.*

Proof. Without loss of generality, assume that $f(t)$ is continuous in $0 < |t - t_0| < \delta$ and $f(t_0 + 0) > f(t_0 - 0)$. Let $\varphi(t)$ be stated as in Example 1.1, and let

$$g(t) = f(t) - \frac{d}{\pi}\varphi(t - t_0), \tag{1.7}$$

where $d = f(t_0 + 0) - f(t_0 - 0) > 0$. By the assumption, we see that $g(t)$ is a 2π-periodic function of bounded variation and continuous in $0 < |t - t_0| < \delta$. According to the Jordan criterion, the partial sums of the Fourier series of $g(t)$ converge and

$$S_n(g; t) \to \frac{1}{2}(g(t_0 + 0) + g(t_0 - 0)) \quad (n \to \infty, 0 < |t - t_0| < \delta).$$

Since $\varphi(0 + 0) = \frac{\pi}{2}$ and $\varphi(0 - 0) = -\frac{\pi}{2}$ (see Example 1.1), it follows from (1.7) that

$$g(t_0 + 0) = f(t_0 + 0) - \frac{d}{2},$$

$$g(t_0 - 0) = f(t_0 - 0) + \frac{d}{2},$$

and so

$$S_n(g; t) \to \frac{1}{2}(f(t_0 + 0) + f(t_0 - 0)), \quad 0 < |t - t_0| < \delta \quad (n \to \infty). \tag{1.8}$$

Now we prove that $S_n(f; t)$ has a double sublimit lying outside the closed interval $[f(t_0 - 0), f(t_0 + 0)]$ as $n \to \infty, t \to t_0$.

Denote the partial sums of the Fourier series of $\varphi(t)$ by $S_n(\varphi; t)$. By (1.7), it follows that

$$S_n(f; t) = S_n(g; t) + \frac{d}{\pi}S_n(\varphi; t - t_0).$$

Take $t - t_0 = t_n = \frac{a}{n}$, where a is any real number. Then

$$S_n(f; t_0 + t_n) = S_n(g; t_0 + t_n) + \frac{d}{\pi}S_n(\varphi; t_n).$$

By Example 1.1,

$$S_n(\varphi; t_n) \to I(a) \quad (n \to \infty, t \to t_0),$$

where $I(a) = \int_0^a \frac{\sin u}{u}\, du$. Denote $f(t_0) = \frac{1}{2}(f(t_0 + 0) + f(t_0 - 0))$. By (1.8),

$$S_n(g; t_0 + t_n) \to f(t_0) \quad (n \to \infty, t \to t_0).$$

Therefore,

$$S_n(f; t_0 + t_n) \to f(t_0) + \frac{d}{\pi}I(a) \quad (n \to \infty, t \to t_0),$$

i.e., $S_n(f;t)$ has double sublimits $f(t_0) + \frac{d}{\pi}I(a)$ as $n \to \infty, t \to t_0$. Since a can be any real number, all values of $f(t_0) + \frac{d}{\pi}I(a)$ consist of the closed interval $\left[f(t_0) + \frac{d}{\pi}I(-\pi), f(t_0) + \frac{d}{\pi}I(\pi)\right]$. Noticing that $I(\pi) > \frac{\pi}{2}$ and $I(-\pi) < -\frac{\pi}{2}$, we have

$$\left[f(t_0) + \frac{d}{\pi}I(-\pi), f(t_0) + \frac{d}{\pi}I(\pi)\right] \supset \left[f(t_0) - \frac{d}{2}, f(t_0) + \frac{d}{2}\right].$$

From $f(t_0) = \frac{1}{2}(f(t_0 + 0) + f(t_0 - 0))$ and $d = f(t_0 + 0) - f(t_0 - 0)$, it follows that

$$\left[f(t_0) + \frac{d}{\pi}I(-\pi), f(t_0) + \frac{d}{\pi}I(\pi)\right] \supset [f(t_0 - 0), f(t_0 + 0)].$$

Therefore, for the sequence of partial sums of the Fourier series of $f(t)$ the Gibbs phenomenon occurs at t_0.   □

1.4   POISSON SUMMATION FORMULAS AND SHANNON SAMPLING THEOREM

We will introduce three important theorems: the Poisson summation formula in $L(\mathbb{R})$, the Poisson summation formula in $L^2(\mathbb{R})$, and the Shannon sampling theorem. In signal processing, the Poisson summation formula leads to the Shannon sampling theorem and the discrete-time Fourier transform.

To prove the Poisson summation formula in $L(\mathbb{R})$, we first give a relation between Fourier transforms in $L(\mathbb{R})$ and Fourier coefficients in $L_{2\pi}$.

Lemma 1.1.   *Let $f \in L(\mathbb{R})$. Then*

(i)   *the series $\sum_n f(t + 2n\pi)$ is absolutely convergent almost everywhere. Denote its sum by $F(t)$;*

(ii) $F(t) \in L_{2\pi}$;

(iii)   *for any integer n,*

$$c_n(F) = \frac{1}{2\pi}\widehat{f}(n),$$

where $c_n(F)$ is the Fourier coefficient of $F(t)$ and $\widehat{f}(\omega)$ is the Fourier transform of $f(t)$.

Proof.   Consider the series $\sum_n f(t + 2n\pi)$. By the assumption that $f \in L(\mathbb{R})$, we have

$$\left| \int_0^{2\pi} \sum_n f(t + 2n\pi)\, dt \right| \leq \sum_n \int_0^{2\pi} |f(t + 2n\pi)|\, dx$$

$$= \sum_n \int_{2n\pi}^{2(n+1)\pi} |f(y)|\, dy$$

$$= \int_{\mathbb{R}} |f(y)|\, dy < \infty.$$

So the series is integrable over $[0, 2\pi]$. Since

$$\sum_n f((t+2\pi) + 2n\pi) = \sum_n f(t + 2(n+1)\pi) = \sum_n f(t + 2n\pi),$$

the series is a 2π-periodic function. Therefore, the series is absolutely convergent almost everywhere. Denote its sum by $F(t)$, i.e.,

$$F(t) = \sum_n f(t + 2n\pi) \quad \text{almost everywhere},$$

and so $F(t)$ is integrable over $[0, 2\pi]$ and is a 2π-periodic function, i.e., $F \in L_{2\pi}$. By the definition of the Fourier coefficients and $e^{in(2k\pi)} = 1$, we have

$$c_n(F) = \frac{1}{2\pi} \int_0^{2\pi} F(t)e^{-int}\,dt = \frac{1}{2\pi} \int_0^{2\pi} \left(\sum_k f(t + 2k\pi) \right) e^{-int}\,dt$$

$$= \frac{1}{2\pi} \sum_k \int_{2k\pi}^{2(k+1)\pi} f(u)e^{-in(u-2k\pi)}\,du = \frac{1}{2\pi} \int_{\mathbb{R}} f(u)e^{-inu}\,du.$$

However, since $f \in L(\mathbb{R})$, by the definition of the Fourier transform, we have

$$\int_{\mathbb{R}} f(u)e^{-inu}\,du = \widehat{f}(n).$$

Therefore, $c_n(F) = \frac{1}{2\pi}\widehat{f}(n)$. $\qquad\qquad\qquad\qquad\qquad\qquad\qquad\qquad\quad\square$

Poisson Summation Formula I. If $f \in L(\mathbb{R})$ and f satisfies one of the following two conditions:

(i) $f(t)$ is of bounded variation on $\mathbb{R}$ and $f(t) := \frac{1}{2}(f(t+0) + f(t-0))$;

(ii) $|f(t)| \le K_1(1 + |t|)^{-\alpha}$ and $|\widehat{f}(\omega)| \le K_2(1 + |\omega|)^{-\alpha}$, where $\alpha > 1$ and K_1, K_2 are constants,

then

$$\sum_n f(t + 2n\pi) = \frac{1}{2\pi} \sum_n \widehat{f}(n)e^{int} \quad (t \in \mathbb{R}).$$

Specially,

$$\sum_n f(2n\pi) = \frac{1}{2\pi} \sum_n \widehat{f}(n).$$

Proof. Suppose that $f(t)$ satisfies the first condition. Lemma 1.1 has shown that the series $\sum_n f(t + 2n\pi)$ is absolutely convergent almost everywhere. Now we prove that the series $\sum_n f(t + 2n\pi)$ is absolutely, uniformly convergent everywhere on $[0, 2\pi]$.

Take $t_0 \in [0, 2\pi]$ such that $\sum_n f(t_0 + 2n\pi)$ converges. When $0 \le t \le 2\pi$,

$$\left| \sum_{|n|>N} f(t+2n\pi) \right| = \left| \sum_{|n|>N} f(t_0 + 2n\pi) + f(t+2n\pi) - f(t_0 + 2n\pi) \right|$$

$$\leq \left| \sum_{|n|>N} f(t_0 + 2n\pi) \right| + \left| \sum_{|n|>N} (f(t+2n\pi) - f(t_0 + 2n\pi)) \right|$$

$$= I_N(t_0) + \tilde{I}_N(t).$$

Since the series $\sum_n f(t_0 + 2n\pi)$ is convergent and is independent of t,

$$I_N(t_0) \to 0 \quad (N \to \infty)$$

uniformly on $[0, 2\pi]$. Note that $f(t)$ is a function of bounded variation on $\mathbb{R}$. Denote its variation by

$$V_n = \bigvee_{2n\pi}^{2(n+1)\pi} (f).$$

So the total variation is

$$\sum_n V_n = \sum_n \left(\bigvee_{2n\pi}^{2(n+1)\pi} (f) \right) = \bigvee_{-\infty}^{\infty} (f) < \infty,$$

and so for $0 \leq t \leq 2\pi$,

$$\tilde{I}_N(t) \leq \sum_{|n|>N} |f(t+2n\pi) - f(t_0 + 2n\pi)| \leq \sum_{|n|>N} V_n \to 0 \quad (N \to \infty),$$

i.e., $\tilde{I}_N(t) \to 0 (N \to \infty)$ uniformly on $[0, 2\pi]$. Therefore,

$$\left| \sum_{|n|>N} f(t+2n\pi) \right| \to 0 \quad (N \to \infty)$$

uniformly on $[0, 2\pi]$, i.e., the series $\sum_n f(t+2n\pi)$ is absolutely, uniformly convergent everywhere on $[0, 2\pi]$. Denote

$$F(t) = \sum_n f(t+2n\pi) \quad (t \in [0, 2\pi]),$$

where $F(t) := \frac{1}{2}(F(t+0) + F(t-0))$ since $f(t) := \frac{1}{2}(f(t+0) + f(t-0))$. Then $F(t)$ is an integrable periodic function of bounded variation with period 2π and its total variation on $[0, 2\pi]$ is

$$\bigvee_{0}^{2\pi}(F) = \bigvee_{0}^{2\pi}\left(\sum_{n}f(t+2n\pi)\right) \le \sum_{n}\left(\bigvee_{0}^{2\pi}f(t+2n\pi)\right)$$

$$= \sum_{n}\left(\bigvee_{2n\pi}^{2(n+1)\pi}f(t)\right) = \bigvee_{-\infty}^{\infty}(f) < \infty,$$

According to the Jordan criterion, the Fourier series of $F(t)$ converges to $F(t)$, i.e.,

$$F(t) = \sum_{n}c_n(F)e^{int} \quad (t \in \mathbb{R}),$$

where $c_n(F)$ are the Fourier coefficients of F. By Lemma 1.1, we get $c_n(F) = \frac{1}{2\pi}\widehat{f}(n)$, and so

$$F(t) = \frac{1}{2\pi}\sum_{n}\widehat{f}(n)e^{int} \quad (t \in \mathbb{R}).$$

Noticing that $F(t) = \sum_{n}f(t+2n\pi)$, we have

$$\sum_{n}f(t+2n\pi) = \frac{1}{2\pi}\sum_{n}\widehat{f}(n)e^{int} \quad (t \in \mathbb{R}).$$

Let $t = 0$. Then

$$\sum_{n}f(2n\pi) = \frac{1}{2\pi}\sum_{n}\widehat{f}(n),$$

i.e., under condition (i), Poisson summation formula I holds.

Suppose that the function $f(t)$ satisfies condition (ii). Clearly, $f \in L(\mathbb{R})$ and $\widehat{f} \in L(\mathbb{R})$.

Consider the series $\sum_{n}f(t+2n\pi)$. Since $\widehat{f} \in L(\mathbb{R})$ and $2\pi f(-t) = \widehat{\widehat{f}}(t)$ (Property (v) of the Fourier transform), it follows from Theorem 1.1(iii) that $f(t)$ is uniformly continuous on $\mathbb{R}$. Since $|f(t)| \le K_1(1+|t|)^{-\alpha}(\alpha > 1)$, the series $\sum_{n}f(t+2n\pi)$ converges uniformly on $\mathbb{R}$. Denote its sum by $F(t)$, i.e., $F(t) = \sum_{n}f(t+2n\pi)$ on $\mathbb{R}$ uniformly and $F(t)$ is a continuous 2π-periodic function.

Denote the Fourier coefficients of $F(t)$ by $c_n(F)$. Then the Fourier series of $F(t)$ is $\sum_{n}c_n(F)e^{int}$. Since $f \in L(\mathbb{R})$, by Lemma 1.1(iii), $c_n(F) = \frac{1}{2\pi}\widehat{f}(n)$. So the Fourier series of $F(t)$ is $\frac{1}{2\pi}\sum_{n}\widehat{f}(n)e^{int}$.

By the condition (ii), $|\widehat{f}(n)| \le K_2(1+|n|)^{-\alpha}(\alpha > 1)$. So $\widehat{f}(n) \to 0$ monotonously as $n \to \infty$. By use of the Dirichlet criterion in calculus, it follows that $\frac{1}{2\pi}\sum_{n}\widehat{f}(n)e^{int} = F(t)(t \in \mathbb{R})$, i.e.,

$$\sum_{n}f(t+2n\pi) = \frac{1}{2\pi}\sum_{n}\widehat{f}(n)e^{int} \quad (t \in \mathbb{R}).$$

Let $t = 0$. Then

$$\sum_n f(2n\pi) = \frac{1}{2\pi} \sum_n \widehat{f}(n),$$

i.e., under condition (ii), Poisson summation formula I holds. $\qquad\square$

The derivation of the Poisson summation formula in $L^2(\mathbb{R})$ needs the following lemma.

Lemma 1.2 (Convolution in Frequency). *Suppose that* $f, g \in L^2(\mathbb{R})$. *Then*

$$2\pi (fg)^{\wedge}(\omega) = (\widehat{f} * \widehat{g})(\omega).$$

i.e., the convolution of Fourier transforms of two functions is equal to 2π times the Fourier transform of the product of these two functions.

Proof. By $f, g \in L^2(\mathbb{R})$, it follows that $fg \in L(\mathbb{R})$. So the Fourier transform of fg is

$$(fg)^{\wedge}(\omega) = \int_{\mathbb{R}} f(t)g(t) e^{-i\omega t} \, dt.$$

Let $h(t) = \overline{g}(t) e^{i\omega t}$, and then using Theorem 1.3, we get

$$(fg)^{\wedge}(\omega) = \int_{\mathbb{R}} f(t)\overline{h}(t) \, dt = \frac{1}{2\pi} \int_{\mathbb{R}} \widehat{f}(u)\overline{\widehat{h}}(u) \, du.$$

However, by the definition of the Fourier transform, the factor of the integrand on the right-side hand

$$\overline{\widehat{h}}(u) = \overline{\int_{\mathbb{R}} h(t) e^{-iut} \, dt} = \overline{\int_{\mathbb{R}} \overline{g}(t) e^{i\omega t} e^{-iut} \, dt} = \int_{\mathbb{R}} g(t) e^{-i(\omega-u)t} \, dt = \widehat{g}(\omega - u).$$

Therefore,

$$(fg)^{\wedge}(\omega) = \frac{1}{2\pi} \int_{\mathbb{R}} \widehat{f}(u)\widehat{g}(\omega - u) \, du = \frac{1}{2\pi} (\widehat{f} * \widehat{g})(\omega).$$

We get the desired result. $\qquad\square$

On the basis of Lemma 1.2 and Poisson summation formula I, we have

Poisson Summation Formula II. If $f \in L^2(\mathbb{R})$ and f satisfies one of the following two conditions:

(i) $\widehat{f}(\omega)$ is a function of bounded variation on $\mathbb{R}$;
(ii) $|f(t)| \leq K_1 |t|^{-\beta} (\beta > 1)$ and $|\widehat{f}(\omega)| \leq K_2 |\omega|^{-\alpha} (\alpha > \frac{1}{2})$, where K_1 and K_2 are constants, then

$$\sum_n |\widehat{f}(\omega + 2n\pi)|^2 = \sum_n \left(\int_{\mathbb{R}} f(t)\overline{f}(n + t) \, dt \right) e^{in\omega} \quad (\omega \in \mathbb{R}).$$

Proof. Let

$$\varphi(\omega) = |\widehat{f}(\omega)|^2 = \widehat{f}(\omega)\overline{\widehat{f}}(\omega).$$

By the assumption $f \in L^2(\mathbb{R})$ and Definition 1.1, $\hat{f} \in L^2(\mathbb{R})$, and so $\varphi \in L(\mathbb{R})$.

Suppose that $f(t)$ satisfies the first condition. Then φ is a function of bounded variation on $\mathbb{R}$. Define $\varphi(\omega) = \frac{1}{2}(\varphi(\omega + 0) + \varphi(\omega - 0))$. So $\varphi(\omega)$ satisfies the first condition of Poisson summation formula I.

Suppose that $f(t)$ satisfies the second condition. By the assumption $|\hat{f}(\omega)| \leq K_2|\omega|^{-\alpha}(\alpha > \frac{1}{2})$, we get $|\varphi(\omega)| \leq K_2^2|\omega|^{-2\alpha}(2\alpha > 1)$. By using Lemma 1.2, we get

$$\hat{\varphi}(u) = \left(\widehat{\hat{f}\hat{f}}\right)^\wedge (u) = \frac{1}{2\pi}\left(\widehat{\hat{f}} * \widehat{\hat{f}}\right)(u).$$

By Properties (iv) and (v) of the Fourier transform, $\widehat{\hat{f}}(u) = 2\pi f(-u)$ and $\widehat{\hat{f}}(u) = \widehat{\hat{f}}(-u) = 2\pi \bar{f}(u)$, and so

$$\hat{\varphi}(u) = 2\pi f(-u) * \bar{f}(u) = 2\pi \int_{\mathbb{R}} f(t)\bar{f}(u + t)\, dt, \qquad (1.9)$$

which can be rewritten in the form

$$\hat{\varphi}(u) = 2\pi \left(\int_{|t| \leq \frac{|u|}{2}} + \int_{|t| > \frac{|u|}{2}} \right) f(t)\bar{f}(u + t)\, dt = I_1(u) + I_2(u).$$

When $|t| \leq \frac{|u|}{2}$, we have $|u + t| \geq |u| - |t| \geq \frac{|u|}{2}$. From this and the assumption $|f(t)| \leq K_1|t|^{-\beta}(\beta > 1)$, we get

$$|I_1(u)| \leq 2\pi \int_{|t| \leq \frac{|u|}{2}} |f(t)\bar{f}(u + t)|\, dt$$

$$\leq 2\pi K_1^2 \int_{|t| \leq \frac{|u|}{2}} \frac{1}{|t(u + t)|^\beta}\, dt$$

$$\leq 2\pi \frac{2^\beta K_1^2}{|u|^\beta} \int_{\mathbb{R}} \frac{1}{|t|^\beta}\, dt \leq K_3|u|^{-\beta} \quad (\beta > 1),$$

where K_3 is a constant.

When $|t| > \frac{|u|}{2}$, by the assumption $|f(t)| \leq K_1|t|^{-\beta}(\beta > 1)$, we get

$$|I_2(u)| \leq 2\pi \int_{|t| > \frac{|u|}{2}} |f(t)\bar{f}(u + t)|\, dt$$

$$\leq 2\pi K_1^2 \int_{|t| > \frac{u}{2}} \frac{1}{|t(u + t)|^\beta}\, dt$$

$$\leq 2\pi \frac{2^\beta K_1^2}{|u|^\beta} \int_{\mathbb{R}} \frac{1}{|u + t|^\beta}\, dt \leq K_4|u|^{-\beta}, \quad \beta > 1,$$

where K_4 is a constant.

Therefore, $\widehat{\varphi}(u) \leq K|u|^{-\beta}(\beta > 1)$, where K is a constant. Therefore, φ satisfies the second condition of Poisson summation formula I.

Using Poisson summation formula I, we get

$$\sum_n \varphi(\omega + 2n\pi) = \frac{1}{2\pi} \sum_n \widehat{\varphi}(n) \, e^{in\omega}.$$

By (1.9), $\widehat{\varphi}(n) = 2\pi \int_{\mathbb{R}} f(t)\bar{f}(n + t) \, dt$, noticing that $\varphi(\omega) = |\widehat{f}(\omega)|^2$, we can rewrite this equality in the form

$$\sum_n |\widehat{f}(\omega + 2n\pi)|^2 = \sum_n \left(\int_{\mathbb{R}} f(t)\bar{f}(n + t) \, dt \right) e^{in\omega}.$$

So Poisson summation formula II holds. $\qquad\square$

The following lemma is used to prove the Shannon sampling theorem.

Lemma 1.3. *Let $X(\omega)$ be the characteristic function of $[-\pi, \pi]$, i.e.,*

$$X(\omega) = \begin{cases} 1, & |\omega| \leq \pi, \\ 0, & |\omega| > \pi. \end{cases}$$

Then the inverse Fourier transform of $X(\omega)e^{-in\omega}(n \in \mathbb{Z})$ is equal to $\frac{\sin \pi(t-n)}{\pi(t-n)}$, i.e.,

$$(X(\omega)e^{-in\omega})^{\vee}(t) = \frac{\sin \pi(t - n)}{\pi(t - n)} \quad (n \in \mathbb{Z}).$$

Proof. It is clear that $X(\omega)e^{-in\omega} \in L(\mathbb{R})$, and its inverse Fourier transform is

$$(X(\omega)e^{-in\omega})^{\vee}(t) = \frac{1}{2\pi} \int_{\mathbb{R}} (X(\omega)e^{-in\omega})e^{it\omega} d\omega$$

$$= \frac{1}{2\pi} \int_{\mathbb{R}} X(\omega)e^{i(t-n)\omega} d\omega.$$

Since $X(\omega) = 1(|\omega| \leq \pi)$ and $X(\omega) = 0(|\omega| > \pi)$, we get

$$(X(\omega)e^{-in\omega})^{\vee}(t) = \frac{1}{2\pi} \int_{-\pi}^{\pi} e^{i(t-n)\omega} d\omega = \frac{1}{2\pi} \frac{e^{i\pi(t-n)} - e^{-i\pi(t-n)}}{i(t - n)}$$

$$= \frac{\sin \pi(t - n)}{\pi(t - n)}.$$

$\qquad\square$

Shannon Sampling Theorem. *Let $f \in L^2(\mathbb{R})$ and its Fourier transform $\widehat{f}(\omega) = 0(|\omega| \geq \pi)$. Then the interpolation formula*

$$f(t) = \sum_n f(n) \frac{\sin \pi(t - n)}{\pi(t - n)} \quad (L^2)$$

holds, and the series $\sum_n f(n) \frac{\sin \pi(t-n)}{\pi(t-n)}$ converges uniformly to a continuous function $g(t)$ in every closed interval on $\mathbb{R}$ and $g(t) = f(t)$ almost everywhere.

Proof. From $\widehat{f}(\omega) = 0(|\omega| \geq \pi)$, it follows that $\widehat{f} \in L^2(\mathbb{R})$ and $\widehat{f} \in L(\mathbb{R})$. Take a 2π-periodic function $f_p(\omega)$ such that $f_p(\omega) = \widehat{f}(\omega)(|\omega| \leq \pi)$. Then $f_p(\omega) \in L_{2\pi}$ and $\widehat{f}(\omega) = f_p(\omega)X(\omega)$, where $X(\omega)$ is the characteristic function of $[-\pi, \pi]$.

We expand $f_p(\omega)$ into the Fourier series

$$f_p(\omega) = \sum_n c_n(f_p)e^{in\omega}(L^2), \tag{1.10}$$

where $c_n(f_p)$ are Fourier coefficients and

$$c_n(f_p) = \frac{1}{2\pi} \int_{-\pi}^{\pi} f_p(\omega)e^{-in\omega}\,d\omega \quad (n \in \mathbb{Z}).$$

By $\widehat{f}(\omega) = f_p(\omega)(|\omega| \leq \pi)$ and the assumption $\widehat{f}(\omega) = 0(|\omega| \geq \pi)$, and $\widehat{\widehat{f}}(t) = 2\pi f(-t)$ (property of the Fourier transform), it follows that

$$c_n(f_p) = \frac{1}{2\pi} \int_{-\pi}^{\pi} \widehat{f}(\omega)e^{-in\omega}\,d\omega = \frac{1}{2\pi} \int_{\mathbb{R}} \widehat{f}(\omega)e^{-in\omega}\,d\omega$$

$$= \frac{1}{2\pi}\widehat{\widehat{f}}(n) = f(-n) \quad (n \in \mathbb{Z}). \tag{1.11}$$

Combining this with (1.9), we get

$$f_p(\omega) = \sum_n f(-n)e^{in\omega} \quad (L^2).$$

Noticing that $\widehat{f}(\omega) = f_p(\omega)X(\omega)$, we get

$$\widehat{f}(\omega) = \sum_n f(-n)X(\omega)e^{in\omega} = \sum_n f(n)X(\omega)e^{-in\omega}.$$

Taking the inverse Fourier transform on both sides, we get

$$f(t) = \sum_n f(n)(X(\omega)e^{-in\omega})^{\vee}(t) \quad (L^2).$$

By Lemma 1.3, we get an interpolation formula:

$$f(t) = \sum_n f(n)\frac{\sin(\pi(t-n))}{\pi(t-n)} \quad (L^2). \tag{1.12}$$

From this, the Riesz theorem shows that the series $\sum_n f(n)\frac{\sin(\pi(t-n))}{\pi(t-n)}$ converges to $f(t)$ almost everywhere.

On the other hand, for Fourier series (1.10), by using Bessel's inequality, we get

$$\sum_n |c_n(f_p)|^2 \leq \frac{1}{2\pi} \int_{-\pi}^{\pi} |f_p(\omega)|^2\,d\omega.$$

By (1.11), $c_n(f_p) = f(-n)(n \in \mathbb{Z})$, the left-hand side is

$$\sum_n |c_n(f_p)|^2 = \sum_n |f(-n)|^2 = \sum_n |f(n)|^2.$$

By $f_p(\omega) = f(\omega)(|\omega| \leq \pi)$ and $\widehat{f}(\omega) = 0(|\omega| \geq \pi)$, the right-hand side is

$$\frac{1}{2\pi} \int_{-\pi}^{\pi} |f_p(\omega)|^2 \, d\omega = \frac{1}{2\pi} \int_{-\pi}^{\pi} |\widehat{f}(\omega)|^2 \, d\omega = \int_{\mathbb{R}} |\widehat{f}(\omega)|^2 \, d\omega.$$

Therefore,

$$\sum_n |f(n)|^2 \leq \frac{1}{2\pi} \int_{\mathbb{R}} |\widehat{f}(\omega)|^2 \, d\omega.$$

From $\widehat{f} \in L^2(\mathbb{R})$, it follows that $\sum_n |f(n)|^2 < \infty$. So the series $\sum_n |f(n)|^2$ converges. Since $\left| \frac{\sin(\pi(t-n))}{\pi(t-n)} \right| \leq \frac{1}{|t-n|}$, the series $\sum_n |\frac{\sin \pi(t-n)}{\pi(t-n)}|^2$ converges uniformly in every closed interval on $\mathbb{R}$.

According to Cauchy's principle of convergence in calculus, for $\epsilon > 0$, there is an $N > 0$ such that when $M \geq m > N$,

$$\sum_{m \leq |k| \leq M} |f(k)|^2 < \epsilon, \qquad \sum_{m \leq |k| \leq M} \left| \frac{\sin \pi(t-k)}{\pi(t-k)} \right|^2 < \epsilon$$

hold simultaneously in every closed interval on $\mathbb{R}$. By using Cauchy's inequality, we have

$$\left| \sum_{m \leq |k| \leq M} f(k) \frac{\sin \pi(t-k)}{\pi(t-k)} \right|^2 \leq \left(\sum_{m \leq |k| \leq M} |f(k)|^2 \right) \left(\sum_{m \leq |k| \leq M} \left| \frac{\sin(\pi(t-k))}{\pi(t-k)} \right|^2 \right).$$

Therefore, for the above $\epsilon > 0$ and $N > 0$, when $M \geq m > N$,

$$\left| \sum_{m \leq |k| \leq M} f(k) \frac{\sin \pi(t-k)}{\pi(t-k)} \right| < \epsilon$$

in every closed interval on $\mathbb{R}$. According to Cauchy's principle of convergence, the series

$$\sum_n f(n) \frac{\sin \pi(t-n)}{\pi(t-n)}$$

converges uniformly in every closed interval on $\mathbb{R}$ to a continuous function, denoted by $g(t)$. By (1.12), we get $g(t) = f(t)$ almost everywhere. $\qquad \square$

1.5   DISCRETE FOURIER TRANSFORM

Discrete Fourier transforms are used in discrete signal or discrete time series. The discrete Fourier transform is defined as follows.

Given an N-point time series $x = (x_0, x_1, \ldots, x_{N-1})$, the *discrete Fourier transform* of x is defined as

$$X_k = \frac{1}{N} \sum_0^{N-1} x_n e^{-in\frac{2\pi k}{N}} \quad (k = 0, 1, \ldots, N-1).$$

In this definition, x_n is called the *sample*, N is called the *number of samples*, $\Delta\omega = \frac{2\pi}{N}$ is called the *sampling frequency interval*, $\omega_k = \frac{2\pi k}{N}$ is called the *discrete frequency*, X_k is called the *frequency coefficient*, and $\{|X_k|^2\}_{k=0,\ldots,N-1}$ is called the *Fourier power spectrum* of x. In detail, the discrete Fourier transform gives a system of equations as follows:

$$X_0 = \frac{1}{N} \sum_0^{N-1} x_n = \frac{1}{N}(x_0 + x_1 + x_2 + \cdots + x_{N-1}),$$

$$X_1 = \frac{1}{N} \sum_0^{N-1} x_n e^{-in\frac{2\pi}{N}} = \frac{1}{N}\left(x_0 + x_1 e^{-i\frac{2\pi}{N}} + x_2 e^{-i\frac{4\pi}{N}} + \cdots + x_{N-1} e^{-i\frac{2(N-1)\pi}{N}}\right),$$

$$\vdots$$

$$X_{N-1} = \frac{1}{N} \sum_0^{N-1} x_n e^{-in\frac{2\pi(N-1)}{N}}$$

$$= \frac{1}{N}\left(x_0 + x_1 e^{-i\frac{2(N-1)\pi}{N}} + x_2 e^{-i\frac{4(N-1)\pi}{N}} + \cdots + x_{N-1} e^{-i\frac{2(N-1)^2\pi}{N}}\right).$$

It can also be rewritten in the matrix form $\mathbf{X} = \frac{1}{N}\mathbf{F}\mathbf{x}$, where

$$\mathbf{X} = \begin{pmatrix} X_0 \\ X_1 \\ \vdots \\ X_{N-1} \end{pmatrix} \quad \mathbf{x} = \begin{pmatrix} x_0 \\ x_1 \\ \vdots \\ x_{N-1} \end{pmatrix}$$

and

$$\mathbf{F} = \begin{pmatrix} 1 & 1 & 1 & \cdots & 1 \\ 1 & e^{-i\frac{2\pi}{N}} & e^{-i\frac{4\pi}{N}} & \cdots & e^{-i\frac{2(N-1)\pi}{N}} \\ \vdots & \vdots & \vdots & \vdots & \vdots \\ 1 & e^{-i\frac{2(N-1)\pi}{N}} & e^{-i\frac{4(N-1)\pi}{N}} & \cdots & e^{-i\frac{2(N-1)^2\pi}{N}} \end{pmatrix} = \left(e^{-in\frac{2\pi k}{N}}\right)_{k,n=0,1,\ldots,N-1}.$$

If all x_n are real, the discrete Fourier transform possesses the property of symmetry:

$$X_{N-k} = \overline{X}_k \quad (k = 0, 1, \ldots, N-1).$$

In fact, by the definition of the discrete Fourier transform,

$$X_{N-k} = \frac{1}{N} \sum_{0}^{N-1} x_n e^{-in\frac{2\pi(N-k)}{N}}.$$

Since $e^{-in\frac{2\pi(N-k)}{N}} = e^{-i(2\pi n - \frac{2n\pi k}{N})} = e^{in\frac{2\pi k}{N}}$, we get

$$X_{N-k} = \frac{1}{N} \sum_{0}^{N-1} x_n e^{in\frac{2\pi k}{N}}.$$

On the other hand, since x_n are real, it follows by the definition that

$$\overline{X}_k = \frac{1}{N} \sum_{0}^{N-1} \overline{x_n e^{-in\frac{2\pi k}{N}}} = \frac{1}{N} \sum_{0}^{N-1} x_n e^{in\frac{2\pi k}{N}}.$$

Therefore, $X_{N-k} = \overline{X}_k (k = 0, 1, \ldots, N-1)$.

Given N frequency coefficients $\{X_k\}_{k=0,\ldots,N-1}$, the *inverse discrete Fourier transform* is defined as

$$x_n = \sum_{0}^{N-1} X_k e^{ik\frac{2\pi n}{N}} \quad (n = 0, \ldots, N-1).$$

In detail, the inverse discrete Fourier transform gives the system of equations as follows:

$$x_0 = \sum_{0}^{N-1} X_k = X_0 + X_1 + X_2 + \cdots + X_{N-1},$$

$$x_1 = \sum_{0}^{N-1} X_k e^{ik\frac{2\pi}{N}} = X_0 + X_1 e^{i\frac{2\pi}{N}} + X_2 e^{i\frac{4\pi}{N}} + \cdots + X_{N-1} e^{i\frac{2(N-1)\pi}{N}},$$

$$\vdots$$

$$x_{N-1} = \sum_{0}^{N-1} X_k e^{ik\frac{2(N-1)\pi}{N}} = X_0 + X_1 e^{i\frac{2(N-1)\pi}{N}} + X_2 e^{i\frac{4(N-1)\pi}{N}} + \cdots + X_{N-1} e^{i\frac{2(N-1)^2\pi}{N}}.$$

It can also be written in the matrix form $\mathbf{x} = \overline{\mathbf{F}}\mathbf{X}$, where

$$\mathbf{x} = \begin{pmatrix} x_0 \\ x_1 \\ \vdots \\ x_{N-1} \end{pmatrix}, \quad \mathbf{X} = \begin{pmatrix} X_0 \\ X_1 \\ \vdots \\ X_{N-1} \end{pmatrix},$$

and

$$\mathbf{F} = \begin{pmatrix} 1 & 1 & 1 & \cdots & 1 \\ 1 & e^{i\frac{2\pi}{N}} & e^{i\frac{4\pi}{N}} & \cdots & e^{i\frac{2\pi(N-1)}{N}} \\ \vdots & \vdots & \vdots & \vdots & \vdots \\ 1 & e^{i\frac{2(N-1)\pi}{N}} & e^{i\frac{4(N-1)\pi}{N}} & \cdots & e^{i\frac{2(N-1)^2\pi}{N}} \end{pmatrix} = \left(e^{in\frac{2\pi k}{N}} \right)_{k,n=0,1,\ldots,N-1}.$$

If all X_n are real, the inverse discrete Fourier transform possesses the property of symmetry:

$$x_{N-k} = \bar{x}_k \quad (k = 0, 1, \ldots, N-1).$$

In fact, the definition of the inverse discrete Fourier transform shows that

$$x_{N-k} = \sum_{0}^{N-1} X_n e^{in\frac{2\pi(N-k)}{N}}.$$

Since $e^{in\frac{2\pi(N-k)}{N}} = e^{i2\pi n}e^{-in\frac{2\pi k}{N}} = e^{-in\frac{2\pi k}{N}}$, we have

$$x_{N-k} = \sum_{0}^{N-1} X_n e^{-in\frac{2\pi k}{N}}.$$

On the other hand, since X_n are real, it follows by the definition that

$$\bar{x}_k = \sum_{0}^{N-1} \overline{X_n e^{in\frac{2\pi k}{N}}} = \sum_{0}^{N-1} X_n e^{-in\frac{2\pi k}{N}}.$$

Therefore, $x_{N-k} = \bar{x}_k (k = 0, 1, \ldots, N-1)$.

Similarly, we consider the two-dimensional case. Given a two-dimensional discrete $M \times N$-point time series,

$$x = \begin{pmatrix} x_{0,0} & x_{0,1} & \cdots & x_{0,N-1} \\ x_{1,0} & x_{1,1} & \cdots & x_{1,N-1} \\ \vdots & \vdots & \vdots & \vdots \\ x_{M-1,0} & x_{M-1,1} & \cdots & x_{M-1,N-1} \end{pmatrix},$$

the *two-dimensional discrete Fourier transform* of x is defined as

$$X_{k,l} = \frac{1}{MN} \sum_{m=0}^{M-1} \left(\sum_{n=0}^{N-1} x_{m,n} e^{-in\frac{2\pi l}{N}} \right) e^{-im\frac{2\pi k}{M}}$$

$$(k = 0, \ldots, M-1; l = 0, \ldots, N-1).$$

In this definition, $x_{m,n}$ is called the *sample* and $X_{k,l}$ are called the *frequency coefficient*. The two-dimensional discrete Fourier transform is a transform from an $M \times N$-point time series to $M \times N$ frequency coefficients.

Conversely, given $M \times N$ frequency coefficients:

$$X_{0,0} \quad X_{0,1} \quad \cdots \quad X_{0,N-1},$$
$$X_{1,0} \quad X_{1,1} \quad \cdots \quad X_{1,N-1},$$
$$\vdots \qquad \vdots \qquad \vdots \qquad \vdots$$
$$X_{M-1,0} \quad X_{M-1,1} \quad \cdots \quad X_{M-1,N-1}.$$

The *two-dimensional inverse discrete Fourier transform* of $\{X_{k,l}\}$ is defined as

$$x_{m,n} = \sum_{k=0}^{M-1} \left(\sum_{l=0}^{N-1} X_{k,l} e^{il\frac{2\pi n}{N}} \right) e^{ik\frac{2\pi m}{M}}$$

$$(m = 0, 1, \ldots, M - 1; n = 0, 1, \ldots, N - 1).$$

The two-dimensional inverse discrete Fourier transform is a transform from $M \times N$ frequency coefficients to an $M \times N$-point time series.

1.6 FAST FOURIER TRANSFORM

The fast Fourier transform is not a new transform, but is a fast algorithm for computing discrete Fourier transform. It is based on the halving trick. This trick halves a given $2N$-point time series into two N-point time subseries, and then discrete Fourier transforms of these two N-point time subseries are used to compute the discrete Fourier transform of the given $2N$-point time series.

Given a $2N$-point time series $z = (z_0, z_1, \ldots, z_{2N-1})$, its discrete Fourier transform is

$$Z_k = \frac{1}{2N} \sum_{0}^{2N-1} z_n e^{-in\frac{2\pi k}{2N}} \quad (k = 0, \ldots, 2N - 1).$$

We use the halving trick to halve the $2N$ frequency coefficients $\{Z_k\}$.

First, we compute the first half: $Z_0, Z_1, \ldots, Z_{N-1}$. We decompose the $2N$-point time series z into two N-point time series x and y as follows:

$$x = (z_0, z_2, z_4, \ldots, z_{2N-2}) =: (x_0, x_1, x_2 \ldots, x_{N-1}),$$
$$y = (z_1, z_3, z_5 \ldots, z_{2N-1}) =: (y_0, y_1, y_2, \ldots, y_{N-1}),$$

i.e., x consists of even samples of $\{z_n\}$ and y consists of odd samples of $\{z_n\}$. So

$$Z_k = \frac{1}{2N} \sum_{0}^{N-1} z_{2n} e^{i2n\frac{2\pi k}{2N}} + \frac{1}{2N} \sum_{0}^{N-1} z_{2n+1} e^{i(2n+1)\frac{2\pi k}{2N}}$$

$$= \frac{1}{2N} \sum_{0}^{N-1} x_n e^{i2n\frac{2\pi k}{2N}} + \frac{1}{2N} \sum_{0}^{N-1} y_n e^{i(2n+1)\frac{2\pi k}{2N}} \quad (k = 0, \ldots, N - 1),$$

where $x_n = z_{2n}$ and $y_n = z_{2n+1}$. Since $e^{-i(2n+1)\frac{2\pi k}{2N}} = e^{-in\frac{2\pi k}{N}}e^{-i\frac{\pi k}{N}}$, we get

$$Z_k = \frac{1}{2N}\sum_0^{N-1} x_n e^{-in\frac{2\pi k}{N}} + e^{-i\frac{\pi k}{N}}\frac{1}{2N}\sum_0^{N-1} y_n e^{-in\frac{2\pi k}{N}}$$

$$= \frac{1}{2}(X_k + e^{-i\frac{\pi k}{N}}Y_k) \quad (k = 0,\ldots,N-1),$$

where

$$X_k = \frac{1}{N}\sum_0^{N-1} x_n e^{-in\frac{2\pi k}{N}} \quad (k = 0,\ldots,N-1),$$

$$Y_k = \frac{1}{N}\sum_0^{N-1} y_n e^{-in\frac{2\pi k}{N}} \quad (k = 0,\ldots,N-1).$$

Clearly, they are just discrete Fourier transforms of $x = (x_0, x_1, \ldots, x_{N-1})$ and $y = (y_0, y_1, \ldots, y_{N-1})$, respectively.

Second, we compute the second half: $Z_N, Z_{N+1}, \ldots, Z_{2N-1}$. Note that

$$Z_k = \frac{1}{2N}\sum_0^{2N-1} z_n e^{-in\frac{2\pi k}{2N}} \quad (k = N,\ldots,2N-1).$$

Taking the substitution $k = j + N$, we get

$$Z_{j+N} = \frac{1}{2N}\sum_0^{2N-1} z_n e^{-in\frac{2\pi(j+N)}{2N}} \quad (j = 0,\ldots,N-1).$$

We decompose Z_{j+N} into two sums according to even samples and odd samples as follows:

$$Z_{j+N} = \frac{1}{2N}\sum_0^{N-1} z_{2n}e^{-i2n\frac{2\pi(j+N)}{2N}} + \frac{1}{2N}\sum_0^{N-1} z_{2n+1}e^{-i(2n+1)\frac{2\pi(j+N)}{2N}}$$

$$(j = 0,\ldots,N-1).$$

Noticing that

$$e^{-in\frac{2\pi(j+N)}{N}} = e^{-in\frac{2\pi j}{N}}e^{-i2\pi n} = e^{-in\frac{2\pi j}{N}},$$

$$e^{-i(2n+1)\frac{2\pi(j+N)}{2N}} = e^{-i(2n+1)\frac{\pi j}{N}}e^{-i\pi(2n+1)} = -e^{-in\frac{2\pi j}{N}}e^{-i\frac{\pi j}{N}},$$

we get

$$Z_{j+N} = \frac{1}{2N}\sum_0^{N-1} z_{2n}e^{-in\frac{2\pi j}{N}} - e^{-i\frac{\pi j}{N}}\frac{1}{2N}\sum_0^{N-1} z_{2n+1}e^{-in\frac{2\pi j}{N}} \quad (j = 0,\ldots,N-1).$$

Noticing that $x_n = z_{2n}$ and $y_n = z_{2n+1}$, and replacing j by k, we get

$$Z_{k+N} = \frac{1}{2N} \sum_0^{N-1} x_n e^{-in\frac{2\pi k}{N}} - e^{-i\frac{\pi k}{N}} \frac{1}{2N} \sum_0^{N-1} y_n e^{-in\frac{2\pi k}{N}}$$

$$= \frac{1}{2}(X_k - e^{-i\frac{\pi k}{N}} Y_k) \quad (k = 0, \ldots, N-1),$$

where X_k and Y_k are stated as above.

Summarizing the above procedure, we see that for the given $2N$-point time series $z = (z_0, z_1, \ldots, z_{2N-1})$, its discrete Fourier transform is

$$Z_k = \frac{1}{2N} \sum_0^{2N-1} z_n e^{-in\frac{2\pi k}{2N}} \quad (k = 0, \ldots, 2N-1).$$

Halving these frequency coefficients $Z_k (k = 0, \ldots, 2N-1)$, we obtain a pair of frequency coefficients:

$$Z_k = \frac{1}{2}(X_k + e^{-i\frac{\pi k}{N}} Y_k) \quad (k = 0, \ldots, N-1),$$

$$Z_{k+N} = \frac{1}{2}(X_k - e^{-i\frac{\pi k}{N}} Y_k) \quad (k = 0, \ldots, N-1),$$

where

$$X_k = \frac{1}{N} \sum_0^{N-1} x_n e^{-in\frac{2\pi k}{N}} \quad (k = 0, \ldots, N-1),$$

$$Y_k = \frac{1}{N} \sum_0^{N-1} y_n e^{-in\frac{2\pi k}{N}} \quad (k = 0, \ldots, N-1),$$

are discrete Fourier transforms of two N-point time subseries x and y which consist of even samples and odd samples of z, respectively.

Now we explain the procedure of the fast Fourier transform.

Given a 2^N-point time series $z = (z_0, z_1, \ldots, z_{2^N-1})$, its discrete Fourier transform is

$$Z_k = \frac{1}{2^N} \sum_0^{2^N-1} z_n e^{-in\frac{2\pi k}{2^N}} \quad (k = 0, 1, \ldots, 2^N - 1).$$

Halving these frequency coefficients $Z_k (k = 0, 1, \ldots, 2^N - 1)$, we obtain a pair of frequency coefficients:

$$Z_k = \frac{1}{2}(X_k + e^{-i\frac{2\pi k}{2^N}} Y_k) \quad (k = 0, \ldots, 2^{N-1} - 1),$$

$$Z_{k+2^{N-1}} = \frac{1}{2}(X_k - e^{-i\frac{2\pi k}{2^N}} Y_k) \quad (k = 0, \ldots, 2^{N-1} - 1),$$

where X_k $(k = 0, \ldots, 2^{N-1} - 1)$ and Y_k $(k = 0, \ldots, 2^{N-1} - 1)$ are the discrete Fourier transforms of two 2^{N-1}-point time subseries x and y which consist of even samples and odd samples of z, respectively.

Again, halving $X_k, Y_k (k = 0, \ldots, 2^{N-1} - 1)$, we obtain two pairs of frequency coefficients:

$$X_k = \frac{1}{2}(X_k' + \mathrm{e}^{-\mathrm{i}\frac{2\pi k}{2^{N-1}}} X_k'') \quad (k = 0, \ldots, 2^{N-2} - 1),$$

$$X_{k+2^{N-2}} = \frac{1}{2}(X_k' - \mathrm{e}^{-\mathrm{i}\frac{2\pi k}{2^{N-1}}} X_k'') \quad (k = 0, \ldots, 2^{N-2} - 1),$$

and

$$Y_k = \frac{1}{2}(Y_k' + \mathrm{e}^{-\mathrm{i}\frac{2\pi k}{2^{N-1}}} Y_k'') \quad (k = 0, \ldots, 2^{N-2} - 1),$$

$$Y_{k+2^{N-2}} = \frac{1}{2}(Y_k' - \mathrm{e}^{-\mathrm{i}\frac{2\pi k}{2^{N-1}}} Y_k'') \quad (k = 0, \ldots, 2^{N-2} - 1),$$

where X_k', X_k'' and $Y_k', Y_k'' (k = 0, \ldots, 2^{N-2} - 1)$ are the discrete Fourier transforms of two 2^{N-2}-point time subseries which consist of even samples and odd samples of x and y, respectively.

If the above procedure is continued again and again, the fast Fourier transform algorithm terminates at the one-point time subseries.

For this fast algorithm, the total number of multiplication operations is equal to $N2^{N-1}$. Let $2^N = M$. Then the total number of multiplication operations is equal to $\frac{1}{2}(\log_2 M)M$. For the original discrete Fourier transform algorithm, the total number of multiplication operations is equal to $2^{2N} = M^2$. So the fast Fourier transform has better computationally efficiency than the discrete Fourier transform.

Zero padding is another trick. It can be used to decrease the frequency interval. The zero padding trick is as follows.

Given an N-point time series $x = (x_0, x_1, \ldots, x_{N-1})$, the discrete Fourier transform of x is

$$X_k = \frac{1}{N} \sum_0^{N-1} x_n \mathrm{e}^{-\mathrm{i}n\frac{2\pi k}{N}} \quad (k = 0, 1, \ldots, N - 1).$$

The sampling frequency interval $\Delta\omega = \frac{2\pi}{N}$, where N is the number of samples. From this, we see that the sampling frequency interval is controlled by the number of samples of the time series.

Let $M > N$. We define a new M-point time series as follows:

$$x^{\mathrm{new}} = (x_0, x_1, \ldots, x_{N-1}, 0, 0, \ldots, 0).$$

The discrete Fourier transform of the new M-point time series is

$$X_k^{\text{new}} = \frac{1}{M} \sum_0^{M-1} x_n e^{-in\frac{2\pi k}{M}} \quad (k = 0, 1, \ldots, M - 1).$$

Note that $x_n = 0(n = N, \ldots, M - 1)$, and the discrete Fourier transform of the new M-point time series is

$$X_k^{\text{new}} = \frac{1}{M} \sum_0^{N-1} x_n e^{-in\frac{2\pi k}{M}} \quad (k = 0, \ldots, M - 1).$$

The new sampling frequency interval $\Delta\omega^{\text{new}} = \frac{2\pi}{M}$. By $M > N$, we see that

$$\Delta\omega^{\text{new}} = \frac{2\pi}{M} < \frac{2\pi}{N} = \Delta\omega.$$

This means that when the zero padding trick is used, the sampling frequency interval decreases.

1.7 HEISENBERG UNCERTAINTY PRINCIPLE

The Heisenberg uncertainty principle is the fundament of time-frequency analysis in Chapter 2. This principle is related to the temporal variance and the frequency variance of signals of finite energy.

Heisenberg Uncertainty Principle. *If $f \in L^2(\mathbb{R})$, then*

$$\left(\int_{\mathbb{R}} t^2 |f(t)|^2 \, dt \right) \left(\int_{\mathbb{R}} \omega^2 |\widehat{f}(\omega)|^2 \, d\omega \right) \geq \frac{\pi}{2} \, \| f \|_2^4 \, .$$

In particular, the necessary and sufficient condition that the sign of equality holds is $f(t) = Ce^{-t^2/4a}$, where C is a constant and $a > 0$.

Proof. By the assumption $f \in L^2(\mathbb{R})$ and Definition 1.1, it is clear that $\widehat{f} \in L^2(\mathbb{R})$. When

$$\int_{\mathbb{R}} t^2 |f(t)|^2 \, dt = \infty \quad \text{or} \quad \int_{\mathbb{R}} \omega^2 |\widehat{f}(\omega)|^2 \, d\omega = \infty,$$

the conclusion holds clearly. Therefore, we may assume that

$$\int_{\mathbb{R}} t^2 |f(t)|^2 \, dt < \infty, \quad \int_{\mathbb{R}} \omega^2 |\widehat{f}(\omega)|^2 \, d\omega < \infty.$$

Based on this assumption and noticing that

$$\left(\int_{\mathbb{R}} |f(t)| \, dt \right)^2 \leq \int_{\mathbb{R}} \frac{1}{1 + |t|^2} \, dt \int_{\mathbb{R}} (1 + |t|^2) |f(t)|^2 \, dt < \infty,$$

it follows that $f \in L(\mathbb{R})$. Similarly, $\widehat{f} \in L(\mathbb{R})$.

Note that

$$\text{Re}(tf(t)\overline{f'}(t)) \le |tf(t)\overline{f'}(t)| = |tf(t)f'(t)|,$$

and using Cauchy's inequality, we get

$$\left| \int_{\mathbb{R}} \text{Re}(tf(t)\overline{f'}(t)) \, dt \right|^2 \le \left(\int_{\mathbb{R}} |tf(t)f'(t)| dt \right)^2$$

$$\le \left(\int_{\mathbb{R}} t^2 |f(t)|^2 \, dt \right) \left(\int_{\mathbb{R}} |f'(t)|^2 \, dt \right).$$

Using Parseval's equality and $\widehat{f'}(\omega) = (i\omega)\widehat{f}(\omega)$ (Property (vi) of the Fourier transform), we find the integral in the second set of brackets on the right-hand side is

$$\int_{\mathbb{R}} |f'(t)|^2 \, dt = \frac{1}{2\pi} \int_{\mathbb{R}} |\widehat{f'}(\omega)|^2 \, d\omega = \frac{1}{2\pi} \int_{\mathbb{R}} \omega^2 |\widehat{f}(\omega)|^2 \, d\omega.$$

Therefore,

$$\left| \int_{\mathbb{R}} \text{Re}(tf(t)\overline{f'}(t)) \, dt \right|^2 \le \frac{1}{2\pi} \left(\int_{\mathbb{R}} t^2 |f(t)|^2 \, dt \right) \left(\int_{\mathbb{R}} \omega^2 |\widehat{f}(\omega)|^2 \, d\omega \right). \quad (1.13)$$

Let $f(t) = u(t) + iv(t)$, where u and v are real functions. Then $u(t) = \text{Re}f(t)$ and $v(t) = \text{Im}f(t)$, and

$$f'(t) = u'(t) + iv'(t),$$

$$\frac{d|f(t)|^2}{dt} = \frac{d}{dt}(u^2(t) + v^2(t)) = 2(u(t)u'(t) + v(t)v'(t)),$$

and so

$$\text{Re}(tf(t)\overline{f'}(t)) = \text{Re}\{t(u(t) + iv(t))(u'(t) - iv'(t))\}$$

$$= t(u(t)u'(t) + v(t)v'(t)) = \frac{t}{2} \frac{d|f(t)|^2}{dt}.$$

Integrating both sides over $\mathbb{R}$, we get

$$\int_{\mathbb{R}} \text{Re}(tf(t)\overline{f'}(t)) \, dt = \int_{\mathbb{R}} \frac{t}{2} \frac{d|f(t)|^2}{dt} \, dt.$$

Using integration by parts and noticing that $\lim_{r\to\infty} \left(r|f(r)|^2 \right) \to 0$ and $\lim_{r\to\infty} \left(r|f(-r)|^2 \right) \to 0$, we obtain for the right-hand side

$$\int_{\mathbb{R}} \frac{t}{2} \frac{d|f(t)|^2}{dt} \, dt = \lim_{r\to\infty} \left(\frac{r}{2} |f(r)|^2 + \frac{r}{2} |f(-r)|^2 \right)$$

$$- \frac{1}{2} \int_{\mathbb{R}} |f(t)|^2 \, dt = -\frac{1}{2} \| f \|_2^2.$$

So

$$\int_{\mathbb{R}} \text{Re}(tf(t)\overline{f'}(t))\,dt = -\frac{1}{2}\,\|f\|_2^2,$$

and so

$$\left|\int_{\mathbb{R}} \text{Re}(tf(t)\overline{f'}(t))\,dt\right|^2 = \frac{1}{4}\,\|f\|_2^4.$$

Combining this with (1.13), we get the desired result:

$$\left(\int_{\mathbb{R}} t^2|f|^2\,dt\right)\left(\int_{\mathbb{R}} \omega^2|\widehat{f}(\omega)|^2\,d\omega\right) \geq \frac{\pi}{2}\,\|f\|_2^4. \tag{1.14}$$

Now we give the necessary and sufficient condition that the sign of equality of (1.14) holds. According to the line of the proof of the inequality (1.14), we need to prove only that the necessary and sufficient condition that the equalities

$$\left|\int_{\mathbb{R}} \text{Re}(tf(t)\overline{f'}(t))\,dt\right|^2 = \left(\int_{\mathbb{R}} |tf(t)f'(t)|\,dt\right)^2$$

$$= \left(\int_{\mathbb{R}} t^2|f(t)|^2\,dt\right)\left(\int_{\mathbb{R}} |f'(t)|^2\,dt\right) \tag{1.15}$$

hold is $f(t) = Ce^{-t^2/4a}$, where C is a constant and $a > 0$.

If the first sign of equality holds, then

$$\int_{\mathbb{R}} \text{Re}(tf(t)\overline{f'}(t))\,dt = \int_{\mathbb{R}} |tf(t)f'(t)|\,dt \quad \text{or}$$

$$-\int_{\mathbb{R}} \text{Re}(tf(t)\overline{f'}(t))\,dt = \int_{\mathbb{R}} |tf(t)f'(t)|\,dt.$$

From $\pm\text{Re}(tf(t)f'(t)) \leq |tf(t)f'(t)|$, it follows that

$$\text{Re}(tf(t)\overline{f'}(t)) = |tf(t)\overline{f'}(t)| \quad \text{or} \quad \text{Re}(tf(t)\overline{f'}(t)) = -|tf(t)\overline{f'}(t)|,$$

and so

$$tf(t)\overline{f'}(t) \geq 0 \quad \text{or} \quad tf(t)\overline{f'}(t) \leq 0.$$

If the second sign of equality holds, then $|tf(t)| = 2a|f'(t)|$ $(a > 0)$, and so $tf(t) = 2af'(t)e^{i\theta(t)}$. Multiplying both sides by $\overline{f'}(t)$, we get

$$tf(t)\overline{f'}(t) = 2a|f'(t)|^2e^{i\theta(t)}.$$

If these two signs of equality hold simultaneously, i.e., (1.15) holds, then the results

$$tf(t)\overline{f'}(t) \geq 0 \quad \text{or} \quad tf(t)\overline{f'}(t) \leq 0,$$

$$tf(t)\overline{f'}(t) = 2a|f'(t)|^2e^{i\theta(t)}$$

hold simultaneously. So $e^{i\theta(t)} = \pm 1$, and so

$$tf(t) = 2af'(t) \quad \text{or} \quad tf(t) = -2af'(t).$$

Solving these two equations, we obtain that $f(t) = Ce^{t^2/4a}$ or $f(t) = Ce^{-t^2/4a}$. Noticing that $e^{t^2/4a} \notin L^2(\mathbb{R})$, we obtain finally that the necessary and sufficient condition that the sign of equality of (1.14) holds is $f(t) = Ce^{-t^2/4a}$, where C is a constant and $a > 0$. $\square$

1.8   CASE STUDY: ARCTIC OSCILLATION INDICES

The Arctic Oscillation (AO) is a key aspect of climate variability in the Northern Hemisphere (see Figure 1.1). The AO indices are defined as the leading empirical orthogonal function of Northern Hemisphere sea level pressure anomalies poleward of 20°N and are characterized by an exchange of atmospheric mass between the Arctic and middle latitudes (Thompson and Wallace, 1998). We research the Fourier power spectrum of AO indices (December to February 1851-1997) with the help of the discrete Fourier transform (see Figure 1.2). The highest peak in the Fourier power spectrum occurs with a period of about 2.2 years.

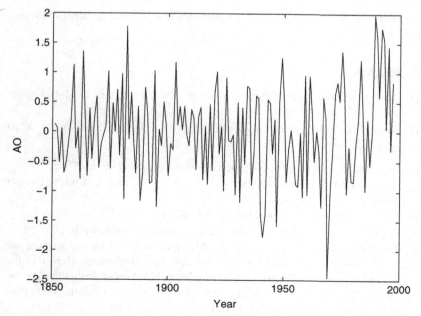

FIGURE 1.1   AO indices.

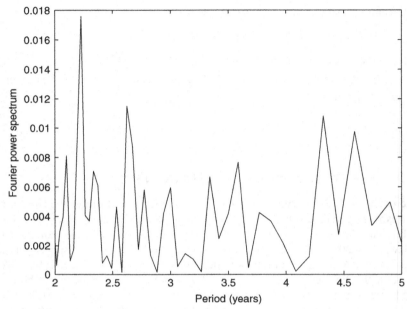

FIGURE 1.2 Fourier power spectrum of AO indices.

In Chapter 7, using the statistical significant test, we will do further research on it.

PROBLEMS

1.1 Let f be a 2π-periodic signal and $f(t) = |t| (t \in [-\pi, \pi])$. Find its Fourier series and Parseval's equality.

1.2 Show that the Legendre polynomials $X_n(t) (n = 0, 1, \ldots)$ satisfy $\int_{-1}^{1} X_n^2(t)\, dt = \frac{2}{2n+1}$.

1.3 Find the Fourier transform of the Gaussian function $f(t) = e^{-t^2/2}$.

1.4 Given a four-point time series $x = (i, 1, -i, 1 + i)$, find its discrete Fourier transform.

1.5 Compute the one-sided Laplace transform of te^{-2t}.

1.6 Let $t = (t_1, t_2)$. Find the two-dimensional Fourier transform of $e^{-|t|^2/2}$.

1.7 The North Atlantic Oscillation (NAO) index is based on the surface sea level pressure difference between the Subtropical (Azores) High and the Subpolar Low. Download the monthly mean NAO index from http://www.cpc.ncep.noaa.gov/products/precip/CWlink/pna/new.nao.shtml and then research the Fourier power spectrum of the NAO index.

BIBLIOGRAPHY

Allen, M.R., Smith, L.A., 1994. Investigating the origins and significance of low-frequency modes of climate variability. Geophys. Res. Lett. 21, 883-886.

Andreo, B., Jimenez, P., Duran, J.J., Carrasco, F., Vadillo, I., Mangin, A., 2006. Climatic and hydrological variations during the last 117-166 years in the south of the Iberian Peninsula, from spectral and correlation analyses and continuous wavelet analyses. J. Hydrol. 324, 24-39.

Chandrasekharan, K., 1989. Classical Fourier Transforms. Springer-Verlag, Berlin.

Duhamel, P., Vetterli, M., 1990. Fast Fourier transforms: a tutorial review and a state of the art. Signal Process. 19, 259-299.

Ghil, M., et al., 2002. Advanced spectral methods for climatic time series. Rev. Geophys. 40, 1003.

Lüdecke, H.-J., Hempelmann, A., Weiss, C.O., 2013. Multi-periodic climate dynamics: spectral analysis of long-term instrumental and proxy temperature records. Clim. Past 9, 447-452.

Lee, H.S., Yamashita, T., Mishima, T., 2012. Multi-decadal variations of ENSO, the pacific decadal oscillation and tropical cyclones in the western North Pacific. Prog. Oceanogr. 105, 67-80.

Murray, R.J., Reason, C.J.C., 2002. Fourier filtering and coefficient tapering at the North Pole in OGCMs. Ocean Model. 4, 1-25.

Nussbaumer, H.J., 1982. Fast Fourier Transform and Convolution Algorithms. Springer-Verlag, Berlin.

Osborne, A.R., 2010. Nonlinear Fourier analysis and filtering of ocean waves. Int. Geophys. 97, 713-744.

Papoulis, A., 1987. The Fourier Integral and its Applications. McGraw-Hill, New York.

Schulz, M., Stattegger, K., 1997. Spectrum: spectral analysis of unevenly spaced paleoclimatic time series. Comput. Geosci. 23, 929-945.

Sleighter, R.L., Hatcher, P.G., 2008. Molecular characterization of dissolved organic matter (DOM) along a river to ocean transect of the lower Chesapeake Bay by ultrahigh resolution electrospray ionization Fourier transform ion cyclotron resonance mass spectrometry. Mar. Chem. 110, 140-152.

Strichartz, R., 1994. A Guide to Distribution Theory and Fourier Transforms. CRC Press, Boca Raton.

Swindles, G.T., Patterson, R.T., Roe, H.M., Galloway, J.M., 2012. Evaluating periodicities in peat-based climate proxy records. Quat. Sci. Rev. 41, 94-103.

Szego, G., 1959. Orthogonal Polynomials, vol. 23. AMS Colloquium Publications, Providence, RI.

Thompson, D.W.J., Wallace, J.M., 1998. The Arctic Oscillation signature in the winter geopotential height and temperature fields. Geophys. Res. Lett. 25, 1297-1300.

Wirth, A., 2005. A non-hydrostatic flat-bottom ocean model entirely based on Fourier expansion. Ocean Model. 9, 71-87.

Yiou, P., Baert, E., Loutre, M.F., 1996. Spectral analysis of climate data. Surv. Geophys. 17, 619-663.

Zhang, Z., Moore, J.C., 2011. New significance test methods for Fourier analysis of geophysical time series. Nonlinear Process. Geophys. 18, 643-652.

Chapter 2

Time-Frequency Analysis

The Fourier transform of a signal can provide only global frequency information. While a time-frequency distribution of a signal can provide information about how the frequency content of the signal evolves with time. This is performed by mapping a one-dimensional time domain signal into a two-dimensional time-frequency representation of the signal. A lot of techniques have been developed to extract local time-frequency information. In this chapter, we introduce basic concepts and theory in time-frequency analysis, including windowed Fourier transform, wavelet transform, multiresolution analysis, wavelet basis, Hilbert transform, instantaneous frequency, Wigner-Ville distribution, and empirical mode decomposition.

2.1 WINDOWED FOURIER TRANSFORM

In order to compute the Fourier transform of a signal, we must have full knowledge of this signal in the whole time domain. However, in practice, since one does not know the information of the signal in the past or in the future, the Fourier transform alone is quite inadequate.

The windowed Fourier transform of $f \in L^2(\mathbb{R})$ is defined as

$$\left(G_b^\alpha f\right)(\omega) = \int_{\mathbb{R}} e^{-it\omega} f(t) g_\alpha(t-b) \, dt,$$

where $g_\alpha(t)$ is the Gaussian function $g_\alpha(t) = \frac{1}{2\sqrt{\pi\alpha}} e^{-(t^2/4\alpha)} (\alpha > 0)$. Since

$$\int_{\mathbb{R}} (G_b^\alpha f)(\omega) \, db = \int_{\mathbb{R}} e^{-it\omega} f(t) \, dt \int_{\mathbb{R}} g_\alpha(t-b) \, db = \int_{\mathbb{R}} f(t) e^{-it\omega} \, dt = \widehat{f}(\omega),$$

the windowed Fourier transform is a nice tool to extract local-frequency information from a signal.

In general, the windowed Fourier transform is defined as

$$(S_b f)(\omega) = \int_{\mathbb{R}} e^{-it\omega} f(t) \overline{W}(t-b) \, dt =: (f, W_{b,\omega}), \tag{2.1}$$

where $W(t)$ is a window function and $W_{b,\omega}(t) = e^{it\omega} W(t-b)$.

The main window functions are as follows:

1. Rectangular window $\chi_{[-\frac{1}{2},\frac{1}{2}]}(t)$;
2. Hamming window $(0.54 + 0.46\cos(2\pi t))\chi_{[-\frac{1}{2},\frac{1}{2}]}(t)$;
3. Gaussian window e^{-18t^2};
4. Hanning window $\cos^2(\pi t)\chi_{[-\frac{1}{2},\frac{1}{2}]}(t)$;
5. Blackman window $(0.42 + 0.5\cos(2\pi t) + 0.08\cos(4\pi t))\chi_{[-\frac{1}{2},\frac{1}{2}]}$,

where $\chi_{[-\frac{1}{2},\frac{1}{2}]}(t)$ is the characteristic function on $[-\frac{1}{2},\frac{1}{2}]$.

From (2.1), we see that the windowed Fourier transform $(S_b f)(\omega)$ is the Fourier transform of $f(t)\overline{W(t-b)}$, i.e.

$$(S_b f)(\omega) = (f(t)\overline{W(t-b)})^{\wedge}(\omega).$$

Let $\| W \|_2 = \left(\int_{\mathbb{R}} |W(t)|^2 \, dt\right)^{1/2}$. Define the center t^* and the radius Δ_W of a window function W as follows:

$$t^* = \frac{1}{\| W \|_2} \int_{\mathbb{R}} t |W(t)|^2 \, dt,$$

$$\Delta_W = \frac{1}{\| W \|_2} \left(\int_{\mathbb{R}} (t - t^*)^2 |W(t)|^2 \, dt\right)^{1/2}.$$

So the windowed Fourier transform gives local-time information of f in the time window:

$$[t^* + b - \Delta_W, t^* + b + \Delta_W].$$

On the other hand, by (2.1) and Theorem 1.3, it follows that

$$(S_b f)(\omega) = \frac{1}{2\pi}(\hat{f}, \widehat{W}_{b,\omega}).$$

So the windowed Fourier transform also gives local-frequency information of f in the frequency window:

$$[\omega^* + \omega - \Delta_{\widehat{W}}, \omega^* + \omega + \Delta_{\widehat{W}}],$$

where ω^* and $\Delta_{\widehat{W}}$ are the center and the radius of $\widehat{W}$, respectively. Furthermore, the windowed Fourier transform possesses a time-frequency window:

$$[t^* + b - \Delta_W, t^* + b + \Delta_W] \times [\omega^* + \omega - \Delta_{\widehat{W}}, \omega^* + \omega + \Delta_{\widehat{W}}]$$

with window area $4\Delta_W \Delta_{\widehat{W}}$. If W is the Gaussian function g_α, then $\Delta_W = \sqrt{\alpha}$ and $\Delta_{\widehat{W}} = \frac{1}{2\sqrt{\alpha}}$. So the window area $4\Delta_W \Delta_{\widehat{W}} = 2$. The Heisenberg uncertainty principle in Section 1.7 shows that it is not possible to construct a window function W such that the window area is less than 2.

Therefore, the windowed Fourier transform with a Gabor function has the smallest time-frequency window.

Theorem 2.1. *Let the window function* W *satisfy* $\| W \|_2 = 1$. *Then, for any* $f, h \in L^2(\mathbb{R})$,

$$\int \int_{\mathbb{R}^2} (S_b f)(\omega)\overline{(S_b h)}(\omega)\, d\omega\, db = 2\pi (f, h).$$

Proof. For any $f, h \in L^2(\mathbb{R})$, by Theorem 1.3, it follows that

$$\int_{\mathbb{R}} (S_b f)(\omega)\overline{(S_b h)}(\omega)\, d\omega = 2\pi \int_{\mathbb{R}} (S_b f)^{\vee}(t)\overline{(S_b h)^{\vee}}(t)\, dt,$$

where $\xi^{\vee}$ is the inverse Fourier transform of ξ. Since

$$(S_b f)^{\vee}(t)\overline{(S_b h)^{\vee}}(t) = f(t)\overline{h}(t)|W(t - b)|^2,$$

it follows that

$$\int_{\mathbb{R}} (S_b f)(\omega)\overline{(S_b h)}(\omega)\, d\omega = 2\pi \int_{\mathbb{R}} f(t)\overline{h}(t)|W(t - b)|^2\, dt.$$

Integrating on both sides over $\mathbb{R}$ with respect to b, we get

$$\int \int_{\mathbb{R}^2} (S_b f)(\omega)\overline{(S_b h)}(\omega)\, d\omega\, db = 2\pi \int_{\mathbb{R}} f(t)\overline{h}(t) \left(\int_{\mathbb{R}} |W(t - b)|^2\, db \right) dt.$$

By $\int_{\mathbb{R}} |W(t - b)|^2\, db = \| W \|_2^2 = 1$, we get the desired result. $\qquad\square$

Taking $h = g_\alpha(\cdot - t)$ in Theorem 2.1, where $g_\alpha(t)$ is the Gaussian function $g_\alpha(t) = \frac{1}{2\sqrt{\pi\alpha}} e^{-(t^2/4\alpha)}$ $(\alpha > 0)$, and then letting $\alpha \to 0+$, we derived the following theorem immediately.

Theorem 2.2. *Under the conditions of Theorem 2.1, we have*

$$f(t) = \frac{1}{2\pi} \int \int_{\mathbb{R}^2} e^{it\omega}(S_b f)(\omega) W(x - b)\, d\omega db.$$

The formula in Theorem 2.2 is called the reconstruction formula of the windowed Fourier transform.

2.2 WAVELET TRANSFORM

The wavelet transform possesses the ability to construct a time-frequency representation of a signal that offers very good time and frequency localization, so wavelet transforms can analyze localized intermittent periodicity of geophysical time series very well.

A wavelet is a function $\psi \in L^2(\mathbb{R})$ with zero-average $\int_{\mathbb{R}} \psi(t)\, dt = 0$. The wavelet transform of $f \in L^2(\mathbb{R})$ is defined as

$$(W_\psi f)(b, a) = \frac{1}{\sqrt{|a|}} \int_{\mathbb{R}} f(t)\overline{\psi}\left(\frac{t - b}{a}\right) dt = (f, \psi_{b,a}) \quad (a \neq 0, b \in \mathbb{R}),$$

$$(2.2)$$

where a is called the *dilation parameter*, b is called the *translation parameter*, and $\psi_{b,a}(t) = \frac{1}{\sqrt{|a|}} \psi \left(\frac{t-b}{a} \right)$.

From Theorem 1.3, it follows that $(f, \psi_{b,a}) = \frac{1}{2\pi} (\hat{f}, \hat{\psi}_{b,a})$, and so

$$(W_\psi f)(b, a) = \frac{1}{2\pi} (\hat{f}, \hat{\psi}_{b,a}) \quad (a > 0, b \in \mathbb{R}),$$

where $\hat{\psi}_{b,a}$ is the Fourier transform of $\psi_{b,a}$ and

$$\hat{\psi}_{b,a}(\omega) = \frac{1}{\sqrt{|a|}} \int_{\mathbb{R}} e^{-it\omega} \psi \left(\frac{t-b}{a} \right) \, dt = \sqrt{|a|} e^{-ib\omega} \hat{\psi}(a\omega).$$

The wavelet transform $(W_\psi f)(b, a)$ possesses the time-frequency window

$$[b + at^* - |a|\Delta_\psi, b + at^* + |a|\Delta_\psi] \times \left[\frac{\omega^*}{a} - \frac{\Delta_{\hat{\psi}}}{|a|}, \frac{\omega^*}{a} + \frac{\Delta_{\hat{\psi}}}{|a|} \right],$$

where t^* and ω^* are the centers of ψ and $\hat{\psi}$, respectively, and $\Delta_{\hat{\psi}}$ and $\Delta_{\hat{\psi}^*}$ are the radii of ψ and $\hat{\psi}$, respectively. This time-frequency window automatically narrows when detecting high-frequency information (i.e., small $|a|$) and widens when detecting low-frequency information (i.e., large $|a|$). Similarly to the Fourier power spectrum, the wavelet power spectrum of a signal f is defined as the square of the modulus of the wavelet transform of the signal, i.e., $|W_\psi f(b, a)|^2$.

To reconstruct the signals from their wavelet transform, we need to assume only that wavelet ψ satisfies the *admissibility condition*:

$$C_\psi = \int_{\mathbb{R}} \frac{|\hat{\psi}(\omega)|^2}{|\omega|} \, d\omega < \infty. \tag{2.3}$$

A wavelet ψ with an admissibility condition is called a *basic wavelet*.

If $\int_{\mathbb{R}} \psi(t) \, dt = 0$ and for some constant K and $\epsilon > 0$,

$$|\psi(t)| \leq K \frac{1}{(1 + |t|)^{1+\epsilon}} \quad (t \in \mathbb{R}),$$

then ψ is a basic wavelet.

Theorem 2.3. *Let ψ be a basic wavelet. Then any signal $f \in L^2(\mathbb{R})$ satisfies*

$$f(t) = \frac{1}{C_\psi} \int \int_{\mathbb{R}^2} (W_\psi f)(b, a) \frac{1}{\sqrt{|a|}} \psi \left(\frac{t-b}{a} \right) \frac{da}{a^2} \, db.$$

The formula in Theorem 2.3 is called the reconstruction formula of the wavelet transform.

Proof. Denote the integral on the right-hand side by $\lambda(t)$. Let $\psi_a(t) = \frac{1}{\sqrt{|a|}} \psi \left(\frac{t}{a} \right)$. Then

$$\lambda(t) = \frac{1}{C_\psi} \int_{\mathbb{R}} \left(\int_{\mathbb{R}} (W_\psi f)(b, a) \psi_a(t - b) \, db \right) \frac{da}{a^2}.$$

The integral in brackets can be represented by a convolution:

$$\int_{\mathbb{R}} (W_\psi f)(b, a)\psi_a(t - b)\, db = ((W_\psi f)(\cdot, a) * \psi_a)(t),$$

and so

$$\lambda(t) = \frac{1}{C_\psi} \int_{\mathbb{R}} ((W_\psi f)(\cdot, a) * \psi_a)(t)\frac{da}{a^2}.$$

However, by (2.2),

$$(W_\psi f)(b, a) = \frac{1}{\sqrt{a}} \int_{\mathbb{R}} f(t)\overline{\psi}\left(\frac{t - b}{a}\right)\, dt = (f * \tilde{\psi}_a)(b),$$

where $\tilde{\psi}_a(t) = \frac{1}{\sqrt{|a|}}\overline{\psi}\left(\frac{-t}{a}\right)$. Therefore,

$$\lambda(t) = \frac{1}{C_\psi} \int_{\mathbb{R}} ((f * \tilde{\psi}_a) * \psi_a)(t)\frac{da}{a^2}.$$

Taking the Fourier transform on both sides, using the convolution property in frequency, we get

$$\hat{\lambda}(\omega) = \frac{1}{C_\psi} \int_{\mathbb{R}} \hat{f}(\omega)\sqrt{|a|}\overline{\hat{\psi}}(a\omega)\sqrt{|a|}\hat{\psi}(a\omega)\frac{da}{a^2}$$

$$= \frac{\hat{f}(\omega)}{C_\psi} \int_{\mathbb{R}} \frac{|\hat{\psi}(a\omega)|^2}{a}da = \frac{\hat{f}(\omega)}{C_\psi} \int_{\mathbb{R}} \frac{|\hat{\psi}(u)|^2}{|u|}\, du.$$

Note that $C_\psi = \int_{\mathbb{R}} \frac{|\hat{\psi}(u)|^2}{|u|}\, du$. Then

$$\hat{\lambda}(\omega) = \hat{f}(\omega).$$

Taking the inverse Fourier transform on both sides, we get the desired result: $\lambda(t) = f(t)$.   □

Let

$$K(b_0, b, a_0, a) = (\psi_{b,a}, \psi_{b_0,a_0}),$$

where $\psi_{b,a}(t) = \frac{1}{\sqrt{|a|}}\psi\left(\frac{t-b}{a}\right)$. A wavelet transform is a redundant representation whose redundancy is characterized by the *reproducing equation*:

$$(W_\psi f)(b_0, a_0) = \frac{1}{C_\psi} \int \int_{\mathbb{R}^2} K(b, b_0, a, a_0)(W_\psi f)(b, a)\frac{da}{a^2}\, db,$$

where $K(b_0, b, a_0, a)$ is called the *reproducing kernel*. It measures the correlation of two wavelets $\psi_{b,a}$ and ψ_{b_0,a_0}. The reproducing equation can be derived directly by Theorem 2.3 and the definition of the wavelet transform.

Example 2.1. In geoscience, the Morlet wavelet and the Mexican hat wavelet are often used. Morlet wavelets consist of a plane wave modulated by a Gaussian function:

$$\psi^{M}(t) = \pi^{-(1/4)} e^{it\theta} e^{-(t^2/2)}.$$

When $\theta \geq 6$, the value of its Fourier transform at the origin approximates to 0, i.e., the Morlet wavelet has zero mean and is localized in both time and frequency space. The Mexican hat wavelet is

$$\psi^{H}(t) = -\frac{1}{\sqrt{\Gamma(2.5)}} (1 - t^2) e^{-(t^2/2)},$$

where $\Gamma(t)$ is the Gamma function.

To measure the degree of uncertainty of a random signal, the continuous wavelet entropy is defined as

$$S(t) = -\int_{0}^{\infty} P(a, b) \log P(a, b) \, da,$$

where $P(a, b) = \frac{|W_{\psi}f(b,a)|^4}{\int_{\mathbb{R}} |W_{\psi}f(\tau,a)|^4 \, d\tau}$. The wavelet entropy of a white noise is maximal.

Theorem 2.3 shows a signal is reconstructed by all the values of wavelet transform $W_{\psi}f(b, a)(a \neq 0, t \in \mathbb{R})$. Since the wavelet transform provides redundant information, a signal may be reconstructed by discretizing the wavelet transform. If a wavelet ψ satisfies the stability condition

$$A \leq \sum_{m} |\widehat{\psi}(2^{-m}\omega)|^2 \leq B \quad (\omega \in \mathbb{R}),$$

where $\widehat{\psi}$ is the Fourier transform of ψ, then the half-discrete values $W_{\psi}f(b, 2^{-m})(b \in \mathbb{R}, m \in \mathbb{Z})$ can reconstruct the signal f. Such a wavelet ψ is called a dyadic wavelet.

Taking $a = 2^{-m}$ and $b = 2^{-m}n$ in $\psi_{b,a}(t) = \frac{1}{\sqrt{a}} \psi(\frac{t-b}{a})$, we get

$$\psi_{m,n}(t) = 2^{m/2} \psi(2^m t - n),$$

where m is the *dilation parameter* and n is the *translation parameter*.

For any signal $f \in L^2(\mathbb{R})$, the discrete values $W_{\psi}f(2^{-m}n, 2^{-m})(m, n \in \mathbb{Z})$ can reconstruct the signal if and only if the wavelet ψ satisfies the *frame condition*:

$$A \| f \|^2 \leq |(f, \psi_{m,n})|^2 \leq B \| f \|^2.$$

The family $\{\psi_{m,n}\}_{m,n\in\mathbb{Z}}$ is called a *wavelet frame* with *upper bound A* and *lower bound B*. If $A = B = 1$, then it is called the *Parseval wavelet frame*.

If $\{\psi_{m,n}\}_{m,n\in\mathbb{Z}}$ is an orthonormal basis, then $\{\psi_{m,n}\}_{m,n\in\mathbb{Z}}$ is called a *wavelet basis* and ψ is called an *orthonormal wavelet*.

2.3   MULTIRESOLUTION ANALYSES AND WAVELET BASES

All orthonormal wavelets can be characterized by their Fourier transforms as follows.

A wavelet $\psi \in L^2(\mathbb{R})$ is an orthonormal wavelet if and only if ψ satisfies the following equations:

$$\| \psi \|^2 = 1, \quad \sum_m |\widehat{\psi}(2^m\omega)|^2 = 1 \quad (\omega \in \mathbb{R})$$

and for each odd integer k,

$$\sum_{m=0}^{\infty} \widehat{\psi}(2^m\omega)\overline{\widehat{\psi}}(2^m(\omega + 2k\pi)) = 0 \quad (\omega \in \mathbb{R}).$$

However, orthonormal wavelets cannot be constructed easily by this characterization.

2.3.1   Multiresolution Analyses

To construct orthonormal wavelets, multiresolution analysis is the most important method.

A sequence of closed subspaces $\{V_m\}_{m\in\mathbb{Z}}$ of $L^2(\mathbb{R})$ is a multiresolution analysis if

 (i) $V_m \subset V_{m+1}$ $(m \in \mathbb{Z})$;
 (ii) $f \in V_m$ if and only if $f(2\cdot) \in V_{m+1}$ $(m \in \mathbb{Z})$;
 (iii) $\overline{\bigcup_m V_m} = L^2(\mathbb{R})$;
 (iv) $\bigcap_m V_m = \{0\}$;
 (v) there exists a function $\varphi \in V_0$ such that $\{\varphi(t - n)\}_{n\in\mathbb{Z}}$ is an orthonormal basis of V_0.

Here the function φ is called a *scaling function* and V_0 is called the center space.

Proposition 2.1. *Let* $\varphi \in L^2(\mathbb{R})$. *Then* $\{\varphi(t - n)\}_{n\in\mathbb{Z}}$ *is an orthonormal system if and only if*

$$\sum_n |\widehat{\varphi}(\omega + 2n\pi)|^2 = 1 \quad (\omega \in \mathbb{R}).$$

Proof. We know that $\{\varphi(t - n)\}_{n\in\mathbb{Z}}$ is an orthonormal system if and only if

$$\int_{\mathbb{R}} \varphi(t)\overline{\varphi}(t - n) \, \mathrm{d}t = \begin{cases} 1, & n = 0, \\ 0, & n \neq 0. \end{cases}$$

However, by Theorem 1.3, it follows that

$$\int_{\mathbb{R}} \varphi(t)\overline{\varphi}(t-n)\,dt = \frac{1}{2\pi}\int_{\mathbb{R}} |\widehat{\varphi}(\omega)|^2 e^{in\omega}\,d\omega = \frac{1}{2\pi}\sum_k \int_{2k\pi}^{2(k+1)\pi} |\widehat{\varphi}(\omega)|^2 e^{in\omega}\,d\omega$$

$$= \frac{1}{2\pi}\sum_k \int_0^{2\pi} |\widehat{\varphi}(\omega+2k\pi)|^2 e^{in\omega}\,d\omega$$

$$= \frac{1}{2\pi}\int_0^{2\pi} \sum_k |\widehat{\varphi}(\omega+2k\pi)|^2 e^{in\omega}\,d\omega.$$

Denote $g(\omega) = \sum_k |\widehat{\varphi}(\omega+2k\pi)|^2$. Then

$$\int_{\mathbb{R}} \varphi(t)\overline{\varphi}(t-n)\,dt = \frac{1}{2\pi}\int_0^{2\pi} g(\omega)e^{in\omega}\,d\omega.$$

Therefore, $\{\varphi(\cdot - n)\}_{n\in\mathbb{Z}}$ is an orthonormal system if and only if

$$\frac{1}{2\pi}\int_0^{2\pi} g(\omega)e^{in\omega}\,d\omega = \begin{cases} 1, & n = 0, \\ 0, & n \neq 0, \end{cases}$$

that is, the Fourier coefficients of $g(\omega)$ vanish at $n \neq 0$ and equal 1 at $n = 0$. So $g(\omega) = 1$, i.e.

$$\sum_k |\widehat{\varphi}(\omega+2k\pi)|^2 = 1.$$

$\square$

By Proposition 2.1 and (v), it follows that $\{\varphi(t-n)\}_{n\in\mathbb{Z}}$ must satisfy $\sum_n |\widehat{\varphi}(\omega+2n\pi)|^2 = 1$. Since $\varphi \in V_0$ and $\frac{1}{2}\varphi(\frac{t}{2}) \in V_{-1} \subset V_0$, we expand $\frac{1}{2}\varphi(\frac{t}{2})$ in terms of the orthonormal basis $\{\varphi(t-n)\}_{n\in\mathbb{Z}}$ as follows:

$$\frac{1}{2}\varphi\left(\frac{t}{2}\right) = \sum_n c_n\varphi(t-n).$$

This equation is called the *bi-scale equation* and $\{c_n\}_{n\in\mathbb{Z}}$ are called *bi-scale coefficients*. Taking the Fourier transform on both sides of the bi-scale equation, we get

$$\widehat{\varphi}(2\omega) = \widehat{\varphi}(\omega)\sum_n c_n e^{-in\omega} = \widehat{\varphi}(\omega)H(\omega),$$

where $H(\omega) = \sum_n c_n e^{-in\omega}$ is called the *transfer function* associated with the scaling function φ. It is clear that $H(\omega)$ is a 2π-periodic function.

Theorem 2.4. *Let $H(\omega)$ be the transfer function associated with the scaling function φ. Then*

$$|H(\omega)|^2 + |H(\omega+\pi)|^2 = 1 \quad (\omega \in [0, 2\pi]).$$

Proof. Since $\widehat{\varphi}(2\omega) = \widehat{\varphi}(\omega)H(\omega)$, it is clear that

$$\widehat{\varphi}(2\omega + 2n\pi) = \widehat{\varphi}(\omega + n\pi)H(\omega + n\pi).$$

Since φ is a scaling function, by Proposition 2.1, $\sum_n |\widehat{\varphi}(\omega + 2n\pi)|^2 = 1$, and so

$$\sum_n |\widehat{\varphi}(2\omega + 2n\pi)|^2 = 1, \quad \sum_n |\widehat{\varphi}(\omega + \pi + 2n\pi)|^2 = 1.$$

Since $H(\omega)$ is a 2π-periodic function, $H(\omega + 2l\pi) = H(\omega)(l \in \mathbb{Z})$. Therefore,

$$1 = \sum_n |\widehat{\varphi}(2\omega + 2n\pi)|^2 = \sum_n |\widehat{\varphi}(\omega + n\pi)|^2 |H(\omega + n\pi)|^2$$

$$= \sum_k |\widehat{\varphi}(\omega + 2k\pi)|^2 |H(\omega + 2k\pi)|^2$$

$$+ \sum_k |\widehat{\varphi}(\omega + (2k + 1)\pi)|^2 |H(\omega + (2k + 1)\pi)|^2$$

$$= |H(\omega)|^2 \sum_k |\widehat{\varphi}(\omega + 2k\pi)|^2 + |H(\omega + \pi)|^2 \sum_k |\widehat{\varphi}(\omega + \pi + 2k\pi)|^2$$

$$= |H(\omega)|^2 + |H(\omega + \pi)|^2.$$

We get Theorem 2.4.   □

Since φ is the scaling function, by (v), $\{\varphi(t - n)\}_{n \in \mathbb{Z}}$ is an orthonormal basis of V_0. Let

$$\varphi_{m,n}(t) = 2^{m/2}\varphi(2^m t - n) \quad (m, n \in \mathbb{Z}).$$

Then $\{\varphi_{m,n}(t)\}_{n \in \mathbb{Z}}$ is the orthonormal basis of V_m.

To construct an orthonormal wavelet by using a multiresolution analysis $\{V_m\}_{m \in \mathbb{Z}}$, we consider the orthogonal complement space W_0 of the center space V_0 in V_1, i.e.

$$V_1 = V_0 \bigoplus W_0,$$

where $\bigoplus$ represents the orthogonal sum. The following theorem gives a construction method for the orthonormal wavelet.

Theorem 2.5. *Suppose that for a multiresolution analysis, φ is the scaling function, H is the transfer function, and $\{c_n\}_{n \in \mathbb{Z}}$ are bi-scale coefficients. Let ψ satisfy $\widehat{\psi}(\omega) = \widetilde{H}\left(\frac{\omega}{2}\right)\widehat{\varphi}\left(\frac{\omega}{2}\right)$, where $\widetilde{H}(\omega) = e^{-i\omega}\overline{H}(\omega + \pi)$, i.e.,*

$$\psi(t) = -2\sum_n (-1)^n \overline{c}_{1-n}\, \varphi(2t - n),$$

Then $\{\psi(t - n)\}_{n \in \mathbb{Z}}$ is an orthonormal basis of W_0 and $\{2^{m/2}\psi(2^m t - n)\}_{m,n \in \mathbb{Z}}$ is an orthonormal basis of $L^2(\mathbb{R})$, i.e., ψ is an orthonormal wavelet.

Theorem 2.5 is called the *existence theorem of orthonormal wavelets*. As an example, let

$$N_1(t) = \begin{cases} 1, & t \in [0, 1], \\ 0, & \text{otherwise.} \end{cases}$$

Define

$$N_k(t) = (N_{k-1} * N_1)(t) = \int_0^1 N_{k-1}(t - x)\, dx \quad (k \geq 2)$$

and call $N_k(t)$ the *k-order cardinal B-spline*. Its Fourier transform is

$$\widehat{N}_k(\omega) = \left(\frac{1 - e^{-i\omega}}{i\omega} \right)^k = \left(\frac{\sin(\omega/2)}{\omega/2} \right)^k e^{-i(k\omega/2)}.$$

A direct computation shows that

$$\sum_l |\widehat{N}_k(2\omega + 2l\pi)|^2 = -\frac{\sin^{2k}\omega}{(2k-1)!} \frac{d^{2k-1}(\cot\omega)}{d\omega^{2k-1}} =: F_k(2\omega).$$

Especially, $F_1(\omega) = 1$ and $F_2(\omega) = \frac{1}{3}\sin^2\frac{\omega}{2} + \cos^2\frac{\omega}{2}$. Let φ_k satisfy the condition

$$\widehat{\varphi}_k(\omega) = \frac{\widehat{N}_k(\omega)}{\left(\sum_l |\widehat{N}_k(\omega + 2l\pi)|^2 \right)^{1/2}} = \left(\frac{\sin(\omega/2)}{\omega/2} \right)^k e^{-i(k\omega/2)} F_k^{-(1/2)}(\omega).$$

Then φ_k is a scaling function. By Theorem 2.5, the corresponding orthonormal wavelet $\psi_k(t)$ satisfies

$$\widehat{\psi}_k(\omega) = \left(\frac{4}{i\omega} \right)^k e^{-i(\omega/2)} \sin^{2k}\frac{\omega}{4} \left(\frac{F_k((\omega/2) + \pi)}{F_k(\omega/2) F_k(\omega)} \right)^{1/2}.$$

The wavelet ψ_k is called the Battle-Lemarié wavelet of order k.

A function f is called a *compactly supported function* if there exists a $c > 0$ such that $f(t) = 0(|t| > c)$. Daubechies constructed a lot of compactly supported orthonormal wavelets and applied them widely in signal processing.

For any $N \in \mathbb{Z}_+$, Daubechies constructed a rational function $P(z) = \sum_{-N+1}^{N-1} c_n z^n$ with real-valued coefficients $c_n \in \mathbb{R}$ such that

$$P(1) = 1, \quad |P(e^{-i\omega})|^2 = \sum_0^{N-1} C_{N+k-1}^k \left(\sin\frac{\omega}{2} \right)^{2k},$$

where $C_m^n = \frac{m!}{n!(m-n)!}$. Denote the filter

$$H_N^D(\omega) := \left(\frac{1 + e^{-i\omega}}{2} \right)^N P(e^{-i\omega}) = \sum_{n=0}^{2N-1} h_{n,N} e^{-in\omega}.$$

On the basis of $\{h_{n,N}\}_{n=0,\dots,2N-1}$, the scaling function φ_N^D can be obtained numerically by the bi-scale equation:

$$\frac{1}{2}\varphi_N^D(t) = \sum_{n=0}^{2N-1} h_{n,N}\varphi_N^D(2t-n),$$

and φ_N^D is compactly supported. By Theorem 2.5, the corresponding orthonormal wavelet is

$$\psi_N^D(t) = -2 \sum_{n=2-2N}^{1} (-1)^n \overline{h}_{1-n,N}\varphi_N^D(2t-n).$$

It is compactly supported. The wavelet $\psi_N^D(t)$ is called the Daubechies wavelet.

Let ψ be an orthonormal wavelet and $\psi_{m,n}(t) = 2^{m/2}\psi(2^m t - n)(m, n \in \mathbb{Z})$. Then any $f \in L^2(\mathbb{R})$ can be expanded into a wavelet series:

$$f = \sum_{m,n} d_{m,n}\psi_{m,n} \quad (L^2(\mathbb{R})),$$

where the coefficients are

$$d_{m,n} = (f, \psi_{m,n}) = \int_{\mathbb{R}} f(t)\,\overline{\psi}_{m,n}(t)\,dt \quad (m, n \in \mathbb{Z})$$

and $d_{m,n}$ are called *wavelet coefficients*, and Parseval's identity $\sum_{m,n} |d_{m,n}|^2 = \|f\|_2^2$ holds. Notice that the coefficient formula can be written as

$$d_{m,n} = 2^{m/2} \int_{\mathbb{R}} f(t)\,\overline{\psi}\left(\frac{t - 2^{-m}n}{2^{-m}}\right) dt.$$

Therefore, when we regard ψ as a basic wavelet, the wavelet coefficients are just the values of wavelet transform at $a = 2^{-m}$ and $b = 2^{-m}n$. If φ is the scaling function corresponding to ψ, then any $f \in L^2(\mathbb{R})$ can also be expanded into another wavelet series:

$$f(t) = \sum_n c_n\varphi(t-n) + \sum_{m=0}^{\infty}\sum_n d_{m,n}\,\psi_{m,n}(t) \quad (L^2(\mathbb{R})),$$

where

$$c_n = (f, \varphi) = \int_{\mathbb{R}} f(t)\overline{\varphi}(t-n)\,dt \quad (n \in \mathbb{Z}),$$

$$d_{m,n} = (f, \psi_{m,n}) = \int_{\mathbb{R}} f(t)\overline{\psi}_{m,n}(t)\,dt \quad (m, n \in \mathbb{Z}).$$

For any $f \in L^2(\mathbb{R})$, since $L^2(\mathbb{R}) = \overline{\bigcup_{m\in\mathbb{Z}} V_m}$, the projection of f on space V_m

$$\text{Proj}_{V_m} f \to f \quad (m \to \infty),$$

that is, $f \approx \text{Proj}_{V_m} f$ when m is sufficiently large. Denote the orthogonal complement space of V_m in V_{m+1} by W_m, i.e., $V_{m+1} = V_m \oplus W_m$. So

$$\text{Proj}_{V_{m+1}} f = \text{Proj}_{V_m} f + \text{Proj}_{W_m} f,$$

where $\text{Proj}_{V_m} f$ and $\text{Proj}_{W_m} f$ are the low-frequency part and the high-frequency part of the projection $\text{Proj}_{V_{m+1}} f$, respectively. Note that

$$\varphi_{m,n}(t) = 2^{m/2} \varphi(2^m t - n) \quad (m, n \in \mathbb{Z}),$$

$$\psi_{m,n}(t) = 2^{m/2} \psi(2^m t - n) \quad (m, n \in \mathbb{Z}).$$

Since $\{\varphi_{m,n}\}_{n \in \mathbb{Z}}$ and $\{\psi_{m,n}\}_{n \in \mathbb{Z}}$ are orthonormal bases of V_m and W_m, respectively,

$$\sum_n c_{m+1,n} \, \varphi_{m+1,n} = \sum_n c_{m,n} \, \varphi_{m,n} + \sum_n d_{m,n} \, \psi_{m,n} \quad (m \in \mathbb{Z}), \qquad (2.4)$$

where $c_{m,n} = (f, \varphi_{m,n})$ and $d_{m,n} = (f, \psi_{m,n})$. This formula is called the *decomposition formula*.

Replacing m by $m - 1$ in (2.4), we get

$$\sum_n c_{m,n} \, \varphi_{m,n} = \sum_n c_{m-1,n} \, \varphi_{m-1,n} + \sum_n d_{m-1,n} \, \psi_{m-1,n},$$

and then substituting it into the first term on the right-hand side of (2.4), we get

$$\sum_n c_{m+1,n} \, \varphi_{m+1,n} = \left(\sum_n c_{m-1,n} \, \varphi_{m-1,n} + \sum_n d_{m-1,n} \, \psi_{m-1,n} \right) + \sum_n d_{m,n} \, \psi_{m,n}.$$

Continuing this procedure l times, when m is sufficiently large, we have

$$f \approx \sum_n c_{m+1,n} \, \varphi_{m+1,n} = \sum_n c_{m-l,n} \, \varphi_{m-l,n} + \sum_{j=m-l}^{m} \sum_n d_{j,n} \, \psi_{j,n}.$$

In application, one often uses such a decomposition.

2.3.2 Discrete Wavelet Transform

To avoid computing each coefficient $c_{m,n}, d_{m,n} (n \in \mathbb{Z})$ in (2.4), by using integrals,

$$c_{m,n} = \int_{\mathbb{R}} f(t) \overline{\varphi}_{m,n}(t) \, dt \quad (n \in \mathbb{Z}),$$

$$d_{m,n} = \int_{\mathbb{R}} f(t) \overline{\psi}_{m,n}(t) \, dt \quad (n \in \mathbb{Z}),$$

the discrete wavelet transform provides a fast algorithm that can compute coefficients $\{c_{m,n}\}$ and $\{d_{m,n}\}$ with the help of $\{c_{m+1,n}\}_{n \in \mathbb{Z}}$.

Now we introduce the discrete wavelet transform.

Let φ be a scaling function and ψ be the corresponding orthonormal wavelet, and

$$c_{m,n} = (f, \varphi_{m,n}), \quad d_{m,n} = (f, \psi_{m,n}),$$

where

$$\varphi_{m,n}(t) = 2^{m/2}\varphi(2^m t - n) \quad (m, n \in \mathbb{Z}),$$

$$\psi_{m,n}(t) = 2^{m/2}\psi(2^m t - n) \quad (m, n \in \mathbb{Z}).$$

From the bi-scale equation and Theorem 2.5, it follows that

$$\varphi(t) = \sum_k p_k \varphi(2t - k),$$

$$\psi(t) = \sum_k q_k \varphi(2t - k), \tag{2.5}$$

where $p_k = 2c_k$ and $q_k = (-1)^{k+1} 2\bar{c}_{1-k}$, and c_k is the bi-scale coefficient. Since $\{\varphi_{m+1,l}\}_{l \in \mathbb{Z}}$ and $\{\psi_{m+1,l}\}_{l \in \mathbb{Z}}$ are an orthonormal basis of V_{m+1} and W_{m+1}, respectively, and $\varphi_{m,n} \in V_m \subset V_{m+1}$, $\psi_{m,n} \in W_m \subset V_{m+1}$,

$$\varphi_{m,n} = \sum_l (\varphi_{m,n}, \varphi_{m+1,l}) \varphi_{m+1,l},$$

$$\psi_{m,n} = \sum_l (\psi_{m,n}, \varphi_{m+1,l}) \varphi_{m+1,l}.$$

By (2.5), it follows that

$$(\varphi_{m,n}, \varphi_{m+1,l}) = \sqrt{2} \int_{\mathbb{R}} \varphi(u - n)\overline{\varphi}(2u - l)\, du = \frac{1}{\sqrt{2}} p_{l-2n},$$

$$(\psi_{m,n}, \varphi_{m+1,l}) = \sqrt{2} \int_{\mathbb{R}} \psi(u - n)\overline{\varphi}(2u - l)\, du = \frac{1}{\sqrt{2}} q_{l-2n}.$$

Therefore,

$$\varphi_{m,n} = \frac{1}{\sqrt{2}} \sum_l p_{l-2n} \varphi_{m+1,l},$$

$$\psi_{m,n} = \frac{1}{\sqrt{2}} \sum_l q_{l-2n} \varphi_{m+1,l}.$$

Noticing that $c_{m,n} = (f, \varphi_{m,n})$ and $d_{m,n} = (f, \psi_{m,n})$, we find

$$c_{m,n} = \frac{1}{\sqrt{2}} \sum_l p_{l-2n} c_{m+1,l},$$

$$d_{m,n} = \frac{1}{\sqrt{2}} \sum_l q_{l-2n} c_{m+1,l}.$$

These formulas are called the *discrete wavelet transform*.

Since the union of $\{\varphi_{m,n}\}_{n\in\mathbb{Z}}$ and $\{\psi_{m,n}\}_{n\in\mathbb{Z}}$ is an orthonormal basis of V_{m+1} and $\varphi_{m+1,n} \in V_{m+1}$,

$$\varphi_{m+1,n} = \sum_l (\varphi_{m+1,n}, \varphi_{m,l})\, \varphi_{m,l} + \sum_l (\varphi_{m+1,n}, \psi_{m,l})\, \psi_{m,l},$$

and so

$$(f, \varphi_{m+1,n}) = \sum_l (\varphi_{m+1,n}, \varphi_{m,l})(f, \varphi_{m,l}) + \sum_l (\varphi_{m+1,n}, \psi_{m,l})(f, \psi_{m,l}),$$

that is, the *inverse discrete wavelet transform* is

$$c_{m+1,n} = \frac{1}{\sqrt{2}} \left(\sum_l \overline{p}_{n-2l}\, c_{m,l} + \sum_l \overline{q}_{n-2l} d_{m,l} \right).$$

2.3.3 Biorthogonal Wavelets, Bivariate Wavelets, and Wavelet Packet

Biorthogonal wavelets are a kind of wavelet that are used often. Their constructions depend on the concept of the Riesz basis. Let $\{g_n\}$ be a basis for $L^2(\mathbb{R})$, and for any sequence $c_n (\sum_n |c_n|^2 < \infty)$ there exists $B \geq A > 0$ such that

$$A \sum_n |c_n|^2 \leq \| \sum_n c_n g_n \|_2^2 \leq B \sum_n |c_n|^2,$$

then $\{g_n\}$ is called a *Riesz basis* for $L^2(\mathbb{R})$.

Let $\psi, \widetilde{\psi} \in L^2(\mathbb{R})$. If their integral translations and dyadic dilations satisfy $(\psi_{m,n}, \widetilde{\psi}_{m',n'}) = \delta_{m,m'}\delta_{n,n'}$ and both $\{\psi_{m,n}\}_{m,n\in\mathbb{Z}}$ and $\{\widetilde{\psi}_{m,n}\}_{m,n\in\mathbb{Z}}$ are Riesz bases of $L^2(\mathbb{R})$, then $\{\psi, \widetilde{\psi}\}$ is called a *pair of biorthogonal wavelets*, where $\delta_{k,l} = 0(k \neq l)$, $\delta_{k,l} = 1(k = l)$.

If $\{\psi, \widetilde{\psi}\}$ is a pair of biorthogonal wavelets, then, for $f \in L^2(\mathbb{R})$, the *reconstruction formula* holds:

$$f = \sum_{m,n}(f, \psi_{m,n})\widetilde{\psi}_{m,n} = \sum_{m,n}(f, \widetilde{\psi}_{m,n})\, \psi_{m,n}.$$

Symmetric or antisymmetric compactly supported spline biorthogonal wavelets are applied widely. The construction method is as follows. First, a pair of trigonometric polynomials $H(\omega)$ and $\widetilde{H}(\omega)$ are defined as

$$H(\omega) = e^{-i(\epsilon \, \omega/2)} \left(\cos \frac{\omega}{2} \right)^p L(\omega),$$

$$\widetilde{H}(\omega) = e^{-i(\epsilon \, \omega/2)} \left(\cos \frac{\omega}{2} \right)^{\widetilde{p}} \widetilde{L}(\omega),$$

where $\epsilon = 0$ for even numbers p and $\widetilde{p}$, and $\epsilon = 1$ for odd numbers p and $\widetilde{p}$, and

$$L(\cos \omega)\widetilde{L}(\cos \omega) = \sum_0^{q-1} C_{q-1+k}^k \sin^{2k} \frac{\omega}{2}, \quad \left(q = \frac{1}{2}(p + \widetilde{p}) \right).$$

Next, the bi-scale coefficients $\{h_n\}$ and $\{\widetilde{h}_n\}$ are computed using

$$H(\omega) = \sum_{-p}^{p} h_n e^{-in\omega},$$

$$\widetilde{H}(\omega) = \sum_{-\widetilde{p}}^{\widetilde{p}} \widetilde{h}_n e^{-in\omega}.$$

For example, let $p = 2$, $\widetilde{p} = 4$, and $L(\omega) = 1$. Then

$$h_2 = h_{-2} = 0, \quad h_1 = h_{-1} = 0.35355, \quad h_0 = 1$$

and

$$\widetilde{h}_0 = 0.9944, \quad \widetilde{h}_{-1} = \widetilde{h}_1 = 0.4198, \quad \widetilde{h}_{-2} = \widetilde{h}_2 = -0.1767,$$

$$\widetilde{h}_{-3} = \widetilde{h}_3 = -0.0662, \quad \widetilde{h}_{-4} = \widetilde{h}_4 = 0.0331.$$

From this, with use of bi-scale equations,

$$\varphi(t) = \sum_{-p}^{p} 2h_n \, \varphi(2t - n),$$

$$\widetilde{\varphi}(t) = \sum_{-\widetilde{p}}^{\widetilde{p}} 2\widetilde{h}_n \, \widetilde{\varphi}(2t - n),$$

the scaling functions $\varphi(t)$ and $\widetilde{\varphi}(t)$ can be solved numerically. Finally, the corresponding biorthogonal wavelets $\psi(t)$ and $\widetilde{\psi}(t)$ are obtained.

If φ is a scaling function and ψ is the corresponding wavelet, define

$$\psi^{(1)}(t) = \varphi(t_1)\psi(t_2),$$

$$\psi^{(2)}(t) = \psi(t_1)\varphi(t_2),$$

$$\psi^{(3)}(t) = \psi(t_1)\psi(t_2).$$

Denote $\psi_{m,n}^{(k)}(t) = \frac{1}{2^m} \psi^{(k)}(2^m t - n)$, where $m \in \mathbb{Z}, n \in \mathbb{Z}^2$ and $k = 1, 2, 3$. Then

$$\{\psi_{m,n}^{(1)}, \psi_{m,n}^{(2)}, \psi_{m,n}^{(3)}\}_{(m,n)\in\mathbb{Z}^3}$$

forms an orthonormal basis of $L^2(\mathbb{R}^2)$. Such a basis is called a *bivariate wavelet basis*.

A multiresolution analysis can generate not only an orthogonal basis but also a library of functions, called a *wavelet packet*, from which infinitely many wavelet packet bases can be constructed. The Heisenberg uncertainty principle considers only the minimal area of time-frequency windows and does not mention their shapes. For a wavelet basis, the shape of the time-frequency window has been predetermined by the choice of the wavelet function. However, in a wavelet packet, the time-frequency windows are rectangular with arbitrary aspect ratios.

For a multiresolution analysis, let $\varphi(t)$ be the scaling function, $H(\omega)$ be the transfer function, and ψ be the corresponding wavelet. Define $\mu_0 = \varphi, \mu_1 = \psi$, and

$$\widehat{\mu_{2l}}(\omega) = H\left(\frac{\omega}{2}\right) \widehat{\mu_l}\left(\frac{\omega}{2}\right),$$

$$\widehat{\mu_{2l+1}}(\omega) = e^{-i(\omega/2)} \overline{H}\left(\frac{\omega}{2} + \pi\right) \widehat{\mu_l}\left(\frac{\omega}{2}\right) \quad (l = 0, 1, \ldots).$$

The sequence $\{\mu_l\}_{l=0,1,\ldots}$ is called the *wavelet packet* determined by the scale function φ, where l is called the *modulation parameter*. The integral translations and dyadic dilations of all wavelet packet functions,

$$\mu_{l,m,n} = 2^{\frac{m}{2}} \mu_l(2^m t - n) \quad (l = 0, 1, \ldots; m, n \in \mathbb{Z}),$$

are called the *dictionary*. The choice of the modulation parameter l and the dilation parameter m, and the translation parameter n can give a lot of orthonomal bases. These orthonomal bases are called *wavelet packet bases*. A signal f can be expanded into an orthogonal series with respect to a wavelet packet basis of order $k(0 \le k \le j_0)$ as follows:

$$f(t) = P_{j_0}f + \sum_{j=j_0}^{\infty} \sum_{m=0}^{2^k-1} \sum_n c_{j,k,m,n} \mu_{2^k+m}(2^{j-k}t - n),$$

where $P_{j_0}f$ is the projection of f on the space V_{j_0}, and $c_{j,k,m,n} = \langle f, \mu_{2^k+m}(2^{j-k}t - n)\rangle$.

Recently, great advances in wavelet analysis have resulted from the study of Parseval wavelet frames (see Section 2.2).

The Parseval wavelet frame has now become an alternative to the wavelet basis and it is anticipated that Parseval wavelet frames will soon be applied in the analysis of geophysical processes. For any signal f of finite energy, if $\{\psi_{m,n}\}_{m,n\in\mathbb{Z}}$ is a Parseval wavelet frame, then

$$f = \sum_{m,n} d_{m,n}\psi_{m,n}, \quad \text{where} \quad d_{m,n} = (f, \psi_{m,n}).$$

This is similar to the orthogonal expansion of a signal with respect to a wavelet basis. However, Parseval wavelet frames $\{\psi_{m,n}\}_{m,n\in\mathbb{Z}}$ may not be orthogonal or linear independent. Their construction is easier than that of wavelet bases. It is well known that a univariate wavelet basis is generated by one function, and a bivariate wavelet basis is generated by three functions. However, the number of functions generating a Parseval wavelet frame may be arbitrary. Their construction method is based on the following unitary extension principle.

Let a function φ satisfy $\widehat{\varphi}(\omega) = P\left(\frac{\omega}{2}\right)\widehat{\varphi}\left(\frac{\omega}{2}\right)$, where P is a trigonometric polynomial. One constructs r trigonometric polynomials $\{Q_j\}_{j=1,\dots,r}$ such that

$$P(\omega)\overline{P}(\omega + l) + \sum_{1}^{r} Q_j(\omega)\,\overline{Q}_j(\omega + l) = \begin{cases} 1, & l = 0, \\ 0, & l = 1, \end{cases}$$

and then defines $\{\psi_j\}_{j=1,\dots,r}$ as

$$\widehat{\psi}_j(\omega) = Q_j\left(\frac{\omega}{2}\right)\widehat{\varphi}\left(\frac{\omega}{2}\right).$$

The integral translations and dyadic dilations of these functions form a Parseval wavelet frame.

2.4 HILBERT TRANSFORM, ANALYTICAL SIGNAL, AND INSTANTANEOUS FREQUENCY

For a function $f(t)(t \in \mathbb{R})$, if the Cauchy principal value

$$\text{p.v.}\frac{1}{\pi}\int_{\mathbb{R}}\frac{f(\tau)}{t-\tau}\,d\tau = \frac{1}{\pi}\lim_{\epsilon\to 0}\int_{|t-\tau|>\epsilon}\frac{f(\tau)}{t-\tau}\,d\tau$$

exists, then it is called the *Hilbert transform* of $f(t)$, denoted by $\widetilde{f}(t)$, i.e.

$$\widetilde{f}(t) = \text{p.v.}\frac{1}{\pi}\int_{\mathbb{R}}\frac{f(\tau)}{t-\tau}\,d\tau.$$

Hilbert transforms have the following properties:

(i) (Linearity). Let $F = \alpha f_1 + \beta f_2$, where α, β are constants. Then $\widetilde{F} = \alpha\widetilde{f_1} + \beta\widetilde{f_2}$.

(ii) (Translation). Let $F(t) = f(t - \alpha)$. Then the Hilbert transform $\widetilde{F} = \widetilde{f}(t - \alpha)$.

  In fact,

$$\widetilde{F}(t) = \text{p.v.}\frac{1}{\pi}\int_{\mathbb{R}}\frac{f(\tau - \alpha)}{t-\tau}\,d\tau = \text{p.v.}\frac{1}{\pi}\int_{\mathbb{R}}\frac{f(u)}{t-\alpha-u}\,du = \widetilde{f}(t - \alpha).$$

(iii) (Dilation). Let $F(t) = f(\lambda t)$, where λ is a real number. Then the Hilbert transform

$$\widetilde{F}(t) = \widetilde{f}(\lambda t)\, \mathrm{sgn}\lambda,$$

where $\mathrm{sgn}\lambda = 1 (\lambda > 0)$ and $\mathrm{sgn}\lambda = -1 (\lambda < 0)$, and $\mathrm{sgn}\, 0 = 0$.
For $\lambda > 0$,

$$\widetilde{F}(t) = \mathrm{p.v.}\frac{1}{\pi}\int_{\mathbb{R}}\frac{f(\lambda\tau)}{t-\tau}\,d\tau = \mathrm{p.v.}\frac{1}{\pi}\int_{\mathbb{R}}\frac{f(u)}{\lambda t-u}\,du = \widetilde{f}(\lambda t) = \widetilde{f}(\lambda t)\,\mathrm{sgn}\lambda,$$

and for $\lambda < 0$,

$$\widetilde{F}(t) = \mathrm{p.v.}\frac{1}{\pi}\int_{\mathbb{R}}\frac{f(\lambda\tau)}{t-\tau}\,d\tau$$

$$= -\mathrm{p.v.}\frac{1}{\pi}\int_{\mathbb{R}}\frac{f(u)}{\lambda t-u}\,du = -\widetilde{f}(\lambda t) = \widetilde{f}(\lambda t)\,\mathrm{sgn}\lambda.$$

The following theorem shows that the Hilbert transform of a harmonic wave is also a harmonic wave.

Theorem 2.6. *Let $f(t)$ be a periodic signal with period 2π. Then its Hilbert transform is*

$$\widetilde{f}(t) = -\frac{1}{2\pi}\lim_{\epsilon\to0}\int_{\epsilon}^{\pi}\frac{f(t+\tau)-f(t-\tau)}{\tan(\tau/2)}\,d\tau. \tag{2.6}$$

Especially, if $f(t) = \cos t$, then $\widetilde{f}(t) = \sin t$; if $f(t) = \sin t$, then $\widetilde{f}(t) = -\cos t$.

Proof. The Hilbert transform of f is

$$\widetilde{f}(t) = \mathrm{p.v.}\frac{1}{\pi}\int_{\mathbb{R}}\frac{f(\tau)}{t-\tau}\,d\tau = \lim_{N\to\infty}\mathrm{p.v.}\frac{1}{\pi}\sum_{-N}^{N}\int_{(2k-1)\pi}^{(2k+1)\pi}\frac{f(\tau)}{t-\tau}\,d\tau$$

$$= \lim_{N\to\infty}\mathrm{p.v.}\frac{1}{\pi}\int_{-\pi}^{\pi}\sum_{-N}^{N}\frac{f(u+2k\pi)}{t-(u+2k\pi)}\,du.$$

From $f(u+2k\pi) = f(u)$, it follows that

$$\widetilde{f}(t) = \lim_{N\to\infty}\mathrm{p.v.}\frac{1}{\pi}\int_{-\pi}^{\pi}\sum_{-N}^{N}\frac{f(u)}{(t-u)-2k\pi}\,du$$

$$= \mathrm{p.v.}\frac{1}{\pi}\int_{-\pi}^{\pi}f(u)\left(\lim_{N\to\infty}\sum_{-N}^{N}\frac{1}{(t-u)-2k\pi}\right)du.$$

By using the known formula $\frac{1}{2\tan(t/2)} = \lim_{N\to\infty}\sum_{-N}^{N}\frac{1}{t-2k\pi}$, we find the right-hand side is equal to

$$\text{p.v.}\frac{1}{2\pi}\int_{-\pi}^{\pi}\frac{f(u)}{\tan((t-u)/2)}\,du = -\frac{1}{2\pi}\lim_{\epsilon\to 0}\left(\int_{-\pi}^{-\epsilon}+\int_{\epsilon}^{\pi}\right)\frac{f(t-\tau)}{\tan(\tau/2)}d\tau$$

$$= -\frac{1}{2\pi}\lim_{\epsilon\to 0}\int_{\epsilon}^{\pi}\frac{f(t+\tau)-f(t-\tau)}{\tan(\tau/2)}\,d\tau.$$

So we get (2.6).

If $f(t) = \cos t$, from (2.6), it follows that

$$\widetilde{f}(t) = -\frac{1}{2\pi}\lim_{\epsilon\to 0}\int_{\epsilon}^{\pi}\frac{\cos(t+\tau)-\cos(t-\tau)}{\tan(\tau/2)}d\tau$$

$$= \frac{2}{\pi}\sin t\int_{0}^{\pi}\cos^2\frac{\tau}{2}\,d\tau = \sin t.$$

If $f(t) = \sin t$, then $f(t) = -\cos(t+\frac{\pi}{2})$, and so $\widetilde{f}(t) = -\sin(t+\frac{\pi}{2}) = -\cos t$. $\qquad\square$

By Theorem 2.6 and the properties of the Hilbert transform, if a signal $f(t)$ is a trigonometric polynomial and

$$f(t) = \sum_{0}^{N}(c_n\cos(nt)+d_n\sin(nt)),$$

then its Hilbert transform is also a trigonometric polynomial and

$$\widetilde{f}(t) = \sum_{0}^{N}(c_n\sin(nt)-d_n\cos(nt)).$$

If a signal $f \in L_{2\pi}$ can be expanded into a Fourier series,

$$f(t) = \sum_{n}c_n(f)e^{int} = \frac{a_0}{2}(f)+\sum_{1}^{\infty}(a_n(f)\cos(nt)+b_n(f)\sin(nt)),$$

then its Hilbert transform satisfies

$$\widetilde{f}(t) = \sum_{n}-ic_n(f)\,\mathrm{sgn}\,ne^{int} = \sum_{1}^{\infty}(a_n(f)\sin(nt)-b_n(f)\cos(nt)),$$

where the series on the right-hand side is called the *conjugate Fourier series*. So

$$c_n(\widetilde{f}) = -i\,c_n(f)\,\mathrm{sgn}\,n \quad (n\in\mathbb{Z}),$$

$$f(t)+i\widetilde{f}(t) = c_0(f)+\sum_{1}^{\infty}2c_n(f)\,z^n \quad (z=e^{it}).$$

From this, we get the following theorem.

Theorem 2.7. *Let $f \in L_{2\pi}$ and $\widetilde{f}$ be its Hilbert transform, and $c_n(f)$, $c_n(\widetilde{f})$ be their Fourier coefficients. Then $c_n(\widetilde{f}) = -i\,c_n(f)\,\mathrm{sgn}\,n(n\in\mathbb{Z})$ and $f(t)+i\widetilde{f}(t) = c_0(f)+\sum_{1}^{\infty}2c_n(f)\,z^n(z=e^{it})$.*

From Theorem 2.7, we see that for a real-valued periodic signal f, adding the Hilbert transform $\widetilde{f}$ as the imaginary part, we obtain an analytic function in the unit disk $f_\alpha(z) = c_0(f) + \sum_1^\infty 2c_n(f) z^n (|z| < 1)$.

For a nonperiodic signal of finite energy, replacing Fourier coefficients by Fourier transforms, we obtain a result similar to Theorem 2.7, as follows.

Theorem 2.8. *Let $f \in L^2(\mathbb{R})$ and $\widetilde{f}$ be its Hilbert transform. Then their Fourier transforms satisfy*

$$\widehat{\widetilde{f}}(\omega) = -\mathrm{i}\widehat{f}(\omega) \operatorname{sgn} \omega.$$

Proof. Denote

$$K_{\delta,\eta}(t) = \begin{cases} \frac{1}{t}, & 0 < \delta \le |t| \le \eta < \infty, \\ 0, & \text{otherwise,} \end{cases}$$

and

$$\widetilde{f}_{\delta,\eta}(t) = \frac{1}{\pi} \int_{\delta \le |u| \le \eta} \frac{f(t-u)}{u} \, du.$$

From these two representation, it follows that

$$\widetilde{f}_{\delta,\eta}(t) = \frac{1}{\pi} \int_{\mathbb{R}} f(t-u) \, K_{\delta,\eta}(u) \, du = \frac{1}{\pi} (f * K_{\delta,\eta})(t).$$

By the convolution property of the Fourier transform, we get

$$\widehat{\widetilde{f}}_{\delta,\eta}(\omega) = \frac{1}{\pi} (f * K_{\delta,\eta})^\wedge(\omega) = \frac{1}{\pi} \widehat{f}(\omega) \widehat{K}_{\delta,\eta}(\omega). \tag{2.7}$$

With use of the Euler formula, $\mathrm{e}^{-\mathrm{i}v} - \mathrm{e}^{\mathrm{i}v} = -2\mathrm{i} \sin v$, the Fourier transform of $K_{\delta,\eta}$ is

$$\widehat{K}_{\delta,\eta}(\omega) = \int_{\delta < |t| \le \eta} \frac{1}{t} \mathrm{e}^{-\mathrm{i}t\omega} \, dt = \left(\int_{-\eta}^{-\delta} + \int_{\delta}^{\eta} \right) \frac{\mathrm{e}^{-\mathrm{i}t\omega}}{t} dt$$

$$= \int_{\delta}^{\eta} \frac{\mathrm{e}^{-\mathrm{i}t\omega} - \mathrm{e}^{\mathrm{i}t\omega}}{t} \, dt = -2\mathrm{i} \int_{\delta\omega}^{\eta\omega} \frac{\sin u}{u} \, du.$$

By the formula $\int_0^\infty \frac{\sin u}{u} \, du = \frac{\pi}{2}$, we deduce that for $\omega > 0$,

$$\lim_{\substack{\delta \to 0 \\ \eta \to \infty}} \widehat{K}_{\delta,\eta}(\omega) = \lim_{\substack{\delta \to 0 \\ \eta \to \infty}} \left(-2\mathrm{i} \int_{\delta\omega}^{\eta\omega} \frac{\sin u}{u} \, du \right) = -2\mathrm{i} \int_0^\infty \frac{\sin u}{u} \, du = -\pi \mathrm{i}.$$

Similarly, for $\omega < 0$, we can deduce that $\lim_{\substack{\delta \to 0 \\ \eta \to \infty}} \widehat{K}_{\delta,\eta}(\omega) = \pi \mathrm{i}$. Therefore,

$$\lim_{\substack{\delta \to 0 \\ \eta \to \infty}} \widehat{K}_{\delta,\eta}(\omega) = -\pi \mathrm{i} \operatorname{sgn} \omega.$$

From this and

$$\widetilde{f}(t) = \text{p.v.} \frac{1}{\pi} \int_{\mathbb{R}} \frac{f(\tau)}{t - \tau} \, d\tau = \lim_{\substack{\delta \to 0 \\ \eta \to \infty}} \widetilde{f}_{\delta,\eta}(t),$$

by (2.7), we get

$$\widehat{\widetilde{f}}(\omega) = \lim_{\substack{\delta \to 0 \\ \eta \to \infty}} \widehat{\widetilde{f}}_{\delta,\eta}(\omega) = \lim_{\substack{\delta \to 0 \\ \eta \to \infty}} \left(\frac{1}{\pi} \widehat{f}(\omega) \widehat{K}_{\delta,\eta}(\omega) \right) = -i\widehat{f}(\omega) \, \text{sgn} \, \omega.$$

$\square$

From Theorem 2.8, it follows that

$$\widehat{f}(\omega) + i\widehat{\widetilde{f}}(\omega) = \widehat{f}(\omega) + \widehat{f}(\omega) \, \text{sgn} \, \omega = \begin{cases} 2\widehat{f}(\omega), & \omega > 0, \\ 0, & \omega < 0 \end{cases} =: F(\omega).$$

So

$$f(t) + i\widetilde{f}(t) = F^{\vee}(t),$$

where $F^{\vee}(t)$ is the inverse Fourier transform of $F(\omega)$. This implies that $\widetilde{f}(t) = \text{Im} \, F^{\vee}(t)$.

Corollary 2.1. *Let $f \in L^2(\mathbb{R})$ and $\widetilde{f}$ be the Hilbert transform of f. Then $\widetilde{\widetilde{f}}(t) = -f(t)$.*

Proof. Let $\varphi(t) = \widetilde{f}(t)$. Then, by Theorem 2.8, we have

$$\widehat{\varphi}(\omega) = -i\widehat{\varphi}(\omega) \, \text{sgn} \, \omega = -i\widehat{\widetilde{f}}(\omega) \, \text{sgn} \, \omega = (-i \, \text{sgn} \, \omega)^2 \widehat{f}(\omega) = -\widehat{f}(\omega),$$

and so $\widetilde{\widetilde{f}}(t) = \widetilde{\varphi}(t) = -f(t)$.

$\square$

Bedrosian studied the Hilbert transform of products of two signals as follows.

Bedrosian Identity. *Let $f, g \in L^2(\mathbb{R})$ and the Fourier transforms of f, g satisfy $\widehat{f}(\omega) = 0 (\omega \in \mathbb{R} \setminus (-a, a))$ and $\widehat{g}(\omega) = 0 (\omega \in [-a, a])$ for some $a > 0$. Then $\widetilde{fg} = f\widetilde{g}$.*

Proof. By the assumption and the convolution property in frequency, it follows that

$$\widehat{fg}(\omega) = \frac{1}{2\pi} (\widehat{f} * \widehat{g})(\omega) = \frac{1}{2\pi} \int_{-a}^{a} \widehat{f}(u) \widehat{g}(\omega - u) \, du.$$

By Theorem 2.8, $\widehat{\widetilde{fg}} = -i \, \text{sgn} \, \omega \widehat{fg}(\omega)$, and so

$$\widetilde{fg}(t) = \frac{1}{2\pi} \int_{\mathbb{R}} (-i \, \text{sgn} \, \omega) \widehat{fg}(\omega) e^{it\omega} \, d\omega$$

$$= \frac{1}{(2\pi)^2} \int_{\mathbb{R}} (-i \, \text{sgn} \, \omega) e^{it\omega} \int_{-a}^{a} \widehat{f}(u) \widehat{g}(\omega - u) \, du \, d\omega$$

$$= \frac{1}{(2\pi)^2} \int_{-a}^{a} \widehat{f}(u) \left(\int_{\mathbb{R}} \widehat{g}(v)(-i \, \text{sgn}(u + v)) e^{it(u+v)} \, dv \right) du.$$

Consider $\text{sgn}(u + v)$. Note that $-a \le u \le a$. If $v \ge a$, then $u + v \ge 0, v \ge 0$, and so $\text{sgn}(u + v) = \text{sgn } v$. If $v \le -a$, then $u + v \le 0, v \le 0$, and so $\text{sgn}(u + v) = \text{sgn } v$. Note that $\widehat{g}(\omega) = 0 (\omega \in [-a, a])$. The integral in brackets is equal to

$$\left(\int_{-\infty}^{-a} + \int_{a}^{\infty} \right) \widehat{g}(v)(-i \, \text{sgn}(u + v)) e^{it(u+v)} \, dv$$

$$= -i \int_{\mathbb{R} \setminus [-a,a]} \widehat{g}(v) \, \text{sgn } v e^{it(u+v)} \, dv = -i \int_{\mathbb{R}} \widehat{g}(v) \, \text{sgn } v e^{it(u+v)} \, dv.$$

Therefore,

$$\widetilde{fg}(t) = \frac{1}{(2\pi)^2} \int_{-a}^{a} \widehat{f}(u) e^{itu} \, du \int_{\mathbb{R}} \widehat{g}(v)(-i \, \text{sgn } v) e^{ivt} \, dv = f(t)\widetilde{g}(t).$$

$\square$

A signal of finite energy is called an *analytic signal* if its Fourier transform is zero for negative frequency.

Proposition 2.2. *Let* $f \in L^2(\mathbb{R})$ *and* $\widetilde{f}$ *be the Hilbert transform of* f. *Then* $f_\alpha(t) = f(t) + i\widetilde{f}(t)$ *is an analytic signal.*

Proof. By Theorem 2.8: $\widehat{\widetilde{f}} = -i\widehat{f}(\omega) \, \text{sgn } \omega$, it follows that

$$\widehat{f_\alpha}(\omega) = \widehat{f}(\omega) + i\widehat{\widetilde{f}}(\omega) = \widehat{f}(\omega) + \widehat{f}(\omega)\text{sgn } \omega = \begin{cases} 2\widehat{f}(\omega), & \omega \ge 0, \\ 0, & \omega < 0. \end{cases}$$

that is, $f_\alpha(t)$ is an analytic signal. $\square$

Complex analysis shows that $f_\alpha(t) = f(t) + i\widetilde{f}(t)$ can be extended to an analytic function $f_\alpha(z)$ on the upper-half plane. Denote $f_\alpha(t) = A(t)e^{i\theta(t)}$. Then

$$A(t) = ((f(t))^2 + (\widetilde{f}(t))^2)^{1/2},$$

$$\theta(t) = \tan^{-1}\left(\frac{\widetilde{f}(t)}{f(t)} \right),$$

where $A(t)$ and $\theta(t)$ are called the *modulus* and *argument* of $f_\alpha(t)$, respectively.

Definition 2.1. Let $f \in L^2(\mathbb{R})$ be a real signal and

$$f_\alpha(t) = f(t) + i\widetilde{f}(t) = A(t)e^{i\theta(t)} \quad (A(t) \ge 0).$$

Then $\theta'(t)$ is called the *instantaneous frequency* of $f(t)$.

Example 2.2. Let $f(t) = a(t) \cos(\omega_0 t + \varphi)$, where $a(t) \in L^2(\mathbb{R})$ and $a(t) > 0$, $\omega_0 > 0$, and $\widehat{a}(\omega) = 0 \ (|\omega| > \omega_0)$. Then the instantaneous frequency of f is ω_0.

Let $g(t) = \cos(\omega_0 t + \varphi)$. By Theorem 2.6 and properties of the Hilbert transform, it follows that

$$\widetilde{g}(t) = \sin(\omega_0 t + \varphi) \, \text{sgn } \omega_0 = \sin(\omega_0 t + \varphi) \quad (\omega_0 > 0).$$

Noticing that $\widehat{a}(\omega) = 0(|\omega| > \omega_0)$ and $\operatorname{supp}\widehat{g}(\omega) = \{\omega_0, -\omega_0\}$, by the Bedrosian identity, we find that

$$\widetilde{f}(t) = a(t)\widetilde{g}(t) = a(t)\sin(\omega_0 t + \varphi).$$

Therefore,

$$f_\alpha(t) = f(t) + \mathrm{i}\widetilde{f}(t) = a(t)(\cos(\omega_0 t + \varphi) + \mathrm{i}\sin(\omega_0 t + \varphi)) = a(t)\mathrm{e}^{\mathrm{i}(\omega_0 t + \varphi)}.$$

By Definition 2.1, the instantaneous frequency is ω_0.

Let a signal f be the sum of two cosine waves with the same amplitude:

$$f(t) = a\cos(\omega_1 t) + a\cos(\omega_2 t) \quad (\omega_1 > \omega_2 > 0).$$

Then its Hilbert transform is

$$\widetilde{f}(t) = a\sin(\omega_1 t)\operatorname{sgn}\omega_1 + a\sin(\omega_2 t)\operatorname{sgn}\omega_2 = a\sin(\omega_1 t) + a\sin(\omega_2 t).$$

Then the corresponding analytic signal is

$$f_\alpha(t) = f(t) + \mathrm{i}\widetilde{f}(t) = a(\mathrm{e}^{\mathrm{i}\omega_1 t} + \mathrm{e}^{\mathrm{i}\omega_2 t}) = 2a\cos\frac{(\omega_1 - \omega_2)t}{2}\mathrm{e}^{\mathrm{i}((\omega_1 + \omega_2)t/2)}.$$

By Definition 2.1, the instantaneous frequency is $\frac{1}{2}(\omega_1 + \omega_2)$. This does not reveal that the signal includes two cosine waves with frequency ω_1 and ω_2, respectively, so the Hilbert transform can deal only with narrow-band signals

2.5   WIGNER-VILLE DISTRIBUTION AND COHEN'S CLASS

The windowed Fourier transform and the wavelet transform analyze the time-frequency structure by using a window function, while the Wigner-Ville distribution analyzes the time-frequency structure by translations. The Wigner-Ville distribution is defined as

$$W_V f(u, \omega) = \int_{\mathbb{R}} f\left(u + \frac{\tau}{2}\right)\overline{f}\left(u - \frac{\tau}{2}\right)\mathrm{e}^{-\mathrm{i}\tau\omega}\,\mathrm{d}\tau \quad (f \in L^2(\mathbb{R})).$$

If $f(t) = \mathrm{e}^{\mathrm{i}bt}$, then $W_V f(u, \omega) = \frac{1}{2\pi}\delta(\omega - b)$, where δ is the Dirac function.

The Wigner-Ville distribution possesses the following properties:

(i) (Phase translation). If $f(t) = \mathrm{e}^{\mathrm{i}\varphi}g(t)$, then $W_V f(u, \omega) = W_V g(u, \omega)$.

(ii) (Time translation). If $f(t) = g(t - u_0)$, then $W_V f(u, \omega) = W_V g(u - u_0, \omega)$.

(iii) (Frequency translation). If $f(t) = \mathrm{e}^{\mathrm{i}t\omega_0}g(t)$, then $W_V f(u, \omega) = W_V g(u, \omega - \omega_0)$.

(iv) (Scale dilation). If $f(t) = \frac{1}{\sqrt{s}}g\left(\frac{t}{s}\right)$, then $W_V f(u, \omega) = W_V f(\frac{u}{s}, s\omega)$.

The Wigner-Ville distribution can localize the time-frequency structure of the signal f. In the Wigner-Ville distribution, time and frequency have a symmetrical role, i.e., the following proposition holds.

Proposition 2.3. $W_V f(u, \omega) = \frac{1}{2\pi}\int_{\mathbb{R}}\widehat{f}(\omega + \frac{r}{2})\overline{\widehat{f}}(\omega - \frac{r}{2})\mathrm{e}^{\mathrm{i}ru}\,\mathrm{d}r.$

Proof. Denote

$$\varphi(\tau) = f\left(u + \frac{\tau}{2}\right)e^{-i\tau\xi}, \quad g(\tau) = f\left(u - \frac{\tau}{2}\right).$$

By Theorem 1.3, the Wigner-Ville distribution is

$$W_V f(u, \xi) = \int_{\mathbb{R}} \varphi(\tau) \overline{g}(\tau)\, d\tau = \frac{1}{2\pi} \int_{\mathbb{R}} \widehat{\varphi}(\omega) \overline{\widehat{g}}(\omega)\, d\omega.$$

Let $u + \frac{\tau}{2} = t$. Then

$$\widehat{\varphi}(\omega) = \int_{\mathbb{R}} f\left(u + \frac{\tau}{2}\right) e^{-i\tau\xi} e^{-i\tau\omega} d\tau$$

$$= 2 \int_{\mathbb{R}} f(t) e^{-2i(t-u)(\xi+\omega)}\, dt = 2\widehat{f}(2\omega + 2\xi)e^{2iu(\xi+\omega)}$$

and

$$\overline{\widehat{g}}(\omega) = \overline{\int_{\mathbb{R}} f\left(u - \frac{\tau}{2}\right) e^{-i\tau\omega}\, d\tau} = 2\overline{\widehat{f}}(-2\omega)e^{2iu\omega}.$$

Using the substitution $\xi + 2\omega = \frac{r}{2}$, we have

$$W_V f(u, \xi) = \frac{2}{\pi} \int_{\mathbb{R}} \widehat{f}(2(\omega + \xi)) \overline{\widehat{f}}(-2\omega) e^{2iu(\xi+2\omega)} d\omega$$

$$= \frac{1}{2\pi} \int_{\mathbb{R}} \widehat{f}\left(\xi + \frac{r}{2}\right) \overline{\widehat{f}}\left(\xi - \frac{r}{2}\right) e^{iru}\, dr.$$

$\square$

From Proposition 2.3 and the definition of the Wigner-Ville distribution, we get the following proposition.

Proposition 2.4. (i) *If* $\operatorname{supp} f(u) = [u_0 - \mu, u_0 + \mu]$, *then* $\operatorname{supp} W_V f(\cdot, \omega) \subset [u_0 - \mu, u_0 + \mu]$ ($\omega \in \mathbb{R}$).

(ii) *If* $\operatorname{supp} \widehat{f}(\omega) = [\omega_0 - \eta, \omega_0 + \eta]$, *then* $\operatorname{supp} W_V f(u, \cdot) \subset [\omega_0 - \eta, \omega_0 + \eta]$ ($u \in \mathbb{R}$).

Proof. Let $g(t) = f(-t)$. The Wigner-Ville distribution is written in the form

$$W_V f(u, \omega) = \int_{\mathbb{R}} f\left(\frac{\tau + 2u}{2}\right) \overline{g}\left(\frac{\tau - 2u}{2}\right) e^{-i\tau\omega}\, d\tau.$$

Since $\operatorname{supp} f = [u_0 - \mu, u_0 + \mu]$ and $\operatorname{supp} g = [-u_0 - \mu, -u_0 + \mu]$, it follows that

$$\operatorname{supp} f\left(\frac{\tau + 2u}{2}\right) = [2(u_0 - u) - 2\mu, 2(u_0 - u) + 2\mu],$$

$$\operatorname{supp} g\left(\frac{\tau - 2u}{2}\right) = [-2(u_0 - u) - 2\mu, -2(u_0 - u) + 2\mu].$$

Therefore, $W_V f(\cdot, \omega) \neq 0$ only if these two intervals overlap. This is equivalent to $|u_0 - u| \leq \mu$. So we get (i). Similarly, by Proposition 2.3, we can get (ii). □

Since the Fourier transform of a Gaussian function is still a Gaussian function, a direct computation shows that the Wigner-Ville distribution of Gaussian function $f(t) = (\sigma^2 \pi)^{-(1/4)} e^{-(t^2/2\sigma^2)}$ is a bivariate Gaussian function,

$$W_V f(u, \omega) = \frac{1}{\pi} e^{-(u^2/\sigma^2) - \sigma^2 \omega^2} \quad \text{i.e.,} \quad W_V f(u, \omega) = |f(u)|^2 |\widehat{f}(\omega)|^2.$$

For a signal $f(t)$, we know that $f_a(t) = f(t) + i\widetilde{f}(t) = A(t) e^{i\theta(t)} (A(t) \geq 0)$ is an analytic signal and $\theta'(t)$ is the instantaneous frequency of $f(t)$. The formula

$$\theta'(u) = \frac{\int_{\mathbb{R}} \omega W_V f_\alpha(u, \omega) \, d\omega}{\int_{\mathbb{R}} W_V f_\alpha(u, \omega) \, d\omega}$$

gives an *equivalent definition of the instantaneous frequency* computed by the Wigner-Ville distribution. This shows that the instantaneous frequency is the average frequency. Moreover, the Wigner-Ville distribution is a unitary transform which can imply the energy conservation property.

Theorem 2.9. *For* $f, g \in L^2(\mathbb{R})$,

$$\left| \int_{\mathbb{R}} f(t) \overline{g}(t) \, dt \right|^2 = \frac{1}{2\pi} \int \int_{\mathbb{R}^2} W_V f(u, \omega) W_V g(u, \omega) \, du \, d\omega.$$

Proof. Note that

$$A = \int \int_{\mathbb{R}^2} W_V f(u, \omega) W_V g(u, \omega) \, du \, d\omega$$

$$= \int \int_{\mathbb{R}^2} \left(\int_{\mathbb{R}} f\left(u + \frac{\tau}{2}\right) \overline{f}\left(u - \frac{\tau}{2}\right) e^{-i\omega\tau} \, d\tau \right) \left(\int_{\mathbb{R}} g\left(u + \frac{\tau'}{2}\right) \right.$$

$$\left. \overline{g}\left(u - \frac{\tau'}{2}\right) e^{-i\omega\tau'} \, d\tau' \right) \, du \, d\omega.$$

By Formula 1.1, the Fourier transform of 1 is $2\pi\delta$, we get

$$\int_{\mathbb{R}} e^{-i\omega(\tau+\tau')} \, d\omega = 2\pi\delta(\tau + \tau').$$

Moreover, we have

$$A = 2\pi \int \int_{\mathbb{R}^2} f\left(u + \frac{\tau}{2}\right) \overline{f}\left(u - \frac{\tau}{2}\right) \left(\int_{\mathbb{R}} \delta(\tau + \tau') g\left(u - \frac{\tau'}{2}\right) \overline{g}\left(u + \frac{\tau'}{2}\right) \, d\tau' \right) \, d\tau \, du$$

$$= 2\pi \int \int_{\mathbb{R}^2} f\left(u + \frac{\tau}{2}\right) \overline{f}\left(u - \frac{\tau}{2}\right) g\left(u - \frac{\tau}{2}\right) \overline{g}\left(u + \frac{\tau}{2}\right) \, d\tau \, du$$

Let $t = u + (\tau/2)$ and $s = u - (\tau/2)$. Then

$$A = 2\pi \int \int_{\mathbb{R}^2} f(t) \overline{f}(s) g(s) \overline{g}(t) \, dt \, ds = 2\pi \left| \int_{\mathbb{R}} f(t) \overline{g}(t) \, dt \right|^2.$$

□

Proposition 2.5. *The Wigner-Ville distribution satisfies*

$$\int_{\mathbb{R}} W_V f(u, \omega)\, du = |\widehat{f}(\omega)|^2 \quad \text{and} \quad \frac{1}{2\pi} \int_{\mathbb{R}} W_V f(u, \omega)\, d\omega = |f(u)|^2.$$

Proof. Let $g_\omega(u) = (W_V f)(u, \omega)$. Note that $\widehat{g}_\omega(0) = \int_{\mathbb{R}} g_\omega(u) e^{-i0u}\, du = \int_{\mathbb{R}} g_\omega(u)\, du$. Then

$$\int_{\mathbb{R}} (W_V f)(u, \omega)\, du = \int_{\mathbb{R}} g_\omega(u)\, du = \widehat{g}_\omega(0).$$

By Proposition 2.3, the Fourier transform of g_ω is $\widehat{g}_\omega(r) = \widehat{f}(\omega + \frac{r}{2})\overline{\widehat{f}}(\omega - \frac{r}{2})$. Therefore,

$$\int_{\mathbb{R}} (W_V f)(u, \omega)\, du = |\widehat{f}(\omega)|^2.$$

Similarly, let $h_u(\omega) = W_V f(u, \omega)$. Then

$$\int_{\mathbb{R}} W_V f(u, \omega)\, d\omega = \int_{\mathbb{R}} h_u(\omega)\, d\omega = \widehat{h}_u(0).$$

By the definition of the Wigner-Ville distribution, the Fourier transform $\widehat{h}_u(\tau) = 2\pi f(u + \frac{\tau}{2})\overline{f}(u - \frac{\tau}{2})$. Therefore,

$$\int_{\mathbb{R}} W_V f(u, \omega)\, d\omega = 2\pi |f(u)|^2.$$

$\square$

However, the Wigner-Ville distribution may take negative values. For example, let $f = \chi_{[-T,T]}$. Since f is a real even function,

$$(W_V f)(u, \omega) = \int_{\mathbb{R}} f\left(\frac{\tau}{2} + u\right) f\left(\frac{\tau}{2} - u\right) e^{-i\tau\omega} d\tau$$

$$= 2 \int_{\mathbb{R}} f(\tau + u) f(\tau - u) e^{-2i\tau\omega}\, d\tau$$

and $f(\tau + u)f(\tau - u) = \chi_{[-T+|u|, T-|u|]}(\tau)$, and its Fourier transform

$$(f(\tau + u)f(\tau - u))^{\wedge}(\omega) = \int_{-T+|u|}^{T-|u|} e^{-it\omega} dt$$

$$= \frac{e^{-i\omega(T-|u|)} - e^{i\omega(T-|u|)}}{-i\omega} = \frac{2\sin((T - |u|)\omega)}{\omega}.$$

Note that $(W_V f)(u, \omega) = 0 (|u| > \frac{T}{2})$. Then

$$\frac{1}{2} W_V f(u, \omega) = (f(\tau + u)f(\tau - u))^{\wedge}(2\omega) = \frac{2\sin(2(T - |u|)\omega)}{\omega} \chi_{\left[-\frac{T}{2}, \frac{T}{2}\right]}(u).$$

Clearly, $W_V f(u, \omega)$ takes negative values. A Gaussian function is the only function whose Wigner-Ville distribution remains positive.

To obtain a positive energy distribution, one needs to average the Wigner-Ville distribution and introduce the *Cohen's class distributions* as follows

$$Kf(u, \omega) := \int \int_{\mathbb{R}^2} W_V f(u', \omega') \, k(u - u', \omega - \omega') \, du' \, d\omega',$$

where $k(u, v)$ is a smooth kernel function. The windowed Fourier transform belongs to Cohen's class distributions, and the corresponding smooth kernel is

$$k(u, \omega) = \frac{1}{2\pi} W_V g(u, \omega),$$

where $g(t)$ is a window function.

2.6 EMPIRICAL MODE DECOMPOSITIONS

Spline functions play a key role in the empirical mode decomposition (EMD) algorithm. If f is a polynomial of degree $k - 1$ on each interval $[x_n, x_{n+1}] (n \in \mathbb{Z})$ and f is a $k - 2$-order continuously differentiable function on $\mathbb{R}$, then f is called a *spline function* of degree $k(k \geq 2)$ with knots $\{x_n\}_{n \in \mathbb{Z}}$.

Let a function f on $\mathbb{R}$ have local maximal values on $\{\alpha_n\}$:

$$\cdots < \alpha_{-1} < \alpha_0 < \alpha_1 < \alpha_2 < \cdots.$$

Define the *upper envelope $M(f)$* of f as follows:

(i) $M(f)(\alpha_n) = f(\alpha_n)(n \in \mathbb{Z})$;
(ii) $M(f)$ is a 3-order spline function with knots $\{\alpha_n\}$.

Let a function f on $\mathbb{R}$ have local minimal values on $\{\beta_n\}$:

$$\cdots < \beta_{-1} < \beta_0 < \beta_1 < \beta_2 < \cdots.$$

Define the *lower envelope $m(f)$* of f as follows:

(i) $m(f)(\beta_n) = f(\beta_n)(n \in \mathbb{Z})$;
(ii) $m(f)$ is 3-order spline function with knots $\{\beta_n\}$.

The *local mean* of a function f on $\mathbb{R}$ is defined as

$$V(f)(t) = \frac{1}{2}(M(f)(t) + m(f)(t)).$$

For example, $f(t) = 3 \sin(2t + \frac{\pi}{4})$ attains the maximal values on $\alpha_n = \frac{1}{2}$ $((2n + \frac{1}{2})\pi - \frac{\pi}{4})(n \in \mathbb{Z})$ and attains the minimal values on $\beta_n = \frac{1}{2}((2n - \frac{1}{2})\pi - \frac{\pi}{4})(n \in \mathbb{Z})$, and attains the crossing zeros on $\gamma_n = \frac{1}{2}(n\pi - \frac{\pi}{4})(n \in \mathbb{Z})$. Clearly,

$$\cdots < \gamma_{2n-1} < \beta_n < \gamma_{2n} < \alpha_n < \gamma_{2n+1} < \beta_{n+1} < \cdots.$$

So $f(t)$ has the upper envelope $M(f)(t) = 3$ and the lower envelope $m(f)(t) = -3$, and its local mean $V(f)(t) = 0$.

A function f is called an *intrinsic mode function* (IMF) if it satisfies the following conditions:

(i) The number of extrema and the number of crossing zeros are equal or differ at most by one.
(ii) Its local mean is zero.

Empirical mode decomposition is used to decompose a signal f into several IMFs. If a discrete signal $f(t)$ has more than one oscillatory mode, then it can be decomposed into a sum of several IMFs and a monotonic signal as follows:

(i) Take the upper envelope $M(f)$ and lower envelope $m(f)$ of $f(t)$.
(ii) Compute the mean $V(f)(t) = \frac{1}{2}(M(f)(t) + m(f)(t))$ and the residual $r(t) = f(t) - V(f)(t)$.
(iii) Let $r(t)$ be the new signal. Follow this procedure until the local mean of $r(t)$ is equal to zero.
(iv) Once we have the zero-mean $r(t)$, it is designated as the first IMF, $c_1(t)$.
(v) Denote $f_1(t) = f(t) - c_1(t)$. We start from $f_1(t)$. Repeating the procedure from (i) to (iv), we get the second IMF, $c_2(t)$.
(vi) Continuing this procedure, we get $c_1(t), c_2(t), \ldots, c_n(t)$.

This process is stopped when the residual $r_n(t)$ is a monotonic function.

The procedure from (i) to (vi) gives an empirical mode decomposition of the signal $f(t)$ as follows:

$$f(t) = \sum_1^n c_k(t) + r_n(t),$$

where each $c_k(t)$ is an IMF and $r_n(t)$ is monotonic.

Let $\tilde{c}_k(t)$ be the Hilbert transform of $c_k(t)$:

$$\tilde{c}_k(t) = \text{p.v.} \frac{1}{\pi} \int_{\mathbb{R}} \frac{c_k(u)}{t - u} \, du.$$

Then $Z_k(t) := c_k(t) + i\tilde{c}_k(t) = A_k(t)e^{i\theta_k(t)}$ is an analytic signal, where

$$A_k(t) = (c_k^2(t) + \tilde{c}_k^2(t))^{1/2}, \quad \theta_k(t) = \arctan\left(\frac{\tilde{c}_k(t)}{c_k(t)}\right).$$

Denote by $\omega_k(t)$ the instantaneous frequency of $c_k(t)$. Then the instantaneous frequency $\omega_k(t) = \theta_k'(t)$. This process is also called the *Hilbert-Huang transform*.

PROBLEMS

2.1 Let $f(t) = e^{-t^2}$. Compute its Gabor transform $\left(G_0^\alpha f\right)(\omega)$.
2.2 Compare the time-frequency window of the windowed Fourier transform with that of the wavelet transform.

2.3 Download the monthly mean North Atlantic Oscillation index from http://www.cpc.ncep.noaa.gov/products/precip/CWlink/pna/new.nao.shtml and then research the wavelet power spectrum of the North Atlantic Oscillation index at different scales.

2.4 Let $\chi(x)$ be a Haar wavelet, i.e.

$$\chi(t) = \begin{cases} -1, & 0 \le t < \frac{1}{2}, \\ 1, & \frac{1}{2} < t < 1, \\ 0, & \text{otherwise.} \end{cases}$$

Prove $\{2^{m/2}\chi(2^m - n)\}_{m,n\in\mathbb{Z}}$ is a wavelet basis of $L^2(\mathbb{R})$.

2.5 Given a multiresolution analysis $\{V_m\}$,

$$V_m = \{f \in L^2(\mathbb{R}), \ \widehat{f}(\omega) = 0, \ |\omega| \ge 2^m\pi\},$$

try to find the scaling function and the corresponding orthonormal wavelet.

2.6 Let $H(\omega)$ be the filter of a scaling function and

$$H(\omega) = \sum_n a_n e^{-in\omega}.$$

Prove that

(i) $\sum_n a_{2n} = \sum_n a_{2n+1} = \frac{1}{2}$;

(ii) $\sum_n a_n \bar{a}_{n-2k} = \begin{cases} 0, & k \ne 0, k \in \mathbb{Z}, \\ \frac{1}{2}, & k = 0. \end{cases}$

2.7 Perform empirical mode decomposition of local temperature data and analyze when significant warming occurs.

BIBLIOGRAPHY

Amirmazlaghani, M., Amindavar, H., 2013. Statistical modeling and denoising Wigner-Ville distribution. Digital Signal Process. 23, 506-513.

Andreo, B., Jimenez, P., Duran, J.J., Carrasco, F., Vadillo, I., Mangin, A., 2006. Climatic and hydrological variations during the last 117–166 years in the south of the Iberian Peninsula, from spectral and correlation analyses and continuous wavelet analyses. J. Hydrol. 324, 24-39.

Barnhart, B.L., Eichinger, W.E., 2011. Empirical mode decomposition applied to solar irradiance, global temperature, sunspot number, and CO_2 concentration data. J. Atmos. Sol.-Terr. Phys. 73, 1771-1779.

Bedrosian, E., 1963. A product theorem for Hilbert transform. Proc. IEEE 51, 868-869.

Boashash, B., 1992. Estimating and interpreting the instantaneous frequency of a signal-Part I: fundamentals. Proc. IEEE 80, 519-538.

Cherneva, Z., Guedes Soares, C., 2008. Non-linearity and non-stationary of the New Year abnormal wave. Appl. Ocean Res. 30, 215-220.

Chui, C.K., 1992. An Introduction to Wavelet. Academic Press, Inc., San Diego, CA.

Cohen, L., 1995. Time-Frequency Analysis. Prentice-Hall, New York.

Daubechies, I., 1992. Ten Lectures on Wavelets, vol. 6. CBMS-Conference Lecture Notes. SIAM, Philadelphia.

Fang, K., Frank, D., Gou, X., Liu, C., Zhou, F., Li, J., Li, Y., 2013. Precipitation over the past four centuries in the Dieshan Mountains as inferred from tree rings: an introduction to an HHT-based method. Global Planet. Change 107, 109-118.

Flandrin, P., 1999. Time-Frequency/Time-Scale Analysis, Wavelet Analysis and Its Applications, vol. 10. Academic Press, San Diego.

Galloway, J.M., Wigston, A., Patterson, R.T., Swindles, G.T., Reinhardt, E., Roe, H.M., 2013. Climate change and decadal to centennial-scale periodicities recorded in a late holocene NE Pacific marine record: examining the role of solar forcing. Palaeogeogr. Palaeoclimatol. Palaeoecol. 386, 669-689.

Hu, W., Biswas, A., Si, B.C., 2014. Application of multivariate empirical mode decomposition for revealing scale- and season-specific time stability of soil water storage. CATENA 113, 377-385.

Huang, Y., Schmitt, F.G., 2014. Time dependent intrinsic correlation analysis of temperature and dissolved oxygen time series using empirical mode decomposition. J. Mar. Syst. 130, 90-100.

Huang, Y., Schmitt, F.G., Lu, Z., Liu, Y., 2009. Analysis of daily river flow fluctuations using empirical mode decomposition and arbitrary order Hilbert spectral analysis. J. Hydrol. 373, 103-111.

Jevrejeva, S., Moore , J.C., Woodworth, P.L., Grinsted, A., 2005. Influence of large scale atmospheric circulation on European sea level: results based on the wavelet transform method. Tellus 57A, 183-193.

Karthikeyan, L., Nagesh Kumar, D., 2013. Predictability of nonstationary time series using wavelet and EMD based ARMA models. J. Hydrol. 502, 103-119.

Mallat, S., 1998. A Wavelet Tour of Signal Processing. Academic Press, San Diego, CA.

Massei, N., Fournier, M., 2012. Assessing the expression of large-scale climatic fluctuations in the hydrological variability of daily Seine river flow (France) between 1950 and 2008 using Hilbert-Huang transform. J. Hydrol. 448-449, 119-128.

Moosavi, V., Malekinezhad, H., Shirmohammadi, B., 2014, Fractional snow cover mapping from MODIS data using wavelet-artificial intelligence hybrid models. J. Hydrol. 511, 160-170.

Nalley, D., Adamowski, J., Khalil, B., Ozga-Zielinski, B., 2013. Trend detection in surface air temperature in Ontario and Quebec, Canada during 1967–2006 using the discrete wavelet transform. Atmos. Res. 132-133, 375-398.

Narasimhan, S.V., Nayak, M.B., 2003. Improved Wigner-Ville distribution performance by signal decomposition and modified group delay. Signal Process. 83, 2523-2538.

Olsen, L.R., Chaudhuri, P., Godtliebsen, F., 2008. Multiscale spectral analysis for detecting short and long range change points in time series. Comput. Stat. Data Anal. 52, 3310-3330.

Rossi, A., Massei, N., Laignel, B., Sebag, D., Copard, Y., 2009. The response of the Mississippi River to climate fluctuations and reservoir construction as indicated by wavelet analysis of streamflow and suspended-sediment load, 1950-1975. J. Hydrol. 377, 237-244.

Rossi, A., Massei, N., Laignel, B. 2011. A synthesis of the time-scale variability of commonly used climate indices using continuous wavelet transform. Global Planet. Change 78, 1-13.

Sang, Y.-F., Wang, Z., Liu, C., 2014. Comparison of the MK test and EMD method for trend identification in hydrological time series. J. Hydrol. 510, 293-298.

Thakur, G., Brevdo, E., Fukar, N.S., Wu, H.-T., 2013. The synchrosqueezing algorithm for time-varying spectral analysis: robustness properties and new paleoclimate applications. Signal Process. 93, 1079-1094.

Yi, H., Shu, H., 2012. The improvement of the Morlet wavelet for multi-period analysis of climate data. C. R. Geosci. 344, 483-497.

Zhu, Z., Yang, H. 2002. Discrete Hilbert transformation and its application to estimate the wind speed in Hong Kong. J. Wind Eng. Ind. Aerodyn. 90, 9-18.

Chapter 3

Filter Design

The purpose of filtering is to extract the information of geophysical signals for a given frequency band or restore the original signal details as much as possible by removing the unwanted noise produced by measurement imperfections. A lot of filters have been proposed, each of which has its own advantages and limitations. Implementation of these filters is easy, fast, and cost-effective by using a linear time-invariant system. In this chapter, we first focus on continuous linear time-invariant systems and the corresponding analog filters, including Butterworth filters, Chebeshev filters, and elliptic filters. Then we turn to discrete linear time-invariant systems, finite impulse response (FIR) filters, infinite impulse response (IIR) filters, and conjugate mirror filters.

3.1 CONTINUOUS LINEAR TIME-INVARIANT SYSTEMS

Linear time-invariant systems play a key role in the construction of filters. To explain this concept, we use the notation $y(t) = T[x(t)]$ to represent a system, where $x(t)$ is the input to the system and $y(t)$ is the output from the system.

If, for arbitrary constants a and b,

$$T[\alpha x_1(t) + \beta x_2(t)] = \alpha T[x_1(t)] + \beta T[x_2(t)],$$

then the system $y(t) = T[x(t)]$ is called a *linear system*. If, for $x_n(t) \rightarrow x(t) \ (L^2)$,

$$T[x_n(t)] \rightarrow T[x(t)] \quad (t \in \mathbb{R}),$$

then the system $y(t) = T[x(t)]$ is *continuous*.

Let $y(t) = T[x(t)]$ be a linear system and τ be a constant. If $y(t - \tau) = T[x(t - \tau)]$ for any τ, then this linear system is called a *linear time-invariant system*.

In order to study linear time-invariant systems, we first define convolution.

Let $g(t) \in L^2(\mathbb{R})$ and $x(t) \in L^2(\mathbb{R})$ be two continuous signals. The *convolution* of $g(t)$ and $x(t)$ is

$$(g * x)(t) = \int_{\mathbb{R}} g(t - u) x(u) \, du \quad (t \in \mathbb{R}).$$

It has the following properties:

$$(g * (c x + d y))(t) = c (g * x)(t) + d (g * y)(t) \quad (t \in \mathbb{R}),$$

$$(g * x)(t) = (x * g)(t), \quad (g * (x * y))(t) = ((g * x) * y)(t) \quad (t \in \mathbb{R}),$$

where $g(t)$, $x(t)$, and $y(t)$ are continuous signals and c and d are constants.

Proposition 3.1. *Let $g \in L^2(\mathbb{R})$. A system $y(t) = T[x(t)]$ determined by the convolution*

$$y(t) = (g * x)(t) = \int_{\mathbb{R}} g(t - u) x(u) \, du \quad (x \in L^2(\mathbb{R}))$$

is a linear time-invariant system and is continuous. Here g is often called a filter.

Proof. Take $y_1(t) = T[x_1(t)]$ and $y_2(t) = T[x_2(t)]$. For any two constants α and β, it is clear that

$$\alpha y_1(t) + \beta y_2(t) = \alpha T[x_1(t)] + \beta T[x_2(t)].$$

On the other hand, by the assumption

$$y_1(t) = (g * x_1)(t) = \int_{\mathbb{R}} g(t - u) x_1(u) \, du,$$

$$y_2(t) = (g * x_2)(t) = \int_{\mathbb{R}} g(t - u) x_2(u) \, du,$$

it follows that

$$\begin{aligned}
\alpha y_1(t) + \beta y_2(t) &= \alpha \int_{\mathbb{R}} g(t - u) x_1(u) \, du + \beta \int_{\mathbb{R}} g(t - u) x_2(u) \, du \\
&= \int_{\mathbb{R}} g(t - u) (\alpha x_1(u) + \beta x_2(u)) \, du \\
&= T[\alpha x_1(t) + \beta x_2(t)].
\end{aligned}$$

Therefore,

$$T[\alpha x_1(t) + \beta x_2(t)] = \alpha T[x_1(t)] + \beta T[x_2(t)],$$

i.e., the system T is a linear system.

Let $x_n(t) \to x(t) \, (L^2)$. By the Schwarz inequality,

$$\begin{aligned}
|T[x_n(t)] - T[x(t)]| &= \left| \int_{\mathbb{R}} g(t - u)(x_n(u) - x(u)) \, du \right| \\
&\le \left(\int_{\mathbb{R}} |g(t - u)|^2 \, du \int_{\mathbb{R}} |x_n(u) - x(u)|^2 \, du \right)^{1/2} \\
&= \left(\int_{\mathbb{R}} |g(u)|^2 \, du \right)^{1/2} \left(\int_{\mathbb{R}} |x_n(u) - x(u)|^2 \, du \right)^{1/2} \\
&\to 0 \quad (n \to \infty).
\end{aligned}$$

So $T[x_n(t)] \to T[x(t)] (t \in \mathbb{R})$, i.e., the system is continuous.

By the assumption and $g * x = x * g$, it follows that

$$y(t) = (g * x)(t) = (x * g)(t) = \int_{\mathbb{R}} x(t - u)g(u)\, du,$$

and so, for any τ,

$$y(t - \tau) = \int_{\mathbb{R}} x(t - \tau - u)g(u)\, du = T[x(t - \tau)],$$

i.e., the system $y(t) = T[x(t)]$ determined by the convolution is a time-invariant system. □

For a continuous linear time-invariant system, the inverse proposition of Proposition 3.1 holds.

Proposition 3.2. *If a linear time-invariant system T is continuous, then there exists a filter $g(t)$ such that the input $x(t)$ and the output $y(t)$ of the system satisfy*

$$y(t) = \int_{\mathbb{R}} g(t - u)x(u)\, du = (g * x)(t). \tag{3.1}$$

Proof. Since the system is a linearly continuous system, from

$$x(t) = \, < \delta(t - u), x(u) > \, = \int_{\mathbb{R}} \delta(t - u)x(u)\, du,$$

where δ is the Dirac function, it follows that

$$T[x(t)] = \int_{\mathbb{R}} T[\delta(t - u)]x(u)\, du.$$

Let $g(t) = T[\delta(t)]$. Since T is a time-invariant system, $T[\delta(t - u)] = g(t - u)$, and so

$$y(t) = \int_{\mathbb{R}} g(t - u)x(u)\, du.$$

□

Propositions 3.1 and 3.2 state that a continuous system T is a linear time-invariant system if and only if T can be represented by a convolution form, i.e., $y(t) = T[x(t)] = (g * x)(t)$, where the filter g is the response of the Dirac impulse, i.e., $g(t) = T[\delta(t)]$.

A linear time-invariant system T is *causal* if the output $y(t)$ depends only on the input $x(u)$ ($u \le t$). Proposition 3.2 shows that T is causal if and only if the filter $g(u) = 0$ ($u < 0$). A linear time-invariant system T is *stable* if any bounded input produces a bounded output. By (3.1), we have

$$|y(t)| \le \sup_{u \in \mathbb{R}} |x(u)| \int_{\mathbb{R}} |g(u)|\, du.$$

So T is stable if and only if $\int_{\mathbb{R}} |g(u)|\, du < \infty$. Suppose that T is a continuous linear time-invariant system with the filter g. For complex exponent $e^{i\Omega t}$, the output of the system T is

$$T[e^{i\Omega t}] = \int_{\mathbb{R}} g(u) e^{i\Omega(t-u)} du = e^{i\Omega t} \int_{\mathbb{R}} g(u) e^{-i\Omega u} du = e^{i\Omega t} \hat{g}(\Omega), \qquad (3.2)$$

so $\hat{g}(\Omega)$ is called the frequency response of the system T. If T is regarded as a linear continuous operator, then each $e^{i\Omega t}$ is the eigenfunction of T corresponding to the eigenvalue $\hat{g}(\Omega)$.

Now we introduce an ideal low-pass filter $g_d(t)$ which passes low-frequency signals and completely eliminates all high-frequency information.

Let

$$G_d(\omega) = \chi_{[-\Omega_c, \Omega_c]}(\omega) =: \begin{cases} 1, & |\omega| \le \Omega_c, \\ 0, & |\omega| > \Omega_c \end{cases}$$

and the filter $g_d(t)$ be the inverse Fourier transform of $G_d(\omega)$, i.e.,

$$g_d(t) = \frac{1}{2\pi} \int_{-\Omega_c}^{\Omega_c} e^{i\omega t} d\omega = \frac{e^{it\Omega_c} - e^{-it\Omega_c}}{2\pi i t} = \frac{\sin(t\Omega_c)}{\pi t}.$$

Define a linear time-invariant system T by $y(t) = T[x(t)] = (g_d * x)(t)$. Taking Fourier transforms on both sides, by the convolution property of the Fourier transform, we get

$$Y(\omega) = G_d(\omega)X(\omega) = \begin{cases} X(\omega), & |\omega| \le \Omega_c, \\ 0, & |\omega| > \Omega_c, \end{cases}$$

where $X(\omega)$ and $Y(\omega)$ are Fourier transforms of the input $x(t)$ and the output $y(t)$, respectively. This equality states that the frequency spectrum of low-frequency waves remains invariant, while that of high-frequency waves vanishes. Therefore, $g_d(t)$ is called an *ideal low-pass filter* and $e^{i\omega t}\chi_{[-\Omega_c, \Omega_c]}(\omega)$ is the frequency response. However, the continuous linear time-invariant system with an ideal low-pass filter is not stable, and this implies that bounded input does not imply bounded output, moreover, it is also not causal, so it cannot be used in practice.

3.2 ANALOG FILTERS

Three classical analog filters are follows:

(i) *Butterworth filter.* A Butterworth filter $g_b(t)$ is a filter whose Laplace transform $G_b(s)$ satisfies

$$|G_b(i\Omega)|^2 = \frac{1}{1 + (\Omega/\Omega_c)^{2N}} = \frac{\Omega_c^{2N}}{\Omega^{2N} + \Omega_c^{2N}}, \qquad (3.3)$$

where Ω_c is the *width* of the passband, N is an integer, and N is the *order* of the filter.

When $\Omega = \Omega_c$, $|G_b(i\Omega)|^2 = \frac{1}{2}$. When N is increasing, $|G_b(i\Omega)|$ approximates to an ideal low-pass filter. It does not have a zero. Its poles s_p are determined as follows.

From $\Omega^{2N} + \Omega_c^{2N} = 0$, it follows that $\Omega^{2N} = -\Omega_c^{2N}$, and so $\Omega = (-1)^{\frac{1}{2N}}\Omega_c$. Since $(-1)^{\frac{1}{2N}}$ has $2N$ values, $\{e^{i\frac{(2k+1)\pi}{2N}}\}_{k=0,...,2N-1}$. Therefore,

$$|G_b(i\Omega)|^2 = \frac{\Omega_c^{2N}}{\prod_0^{2N-1}(\Omega - s_k)} \quad (\Omega \in \mathbb{R}),$$

where $s_k = e^{i\frac{(2k+1)\pi}{2N}}\Omega_c$. Let $s = i\Omega$. Then

$$\frac{\Omega_c^{2N}}{\prod_0^{2N-1}(\Omega - s_k)} = \frac{(-1)^N\Omega_c^{2N}}{\prod_0^{2N-1}(s - is_k)}.$$

These poles $\{is_k\}_{k=0,...,2N-1}$ are symmetric about the origin. Let $p_1, \ldots, p_N$ lie in the left-half plane. Then the other poles are $-p_1, \ldots, -p_N$. So

$$\frac{\Omega_c^{2N}}{\prod_0^{2N-1}(\Omega - s_k)} = \frac{\Omega_c^N}{\prod_1^N(s - p_k)}\frac{(-1)^N\Omega_c^N}{\prod_1^N(s + p_k)}.$$

Noticing that $|G_b(i\Omega)|^2 = G_b(i\Omega)G_b(-i\Omega)$ for a real filter $g_b(t)$, we know that

$$G_b(s) = \frac{\Omega_c^N}{\prod_1^N(s - p_k)}.$$

Taking the inverse Laplace transform, we have $g_b(t) = L^{-1}(G_b(s))$.

(ii) *Chebeshev filter.* The Chebyshev polynomial of order N is defined as

$$T_N(x) = \begin{cases} \cos(N\theta), & \theta = \cos^{-1}x \quad (|x| \leq 1), \\ \cosh(N\tau), & \tau = \cosh^{-1}x \quad (|x| > 1). \end{cases}$$

Especially,

$$T_0(x) = 1, \quad T_1(x) = x.$$

From this and the recurrence formula $T_{N+1}(x) = 2x\,T_n(x) - T_{N-1}(x)$, it follows that

$$T_2(x) = 2x^2 - 1, T_3(x) = 4x^3 - 3x, \ldots.$$

A type I Chebyshev filter $g_c^1(t)$ is a filter whose Laplace transform satisfies

$$|G_c^1(i\Omega)|^2 = \frac{1}{1 + \epsilon^2\,T_N^2(\Omega/\Omega_c)},$$

where N is the order of the filter and $0 < \epsilon < 1$. The larger ϵ is, the larger the ripple is. Since all zeros of Chebyshev polynomials $T_N(x)$ lie in $[-1, 1]$, when $0 \leq \Omega \leq \Omega_c$,

$$\frac{1}{1 + \epsilon^2} \leq |G_c^1(i\Omega)|^2 \leq 1.$$

When $\Omega \geq \Omega_c$, $|G_c^1(i\Omega)|^2$ increases monotonically as Ω increases.

We compute the bandwidth Ω_A, which is defined as $|G_c^1(i\Omega_A)|^2 = \frac{1}{2}$. It is clear that $|G_c^1(i\Omega_A)|^2 = \frac{1}{2}$ is equivalent to $\epsilon^2 T_N^2(\Omega_A/\Omega_c) = 1$. If $\Omega_A \leq \Omega_c$, then $T_N^2(\Omega_A/\Omega_c) \leq 1$, and so $\epsilon \geq 1$. This is contrary to $0 < \epsilon < 1$. Therefore, $\Omega_A > \Omega_c$, i.e., $\frac{\Omega_A}{\Omega_c} > 1$. From this and $T_N(\Omega_A/\Omega_c) = \frac{1}{\epsilon}$, we get

$$\cosh\left[N \cosh^{-1}\left(\frac{\Omega_A}{\Omega_c} \right) \right] = \frac{1}{\epsilon}.$$

Therefore,

$$\Omega_A = \Omega_c \cosh\left(\frac{1}{N} \cosh^{-1} \frac{1}{\epsilon} \right).$$

The poles of $|G_c^1(i\Omega)|^2$ are

$$p_k = \sigma_k + i\tau_k,$$

where $\sigma_k = -\Omega_c \sinh \zeta \sin \frac{(2k-1)\pi}{2N}$ and $\tau_k = \Omega_c \cosh \zeta \cos \frac{(2k-1)\pi}{2N}$, and $\zeta = \frac{1}{N} \sinh^{-1} \frac{1}{\epsilon}$. This implies that each pole p_k of the type I Chebyshev filter lies on the ellipse $\frac{\sigma^2}{a^2} + \frac{\tau^2}{b^2} = 1$, where $a = \Omega_c \sinh \zeta$ and $b = \Omega_c \cosh \zeta$. Similarly to the Butterworth filter, with the help of these poles, we can obtain the type I Chebyshev filter $g_c^1(t)$ and its Laplace transform $G_c^1(s)$.

The type II Chebeshev filter $g_c^2(t)$ is a filter whose Laplace transform $G_c^2(s)$ satisfies

$$|G_c^2(i\Omega)|^2 = \frac{1}{1 + \epsilon^2 \left(\frac{T_N(\Omega_r/\Omega_c)}{T_N(\Omega_r/\Omega)} \right)^2} \qquad (0 < \epsilon < 1).$$

It is clear that it decreases monotonically in the passband and $|G(0)| = 1$, and $G(i\Omega_c) = \frac{1}{1+\epsilon^2}$, and it has equirriple in $\Omega \geq \Omega_r$.

$|G_c^2(i\Omega)|^2$ has $2N$ zeros $z_k = \frac{i\Omega_r}{\cos((2k-1)\pi/2N)}$ $(k = 1, \ldots, 2N)$ and $2N$ poles $p_k = \sigma_k + i\tau_k$ $(k = 1, \ldots, 2N)$, where

$$\sigma_k = \frac{\Omega_r \alpha_k}{\alpha_k^2 + \beta_k^2}, \qquad \tau_k = -\frac{\Omega_r \beta_k}{\alpha_k^2 + \beta_k^2},$$

and

$$\alpha_k = -\sinh \xi \sin \frac{(2k-1)\pi}{2N},$$
$$\beta_k = \cosh \xi \cos \frac{(2k-1)\pi}{2N},$$
$$\xi = \frac{1}{N} \sinh^{-1}\left(\epsilon T_N \left(\frac{\Omega_r}{\Omega_c} \right) \right).$$

Similarly to the Butterworth filter, with the help of these poles and zeros, we can obtain the type II Chebyshev filter $g_c^2(t)$ and its Laplace transform $G_c^1(s)$.

(iii) *Elliptic filter.* An elliptic filter $g_e(t)$ is a filter whose Laplace transform $G_e(s)$ satisfies

$$|G_e(i\Omega)|^2 = \frac{1}{1 + \epsilon^2 J_N^2(\Omega)},$$

where $J_N(\Omega)$ is the Jacobian ellipse function of order N. The elliptic filter allows equiripple for both the passband and the stopband.

3.3   DISCRETE LINEAR TIME-INVARIANT SYSTEMS

A discrete signal comes from sampling or discretization of a continuous signal. A discrete signal is also called a digital signal. If the input and output signals of a system are both discrete signals, then the system is called a *discrete system.*

3.3.1   Discrete Signals

A one-dimensional *discrete signal* is a sequence $\{x(n)\}_{n\in\mathbb{Z}}$. For example, the unit step signal is

$$u(n) = \begin{cases} 1, & n \geq 0, \\ 0, & n < 0. \end{cases}$$

The rectangular signal is

$$r_N(n) = \begin{cases} 1, & 0 \leq n \leq N - 1, \\ 0, & \text{otherwise.} \end{cases}$$

An exponential signal is expressed by $e(n) = a^n u(n)$, where a is a real constant.
The two-dimensional unit step signal is

$$u(n_1, n_2) = \begin{cases} 1, & n_1 \geq 0, n_2 \geq 0, \\ 0, & \text{otherwise.} \end{cases}$$

The two-dimensional exponential signal is $\{a^{n_1} b^{n_2}\}_{n_1,n_2\in\mathbb{Z}}$, where a and b are real constants. The two-dimensional sinusoidal sequence is $\{A\cos(n_1\omega_1 + \theta_1)\cos(n_2\omega_2 + \theta_2)\}_{n_1,n_2\in\mathbb{Z}}$.
Now we discuss frequency domain representations of discrete signals.
The Fourier transform of a one-dimensional discrete signal $\{x(n)\}_{n\in\mathbb{Z}}$ is defined as

$$F[x(n)] := X(e^{i\omega}) = \sum_k x(k)\, e^{-ik\omega}.$$

If the series on the right-hand side is absolutely convergent, i.e., $\sum_k |x(k)| < \infty$, then the Fourier transform $X(e^{i\omega})$ is a periodic continuous function with period 2π. The inverse Fourier transform is defined as

$$F^{-1}(X(e^{i\omega})) := \{x(n)\},$$

where $x(n) = \frac{1}{2\pi} \int_{-\pi}^{\pi} X(e^{i\omega}) e^{in\omega} d\omega$.

For example, the Fourier transform of the rectangular sequence $\{r_N(n)\}$ is

$$F[r_N(n)] = \sum_{0}^{N-1} e^{-in\omega} = \frac{1 - e^{-iN\omega}}{1 - e^{-i\omega}} = e^{-i\frac{(N-1)\omega}{2}} \frac{\sin(N\omega/2)}{\sin(\omega/2)},$$

and the inverse Fourier transform $F^{-1}\left(e^{-i\frac{(N-1)\omega}{2}} \frac{\sin(N\omega/2)}{\sin(\omega/2)}\right) = \{r_N(n)\}$.

Fourier transforms of discrete signals have the following properties.

Property. Let $\{x(n)\}_{n \in \mathbb{Z}}$ and $\{y(n)\}_{n \in \mathbb{Z}}$ be two discrete signals and c and d be two constants. Then

(i) $F[cx(n) + dy(n)] = cF[x(n)] + dF[y(n)]$;

(ii) $F[(x * y)(n)] = F[x(n)]F[y(n)]$.

The Fourier transform of a two-dimensional discrete signal $\{x(n_1, n_2)\}_{n_1, n_2 \in \mathbb{Z}}$ is defined as

$$F[x(n_1, n_2)] := X(e^{i\omega_1}, e^{i\omega_2}) = \sum_{n_1} \sum_{n_2} x(n_1, n_2) e^{-in_1\omega_1} e^{-in_2\omega_2}.$$

If $\sum_{n_1} \sum_{n_2} |x(n_1, n_2)| < \infty$, then the Fourier transform $X(e^{i\omega_1}, e^{i\omega_2})$ is a continuous function of ω_1, ω_2 and

$$X(e^{i\omega_1}, e^{i\omega_2}) = X(e^{i(\omega_1 + 2k\pi)}, e^{i(\omega_2 + 2l\pi)}) \quad (k, l \in \mathbb{Z}).$$

The inverse Fourier transform is defined as

$$F^{-1}[X(e^{i\omega_1}, e^{i\omega_2})] := \{x(n_1, n_2)\},$$

where $x(n_1, n_2) = \frac{1}{4\pi^2} \int_{-\pi}^{\pi} \int_{-\pi}^{\pi} X(e^{i\omega_1}, e^{i\omega_2}) e^{in_1\omega_1} e^{in_2\omega_2} d\omega_1 d\omega_2$.

3.3.2 Discrete Convolution

Let $h = \{h(n)\}_{n \in \mathbb{Z}}$ and $x = \{x(n)\}_{n \in \mathbb{Z}}$ be two infinite discrete signals. The *discrete convolution* of $h(n)$ and $x(n)$ is defined as

$$(h * x)(k) = \sum_{n} h(k - n)x(n) \quad (k \in \mathbb{Z}).$$

If $h = \{h(n)\}_{n=0,...,N_h-1}$ and $x = \{x(n)\}_{n=0,...,N_x-1}$ are two finite signals with lengths N_h and N_x, respectively, then the length of discrete convolution $h * x$ is $N_x + N_h - 1$.

The discrete convolution has the following properties:

$$(h * (cx + dy))(k) = c(h * x)(k) + d(h * y)(k) \quad (k \in \mathbb{Z}),$$

$$(h * x)(k) = (x * h)(k), \quad (h * (x * y))(k) = ((h * x) * y)(k) \quad (k \in \mathbb{Z}),$$

where $h = \{h(n)\}_{n \in \mathbb{Z}}$, $x = \{x(n)\}_{n \in \mathbb{Z}}$, and $y = \{y(n)\}_{n \in \mathbb{Z}}$ are discrete signals and c and d are constants.

Let $h = \{h(n_1, n_2)\}_{n_1, n_2 \in \mathbb{Z}}$ and $x = \{x(n_1, n_2)\}_{n_1, n_2 \in \mathbb{Z}}$ be the two-dimensional discrete signals. Then the convolution of $h(n_1, n_2)$ and $x(n_1, n_2)$ is defined as

$$(h * x)(n_1, n_2) = \sum_{m_1} \sum_{m_2} h(n_1 - m_1, n_2 - m_2)x(m_1, m_2) \quad (n_1, n_2 \in \mathbb{Z}).$$

The following convolution properties also hold in the two-dimensional case:

$$(h * (cx + dy))(n_1, n_2) = c(h * x)(n_1, n_2) + d(h * y)(n_1, n_2) \quad (n_1, n_2 \in \mathbb{Z}),$$
$$(h * x)(n_1, n_2) = (x * h)(n_1, n_2), (h * (x * y))(n_1, n_2) = ((h * x) * y)(n_1, n_2)$$
$$(n_1, n_2 \in \mathbb{Z}).$$

3.3.3 Discrete System

To define discrete time-invariant systems, we use the notation $y(n) = T[x(n)]$ $(n \in \mathbb{Z})$ to represent a discrete system, where $\{x(n)\}_{n \in \mathbb{Z}}$ is the input sequence and $\{y(n)\}_{n \in \mathbb{Z}}$ is the output sequence.

If, for arbitrary constants α and β,

$$T[\alpha x_1(n) + \beta x_2(n)] = \alpha T[x_1(n)] + \beta T[x_2(n)] \quad (n \in \mathbb{Z}),$$

then the system $y(n) = T[x(n)]$ $(n \in \mathbb{Z})$ is called a *discrete linear system*.

Let $y(n) = T[x(n)]$ $(n \in \mathbb{Z})$ be a discrete linear system. If

$$y(n - k) = T[x(n - k)] \, (k \in \mathbb{Z}),$$

then the system $y(n) = T[x(n)]$ $(n \in \mathbb{Z})$ is called a *discrete time-invariant system*.

The sequence $\{\delta(n)\}_{n \in \mathbb{Z}}$, where $\delta(n) = 0$ $(n \neq 0)$ and $\delta(0) = 1$, is called the *unit impulse*. The unit impulse response $h(n) = T[\delta(n)]$ $(n \in \mathbb{Z})$ is called the *filter* of the system T.

Proposition 3.3. *Any discrete linear time-invariant system* $y(n) = T[x(n)]$ $(n \in \mathbb{Z})$ *can be represented by the discrete convolution of the input and the unit impulse response, i.e.,* $y(n) = (h * x)(n)(n \in \mathbb{Z})$.

Proof. Note that $\delta(n) = 0$ $(n \neq 0)$ and $\delta(0) = 1$. Any input $\{x(n)\}_{n \in \mathbb{Z}}$ can be represented by

$$x(n) = \sum_k x(k)\delta(n - k) \quad (n \in \mathbb{Z}).$$

The system T is a linear system, so

$$y(n) = T[x(n)] = \sum_k x(k)T[\delta(n - k)].$$

Note that $h(n) = T[\delta(n)]$. Since the system T is time invariant, we get $h(n - k) = T[\delta(n - k)]$, and so

$$y(n) = \sum_k x(k)h(n - k) = (h * x)(n).$$

$\square$

For a linear time-invariant system $y(n) = (h * x)(n)$ $(n \in \mathbb{Z})$, if the output $y(n)$ depends only on the input $x(k)$ $(k \leq n)$, then this system is called *causal*.

Proposition 3.4. *A system* $y(n) = (h * x)(n)$ $(n \in \mathbb{Z})$ *is causal if and only if* $h(n) = 0$ $(n < 0)$.

Proof. Assume that $h(n) = 0$ $(n < 0)$. Then $h(n - k) = 0$ $(k > n)$, and so

$$y(n) = \sum_{k \leq n} x(k)h(n - k).$$

Therefore, the output $y(n)$ depends only on the input $x(k)$ $(k \leq n)$, i.e., the system is causal.

Assume that the system is causal. If $h(-l) \neq 0$ for some $l \in \mathbb{Z}_+$, we take $x(k) = 0$ $(k \neq n + l)$ and $x(n + l) \neq 0$, then

$$y(n) = \sum_k x(k)h(n - k) = x(n + l)h(-l) \neq 0,$$

so the output $y(n)$ cannot be determined by the input $x(k)$ $(k \leq n)$. This is contrary to the assumption. Hence, $h(n) = 0$ $(n < 0)$. $\square$

If a linear time-invariant system is such that any bounded input products a bounded output, then this system is called a *stable system*.

Proposition 3.5. *A linear time-invariant system*

$$y(n) = (h * x)(n) \quad (n \in \mathbb{Z})$$

is stable if and only if $\sum_n |h(n)| < \infty$.

Proof. Assume that $\sum_n |h(n)| = M < \infty$. If $|x(n)| < A$ $(n \in \mathbb{Z})$, then

$$|y(n)| = \left| \sum_k x(k)h(n - k) \right| \leq A \sum_k |h(n - k)| = A \sum_n |h(n)| \leq AM,$$

and so the system is stable.

Assume that $\sum_n |h(n)| = \infty$. Take

$$x(n) = \begin{cases} \dfrac{\overline{h}(-n)}{|h(-n)|}, & h(-n) \neq 0, \\ 0, & h(-n) = 0. \end{cases}$$

Then

$$|y(0)| = \left| \sum_k x(k)h(-k) \right| = \sum_{h(-k) \neq 0} |h(-k)| = \sum_k |h(k)| = \infty,$$

and so the output is unbounded. This is contrary to the assumption. Hence $\sum_n |h(n)| < \infty$. □

For a linear time-invariant system with the unit impulse response h, we consider the frequency response. If the input is a complex exponent sequence with frequency ω: $x(n) = e^{in\omega}$ ($n \in \mathbb{Z}$), then its output response is

$$y(n) = (h * x)(n) = \sum_k h(k)e^{i(n-k)\omega} = e^{in\omega}H(e^{i\omega}), \qquad (3.4)$$

where $H(e^{i\omega}) = \sum_k h(k)e^{-ik\omega}$. The function $H(e^{i\omega})$ is called the *frequency response* of the system. It is clear that $H(e^{i\omega})$ is a 2π-periodic function. If the system is stable, then $\sum_n |h(n)| < \infty$. So $H(e^{i\omega})$ exists and is continuous, and is the Fourier transform of the filter h. The inverse Fourier transform is

$$h(n) = \frac{1}{2\pi} \int_{-\pi}^{\pi} H(e^{i\omega})\, e^{in\omega}\, d\omega.$$

Now consider the two-dimensional case. For a two-dimensional linear system $y(n_1, n_2) = T[x(n_1, n_2)]$. Noticing that

$$\delta(n_1, n_2) = \begin{cases} 1, & n_1 = n_2 = 0, \\ 0, & n_1 \text{ or } n_2 \neq 0, \end{cases}$$

we can represent the input x by

$$x(n_1, n_2) = \sum_k \sum_l x(k, l)\delta(n_1 - k, n_2 - l),$$

so

$$y(n_1, n_2) = \sum_k \sum_l x(k, l)\, T[\delta(n_1 - k, n_2 - l)].$$

Let h be the response of the two-dimensional unit impulse δ, i.e.,

$$h(n_1, n_2) = T[\delta(n_1, n_2)].$$

The unit impulse response h is also called the *filter* of the system. If T is a time-invariant system, then

$$T[\delta(n_1 - k, n_2 - l)] = h(n_1 - k, n_2 - l),$$

and so

$$y(n_1, n_2) = \sum_k \sum_l x(k, l)h(n_1 - k, n_2 - l) = (h * x)(n_1, n_2), \qquad (3.5)$$

i.e., any two-dimensional linear time-invariant system $y(n_1, n_2) = T[x(n_1, n_2)]$ can be represented by the two-dimensional discrete convolution of the input and the filter. If the filter h of a two-dimensional discrete time-invariant system satisfies $\sum_{n_1} \sum_{n_2} |h(n_1, n_2)| < \infty$, then the system is called *a stable system*.

If the output $y(n_1, n_2)$ depends only on the input $x(k, l)$ $(k \le n_1, l \le n_2)$, then the system is *causal*. Similarly, a two-dimensional time-invariant system is causal if and only if its filter $h(n_1, n_2) = 0$ $(n_1 < 0, n_2 < 0)$.

3.3.4   Ideal Digital Filters

For a discrete linear time-invariant system, let h be the filter and $H(e^{i\omega}) = \sum_n h(n) e^{-in\omega}$ be its frequency response.

Case 1. If $H(e^{i\omega}) = 0$ $(|\omega_c| < |\omega| \le \pi)$, then the filter h is called a *low-pass filter*. Let

$$Y(e^{i\omega}) = \sum_n y(n) e^{-in\omega},$$

$$X(e^{i\omega}) = \sum_n x(n) e^{-in\omega}.$$

Then, by the property of the Fourier transform of discrete signals,

$$Y(e^{i\omega}) = H(e^{i\omega}) X(e^{i\omega}).$$

So a low-pass filter only passes low-frequency signals. The inverse Fourier transform gives

$$h(n) = \frac{1}{2\pi} \int_{-\pi}^{\pi} H(e^{i\omega}) e^{in\omega} \, d\omega = \frac{1}{2\pi} \int_{-\omega_c}^{\omega_c} e^{in\omega} \, d\omega = \begin{cases} \frac{\sin(n\omega_c)}{\pi n}, & n \neq 0, \\ \frac{\omega_c}{\pi}, & n = 0. \end{cases}$$

Case 2. If $H(e^{i\omega}) = 0$ $(|\omega| \le \omega_c < \pi)$, then the filter h is called a *high-pass filter*. We can see that a high-pass filter passes only high-frequency signals. The inverse Fourier transform gives

$$h(n) = \frac{1}{2\pi} \int_{-\pi}^{\pi} H(e^{i\omega}) e^{in\omega} \, d\omega = \begin{cases} -\frac{\sin(n\omega_c)}{\pi n}, & n \neq 0, \\ 1 - \frac{\omega_c}{\pi}, & n = 0. \end{cases}$$

Case 3. If $H(e^{i\omega}) = 0$ $(0 < |\omega_c| \le |\omega| \le |\omega_d| < \pi)$, then the filter h is called a *band-pass filter*.

3.3.5   *Z*-Transforms

For a discrete signal $x = \{x(n)\}_{n \in \mathbb{Z}}$, its *Z-transform* is defined as

$$X(z) = \sum_n x(n) z^{-n}.$$

It is sometimes called the two-sided Z-transform. Denote it by $Z\{x(n)\}$, i.e., $Z\{x(n)\} = X(z)$.

If the limits

$$r_1 = \lim_{n \to \infty} \sqrt[n]{|x(n)|},$$

$$r_2 = \lim_{n \to -\infty} \sqrt[n]{|x(n)|}$$

exist and $\frac{1}{r_1} < r_2$, then the convergence domain of its Z-transform is the annular region $\frac{1}{r_1} < |z| < r_2$. Let $z = re^{i\theta}$ ($\frac{1}{r_1} < r < r_2$). Then the Z-transform of $\{x(n)\}_{n \in \mathbb{Z}}$ can be rewritten as

$$X(re^{i\theta}) = \sum_n x(n) r^{-n} e^{-in\theta}.$$

By the orthogonality of exponential sequence $\{e^{-in\theta}\}_{n \in \mathbb{Z}}$, it follows that

$$\int_{-\pi}^{\pi} X(re^{i\theta}) e^{in\theta} \, d\theta = \sum_k x(k) r^{-k} \int_{-\pi}^{\pi} e^{i(n-k)\theta} \, d\theta = 2\pi x(n) r^{-n}.$$

So the inverse Z-transform of $X(z)$ is

$$x(n) = Z^{-1}[X(z)] = \frac{r^n}{2\pi} \int_{-\pi}^{\pi} X(re^{i\theta}) e^{in\theta} \, d\theta.$$

If the Z-transform $X(z)$ is a rational function which has only simple poles p_k ($k = 1, \ldots, N$), then $X(z)$ can be decomposed into a sum of partial fractions and a polynomial $p(z)$, i.e.,

$$X(z) = \sum_{1}^{N} \frac{A_k}{z - p_k} + p(z),$$

where $A_k = \lim_{z \to p_k} X(z)(z - p_k)$. Expanding each $1/(z - p_k)$ into the positive power series or the negative power series, we can also obtain the inverse Z-transform of $X(z)$.

The Z-transforms have the following properties.

Property. Let $\{x(n)\}_{n \in \mathbb{Z}}$ and $\{y(n)\}_{n \in \mathbb{Z}}$ be two discrete signals. Denote their Z-transforms by $X(z)$ and $Y(z)$, respectively. Then

(i) $Z\{ax(n) + by(n)\}$ is $aX(z) + bY(z)$;
(ii) $Z\{x(n - n_0)\}$ is $z^{-n_0} X(z)$;
(iii) $Z\{a^n x(n)\}$ is $X\left(\frac{z}{a}\right)$;
(iv) $Z\{\overline{x}(n)\}$ is $\overline{X}(\overline{z})$; and
(v) $Z\{nx(n)\}$ is $-zX'(z)$.

Proposition 3.6. *For a discrete linear time-invariant system, let $X(z)$, $H(z)$, and $Y(z)$ be the Z-transforms of the input x, the output y, and the filter h, respectively. Then $Y(z) = H(z)X(z)$.*

The Z-transform of the filter h is called the *transfer function* of the system.

Proof. By Proposition 3.4, we have $y(k) = (h * x)(k) = \sum_n h(k - n)x(n)$. It follows that

$$Y(z) = \sum_k y(k) z^{-k} = \sum_n x(n)z^{-n} \left(\sum_k h(k - n)z^{-(k-n)} \right).$$

Since $\sum_k h(k - n)z^{-(k-n)} = \sum_k h(k)z^{-k}$, we get

$$Y(z) = \left(\sum_k h(k)z^{-k} \right) \left(\sum_n x(n)z^{-n} \right) = H(z)X(z).$$

$\square$

The concept of the one-dimensional Z-transform may be generalized to the two-dimensional case.

Let $\{x(m, n)\}_{m,n \in \mathbb{Z}}$ be a two-dimensional discrete signal. Then its Z-transform is defined as

$$X(z_1, z_2) = \sum_{m,n} x(m, n)z_1^{-m}z_2^{-n}.$$

Let $z_1 = r_1 e^{-i\omega_1}$ and $z_2 = r_2 e^{i\omega_2}$. Then

$$X(r_1 e^{i\omega_1}, r_2 e^{i\omega_2}) = \sum_{m,n} x(m, n) \, r_1^{-m} r_2^{-n} \, e^{-im\omega_1} \, e^{-in\omega_2}.$$

If $\sum_{m,n} |x(m, n)| r_1^{-m} r_2^{-n} < \infty$, then the series on the right-hand side converges. The inverse Z-transform of $X(z_1, z_2)$ is defined as

$$x(m, n) = \frac{1}{4\pi^2} \int_{-\pi}^{\pi} \int_{-\pi}^{\pi} X(r_1 \, e^{i\omega_1}, r_2 \, e^{i\omega_2}) \, r_1^m r_2^n \, e^{im\omega_1} \, e^{in\omega_2} \, d\omega_1 \, d\omega_2 (m, \, n \in \mathbb{Z}).$$

3.3.6 Linear Difference Equations

We will discuss the discrete linear time-invariant system which can be represented by a linear difference equation:

$$\sum_0^N b(k) \, y(u - k) = \sum_0^N a(k) \, x(u - k), \tag{3.6}$$

where $x(u)$ is the input signal, $y(u)$ is the output signal, and $a(k)$ and $b(k)$ are constants. Taking the Z-transform on both sides of (3.6), we get

$$\sum_0^N b(k) \, z^{-k} \, Y(z) = \sum_0^N a(k) \, z^{-k} \, X(z),$$

where $X(z) = \sum_0^N x(n) \, z^{-n}$ and $Y(z) = \sum_0^N y(n) \, z^{-n}$, and so

$$Y(z) = H(z) \, X(z), \tag{3.7}$$

where the transfer function

$$H(z) = \frac{\sum_0^N a(k)\, z^{-k}}{\sum_0^N b(k)\, z^{-k}}.$$

Expand $H(z)$ into the two-sided power series $H(z) = \sum_n h(n) z^n$. By the convolution property of the Z-transform, it follows from (3.7) that

$$y(n) = (h * x)(n).$$

If there exist infinitely many nonzero terms in $\{h(n)\}_{n \in \mathbb{Z}}$, then it is called an *infinite impulse response* (IIR) *filter*. Otherwise, if there exist only finitely many nonzero terms in $\{h(n)\}_{n \in \mathbb{Z}}$, then it is called a *finite impulse response* (FIR) *filter*.

3.4 LINEAR-PHASE FILTERS

Let T be a discrete linear time-variant system with the FIR filter h. The input $x = \{x(n)\}_{n \in \mathbb{Z}}$ and the output $y = \{y(n)\}_{n \in \mathbb{Z}}$ of the system satisfy

$$y(n) = (h * x)(n) \quad (n \in \mathbb{Z}).$$

Without loss of generality, we assume that $h(n) = 0 \ (n \neq 0, \ldots, N-1)$. The transfer function

$$H(z) = \sum_0^{N-1} h(n)\, z^{-n}$$

is an $N-1$ degree polynomial of z^{-1}. This is a causal stable discrete system.

Let $z = e^{i\omega}$. The *frequency response* is

$$H(e^{i\omega}) = \sum_0^{N-1} h(n)\, e^{-in\omega}. \tag{3.8}$$

Clearly, this is a 2π-periodic function. It can be expressed as

$$H(e^{i\omega}) = |H(e^{i\omega})|\, e^{i\theta(\omega)}, \quad \text{where } \tan\theta(\omega) = \frac{\text{Im}(H(e^{i\omega}))}{\text{Re}(H(e^{i\omega}))}. \tag{3.9}$$

Here $|H(e^{i\omega})|$ is called the *frequency spectrum* and $\theta(\omega)$ is called the *phase*.

When $\theta(\omega) = -\tau\omega$, where τ is a constant, we say the filter has a rigorous linear phase. When $\theta(\omega) = b - \tau\omega$, where τ and b are constants, we say the filter has a generalized linear phase. Now we study the FIR filter with a linear phase. It is very important in geophysical signal processing.

From (3.8), it follows that

$$\text{Im}(H(e^{i\omega})) = -\sum_0^{N-1} h(n) \sin(n\omega),$$

$$\mathrm{Re}(H(e^{i\omega})) = \sum_0^{N-1} h(n) \cos(n\omega).$$

Combining this with (3.9), we have

$$\tan\theta(\omega) = -\frac{\sum_0^{N-1} h(n) \sin(n\,\omega)}{\sum_0^{N-1} h(n) \cos(n\,\omega)}.$$

This implies that for any ω, $\theta(\omega) = -\tau\omega$ if and only if

$$\sum_0^{N-1} h(n)(\cos(n\omega)\sin(\tau\omega) - \sin(n\omega)\cos(\tau\omega)) = 0,$$

i.e., for any ω, $\theta(\omega) = -\tau\omega$ if and only if $\sum_0^{N-1} h(n) \sin(\tau - n)\omega = 0$.

Similarly, we can deduce that for any ω, $\theta(\omega) = \frac{\pi}{2} - \tau\omega$ if and only if $\sum_0^{N-1} h(n) \cos(\tau - n)\omega = 0$.

Proposition 3.7. *Let the filter $h = \{h(n)\}_{n=0,\ldots,N-1}$ be an FIR digital filter.*

(i) *If*

$$h(n) = h(N - 1 - n) \quad (n = 0, \ldots, N - 1),$$

then the filter h is a rigorous linear-phase filter and $\arg H(e^{i\omega}) = -\frac{N-1}{2}\omega$.

(ii) *If*

$$h(n) = -h(N - 1 - n) \quad (n = 0, \ldots, N - 1),$$

then the filter h is a generalized linear-phase filter and $\arg H(e^{i\omega}) = \frac{\pi}{2} - \frac{N-1}{2}\omega$.

Proof. Let $\tau = \frac{N-1}{2}$.

(i) By the assumption

$$h(n) = h(N - 1 - n) \quad (n = 0, \ldots, N - 1),$$

it follows that $\{h(n)\}_{n=0,\ldots,N-1}$ is an even symmetric sequence with center τ. So $\{\sin\omega(\tau - n)\}_{n=0,\ldots,N-1}$ is an odd sequence with center τ, and so $h(n)\sin\omega(\tau - n)$ is an odd sequence with center τ. This implies that

$$\sum_0^{N-1} h(n) \sin(\tau - n)\omega = 0,$$

which is equivalent to $\theta(\omega) = -\tau\omega$, so the filter h is a rigorous linear phase filter and $\arg H(e^{i\theta}) = -\tau\omega$.

(ii) By the assumption

$$h(n) = -h(N - 1 - n) \quad (n = 0, \ldots, N - 1),$$

it follows that $\{h(n)\}_{n=0,\ldots,N-1}$ is an odd symmetric sequence with center τ. Since $\{\cos(\tau - n)\omega\}_{n=0,\ldots,N-1}$ is an even sequence with center τ,

$$\sum_{0}^{N-1} h(n)\cos(\tau - n)\omega = 0,$$

which is equivalent to $\theta(\omega) = \frac{\pi}{2} - \tau\omega$. So the filter h is a generalized linear phase filter and $\arg H(e^{i\omega}) = \frac{\pi}{2} - \tau\omega$.

$\square$

3.4.1 Four Types of Linear-Phase Filters

Assume that $\{h(n)\}_{n=0,\ldots,N-1}$ is an FIR filter and its frequency response $H(e^{i\omega}) = \sum_{0}^{N-1} h(n)\,e^{-in\omega}$.

(i) $\{h(n)\}_{n=0,\ldots,N-1}$ has even symmetry and N is odd.
Its frequency response is

$$H(e^{i\omega}) = \sum_{0}^{N-1} h(n)\,e^{-in\omega} = \sum_{0}^{\frac{N-3}{2}} h(n)\,e^{-in\omega}$$

$$+ h\left(\frac{N-1}{2}\right) e^{-i\frac{N-1}{2}\omega} + \sum_{\frac{N+1}{2}}^{N-1} h(n)\,e^{-in\omega}.$$

By $h(n) = h(N-1-n)$, the third term on the right-hand side becomes

$$\sum_{0}^{\frac{N-3}{2}} h(n)\,e^{in\omega}\,e^{-i(N-1)\omega}.$$

So

$$H(e^{i\omega}) = e^{-i\frac{N-1}{2}\omega}\left\{ \sum_{0}^{\frac{N-3}{2}} 2h(n)\cos\left(\frac{N-1}{2} - n\right)\omega + h\left(\frac{N-1}{2}\right) \right\}$$

$$= e^{-i\frac{N-1}{2}\omega}\left\{ \sum_{1}^{\frac{N-1}{2}} 2h\left(\frac{N-1}{2} - m\right)\cos(m\omega) + h\left(\frac{N-1}{2}\right) \right\}.$$

Let $a(0) = h\left(\frac{N-1}{2}\right)$ and $a(m) = 2h\left(\frac{N-1}{2} - m\right)\left(m = 1,\ldots,\frac{N-1}{2}\right)$. Then

$$H(e^{i\omega}) = e^{-i\frac{N-1}{2}\omega}\left\{ \sum_{1}^{\frac{N-1}{2}} a(m)\cos(m\omega) + a(0) \right\} = e^{-i\frac{N-1}{2}\omega} \sum_{0}^{\frac{N-1}{2}} a(n)\cos(n\omega).$$

(ii) $\{h(n)\}_{n=0,\dots,N-1}$ has even symmetry and N is even.

Similarly to the argument in (i), its frequency response is

$$H(e^{i\omega}) = e^{-i\frac{N-1}{2}\omega} \sum_{1}^{\frac{N}{2}} b(n) \cos\left(n - \frac{1}{2}\right)\omega,$$

where $b(n) = 2h(\frac{N}{2} - n)(n = 1, 2, \dots, \frac{N}{2})$.

(iii) $\{h(n)\}_{n=0,\dots,N-1}$ has odd symmetry and N is odd.

$\{h(n)\}_{n=0,\dots,N-1}$ has odd symmetry with center $\frac{N-1}{2}$, so $h(\frac{N-1}{2}) = 0$.
Similarly to the argument in (i), its frequency response is

$$H(e^{i\omega}) = e^{i(\frac{\pi}{2} - \frac{N-1}{2}\omega)} \sum_{1}^{\frac{N-1}{2}} c(n) \sin(n\omega),$$

where $c(n) = 2h(\frac{N-1}{2} - n)(n = 1, 2, \dots, \frac{N-1}{2})$.

(iv) $\{h(n)\}_{n=0,\dots,N-1}$ has odd symmetry and N is even.

Its frequency response is

$$H(e^{i\omega}) = e^{i(\frac{\pi}{2} - \frac{N-1}{2}\omega)} \sum_{1}^{\frac{N}{2}} d(n) \sin\frac{2n-1}{2}\omega,$$

where $d(n) = 2h(\frac{N}{2} - n)(n = 1, 2, \dots, \frac{N}{2})$.

3.4.2 Structure of Linear-Phase Filters

For an FIR digital filter with a rigorous linear phase, $\{h(n)\}_{n=0,\dots,N-1}$, its transfer function

$$H(z) = \sum_{0}^{N-1} h(n) z^{-n},$$

where $h(n) = h(N - 1 - n)$ and $\arg H(e^{i\omega}) = -\tau\omega$, and $\tau = \frac{N-1}{2}$. Therefore,

$$H(z) = \sum_{0}^{N-1} h(N - 1 - n) z^{-n} = z^{-N+1} \sum_{0}^{N-1} h(n) z^{n} = z^{-(N-1)} H(z^{-1}).$$

From this, we see that if z_k is a zero of $H(z)$, then $H(z_k^{-1}) = z_k^{N-1} H(z_k) = 0$. Since each $h(n)$ is real,

$$\overline{H}(z) = \sum_{0}^{N-1} h(n) \overline{z}^{n} = H(\overline{z}).$$

From this, we see that if z_k is a zero of $H(z)$, then $H(\overline{z}_k) = 0$. Therefore, we obtain the following conclusion.

Suppose that z_k is a zero of $H(z)$:

(i) If $|z_k| < 1$ and z_k is not real, then z_k, z_k^{-1}, $\bar{z}_k$, and $\bar{z}_k^{-1}$ are four different zeros of $H(z)$. This constitutes a system of order 4. Denote it by $H_k(z)$, i.e.,

$$H_k(z) = (1 - z^{-1}z_k)(1 - z^{-1}\bar{z}_k)(1 - z^{-1}z_k^{-1})(1 - z^{-1}\bar{z}_k^{-1}).$$

Denote $z_k = r_k \, e^{i\theta_k}$. This equality can be expanded into

$$H_k(z) = 1 - 2z^{-1}\left(r_k + \frac{1}{r_k}\right)\cos\theta_k + z^{-2}\left(r_k^2 + \frac{1}{r_k^2} + 4\cos^2\theta_k\right)$$

$$-2z^{-3}\left(r_k + \frac{1}{r_k}\right)\cos\theta_k + z^{-4}.$$

(ii) If $|z_k| < 1$ and $z_k = r_k$ is real, then r_k and r_k^{-1} are two different zeros of $H(z)$. This constitutes a system of order 2. Denote it by $H_m(z)$, i.e.,

$$H_m(z) = (1 - z^{-1}r_k)\left(1 - \frac{z^{-1}}{r_k}\right) = 1 - z^{-1}\left(r_k + \frac{1}{r_k}\right) + z^{-2}.$$

(iii) If $|z_k| = 1$ and z_k is not real, then $z_k = z_k^{-1}$ and $\bar{z}_k = \bar{z}_k^{-1}$. So z_k and $\bar{z}_k$ are two different zeros of $H(z)$. This also constitutes a system of order 2. Denote it by $H_l(z)$, i.e.,

$$H_l(z) = (1 - z_k z^{-1})(1 - \bar{z}_k z^{-1}) = 1 + z^{-1}(z_k + \bar{z}_k) + z_k\bar{z}_k \, z^{-2}$$

$$= 1 + 2z^{-1}\operatorname{Re}(z_k) + z^{-2}.$$

(iv) If $|z_k| = 1$ and z_k is real, then $z_k = z_k^{-1} = \bar{z}_k = \bar{z}_k^{-1}$. So only z_k is a zero of $H(z)$. This constitutes the simplest system of order 1. Denote it by $H_s(z)$, i.e.,

$$H_s(z) = 1 - z_k z^{-1}.$$

In this way, for the FIR digital filter with a linear phase, its transfer function $H(z)$ can be expressed as

$$H(z) = \left(\prod_k H_k(z)\right)\left(\prod_m H_m(z)\right)\left(\prod_l H_l(z)\right)\left(\prod_s H_s(z)\right),$$

where H_k, H_m, H_l, and H_s are subsystems with a rigorous linear phase.

3.5 DESIGNS OF FIR FILTERS

Now we give three methods for designing FIR digital filters.

3.5.1 Fourier Expansions

From Section 3.3.4, for some $0 < \omega_c < \pi$,

$$h_d(n) = \begin{cases} \frac{\sin(n\omega_c)}{\pi n}, & n \neq 0, \\ \frac{\omega_c}{\pi}, & n = 0 \end{cases}$$

is an ideal low-pass filter, and the corresponding frequency response is

$$H_d(e^{i\omega}) = \sum_n h_d(n)\, e^{-in\omega} = \begin{cases} 1, & |\omega| \leq \omega_c, \\ 0, & \omega_c < |\omega| \leq \pi. \end{cases}$$

The ideal low-pass filter is noncausal and infinite in duration. Clearly, it cannot be implemented in practice. In order to obtain an FIR filter, we may approximate to $H_d(e^{i\omega})$ by $H_{d,N}(e^{i\omega})$:

$$H_{d,N}(e^{i\omega}) = \sum_{-\tau}^{\tau} h_d(n)\, e^{-in\omega} \quad \left(\tau = \frac{N-1}{2}\right). \tag{3.10}$$

By Parseval's identity of Fourier series, the approximation error is

$$r_N = \frac{1}{2\pi} \int_{-\pi}^{\pi} |H_d(e^{i\omega}) - H_{d,N}(e^{i\omega})|^2 \, d\omega = \sum_{|n|>\tau} |h_d(n)|^2 \quad \left(\tau = \frac{N-1}{2}\right).$$

Take an odd number N large enough such that the error $\sum_{|n|>\tau} |h_d(n)|^2 < \epsilon$.

To avoid the noncausal problem, we multiply both sides of (3.10) by $e^{-i\tau\omega}$ to get a new filter:

$$H(e^{i\omega}) = e^{-i\tau\omega} H_{d,N}(e^{i\omega}) = e^{-i\tau\omega} \sum_{-\tau}^{\tau} h_d(n)\, e^{-in\omega} = \sum_{0}^{2\tau} h_d(n-\tau)\, e^{-in\omega}.$$

Let $h(n) = h_d(n - \tau)$. Then the frequency response $H(e^{i\omega}) = \sum_0^{N-1} h(n)\, e^{-in\omega}$, where

$$h(n) = \frac{\sin(n-\tau)\omega_c}{\pi(n-\tau)} \quad (n = 0, \ldots, N-1; n \neq \tau), \quad h(\tau) = \frac{\omega_c}{\pi}, \tag{3.11}$$

and $H(e^{i\omega})$ has rigorous linear phase $\arg H(e^{i\omega}) = -\tau\omega$, where $\tau = \frac{N-1}{2}$. By $\sum_{|n|>\tau} |h_d(n)|^2 < \epsilon$, it follows that the filter $\{h(n)\}_{n=0,\ldots,N-1}$ is a linear-phase filter and approximates to an ideal low-pass filter.

Since $h(n + \tau) = h(\tau - n)$ and $e^{in\omega} + e^{-in\omega} = 2\cos(n\omega)$, by (3.11), the frequency response $H(e^{i\omega})$ can be rewritten in the real form

$$H(e^{i\omega}) = e^{-i\tau\omega} \left(h(\tau) + 2\sum_1^{\tau} h(n+\tau)\cos(n\omega) \right)$$

$$= e^{-i\tau\omega} \left(\frac{\omega_c}{\pi} + 2\sum_1^{\tau} \frac{\sin(n\omega_c)}{\pi n}\cos(n\omega) \right).$$

Since the frequency response of the ideal digital filter has points of discontinuity, the convergence rate of partial sums of its Fourier series is low and truncating the ideal digital filter introduces undesirable ripples and overshoots in the frequency response. Therefore, the filter constructed as above cannot approximate well to the ideal filter and makes the Gibbs phenomenon occur. To solve this problem, window functions are introduced.

3.5.2 Window Design Method

Suppose that $\{h_d(n)\}$ is an ideal digital filter and $H_d(e^{i\omega})$ is its frequency response. To reduce the Gibbs phenomenon, we need to choose a window sequence ω_n with finite length and then multiply $h_d(n)$ by ω_n, i.e.,

$$h(n) = h_d(n)\,\omega_n.$$

Denote the frequency responses corresponding to $h(n)$, $h_d(n)$, and ω_n by $H(e^{i\omega})$, $H_d(e^{i\omega})$, and $W(e^{i\omega})$, respectively. By using the convolution theorem, we get

$$H(e^{i\omega}) = H_d(e^{i\omega}) * W(e^{i\omega}) = \frac{1}{2\pi}\int_{-\pi}^{\pi} H_d(e^{i\theta})\,W(e^{i(\omega-\theta)})\,d\theta.$$

This shows that the frequency response equals the convolution of the frequency response of the ideal digital filter and the frequency response of the window sequence. We choose window sequence $\{\omega_n\}$ such that $H(e^{i\omega})$ is smooth and approximates well to $H_d(e^{i\omega})$. Several window sequences are often used, as follows:

(i) Rectangular window

$$\omega_{R,n} = \begin{cases} 1, & n = 0,\ldots,N-1, \\ 0, & \text{otherwise.} \end{cases}$$

Its frequency response is

$$W_R(e^{i\omega}) = \sum_{0}^{N-1} e^{-in\omega} = \frac{1-e^{-iN\omega}}{1-e^{-i\omega}} = e^{-i\frac{N-1}{2}\omega}\frac{\sin(N\omega/2)}{\sin(\omega/2)}.$$

From this equality, we see that its phase is linear.

(ii) Bartlett window

$$\omega_{B,n} = \begin{cases} \frac{2n}{N-1}, & n = 0,\ldots,\frac{N-1}{2}, \\ 2-\frac{2n}{N-1}, & n = \frac{N-1}{2},\ldots,N-1. \end{cases}$$

(iii) Hanning window

$$\omega_{H,n} = \begin{cases} 0.5 - 0.5\cos\frac{2\pi n}{N-1}, & n = 0,\ldots,N-1, \\ 0, & \text{otherwise.} \end{cases}$$

(iv) Hamming window

$$\omega_{\mathrm{HM},n} = \begin{cases} 0.54 - 0.46 \cos \frac{2\pi n}{N-1}, & n = 0, \ldots, N-1, \\ 0, & \text{otherwise.} \end{cases}$$

(v) Blackman window

$$\omega_{\mathrm{BL},n} = 0.42 - 0.5 \cos \frac{2\pi n}{N-1} + 0.08 \cos \frac{4\pi n}{N-1}, \quad n = 0, \ldots, N-1.$$

3.5.3 Sampling in the Frequency Domain

Suppose that $H_{\mathrm{d}}(z)$ and $H_{\mathrm{d}}(e^{i\omega})$ are the frequency response and the transfer function of an ideal digital filter, respectively. We want to design an even symmetric filter $H(z)$ with linear phase such that $H(z) = H_{\mathrm{d}}(z)$ at $z = e^{\frac{2\pi k}{N}}$ ($k = 0, 1, \ldots, N-1$), where N is odd.

Proposition 3.8. *Let*

$$H(z) = \frac{1 - z^N}{N} \sum_{0}^{N-1} \frac{H_{\mathrm{d}}\left(e^{i\frac{2\pi l}{N}}\right)}{1 - e^{-i\frac{2\pi l}{N}} z}. \tag{3.12}$$

Then the following interpolation formula holds:

$$H\left(e^{i\frac{2\pi k}{N}}\right) = H_{\mathrm{d}}\left(e^{i\frac{2\pi k}{N}}\right) \quad (k = 0, 1, \ldots, N-1), \quad N \text{ is odd.}$$

Proof. By

$$\frac{1 - z^N}{1 - e^{-i\frac{2\pi l}{N}} z} = \sum_{0}^{N-1} \left(e^{-i\frac{2\pi l}{N}} z\right)^n,$$

we have

$$H(z) = \frac{1}{N} \sum_{l=0}^{N-1} H_{\mathrm{d}}\left(e^{i\frac{2\pi l}{N}}\right) \sum_{n=0}^{N-1} \left(e^{-i\frac{2\pi l}{N}} z\right)^n$$

$$= \sum_{n=0}^{N-1} \left(\frac{1}{N} \sum_{l=0}^{N-1} H_{\mathrm{d}}\left(e^{i\frac{2\pi l}{N}}\right) e^{-il\frac{2\pi n}{N}}\right) z^n.$$

When $z = e^{i\frac{2\pi k}{N}}$, the sum on the right-hand side is used to find the discrete Fourier transform of $\{H_{\mathrm{d}}(e^{i\frac{2\pi k}{N}})\}_{k=0,1,\ldots,N-1}$, and then to find the inverse Fourier transform. So

$$H\left(e^{i\frac{2\pi k}{N}}\right) = \sum_{n=0}^{N-1}\left(\frac{1}{N}\sum_{l=0}^{N-1}H_d\left(e^{i\frac{2\pi l}{N}}\right)e^{-il\frac{2n\pi}{N}}\right)e^{in\frac{2\pi k}{N}}$$

$$= H_d\left(e^{i\frac{2\pi k}{N}}\right) \quad (k = 0,\ldots,N-1).$$

□

In (3.12), let $z = e^{i\omega}$. Then

$$H(e^{i\omega}) = \frac{1-e^{iN\omega}}{N}\sum_{0}^{N-1}\frac{H_d\left(e^{i\frac{2\pi l}{N}}\right)}{1-e^{-i\frac{2\pi l}{N}}e^{i\omega}} = \sum_{0}^{N-1}H_d\left(e^{i\frac{2\pi l}{N}}\right)\frac{1-e^{iN\left(\omega-\frac{2\pi l}{N}\right)}}{N\left(1-e^{i\left(\omega-\frac{2\pi l}{N}\right)}\right)}.$$

Denote

$$\varphi(\omega) = \frac{1-e^{iN\omega}}{N\left(1-e^{i\omega}\right)} = \frac{e^{i\frac{N\omega}{2}}\left(e^{-i\frac{N\omega}{2}}-e^{i\frac{N\omega}{2}}\right)}{Ne^{i\frac{\omega}{2}}\left(e^{-i\frac{\omega}{2}}-e^{i\frac{\omega}{2}}\right)} = e^{i\frac{N-1}{2}\omega}\frac{\sin\frac{N\omega}{2}}{N\sin\frac{\omega}{2}}.$$

Then

$$H(e^{i\omega}) = \sum_{0}^{N-1}H_d\left(e^{i\frac{2\pi l}{N}}\right)\varphi\left(\omega-\frac{2\pi l}{N}\right).$$

3.6 IIR FILTERS

In this section, we will discuss how to design IIR filter by using the analog filters described in Section 3.2.

3.6.1 Impulse Invariance Method

Suppose that the Laplace transform of an analog filter $h(t)$ can be written as

$$G(s) = \frac{\sum_0^M c(k)s^k}{\sum_0^N d(k)s^k},$$

where $M \leq N$ and $G(s)$ only has simple poles. Then $G(s)$ can be decomposed to partial fractions:

$$G(s) = \sum_{1}^{N}\frac{A_k}{s-s_k},$$

where $A_k = \lim_{s\to s_k}(s-s_k)G(s)$. Its inverse Laplace transform,

$$h(t) = L^{-1}[G(s)] = \sum_{1}^{N}L^{-1}\left[\frac{A_k}{s-s_k}\right] = \begin{cases}\sum_{1}^{N}A_k e^{s_k t}, & t \geq 0,\\ 0, & t < 0.\end{cases}$$

Furthermore, we can construct a digital filter $\{h(nT)\}$ as follows:

$$h(nT) = \begin{cases} \sum_{1}^{N} A_k \, e^{s_k nT}, & n \geq 0, \\ 0, & n < 0, \end{cases}$$

where we take T such that $G(i\Omega) = 0 \, (\Omega > \frac{\pi}{T})$. By the Poisson summation formula in Section 1.4, it is clear that $\{h(nT)\}$ is a low-pass filter. In addition, the Z-transform of the filter $\{h(nT)\}$ is

$$H(z) = \sum_{0}^{\infty} h(nT) \, z^{-n} = \sum_{n=0}^{\infty} \sum_{k=1}^{N} A_k \, e^{s_k nT} z^{-n} = \sum_{k=1}^{N} A_k \sum_{n=0}^{\infty} (e^{s_k T} z^{-1})^n$$

$$= \sum_{1}^{N} \frac{A_k}{1 - e^{s_k T} z^{-1}},$$

i.e., $H(z)$ is a rational function. This implies that filtering a noisy signal by IIR filter $\{h(nT)\}$ can be implemented through the linear difference equation in Section 3.3.

Example 3.1. Let $G(s) = \frac{s-c}{(s+a)(s+b)}$. Then

$$G(s) = \frac{s - c}{(s + a)(s + b)} = \frac{A}{s + a} + \frac{B}{s + b} \quad (a > 0, b > 0),$$

where

$$A = \lim_{s \to -a} \frac{s - c}{(s + a)(s + b)} \cdot (s + a) = -\frac{a + c}{b - a},$$

$$B = \lim_{s \to -b} \frac{s - c}{(s + a)(s + b)} \cdot (s + b) = \frac{b + c}{b - a}.$$

So

$$G(s) = \left(\frac{a + c}{a - b}\right) \frac{1}{s + a} - \left(\frac{b + c}{a - b}\right) \frac{1}{s + b}.$$

By using the impulse invariance method, we get the Z-transform of the digital filter:

$$H(z) = \frac{1}{a - b} \left(\frac{a + c}{1 - e^{-aT} z^{-1}} - \frac{b + c}{1 - e^{-bT} z^{-1}}\right).$$

Since $a > 0$ and $b > 0$, we have $|e^{-aT}| < 1$ and $|e^{-bT}| < 1$. So the filter is stable.

The advantage of the impulse invariance method is that it preserves the order and stability of the analog filter. The disadvantage is that there is a distortion of the shape of the frequency response because of aliasing.

3.6.2   Matched *Z*-Transform Method

For the analog filter

$$G(s) = \frac{\prod_1^M (s - z_k)}{\prod_1^N (s - p_k)},$$

letting $s - z_k = 1 - e^{z_k T} z^{-1}$, $s - p_k = 1 - e^{p_k T} z^{-1}$, we get the Z-transform of the desired IIR filter $\{h(n)\}$:

$$H(z) = \frac{\prod_1^M (1 - e^{z_k T} z^{-1})}{\prod_1^N (1 - e^{p_k T} z^{-1})}.$$

3.6.3   Bilinear Transform Method

The bilinear transform method can solve the distortion problem caused by the impulse invariance method. First, we need to establish a one-to-one map from the *s*-plane to the *z*-plane. The bilinear transform

$$s = \psi(z) = \frac{1 - z^{-1}}{1 + z^{-1}}$$

is a one-to-one transform from the *z*-plane to the *s*-plane.

Its inverse transform

$$z = \psi^{-1}(s) = \frac{1 + s}{1 - s}$$

maps the left-half *s*-plane onto the interior of the unit circle in the *z*-plane. Let $s = \sigma + i\Omega$. Then

$$|z|^2 = \left| \frac{1 + \sigma + i\Omega}{1 - \sigma - i\Omega} \right|^2 = \frac{(1 + \sigma)^2 + \Omega^2}{(1 - \sigma)^2 + \Omega^2}.$$

When $\sigma = 0$, $|z|^2 = 1$. This shows that the imaginary axis in the *s*-plane maps onto the unit circle in the *z*-plane, one-to-one. When $\sigma < 0$, $|z| < 1$. This shows that the left-half *s*-plane maps onto the interior of the unit circle. Since

$$s = \frac{z - 1}{z + 1} = \frac{e^{i\omega} - 1}{e^{i\omega} + 1} = \frac{e^{i\frac{\omega}{2}} - e^{-i\frac{\omega}{2}}}{e^{i\frac{\omega}{2}} + e^{-i\frac{\omega}{2}}} = i\frac{\sin\frac{\omega}{2}}{\cos\frac{\omega}{2}} = i\tan\frac{\omega}{2},$$

the point $s = i\Omega$ maps to the point $z = e^{i\omega}$, where $\Omega = \tan\frac{\omega}{2}$. From this, we see that the mapping $\Omega \leftrightarrow \omega$ is monotonic, $\Omega = 0$ maps to $\omega = 0$, and $\Omega = \infty$ maps to $\omega = \pi$.

On the basis of the cutoff frequency ω_k of the digital filter, let

$$\Omega_k = \tan\frac{\omega_k}{2}.$$

We design an analog filter with cutoff frequencies Ω_k. Then, applying the bilinear transform

$$s = \frac{1 - z^{-1}}{1 + z^{-1}}$$

to the Laplace transform $G(s)$ of the analog filter, we get the Z-transform $H(z)$ of the digital filter.

3.7 CONJUGATE MIRROR FILTERS

A conjugate mirror filter can decompose a signal into low-pass and high-pass components. It is closely related to wavelet theory (see Chapter 2). We first introduce the following notation for zero insertion and the subsample.

For a signal $b = (b(n))_{n \in \mathbb{Z}}$, its Fourier transform is

$$B(\omega) = \sum_n b(n) \, e^{-in\omega}.$$

The insertion zero signal of b is defined by $b^0 = (b^0(n))_{n \in \mathbb{Z}}$, where

$$b^0(n) = \begin{cases} b(p), & n = 2p, \\ 0, & n = 2p + 1 (p \in \mathbb{Z}). \end{cases}$$

The Fourier transform of b^0 is

$$B^0(\omega) := \sum_n b^0(n) \, e^{-in\omega} = \sum_n b(n) \, e^{-i2n\omega} = B(2\omega). \qquad (3.13)$$

The subsampled signal of b is defined by $b^1 = (b^1(n))_{n \in \mathbb{Z}}$, where $b^1(n) = b(2n)(n \in \mathbb{Z})$. The Fourier transform of b^1 is

$$B^1(\omega) := \sum_n b^1(n) \, e^{-in\omega} = \sum_n b(2n) \, e^{-in\omega},$$

and so

$$B^1(2\omega) = \sum_n b(2n) \, e^{-i2n\omega}.$$

Noticing that $B(\omega) = \sum_n b(n) \, e^{-in\omega}$, we have $B(\omega + \pi) = \sum_n (-1)^n b(n) \, e^{-in\omega}$. So

$$B(\omega) + B(\omega + \pi) = 2 \sum_n b(2n) \, e^{-i2n\omega} = 2B^1(2\omega). \qquad (3.14)$$

Now we give the decomposition and the reconstruction of signals by using low-pass and high-pass filters.

Let $h_1 = \{h_1(n)\}_{n \in \mathbb{Z}}$ be a real low-pass filter and $g_1 = \{g_1(n)\}_{n \in \mathbb{Z}}$ be a real high-pass filter, and let $\widetilde{h}_1 = \{\widetilde{h}_1(n)\}_{n \in \mathbb{Z}}$ and $\widetilde{g}_1 = \{\widetilde{g}_1(n)\}_{n \in \mathbb{Z}}$ satisfy

$\widetilde{h}_1(n) = h_1(-n)$ $(n \in \mathbb{Z})$ and $\widetilde{g}_1(n) = g_1(-n)$ $(n \in \mathbb{Z})$, respectively. For an input signal $x = \{x(n)\}_{n \in \mathbb{Z}}$, define $c_1 = \{c_1(n)\}_{n \in \mathbb{Z}}$ and $d_1 = \{d_1(n)\}_{n \in \mathbb{Z}}$ by

$$c_1(n) = (\widetilde{h}_1 * x)(2n) = \sum_k h_1(k - 2n)x(k),$$

$$d_1(n) = (\widetilde{g}_1 * x)(2n) = \sum_k g_1(k - 2n)x(k). \tag{3.15}$$

This is a *decomposition formula*. In data analysis, an important problem is whether we can choose $h_2 = \{h_2(n)\}_{n \in \mathbb{Z}}$ and $g_2 = \{g_2(n)\}_{n \in \mathbb{Z}}$ such that the *reconstruction formula*

$$x(n) = (h_2 * c_1^0)(n) + (g_2 * d_1^0)(n) \tag{3.16}$$

holds, where c_1^0 and d_1^0 are insertion zero signals of c_1 and d_1, respectively. The following theorem gives a necessary and sufficient condition such that (3.16) holds.

Theorem 3.1. *The filters h_1, g_1 and h_2, g_2 are such that (3.16) holds for any input signal x if and only if their Fourier transforms satisfy*

$$H_1(\omega)\overline{H}_2(\omega) + G_1(\omega)\overline{G}_2(\omega) = 2, \tag{3.17}$$

$$H_1(\omega + \pi)\overline{H}_2(\omega) + G_2(\omega + \pi)\overline{G}_2(\omega) = 0,$$

where H_1, H_2, G_1, and G_2 are Fourier transforms of h_1, h_2, g_1, and g_2, respectively, i.e.,

$$H_i(\omega) = \sum_n h_i(n) e^{-in\omega} \quad (i = 1, 2),$$

$$G_i(\omega) = \sum_n g_i(n) e^{-in\omega} \quad (i = 1, 2).$$

Proof. Since h_1 and g_1 are real,

$$H_1(-\omega) = \sum_n h_1(n)e^{in\omega} = \overline{\sum_n h_1(n)e^{-in\omega}} = \overline{H}_1(\omega),$$

$$G_1(-\omega) = \sum_n g_1(n)e^{in\omega} = \overline{\sum_n g_1(n)e^{-in\omega}} = \overline{G}_1(\omega),$$

and so the Fourier transforms of $\widetilde{h}_1$ and $\widetilde{g}_1$ are, respectively,

$$\widetilde{H}_1(\omega) = \sum_n h_1(-n) e^{-in\omega} = \sum_n h_1(n) e^{in\omega} = H_1(-\omega) = \overline{H}_1(\omega),$$

$$\widetilde{G}_1(\omega) = \sum_n g_1(-n) e^{-in\omega} = \sum_n g_1(n) e^{in\omega} = \overline{G}_1(-\omega) = \overline{G}_1(\omega).$$

Let the sequences $c_1 = \{c_1(n)\}_{n \in \mathbb{Z}}$ and $d_1 = \{d_1(n)\}_{n \in \mathbb{Z}}$ be stated as in (3.15), and let $C_1(\omega)$ and $D_1(\omega)$ be Fourier transforms of c_1 and d_1, respectively. Similarly to the proof of (3.14), by (3.15), it follows that

$$C_1(2\omega) = \frac{1}{2}(X(\omega)\overline{H}_1(\omega) + X(\omega + \pi)\overline{H}_1(\omega + \pi)),$$

$$D_1(2\omega) = \frac{1}{2}(X(\omega)\overline{G}_1(\omega) + X(\omega + \pi)\overline{G}_1(\omega + \pi)). \tag{3.18}$$

From (3.13), it follows that (3.16) holds if and only if the Fourier transform of x satisfies

$$X(\omega) = C_1(2\omega)H_2(\omega) + D_1(2\omega)G_2(\omega).$$

From this and (3.18), it follows that (3.16) holds if and only if the Fourier transform of x satisfies

$$X(\omega) = \frac{1}{2}X(\omega)\left(\overline{H}_1(\omega)H_2(\omega) + \overline{G}_1(\omega)G_2(\omega)\right) + \frac{1}{2}X(\omega + \pi)$$

$$\left(\overline{H}_1(\omega + \pi)H_2(\omega) + \overline{G}_1(\omega + \pi)G_2(\omega)\right).$$

For any input x, its Fourier transform $X(\omega)$ satisfies this formula if and only if (3.17) holds. $\qquad\square$

Formula (3.17) can be written in matrix form:

$$\begin{pmatrix} H_1(\omega) & G_1(\omega) \\ H_1(\omega + \pi) & G_1(\omega + \pi) \end{pmatrix} \begin{pmatrix} \overline{H}_2(\omega) \\ \overline{G}_2(\omega) \end{pmatrix} = \begin{pmatrix} 2 \\ 0 \end{pmatrix}.$$

From this, we can find out the reconstruction filters H_2 and G_2 as follows:

$$\overline{H}_2(\omega) = \frac{2G_1(\omega + \pi)}{\Delta(\omega)},$$

$$\overline{G}_2(\omega) = -\frac{2H_1(\omega + \pi)}{\Delta(\omega)}, \tag{3.19}$$

where the determinant

$$\Delta(\omega) = \begin{vmatrix} H_1(\omega) & G_1(\omega) \\ H_1(\omega + \pi) & G_1(\omega + \pi) \end{vmatrix} = H_1(\omega)G_1(\omega + \pi) - H_1(\omega + \pi)G_1(\omega).$$

Since H_1, H_2, G_1, and G_2 are 2π-periodic functions,

$$\Delta(\omega + \pi) = H_1(\omega + \pi)G_1(\omega) - H_1(\omega)G_1(\omega + \pi) = -\Delta(\omega),$$

by (3.19), we get

$$G_1(\omega)\overline{H}_2(\omega) - G_1(\omega + \pi)\overline{H}_2(\omega + \pi) =$$

$$G_1(\omega)\frac{2G_1(\omega + \pi)}{\Delta(\omega)} - G_1(\omega + \pi)\frac{2G_1(\omega)}{\Delta(\omega)} = 0,$$

$$H_1(\omega)\overline{G}_2(\omega) + H_1(\omega + \pi)\overline{G}_2(\omega + \pi) =$$

$$H_1(\omega)\left(-\frac{2H_1(\omega + \pi)}{\Delta(\omega)}\right) + H_1(\omega + \pi)\frac{2H_1(\omega)}{\Delta(\omega)} = 0, \qquad (3.20)$$

and

$$G_1(\omega)\overline{G}_2(\omega) = H_1(\omega + \pi)\overline{H}_2(\omega + \pi).$$

Combining this with (3.17), we get

$$\overline{H}_1(\omega)H_2(\omega) + \overline{H}_1(\omega + \pi)H_2(\omega + \pi) = 2,$$

$$G_1(\omega + \pi)\overline{G}_2(\omega + \pi) + G_1(\omega)\overline{G}_2(\omega) = 2. \qquad (3.21)$$

Combining (3.17) with (3.20) and (3.21), we get the following corollary.

Corollary 3.1. *Let*

$$T_i(\omega) = \begin{pmatrix} H_i(\omega) & G_i(\omega) \\ H_i(\omega + \pi) & G_i(\omega + \pi) \end{pmatrix} \quad (i = 1, 2).$$

Then condition (3.17) is equivalent to the condition

$$T_2^*(\omega)T_1(\omega) = \begin{pmatrix} 2 & 0 \\ 0 & 2 \end{pmatrix},$$

*where * means conjugate transpose matrix, i.e.,*

$$\begin{pmatrix} \overline{H}_2(\omega) & \overline{H}_2(\omega + \pi) \\ \overline{G}_2(\omega) & \overline{G}_2(\omega + \pi) \end{pmatrix} \begin{pmatrix} H_1(\omega) & G_1(\omega) \\ H_1(\omega + \pi) & G_1(\omega + \pi) \end{pmatrix} = \begin{pmatrix} 2 & 0 \\ 0 & 2 \end{pmatrix}.$$

Letting $h_1 = h_2 =: h$, $g_1 = g_2 =: g$ in (3.15) and (3.16), we deduce the following result: A low-pass filter h and a high-pass filter g are such that for any signal x, the decomposition

$$c(n) = (\widetilde{h} * x)(2n), \quad d(n) = (\widetilde{g} * x)(2n)$$

and the reconstruction

$$x(n) = (h * c^0)(n) + (g * d^0)(n)$$

hold if and only if the matrix

$$T(\omega) = \frac{1}{\sqrt{2}} \begin{pmatrix} H(\omega) & G(\omega) \\ H(\omega + \pi) & G(\omega + \pi) \end{pmatrix}$$

satisfies

$$T^*(\omega)T(\omega) = \begin{pmatrix} 1 & 0 \\ 0 & 1 \end{pmatrix},$$

i.e., T is an orthogonal matrix, and $|H(\omega)|^2 + |H(\omega + \pi)|^2 = 2$, where

$$H(\omega) = \sum_n h(n) \, e^{-in\omega},$$

$$G(\omega) = \sum_n g(n) \, e^{-in\omega}.$$

The filter h satisfying $|H(\omega)|^2 + |H(\omega + \pi)|^2 = 2$ is called a *conjugate mirror filter*. We know from Chapter 2 that the low-pass filter corresponding to a scale function is a conjugate mirror filter. In the construction process of compactly supported wavelets (see Chapter 2), Daubechies constructed many conjugate mirror filters.

PROBLEMS

3.1 Show that a linear differential equation with constant coefficients $y(t) = T[x(t)] = \sum_0^n a_k x^{(k)}(t)$ is a linear time-invariant system. Moreover, find its filter and frequency response.

3.2 Compute the inverse Z-transform of $X(z) = \frac{1}{(z-1)(z+3)}$ in the domains $1 < |z| < 3$ and $|z| > 3$, respectively.

3.3 Construct an ideal band-pass digital filter h such that its frequency response $H(e^{i\omega})$ is

$$H(e^{i\omega}) = \begin{cases} 1, & 0 < \omega_0 \le |\omega| < \pi, \\ 0, & |\omega| < \omega_0 \text{ or } \omega_1 < |\omega| \le \pi. \end{cases}$$

3.4 Construct a five-point FIR filter by using a Hamming window.

3.5 On the basis of a Butterworth filter of order 2, use the impulse invariance method to construct an IIR filter.

3.6 Using the Daubechies filter to decompose local temperature data into low-frequency signal and high-frequency signal.

BIBLIOGRAPHY

Abo-Zahhad, M., Al-Zoubi, Q., 2006. A novel algorithm for the design of selective FIR filters with arbitrary amplitude and phase characteristics. Digit. Signal Process. 16, 211-224.

Apostolov, P., 2011. Method for FIR filters design with compressed cosine using Chebyshev's norm. Signal Process. 91, 2589-2594.

Elliott, D., 1987. Handbook of Digital Signal Processing. Elsevier/Academic Press, San Diego, CA.

Emery, W.J., Thomson, R.E., 2001: Data Analysis Methods in Physical Oceanography. Elsevier/Academic Press, San Diego, CA.

Evrendilek, F., 2013. Quantifying biosphere-atmosphere exchange of CO_2 using eddy covariance, wavelet denoising, neural networks, and multiple regression models. Agricult. Forest Meteorol. 171-172, 1-8.

Fortin, J.G., Bolinder, M.A., Anctil, F., Ktterer, T., Andren, O., Parent, L.E., 2011. Effects of climatic data low-pass filtering on the ICBM temperature- and moisture-based soil biological activity factors in a cool and humid climate. Ecol. Model. 222, 3050-3060.

Frappart, F., Ramillien, G., Leblanc, M., Tweed, S.O., Bonnet, M.-P., Maisongrande, P., 2011. An independent component analysis filtering approach for estimating continental hydrology in the GRACE gravity data. Remote Sens. Environ. 115, 187-204.

Jeon, J., Kim, D., 2012. Design of nonrecursive FIR filters with simultaneously MAXFLAT magnitude and prescribed cutoff frequency. Digit. Signal Process. 22, 1085-1094.

Li, Y., Trenchea, C., 2014. A higher-order Robert-Asselin type time filter. J. Comput. Phys. 259, 23-32.

Mirin, A.A., Shumaker, D.E., Wehner, M.F., 1998. Efficient filtering techniques for finite-difference atmospheric general circulation models on parallel processors. Parallel Comput. 24, 729-740.

Murray, R.J., Reason, C.J.C., 2002. Fourier filtering and coefficient tapering at the North Pole in OGCMs. Ocean Model. 4, 1-25.

Nielsen, U.D., 2007. Response-based estimation of sea state parameters influence of filtering. Ocean Eng. 34, 1797-1810.

Osborne, A.R., 1995. The inverse scattering transform: tools for the nonlinear Fourier analysis and filtering of ocean surface waves. Chaos Soliton. Fract. 5, 2623-2637.

Pei, S.-C., Tseng, C.-C., 1997. Design of equiripple log FIR and IIR filters using multiple exchange algorithm. Signal Process. 59, 291-303.

Reninger, P.-A., Martelet, G., Deparis, J., Perrin, J., Chen, Y., 2011. Singular value decomposition as a denoising tool for airborne time domain electromagnetic data. J. Appl. Geophys. 75, 264-276.

Rusu, C., Dumitrescu, B., 2012. Iterative reweighted l1 design of sparse FIR filters. Signal Process. 92, 905-911.

Sakamoto, T., Wardlow, B.D., Gitelson, A.A., Verma, S.B., Suyker, A.E., Arkebauer, T.J., 2010. A two-step filtering approach for detecting maize and soybean phenology with time-series MODIS data. Remote Sens. Environ. 114, 2146-2159.

Shan, H., Ma, J., Yang, H., 2009. Comparisons of wavelets, contourlets and curvelets in seismic denoising. J. Appl. Geophys. 69, 103-115.

Simpson, J.J., Gobat, J.I., Frouin, R., 1995. Improved destriping of GOES images using finite impulse response filters. Remote Sens. Environ. 52, 15-35.

Vite-Chavez, O., Olivera-Reyna, R., Ibarra-Manzano, O., Shmaliy, Y.S., Morales-Mendoza, L., 2013. Time-variant forward-backward FIR denoising of piecewise-smooth signals. AEU Int. J. Electron. Commun. 67, 406-413.

Yang, Y., Wilson, L.T., Wang, J., 2010. Development of an automated climatic data scraping, filtering and display system. Comput. Electron. Agricult. 71, 77-87.

Zeri, M., Sa, L.D.A., 2010. The impact of data gaps and quality control filtering on the balances of energy and carbon for a southwest Amazon forest. Agricult. Forest Meteorol. 150, 1543-1552.

Chapter 4

Remote Sensing

Earth remote sensing began with the first Landsat Multispectral Scanner System in 1972. This system provided general-purpose satellite image data directly in digital form for the first time. More remote-sensing systems with different spatial resolutions and different spectral bands have appeared successively, for example, the Landsat Thematic Mapper System, Advanced Very High Resolution Radiometer System, Geostationary Operational Environmental Satellite System, Hyperspectral Imager System, and the Moderate Imaging Spectroradiometer System.

Compared with in situ observation, remote sensing gathers information concerning Earth's surface by using data acquired from aircraft and satellites. The traditional and physics-based concepts are now complemented with signal and image processing concepts, so remote sensing is capable of managing the interface between the signal acquired and the physics of the surface of Earth. Remote-sensing technologies have been applied widely in meteorology, climate change detection, environmental monitoring, flood prediction, agriculture, resource explorations, mapping, and so on. In this chapter, we will introduce solar and thermal radiation, spatial filtering, blurring, distortion correction, image fusion, supervised and unsupervised classification, and the applications in climate change.

4.1 SOLAR AND THERMAL RADIATION

Solar radiation and thermal radiation are two important optical radiation processes. The Sun is a near-perfect blackbody radiator. In the visible region, near-infrared region, and short-wave infrared region, the radiation received by sensors originates from the Sun. Part of the solar radiation has been reflected at Earth's surface, and part has been scattered by the atmosphere without ever reaching the earth. In the thermal infrared region, thermal radiation is emitted directly by materials on the earth and combines with self-emitted thermal radiation in the atmosphere as it propagates upward.

Solar radiation can be divided into three significant components:

$$F_\lambda^s = F_\lambda^{su} + F_\lambda^{sd} + F_\lambda^{sp},$$

Mathematical and Physical Fundamentals of Climate Change
© 2015 Elsevier Inc. All rights reserved.

where F_λ^{su} is the unscattered surface-reflected radiation, F_λ^{sd} is the down-scattered surface-reflected skylight, and F_λ^{sp} is the up-scattered path radiance, where λ is the wavelength.

The atmosphere scatters and absorbs radiation between the Sun and Earth along the solar path, and again between Earth and the sensor along the view path. The fraction of radiation that arrives at Earth's surface is called the *solar path atmospheric transmittance*, while the fraction of radiation transmitted by the atmosphere along the view path from Earth's surface to the sensor is called the *view path atmospheric transmittance*.

For a specified wavelength λ, the at-sensor, unscattered surface-reflected radiation

$$F_\lambda^{su} = \alpha(x, y, \lambda) \frac{\mathcal{T}_s(\lambda) \mathcal{T}_v(\lambda) F_\lambda^0}{\pi} \cos(\mathbf{n} \cdot \mathbf{s}),$$

where α is the diffuse spectral reflectance, $\mathcal{T}_s$ and $\mathcal{T}_v$ are the solar path and view path atmospheric transmittances, respectively, $\mathbf{n}$ is the unit vector normal to the surface and $\mathbf{s}$ is the unit vector pointing to the Sun, and F_λ^0 is solar spectral irradiance at the top of the atmosphere. The sum of solar irradiance F_λ^0 over all wavelengths is called the *solar constant*:

$$F_s = \sum_\lambda F_\lambda^0 = 1370 \, \text{W/m}^2.$$

The at-sensor, down-scattered surface-reflected skylight

$$F_\lambda^{sd} = \beta(x, y) \, \alpha(x, y, \lambda) \frac{\mathcal{T}_v(\lambda) \, F_\lambda^d}{\pi},$$

where $\beta(x, y)$ depends on terrain shape and is the fraction of the sky hemisphere that is visible from the position (x, y), and F_λ^d is the irradiance at the surface due to skylight. Therefore, the total at-sensor, solar radiation

$$F_\lambda^s = \alpha(x, y, \lambda) \frac{\mathcal{T}_v(\lambda)}{\pi} \left(\mathcal{T}_s(\lambda) \cos(\mathbf{n} \cdot \mathbf{s}) F_\lambda^0 + \beta(x, y) F_\lambda^d \right) + F_\lambda^{sp}.$$

Thermal radiation can be emitted by every object at a temperature above absolute zero. It can also be divided into three components:

$$F_\lambda^e = F_\lambda^{eu} + F_\lambda^{ed} + F_\lambda^{ep},$$

where F_λ^{eu} is the surface-emitted radiation from Earth, F_λ^{ed} is the down-emitted surface-reflected radiation from the atmosphere, and F_λ^{ep} is the path-emitted radiance.

For a specified wavelength λ, the at-sensor, surface-emitted radiation from Earth

$$F_\lambda^{eu} = \epsilon(x, y, \lambda) \frac{\mathcal{T}_v(\lambda) \, B_\lambda(T(x, y))}{\pi},$$

where ϵ is the emittance, $\mathcal{T}_v$ is the view path atmospheric transmittance, and $B_\lambda(T)$ is the blackbody spectral radiance at the temperature T. Kirchhoff's law states the relation between emittance and reflectance:

$$\epsilon(x, y, \lambda) = 1 - \alpha(x, y, \lambda).$$

The blackbody spectral radiance can be computed by Planck's law:

$$B_\lambda(T) = \frac{2hc^2}{\lambda^5 \left(e^{\frac{hc}{\lambda k_B T}} - 1 \right)},$$

where h is the Planck constant (6.626×10^{-34} J s), k_B is the Boltzmann constant (1.38×10^{-23} J/K), c is the speed of light, and T is the temperature in Kelvin. The at-sensor, down-emitted surface-reflected radiation from the atmosphere

$$F_\lambda^{ed} = \beta(x, y)\, \alpha(x, y, \lambda) \frac{\mathcal{T}_v(\lambda)\, B_\lambda^a}{\pi},$$

where B_λ^a is the atmospheric-emitted radiance. Therefore, the total at-sensor, thermal radiance

$$F_\lambda^e = \mathcal{T}_v(\lambda) \left(\epsilon(x, y, \lambda) \frac{B_\lambda(T(x, y))}{\pi} + \beta(x, y)\, \alpha(x, y, \lambda) \frac{B_\lambda^a}{\pi} \right) + F_\lambda^{ep}.$$

In summary, for a specified wavelength λ, the sum of the total at-sensor, solar and thermal radiations is

$$F_\lambda = F_\lambda^s + F_\lambda^e.$$

In the visible region, the near-infrared region, and the short-wave infrared region, the second term is negligible, while in the thermal infrared region, the first term is negligible.

4.2   SPECTRAL REGIONS AND OPTICAL SENSORS

The major spectral regions used in Earth remote sensing include the visible region, near-infrared region, short-wave infrared region, mid-wave infrared region, thermal infrared region, and microwave region. The thermal infrared region is also known as the long-wave infrared region. Among them, the visible region, near infrared region, and short-wave infrared region are the solar-reflective spectral ranges. The mid-wave infrared region is a transition zone from solar-reflective to thermal radiation. The thermal infrared region and the microwave region correspond to thermal radiation.

Remote-sensing imaging instruments can be divided into two classes:

(1) Passive optical remote sensing, such as multispectral and hyperspectral sensors, relies on solar radiation as an illumination source. It is mainly focused on the visible, near-infrared, and shortwave infrared spectral regions. Many satellites with several onboard passive sensors are currently flying over our heads, and many are being built or planned for the coming years.

(2) Active remote sensing employs an artificial source of radiation as a probe, and the resulting signal that scatters back to the sensor characterizes the atmosphere or the earth. The Synthetic Aperture Radar system is an active system. It can emit radiation in a beam from a moving sensor and measure the backscattered component returned to the sensor from the ground in the microwave region.

Here we mainly focus on passive remote sensing.

All passive, scanning optical sensors at the satellite measure the emergent radiation from the Earth surface-atmosphere system in the sensor observation direction. The grid of pixels in a remote-sensing image is achieved by a combination of scanning in the cross-track direction and the sensor platform motion along the in-track direction. There are three types of scanner: a line scanner, a whisk broom scanner, and a push broom scanner. A line scanner uses a single detector element to scan the entire scene. A whisk broom scanner (or an across-track scanner) uses several detector elements aligned in-track to achieve parallel scanning. A push broom scanner (or an along-track scanner) uses a linear array of thousands of detector elements aligned cross-track to scan the full width of the scene in parallel. A push broom scanner receives a stronger signal than a whisk broom scanner because it looks at each pixel area for longer.

In remote-sensing images, the *ground sample distance* (GSD) is the distance between pixel centers measured on the ground. The bigger the GSD is, the lower the spatial resolution of the image is and fewer details are visible. The GSD is determined by the altitude of the sensor system, the sensor's focal length, and the interdetector spacing, i.e.,

$$\text{GSD} = \text{interdetector spacing} \times \frac{\text{altitude of the sensor system}}{\text{sensor's focal length}}.$$

The *ground-projected instantaneous field of view* (GIFOV) is the maximum angle of view where a sensor can detect electromagnetic energy. The GIFOV is determined by the altitude of the sensor system, the sensor's focal length, and the single detector width, i.e.,

$$\text{GIFOV} = \text{single detector width} \times \frac{\text{altitude of the sensor system}}{\text{sensor's focal length}}.$$

In application, since the single detector width is always designed the same as the interdetector spacing, GSD is equal to GIFOV. In the process of storing remote-sensing data, the radiance obtained by the sensor at each pixel is converted to an electrical signal and then quantized to a discrete integer value. This discrete integer value is called the *digital number* (DN). A finite number of bits is used to code the continuous data measurements as binary numbers. *Image contrast* is defined as the ratio of the maximum and minimum DNs in the image. Since many original remote-sensing images have poorly contrast, contrast enhancement is frequently referred to as one of the most important issues in remote-sensing image processing. The main contrast enhancement techniques include the maximum-minimum contrast method, the percentage contrast method, the piecewise contrast method, and the histogram equalization method.

4.3 SPATIAL FILTERING

Filtering methods have been extensively used in remote-sensing feature extraction. All filters introduced in Chapter 3 can be used in remote-sensing image processing. In addition, the following spatial filters are also used often.

Mean filters are simple spatial filters that replace each pixel value in an image with the mean value of its neighbors. This kind of filter can preserve the local mean and smooth the input signal: the larger the window, the more the smoothing. The corresponding high-pass (HP) filter can remove the local mean and produce an output which measures the deviation of the input signal from the local mean.

Median filters can do an excellent job of rejecting certain types of noise in remote-sensing images, in particular, shot or impulse noise. In the median filtering, the pixel values in the neighborhood are ranked according to intensity, and the middle value (the median) becomes the output value for the pixel under evaluation. Therefore, the median filter is less sensitive than the mean filter to extreme values.

High-boost (HB) *filters* emphasize high-frequency (HF) components representing the remote-sensing image details without eliminating low-frequency components. This kind of filter can be used in edge enhancement and can be created by adding an HF component of a remote-sensing image to the original remote-sensing image:

$$\text{HB}(x, y; k) = \text{Image}(x, y) + k \cdot \text{HF}(x, y),$$

where k is a constant.

Gradient filters are edge detectors that work by numerically computing the first derivatives of an image, and include the Robert filter, Sobel filter, and Prewitt filter:

Filter   Horizontal direction   Vertical direction

$$\text{Robert} \quad \begin{bmatrix} 0 & 1 \\ -1 & 0 \end{bmatrix} \quad \begin{bmatrix} 1 & 0 \\ 0 & -1 \end{bmatrix}$$

$$\text{Sobel} \quad \begin{bmatrix} 1 & 2 & 1 \\ 0 & 0 & 0 \\ -1 & -2 & -1 \end{bmatrix} \quad \begin{bmatrix} -1 & 0 & 1 \\ -2 & 0 & 2 \\ -1 & 0 & 1 \end{bmatrix}$$

$$\text{Prewitt} \quad \begin{bmatrix} 1 & 1 & 1 \\ 1 & -2 & 1 \\ -1 & -1 & -1 \end{bmatrix} \begin{bmatrix} -1 & 1 & 1 \\ -1 & -2 & 1 \\ -1 & 1 & 1 \end{bmatrix}$$

4.4 SPATIAL BLURRING

Remote sensors are complex systems of optical, mechanical, and electronic components. No sensor can measure a physical signal with infinite precision. The spatial blurring is produced by the sensor's optics, motion, detector, and electronics.

The spatial blurring can be characterized by the net sensor point spread function (PSF). This process can be modeled by the convolution:

$$e(x,y) = \int \int s(\alpha, \beta) \text{PSF}(x - \alpha, y - \beta) \, d\alpha \, d\beta = (s * \text{PSF})(x, y),$$

where e is the output signal and s is the input signal.

The net sensor PSF consists of the optical PSF (PSF_o), image motion PSF (PSF_i), detector PSF (PSF_d), and electronics PSF (PSF_e).

The simplest model for the PSF_o is the two-dimensional Gaussian function:

$$\text{PSF}_o(x,y) = \frac{1}{2\pi ab} e^{-\frac{x^2}{2a^2} - \frac{y^2}{2b^2}},$$

where the parameters a and b determine the width of the PSF_o in the cross-track and in-track directions, respectively.

The PSF_i can be modeled by the rectangular PSF. In a whisk broom scanner, the integration time causes the cross-track blurring. The corresponding PSF_i is defined as

$$\text{PSF}_i(x,y) = \text{rect}\left(\frac{x}{S}\right),$$

where S is equal to the scan velocity times the integration time. The rectangular PSF, $\text{rect}(x/S)$, is a square pulse of width S and amplitude 1. In a push broom scanner, the integration time causes in-track blurring. The corresponding PSF_i is defined as

$$\text{PSF}_i(x,y) = \text{rect}\left(\frac{y}{S}\right),$$

where S is equal to the platform velocity times the integration time.

The PSF_d is the spatial blurring caused by the nonzero spatial area of each detector in the sensor. It can be modeled as the product of two rectangular PSFs, i.e.,

$$\text{PSF}_d = \text{rect}\left(\frac{x}{\text{GIFOV}}\right) \text{rect}\left(\frac{y}{\text{GIFOV}}\right).$$

Therefore,

$$e = (((s * \text{PSF}_o) * \text{PSF}_i) * \text{PSF}_d) * \text{PSF}_e,$$

i.e., four cascaded filters are applied to the input image s. Using the convolution property in Chapter 2, we have

$$e = s * (\text{PSF}_o * \text{PSF}_i * \text{PSF}_d * \text{PSF}_e),$$

and so the net sensor PSF is the convolution of these four components, i.e.,

$$\text{PSF} = \text{PSF}_o * \text{PSF}_i * \text{PSF}_d * \text{PSF}_e.$$

Various deconvolution algorithms have been developed to solve the problem of spatial blurring

4.5 DISTORTION CORRECTION

Any remote-sensing image has various geometric distortions. Platform altitude variation (roll, pitch, and yaw) in combination with sensor focal length and Earth's flatness, and topographic factors can change the pixel spacing, and so it can change the orientation and the shape of images. Platform velocity variations can change the line spacing or create line gaps/overlaps.

In the image distortion correction, one must essentially reposition pixels from their original locations in the data array into a specified reference grid. The process consists of three components: selection of mathematical distortion models, coordinate transformation, and resampling.

A suitable model for image distortion correction is the polynomial distortion model:

$$x = \sum_{i=0}^{N} \sum_{j=0}^{N-i} a_{i,j} u^i v^j,$$

$$y = \sum_{i=0}^{N} \sum_{j=0}^{N-i} b_{i,j} u^i v^j,$$

where x, y are the global coordinates in the distorted image and u, v are the coordinates in the reference image.

Especially, let $N = 2$. The quadratic polynomial distortion model is

$$x = a_{0,0} + a_{1,0} u + a_{0,1} v + a_{1,1} u v + a_{2,0} u^2 + a_{0,2} v^2,$$
$$y = b_{0,0} + b_{1,0} u + b_{0,1} v + b_{1,1} u v + b_{2,0} u^2 + b_{0,2} v^2,$$

where the coefficients of the quadratic polynomial distortion are determined by ground control points (GCPs). The GCPs are defined as points on the surface of Earth of known location. Their quantity, distribution, and accuracy play an important role in correcting remote-sensing images. If n pairs of GCPs are used in the distorted image and reference image coordinate systems, the quadratic polynomial distortion model satisfies

$$x_k = a_{0,0} + a_{1,0} u_k + a_{0,1} v_k + a_{1,1} u_k v_k + a_{2,0} u_k^2 + a_{0,2} v_k^2 \quad (k = 1, 2, \dots, n),$$

$$y_k = b_{0,0} + b_{1,0}\, u_k + b_{0,1}\, v_k + b_{1,1}\, u_k\, v_k + b_{2,0}\, u_k^2 + b_{0,2}\, v_k^2 \quad (k = 1, 2, \ldots, n).$$

These two equations can be written simply in matrix forms:

$$\mathbf{X} = \mathbf{H}\,\mathbf{A},$$
$$\mathbf{Y} = \mathbf{H}\,\mathbf{B},$$

where

$$\mathbf{X} = \begin{pmatrix} x_1 \\ x_2 \\ \vdots \\ x_n \end{pmatrix}, \quad \mathbf{A} = \begin{pmatrix} a_{0,0} \\ a_{1,0} \\ a_{0,1} \\ a_{1,1} \\ a_{2,0} \\ a_{0,2} \end{pmatrix}, \quad \mathbf{Y} = \begin{pmatrix} y_1 \\ y_2 \\ \vdots \\ y_n \end{pmatrix}, \quad \mathbf{B} = \begin{pmatrix} b_{0,0} \\ b_{1,0} \\ b_{0,1} \\ b_{1,1} \\ b_{2,0} \\ b_{0,2} \end{pmatrix},$$

and

$$\mathbf{H} = \begin{pmatrix} 1 & u_1 & v_1 & u_1 v_1 & u_1^2 & v_1^2 \\ 1 & u_2 & v_2 & u_2 v_2 & u_2^2 & v_2^2 \\ \vdots & \vdots & \vdots & \vdots & \vdots & \vdots \\ 1 & u_n & v_n & u_n v_n & u_n^2 & v_n^2 \end{pmatrix}.$$

Therefore, the least-squares solutions for $\mathbf{A}$ and $\mathbf{B}$ are

$$\widehat{\mathbf{A}} = (\mathbf{H}^{\mathrm{T}}\,\mathbf{H})^{-1}\,\mathbf{H}^{\mathrm{T}}\,\mathbf{X},$$
$$\widehat{\mathbf{B}} = (\mathbf{H}^{\mathrm{T}}\,\mathbf{H})^{-1}\,\mathbf{H}^{\mathrm{T}}\,\mathbf{Y},$$

where $\mathbf{H}^{\mathrm{T}}$ is the transpose matrix of $\mathbf{H}$.

The simplest linear polynomial model is

$$x = a_{0,0} + a_{1,0}u + a_{0,1}v,$$
$$y = b_{0,0} + b_{1,0}u + b_{0,1}v.$$

This is an affine transformation which can simultaneously accommodate shift, scale, and rotation. The corresponding matrix form is

$$\mathbf{C} = \mathbf{T}\mathbf{D} + \mathbf{T_0},$$

where

$$\mathbf{C} = \begin{pmatrix} x \\ y \end{pmatrix}, \quad \mathbf{T} = \begin{pmatrix} a_{1,0} & a_{0,1} \\ b_{1,0} & b_{0,1} \end{pmatrix},$$

$$\mathbf{D} = \begin{pmatrix} u \\ v \end{pmatrix}, \quad \mathbf{T_0} = \begin{pmatrix} a_{0,0} \\ b_{0,0} \end{pmatrix}.$$

Once the image distortion correction model has been determined, the coordinate transformation f from the reference frame coordinates (u, v) to the distorted image frame coordinates (x, y) is obtained, say,

$$(x, y) = f(u, v).$$

This transformation is implemented by stepping through the integer coordinates (u, v) one-by-one and calculating the transformed (x, y) values. In general, the (x, y) coordinates are not integer values. Therefore, a new pixel must be estimated between existing pixels by an interpolation process which is called a *resampling*. The main resampling algorithms include the following:

- *Nearest-neighbor assignment* is the zero-order interpolation. For the value of each new pixel at (u, v) in the output image, this algorithm selects the value of the original pixel nearest to (x, y).
- *Bilinear resampling* is the first-order interpolation. This algorithm uses the four input pixels surrounding the point (x, y) to estimate the output pixel. Suppose that the values of the input pixels at four points

$$A = ([x], [y]), \qquad B = ([x] + 1, [y]),$$
$$C = ([x], [y] + 1), \qquad D = ([x] + 1, [y] + 1),$$

surrounding the point (x, y) are DN_A, DN_B, DN_C, and DN_D, respectively. Throughout this book, the notation $[x]$ is the integer part of x. The value of the output pixel at (u, v) is defined as

$$DN_{(u,v)} = [\Delta x\, DN_D + (1 - \Delta x)\, DN_C] \Delta y$$
$$+ [\Delta x\, DN_B + (1 - \Delta x)\, DN_A] (1 - \Delta y),$$

where $\Delta x = x - [x]$ and $\Delta y = y - [y]$.
- *Cubic resampling* is the second-order interpolation. The cubic resampling function is a piecewise cubic polynomial that approximates the sinc function.

4.6 IMAGE FUSION

Modern sensors have a set of multispectral bands and a higher spatial resolution panchromatic band. The goal of image fusion is to obtain a high-resolution multispectral image which combines the spectral characteristic of the low-resolution data with the spatial resolution of the panchromatic image. The process combining these data and producing images with high spatial and high spectral resolution is called *multisensor merging fusion*.

The simplest fusion method is similar to an HB filter and is used to add a weighted, HP-filtered version of the high-resolution panchromatic image to low-resolution multispectral images.

The *HF modulation* method is one of the main traditional fusion techniques. Suppose that PAN is a high-resolution panchromatic image and MS is a low-resolution multispectral image. Then a fused multispectral image in bank k

$$R_{ijk} = \frac{MS_{ijk}\, PAN_{ij}}{LP(PAN)_{ij}},$$

where MS_{ijk} is the lower-resolution multispectral image in bank k and $LP(PAN)_{ij}$ is a low-pass-filtered version of the higher-resolution image PAN_{ij}. Therefore, the fused multispectral image in bank k is proportional to the corresponding high-resolution image at each pixel.

The *empirical orthogonal function (EOF)-based fusion* method is used to replace the first EOF of low-resolution multispectral images with the high-resolution panchromatic image. Such replacement maximizes the effect of the panchromatic image in the fused image. In Chapter 6, we will state the EOF algorithm in detail.

The *wavelet-based fusion* method emerged over the past decade. The panchromatic image and each multispectral image are decomposed using discrete wavelet transform algorithm (see Section 2.3.2), and then the detail images of the multispectral image are replaced with those of the panchromatic image. Finally, the fused image is reconstructed by performing inverse discrete wavelet transform algorithm (see Section 2.3.2).

4.7 SUPERVISED AND UNSUPERVISED CLASSIFICATION

One of the main purposes of satellite remote sensing is to interpret the observed data and classify features. Thematic classification of a remote-sensing image consists of feature extraction/selection, training, and labeling. In order to classify a remote-sensing image into categories of interest, the classification algorithm needs to be trained to distinguish those categories from each other. The classification algorithm can be either supervised or unsupervised.

In supervised classification, the prototype pixel samples (i.e., training data) are already labeled by virtue of ground truth, existing maps, or photographic interpretation. On the basis of them, feature parameters for each category of interest are generated. Then various measures of the separation, such as city block distance, Euclidean distance, or angular distance, are applied in the classification. The city block distance is the L-distance:

$$|\mu_a - \mu_b| = \sum_{k=1}^{K} |\mu_{ak} - \mu_{bk}|,$$

where $\mu_a = (\mu_{a1}, \mu_{a2}, \ldots, \mu_{aK})$ and $\mu_b = (\mu_{b1}, \mu_{b2}, \ldots, \mu_{bK})$. The Euclidean distance is the L^2-distance:

$$\| \mu_a - \mu_b \| = \left(\sum_{k=1}^{K} (\mu_{ak} - \mu_{bk})^2 \right)^{1/2}.$$

The angular distance is the arc-cosine of the normalized inner product of the two vectors:

$$\arccos \left(\frac{\mu_a \cdot \mu_b}{\| \mu_a \| \| \mu_b \|} \right).$$

In addition to these distance measures, the normalized city block measure, the Mahalanobis separability measure, the transformed divergence measure, and the Jeffries-Matusita distance are also used. The normalized city block measure is proportional to the separation of the class means and is inversely proportional to their standard deviations. The Mahalanobis separability measure is a multivariate generalization of the Euclidean measure for normal distributions. The transformed divergence measure is based on the ratio of probabilities for two classes. The Jeffries-Matusita distance depends on the difference between the probability function for two classes.

In unsupervised classification, the prototype pixels are not labeled, but have been determined to have distinguishing intrinsic data characteristics. The K-means algorithm is the popular method used in unsupervised training. In this algorithm, an initial mean vector is first arbitrarily specified for each of K clusters. Each pixel of the training set is then assigned to the class whose mean vector is closest to the pixel vector. Therefore, this forms the first set of decision boundaries. A new set of cluster mean vectors is then calculated from this classification, and the pixels are reassigned accordingly. The iterations are terminated if there is no significant change in pixel assignments from one iteration to the next. The final cluster mean vector may be used to classify the entire remote-sensing image.

4.8 REMOTE SENSING OF ATMOSPHERIC CARBON DIOXIDE

Knowledge of present carbon sources and sinks, including their spatial distribution and temporal variability, is essential for predicting future atmospheric concentrations and associated global climate change. Compared with in situ carbon dioxide measurements, space-based instruments can provide data for estimating atmospheric carbon dioxide levels and their variability over a vast region on continuous spatial and temporal intervals.

Scanning Imaging Absorption Spectrometer for Atmospheric Chartography (SCIAMACHY) onboard *Envisat*, which was launched by the European Space Agency in March 2002, is the first instrument to retrieve the carbon dioxide column density from space. By observing backscattered, reflected, transmitted, or emitted radiation from the atmosphere and Earth's surface, from March 2002 to April 2012, SCIAMACHY acquired a lot of data on the global distribution of atmospheric carbon dioxide.

Thermal and Near Infrared Sensor for Carbon Observation (TANSO) onboard *Greenhouse Gases Observing Satellite (GOSAT)*, which was launched successfully by the Japanese Aerospace Exploration Agency (JAXA) on January 23, 2009, has been operating properly since launch. TANSO consists of a Fourier transform spectrometer (TANSO-FTS) and a cloud and aerosol imager (TANSO-CAI). It enables the precise monitoring of the density of carbon dioxide by combining global observation data sent from space with data obtained on land, and with simulation models.

NASA's *Orbiting Carbon Observatory-2* (*OCO-2*) was launched from the Vandenberg Air Force Base in California on a dedicated Delta II rocket in July 2014. It has a planned operational life of 2 years. It will collect the first space-based global measurements of atmospheric carbon dioxide with the precision, resolution, and coverage needed to characterize its sources and sinks on regional scales and quantify their variability over the seasonal cycle. In addition, a Chinese carbon dioxide observation satellite (TanSat) will be launched in 2015.

In remote sensing of atmospheric carbon dioxide, the continuum interpolated band ratio (CIBR) method is a key technique commonly used for retrieval of atmospheric carbon dioxide concentrations. The CIBR index is defined as the ratio of radiances received by the sensor at absorbing wavelength λ_0 with respect to the linear combination of radiances at two nonabsorbing neighboring wavelengths λ_1 and λ_2:

$$\text{CIBR} = \frac{L_{\lambda_0}}{c_1 L_{\lambda_1} + c_2 L_{\lambda_2}},$$

where

$$c_1 = \frac{\lambda_2 - \lambda_0}{\lambda_2 - \lambda_1},$$

$$c_2 = \frac{\lambda_0 - \lambda_1}{\lambda_2 - \lambda_1},$$

and L_λ are the atmosphere radiance values at wavelength λ. The CIBR index can be used to measure the strength of absorption by the atmospheric carbon dioxide column. Again, combining it with CIBR-concentration curves, one can estimate carbon dioxide concentration.

4.9 MODERATE RESOLUTION IMAGING SPECTRORADIOMETER DATA PRODUCTS AND CLIMATE CHANGE

Moderate Resolution Imaging Spectroradiometer (MODIS) is a sensor operating on the *Terra* and *Aqua* satellites, which were launched by NASA in December 1999 and May 2002, respectively. *Terra*'s orbit around Earth is timed so that it passes from north to south across the equator in the morning, while *Aqua* passes south to north over the equator in the afternoon. The main MODIS data products are available from several sources:

(1) MODIS level 1 data, geolocation, cloud mask, and atmosphere products: http://ladsweb.nascom.nasa.gov/.
(2) MODIS land products: https://lpdaac.usgs.gov/.
(3) MODIS cryosphere products: http://nsidc.org/daac/modis/index.html.
(4) MODIS ocean color and sea surface temperature products: http://oceancolor.gsfc.nasa.gov/.

These data products derived from MODIS observations describe features of the land, oceans, and the atmosphere, so they are being used widely in climate change research. For example, Alpert et al. (2012) used MODIS monthly aerosol data to study recent decadal trends of aerosol optical depth (AOD) over 189 of the largest cities in the world. They revealed the increasing AOD trends over the largest cities in the Indian subcontinent, the Middle East, and northern China can be clearly seen. By contrast, megacities in Europe, the northeast of the USA, and Southeast Asia show mainly declining AOD trends.

PROBLEMS

4.1 For a low-pass filter with the weights

$$\frac{1}{25} \begin{bmatrix} +1 & +1 & +1 & +1 & +1 \\ +1 & +1 & +1 & +1 & +1 \\ +1 & +1 & +1 & +1 & +1 \\ +1 & +1 & +1 & +1 & +1 \\ +1 & +1 & +1 & +1 & +1 \end{bmatrix},$$

find its complementary 5×5 HP filter.

4.2 Under the condition of Problem 4.1, if the HB filter is created by

$$HB(x, y; k) = \text{Image}(x, y) + k \cdot HP(x, y),$$

find the 5×5 HB filters for $k = 1, 2, 3$.

4.3 Give the least-squares solution for the linear polynomial distortion model.

4.4 Compare supervised classification with unsupervised classification.

4.5 Try to download some MODIS data products.

BIBLIOGRAPHY

Alpert, P., Shvainshtein, O., Kishcha, P., 2012. AOD trends over megacities based on space monitoring using MODIS and MISR. Am. J. Clim. Change 1, 117-131.

Amalins, K., Zhang, Y., Dare, P., 2007. Wavelet based image fusion techniques—an introduction, review and comparison. ISPRS J. Photogramm. Remote Sens. 62, 249-263.

Bruegge, C.J., Conel, J.E., Margolis, J.S., Green, R.O., Toon, G., Carrere, V., Holm, R.G., Hoover, G., 1990. In-situ atmospheric water-vapor retrieval in support of AVIRIS validation. SPIE 1298, 150-163.

Crevoisier, C., Chedin, A., Matsueda, H., Machida, T., Armante, R., Scott, N.A., 2009. First year of upper tropospheric integrated content of CO_2 from IASI hyperspectral infrared observations. Atmos. Chem. Phys. 9, 4797-4810.

Duda, R.D., Hart, P.E., 1973. Pattern Classification and Scene Analysis. John Wiley & Sons, New York.

Holliger, J.P., Peirce, J.L., Poe, G.A., 1990. SSM/I instrument evaluation. IEEE Trans. Geosci. Remote Sens. 28, 781-790.

Keeling, R.F., Piper, S.C., Bollenbacher, A.F., Walker, J.S., 2009. Atmospheric CO_2 record from sites in the SIO air sampling network. In trends: a compendium of data on global change;

carbon dioxide information analysis center. Oak Ridge National Laboratory, U.S. Department of Energy, Oak Ridge, USA.

Maddy, E.S., Barnet, C.D., Goldberg, M., Sweeney, C., Liu, X., 2008. CO_2 retrievals from the atmospheric infrared sounder: methodology and validation. J. Geophys. Res. 113, 11301.

Park, S.K., Schowengerdt, R.A., 1983. Image reconstruction by parametric cubic convolution. Comput. Vision Graph. Image Process. 20, 258-272.

Prasad, P., Rastogi, S., Singh, R.P., Panigrahy, S., 2014. Spectral modelling near the 1.6 µm window for satellite based estimation of CO_2. Spectrochim. Acta 117, 330-339.

Raupach, M.R., Marland, G., Ciais, P., Le Quere, C., Canadell, J.G., Klepper, G., Field, C.B., 2007. Global and regional drivers of accelerating CO_2 emissions. Proc. Natl Acad. Sci. USA 104, 10288-10293.

Stephens, B.B., Gurney, K.R., Tans, P.P., Sweeney, C., Peters, W., Bruhwiler, L., Ciais, P., Ramonet, M., Bousquet, P., Nakazawa, T., Aoki, S., Machida, T., Inoue, G., Vinnichenko, N., Lloyd, J., Jordan, A., Heimann, M., Shibistova, O., Langenfeds, R.L., Steele, L.P., Francey, R.J., Denning, A.S., 2007. Weak northern and strong tropical land carbon uptake from vertical profiles of atmospheric CO_2. Science 316, 1732-1735.

Strow, L.L., Hannon, S.E., Machado, S.D., Motteler, H.E., Tobin, D.C., 2006. Validation of the atmospheric infrared sounder radiative transfer algorithm. J. Geophys. Res. Atmos. 111, D09S06.

Tahl, S., Schonermark, M.V., 1998. Determination of the column water vapour of the atmosphere using backscattered solar radiation measured by the modular optoelectronic scanner. Int. J. Remote Sens. 19, 3223-3236.

Wolberg, G., 1990. Digital Image Warping. IEEE Computer Society Press, Los Alamitos, CA.

Chapter 5

Basic Probability and Statistics

Basic probability theory and statistics have a wide application in climate change research, ranging from the mean climate state and uncertainty of climatic parameters to the dynamics of the climate system. They provide powerful tools for climatologists to explain and analyze climatic data as well as to model and predict climate change. In this chapter, we will introduce the basic theory and methods in probability and statistics and their applications.

5.1 PROBABILITY SPACE, RANDOM VARIABLES, AND THEIR DISTRIBUTIONS

Starting from Kolmogorov's axioms of probability, we discuss random variables and their distributions.

A probability space consists of three components Ω, F, P. The component Ω is the sample space. Each element $\omega \in \Omega$ is called an *outcome*. The component F is called the *event space*. It is a set of subsets of Ω and satisfies the following conditions:

(i) $\Omega \in F$;
(ii) If $A \in F$, then $\Omega - A \in F$;
(iii) If $A_1, A_2, \cdots \in F$, then $\bigcup_{k \in \mathbb{Z}_+} A_k \in F$.

The component P is a *probability measure* on F satisfying the following conditions:

(i) $P(A) \geq 0$ for all $A \in F$;
(ii) $P(\Omega) = 1$;
(iii) If $A_1, A_2, \cdots \in F$, then $P(\bigcup_{k \in \mathbb{Z}_+} A_k) = \sum_{k \in \mathbb{Z}_+} P(A_k)$, where $A_i \bigcap A_j = \emptyset (i \neq j)$.

For example, (i) $\Omega = \{1, 2, \ldots, N\}$ and $F = \{$all subsets of $\Omega, \emptyset\}$, and $P(A) = \frac{1}{N} \cdot$ (cardinality of A), where $A \in F$, then (Ω, F, P) is a probability space. (ii) Ω is the interior of the circle with center $(0, 0)$ and radius 1 and $F = \{$all measurable subsets of $\Omega, \emptyset\}$, and $P(A) =$ measure of A, where $A \in F$, then (Ω, F, P) is a probability space.

Mathematical and Physical Fundamentals of Climate Change

Let a probability space (Ω, F, P) be given. A random variable is defined as a function X from Ω to the real axis $\mathbb{R}$ that satisfies $\{\omega \in \Omega, X(\omega) \leq c\} \in F$ for any number $c \in \mathbb{R}$. Its *cumulative distribution function* is defined by

$$F(x) = P(X \leq x) \quad (x \in \mathbb{R}).$$

It satisfies the following properties:

(i) F is nondecreasing.
(ii) $F(x) \to 1 (x \to +\infty)$ and $F(x) \to 0 (x \to -\infty)$.
(iii) F is right continuous.

Conversely, each function F satisfying these three conditions must be a cumulative distribution function of some random variable.

5.1.1 Discrete Random Variables

An integer-valued random variable is called a *discrete random variable*. Let X be a discrete random variable. Then its cumulative distribution function is

$$F(x) = P(X \leq x) = \sum_{k \leq x} p(k) \quad (k \in \mathbb{Z}),$$

where $x \in \mathbb{R}$ and $p(k) = P(X = k)$. $\{p(k)\}_{k \in \mathbb{Z}}$ is called the *probability mass function*. It is clear that

$$p(k) \geq 0 (k \in \mathbb{Z}), \quad \sum_k p(k) = 1.$$

Its expectation (mean) and variance are defined, respectively, as

$$E[X] = \sum_k k p(k), \quad \text{Var}(X) = \sum_k (k - \mu)^2 p(k), \tag{5.1}$$

where $\mu = E[X]$.

Two famous discrete random variables are as follows:

(i) Bernoulli random variable with a parameter $0 < r < 1$. Its probability mass function is defined by

$$p(k) = \begin{cases} \frac{n!}{k!(n-k)!} r^k (1 - r)^{n-k}, & k = 0, 1, \ldots, n, \\ 0, & \text{otherwise}. \end{cases}$$

Then the expectation $E[X] = nr$ and the variance $\text{Var} X = nr(1 - r)$.

(ii) Poisson random variable with a parameter $\lambda > 0$. Its probability mass function is defined by

$$p(k) = \frac{\lambda^k e^{-\lambda}}{k!} \quad (k > 0), \quad p(0) = e^{-\lambda}, \quad p(k) = 0 \quad (k < 0).$$

Then the expectation and the variance are, respectively,

$$E[X] = \sum_0^\infty kp(k) = \sum_1^\infty \frac{k\lambda^k}{k!}e^{-\lambda} = \lambda,$$

$$\text{Var}(X) = \sum_0^\infty (k-\lambda)^2 p(k) = \sum_0^\infty (k-\lambda)^2 \frac{\lambda^k e^{-\lambda}}{k!} = \lambda.$$

5.1.2   Continuous Random Variables

A real-valued random variable is called a *continuous random variable*. Let X be a continuous random variable. Then its cumulative distribution function is $F(x) = P(X \le x)$. If there exists a non-negative integrable function $p(x)$ such that

$$F(x) = \int_{-\infty}^x p(t)\, dt \quad (x \in \mathbb{R}),$$

then $p(x)$ is called the *probability density function*. If $p(x)$ is a continuous function, then $dF/dx = p(x)$ and $p(x)$ satisfies

$$p(x) \ge 0, \quad \int_{\mathbb{R}} p(x)\, dx = 1, \quad P(x_1 < X \le x_2) = \int_{x_1}^{x_2} p(t)\, dt.$$

Its expectation (mean) and variance are defined, respectively, by

$$E[X] = \int_{\mathbb{R}} x p(x)\, dx, \quad \text{Var}(X) = \int_{\mathbb{R}} (x-\mu)^2 p(x)\, dx \quad (\mu = E[X]). \quad (5.2)$$

Several famous continuous random variables are as follows:

(i) Uniform random variable. Its probability density function is defined by

$$p(x) = \begin{cases} \frac{1}{b-a}, & a \le x \le b, \\ 0, & \text{otherwise.} \end{cases}$$

Then the expectation $E[X] = \frac{1}{2}(a+b)$ and the variance $\text{Var}(X) = \frac{1}{12}(b-a)^2$.

(ii) Gaussian random variable with parameters μ and σ. Its probability density function is defined by

$$p(x) = \frac{1}{\sqrt{2\pi\sigma^2}} e^{-\frac{(x-\mu)^2}{2\sigma^2}} \quad (x \in \mathbb{R}).$$

Then the expectation $E[X] = \mu$ and the variance $\text{Var}(X) = \sigma^2$. It is also called the *normal random variable*. Denote it by $N(\mu, \sigma^2)$.

(iii) Gamma random variable with parameters $\alpha > 0$ and $\beta > 0$. Its probability density function is defined by

$$p(x) = \frac{1}{\Gamma(\alpha)\beta^\alpha} x^{\alpha-1} e^{-x/\beta} \quad (0 < x < \infty).$$

Then the expectation $E[X] = \alpha\beta$ and the variance $\text{Var}(X) = \alpha\beta^2$.

5.1.3 Properties of Expectations and Variances

The expectation of a function $f(X)$ with respect to a random variable X is defined as

$$E[f(X)] = \sum_k f(k)p(k) \quad \text{if } X \text{ is discrete,}$$

$$E[f(X)] = \int_{\mathbb{R}} f(x)p(x)\,dx \quad \text{if } X \text{ is continuous.}$$

By (5.1), and (5.2), it follows that the variance can be defined in terms of the expectation as

$$\text{Var}(X) = E[(X - EX)^2].$$

If, for arbitrary constants a and b,

$$P(X \le a, Y \le b) = P(X \le a)P(Y \le b),$$

the random variables X and Y are called *independent*.

Property. Let X and Y be random variables and c and d be constants. Then we have the following:

(i) (Linearity). $E[cX + dY] = c\,E[X] + d\,E[Y]$.
(ii) (Schwarz inequality). $E[XY] \le (E[X^2])^{1/2}(E[Y^2])^{1/2}$.
(iii) (Preservation of order). If $X \le Y$, then $E[X] \le E[Y]$.
(iv) If X and Y are independent, then $E[XY] = E[X]E[Y]$.
(v) $\text{Var}(X) = E[X^2] - (EX)^2$.
(vi) If X and Y are independent, then $\text{Var}(X + Y) = \text{Var}(X) + \text{Var}(Y)$.

Proof. It is easy to prove by (5.1) and (5.2) that (i)-(iv) hold. Here we prove only (v) and (vi). By the definition of the variance and (i),

$$\text{Var}(X) = E[(X - EX)^2] = E[X^2 - 2(EX)X + (EX)^2]$$
$$= E[X^2] - 2(EX)(EX) + (EX)^2.$$

So $\text{Var}(X) = E[X^2] - (EX)^2$, i.e., (v) holds.

By (v), $\text{Var}(X + Y) = E[(X + Y)^2] - (E[X + Y])^2$. However, by (i), we get

$$E[(X + Y)^2] = E[X^2] + 2E[XY] + E[Y^2],$$
$$(E[X + Y])^2 = (EX + EY)^2 = (EX)^2 + 2(EX)(EY) + (EY)^2,$$

and since X and Y are independent, $E[XY] = E[X] \cdot E[Y]$. Therefore,

$$\text{Var}(X + Y) = (E[X^2] - (EX)^2) + (E[Y^2] - (EY)^2).$$

Again by (v), $\text{Var}(X + Y) = \text{Var}(X) + \text{Var}(Y)$, i.e., (vi) holds. $\qquad\square$

5.1.4   Distributions of Functions of Random Variables

Suppose that a random variable Y is a function of a random variable X, i.e., $Y = g(X)$. Then the probability density function of Y can be determined by the probability density function of X.

Example 5.1. Given a random transform $Y = X^2$, if X has probability density function $p_X(x)$, we want to find the probability density function $p_Y(y)$ of Y. Denote the distribution functions of X and Y by $F_X(x)$ and $F_Y(y)$, respectively. For $y > 0$,

$$F_Y(y) = P(Y \leq y) = P(X^2 \leq y) = P(-\sqrt{y} \leq X \leq \sqrt{y}) = F_X(\sqrt{y}) - F_X(-\sqrt{y}).$$

From this and the probability density function of Y, $p_Y(y) = \frac{dF_Y(y)}{dy}$, it follows that

$$p_Y(y) = \frac{1}{2\sqrt{y}}(p_X(\sqrt{y}) + p_X(-\sqrt{y})) \quad (y > 0).$$

For $y < 0$, $F_Y(y) = 0$, and so $p_Y(y) = 0$. Thus,

$$p_Y(y) = \begin{cases} \frac{1}{2\sqrt{y}}(p(\sqrt{y}) + p(-\sqrt{y})), & y \geq 0, \\ 0, & y < 0. \end{cases}$$

When X is a Gaussian random variable $N(0, 1)$, its probability density function is $p_X(x) = \frac{1}{\sqrt{2\pi}} e^{-x^2/2} (x \in \mathbb{R})$. So

$$p_Y(y) = \begin{cases} \frac{1}{\sqrt{2\pi}} y^{-1/2} e^{-y/2}, & y > 0, \\ 0, & y < 0, \end{cases}$$

which is called the χ^2-*distribution* with one degree of freedom.

Theorem 5.1. *Let a random variable X have the probability density function $p_X(x)$ and let $g(x)$ be a differentiable function on $\mathbb{R}$. If $g'(x) > 0 (x \in \mathbb{R})$ and $a = \lim_{x \to -\infty} g(x)$ and $b = \lim_{x \to \infty} g(x)$, then $Y = g(X)$ has the probability density function*

$$p_Y(y) = \begin{cases} p_X(h(y))h'(y), & a < y < b, \\ 0, & otherwise, \end{cases} \tag{5.3}$$

where $h(y)$ is the inverse function of $g(x)$.

Proof. By the assumption, $g(x)$ is a monotone function on $\mathbb{R}$ and $a < g(x) < b (x \in \mathbb{R})$, and its inverse function $h(y)$ exists. Therefore,

$$F_Y(y) = P(g(X) \leq y) = 0 \quad (y \leq a),$$

$$F_Y(y) = P(g(X) \leq y) = 1 \quad (y \geq b),$$

$$F_Y(y) = P(g(X) \leq y) = P(X \leq h(y)) = F_X(h(y)) \quad (a < y < b).$$

From this and $p_Y(y) = \frac{dF_Y(y)}{dy}$, we get (5.3). $\qquad \square$

5.1.5 Characteristic Functions

The characteristic function of a random variable X is defined by

$$\Phi_X(\omega) := E[e^{i\omega X}] = \int_{\mathbb{R}} e^{i\omega x} p(x)\,dx,$$

where $p(x)$ is the probability density function of X.

By the definition of the Fourier transform in Chapter 1, we know that the characteristic function of X is the conjugate of the Fourier transform of $p(x)$, i.e.,

$$\Phi_X(\omega) = \overline{\hat{p}}(\omega).$$

This implies that for two random variables X, Y with probability density functions p_X, p_Y, if $\Phi_X(\omega) = \Phi_Y(\omega)$, then $p_X = p_Y$.

The characteristic functions of some random variables are as follows:

(i) Let X be a uniform random variable with probability density function

$$p(x) = \begin{cases} \frac{1}{b-a}, & a \le x \le b, \\ 0, & \text{otherwise.} \end{cases}$$

Then its characteristic function is

$$\Phi_X(\omega) = \int_{\mathbb{R}} p(x) e^{i\omega x}\,dx = \int_a^b \frac{1}{b-a} e^{i\omega x}\,dx = \frac{e^{i\omega b} - e^{i\omega a}}{i\omega(b-a)} \quad (\omega \ne 0).$$

(ii) Let X be an exponential random variable with probability density function

$$p(x) = \begin{cases} \lambda e^{-\lambda x}, & x \ge 0, \\ 0, & x < 0. \end{cases}$$

Then its characteristic function is

$$\Phi_X(\omega) = \int_{\mathbb{R}} p(x) e^{i\omega x}\,dx = \lambda \int_0^\infty e^{-(\lambda - i\omega)x}\,dx = \frac{\lambda}{\lambda - i\omega} \quad (\omega \in \mathbb{R}).$$

(iii) Let X be a normal random variable with mean μ and variance σ^2. Then its characteristic function is

$$\Phi_X(\omega) = e^{i\mu\omega - \frac{1}{2}\omega^2\sigma^2}. \tag{5.4}$$

Proposition 5.1. *Let X and Y be two independent random variables with probability density functions $p(x)$ and $q(y)$, respectively, and let $Z = X + Y$. Then*

(i) *$\Phi_Z(\omega) = \Phi_X(\omega) \cdot \Phi_Y(\omega)$;*

(ii) *the random variable Z has the probability density function $\gamma(z)$ and $\gamma(z) = (p * q)(z)$.*

(iii) *$\Phi_{cX}(\omega) = \Phi_X(c\omega)$, where c is a constant.*

Proof. The characteristic function of $Z = X + Y$ is

$$\Phi_Z(\omega) = E\left[e^{i\omega(X+Y)}\right] = E[e^{i\omega X} \cdot e^{i\omega Y}].$$

Since X and Y are independent,

$$\Phi_Z(\omega) = E[e^{i\omega X} \cdot e^{i\omega Y}] = \int_{\mathbb{R}} e^{i\omega x} p(x)\, dx \cdot \int_{\mathbb{R}} e^{i\omega y} q(y)\, dy = \Phi_X(\omega) \cdot \Phi_Y(\omega).$$

So we get (i).

By $\Phi_X(\omega) = \overline{\widehat{p}}(\omega)$ and $\Phi_Y(\omega) = \overline{\widehat{q}}(\omega)$ and (i),

$$\Phi_Z(\omega) = \overline{\widehat{p}}(\omega) \cdot \overline{\widehat{q}}(\omega).$$

By the convolution property of the Fourier transform, it follows that

$$\widehat{p}(\omega) \cdot \widehat{q}(\omega) = (p * q)^{\wedge}(\omega) = \int_{\mathbb{R}} (p * q)(z) e^{-i\omega z}\, dz,$$

and so

$$\Phi_Z(\omega) = \int_{\mathbb{R}} (p * q)(z) e^{i\omega z}\, dz.$$

By the definition of the characteristic function, the random variable Z has the probability density function $\gamma(z)$ and $\gamma(z) = (p * q)(z)$. So we get (ii).

Let $W = cX$, Then

$$P(a < W < b) = P\left(\frac{a}{c} \le X \le \frac{b}{c}\right) = \int_{a/c}^{b/c} p(x)\, dx = \frac{1}{c} \int_{a}^{b} p\left(\frac{x}{c}\right)\, dx.$$

Let the probability density function of W be $\beta(x)$. Then, for any $a < b$,

$$\int_{a}^{b} \beta(x)\, dx = \frac{1}{c} \int_{a}^{b} p\left(\frac{x}{c}\right)\, dx.$$

This implies that $\beta(x) = \frac{1}{c} p\left(\frac{x}{c}\right)$ and

$$\Phi_{cX}(\omega) = \int_{\mathbb{R}} e^{i\omega x} \beta(x)\, dx = \int_{\mathbb{R}} e^{i\omega x} \frac{1}{c} p\left(\frac{x}{c}\right)\, dx = \int_{\mathbb{R}} e^{ic\omega x} p(x)\, dx = \Phi_X(c\omega).$$

$\square$

Corollary 5.1. *Let the random variable X_1 be $N(\mu_1, \sigma_1^2)$ and the random variable X_2 be $N(\mu_2, \sigma_2^2)$. If X_1 and X_2 are independent, then the random variable $X = X_1 + X_2$ is $N(\mu, \sigma^2)$, where $\mu = \mu_1 + \mu_2$ and $\sigma^2 = \sigma_1^2 + \sigma_2^2$.*
Proof. By Proposition 5.1(i), $\Phi_X(\omega) = \Phi_{X_1}(\omega)\Phi_{X_2}(\omega)$. By (5.4),

$$\Phi_{X_1}(\omega) = e^{i\mu_1\omega - \frac{\omega^2\sigma_1^2}{2}},$$

$$\Phi_{X_2}(\omega) = e^{i\mu_2\omega - \frac{\omega^2\sigma_2^2}{2}}.$$

Therefore,

$$\Phi_X(\omega) = e^{i(\mu_1+\mu_2)\omega-\frac{1}{2}\omega^2(\sigma_1^2+\sigma_2^2)}.$$

This implies that the random variable X is $N(\mu_1 + \mu_2, \sigma_1^2 + \sigma_2^2)$. □

From Corollary 5.1, it follows that if $X_1, X_2, \ldots, X_n$ are independent Gaussian random variables, then any linear combination $a_1 X_1 + \cdots + a_n X_n$ is a Gaussian random variable.

5.2 JOINTLY DISTRIBUTED RANDOM VARIABLES

The *joint distribution function* of two random variables X, Y is defined by

$$F(x, y) = P(X \le x, Y \le y) \quad (x, y \in \mathbb{R}).$$

It follows immediately from this definition that $F(x, +\infty) = F_X(x)$ and $F(+\infty, y) = F_Y(y)$, where F_X and F_Y are called *marginal distribution functions* of X and Y, respectively. If X and Y are independent, then

$$F(x, y) = F_X(x)F_Y(y).$$

Let X be a discrete random variable with $P(X = k) = p(k)(k \in \mathbb{Z})$ and Y be a discrete random variable with $P(Y = k) = q(k)(k \in \mathbb{Z})$. The *joint probability mass function* of X and Y is defined by $\gamma(k, l) = P(X = k, Y = l)$ $(k, l \in \mathbb{Z})$.

Proposition 5.2. *Let X and Y have joint probability mass function $\gamma(k, l)(k, l \in \mathbb{Z})$. Then*

$$\sum_l \gamma(k, l) = p(k) \quad (k \in \mathbb{Z}), \quad \sum_k \gamma(k, l) = q(l) \quad (l \in \mathbb{Z}).$$

In fact, by the definition,

$$\sum_l \gamma(k, l) = \sum_l P(X = k, Y = l) = P(X = k, Y \in \mathbb{Z}) = p(k).$$

Similarly, $\sum_k \gamma(k, l) = q(l)(l \in \mathbb{Z})$.

If X and Y are independent, then $P(X = k, Y = l) = P(X = k)P(Y = l)$, i.e.,

$$\gamma(k, l) = p(k)q(l) \quad (k, l \in \mathbb{Z}).$$

Now we consider two continuous random variables X and Y with probability density functions $p(x)$ and $q(y)$, respectively.

If a non-negative integrable function $\gamma(x, y)$ exists such that the joint distribution function of X and Y satisfies

$$F(x, y) = \int_{-\infty}^{x} \int_{-\infty}^{y} \gamma(x, y) \, dx \, dy \quad (x, y \in \mathbb{R}),$$

then $\gamma(x, y)$ is called the joint probability density function of X and Y. Given $Y = y$, the conditional probability density function of X is defined as

$$\gamma(x|y) = \frac{\gamma(x,y)}{q(y)} \quad (x \in \mathbb{R}).$$

Given $X = x$, the conditional probability density function of Y is defined as

$$\gamma(y|x) = \frac{\gamma(x,y)}{p(x)} \quad (y \in \mathbb{R}).$$

Let $u(X)$ be a function of X. Given $Y = y$, the conditional expectation of $u(X)$ is defined as

$$E[u(X)|y] = \int_{\mathbb{R}} u(x)\gamma(x|y)\,dx.$$

Especially, $E[X|y] = \int_{\mathbb{R}} x\gamma(x|y)\,dx$.

Proposition 5.3. *Let X and Y have the joint probability density function $\gamma(x,y)$. Then*

$$p(x) = \int_{\mathbb{R}} \gamma(x,y)\,dy \quad (x \in \mathbb{R}), \qquad q(y) = \int_{\mathbb{R}} \gamma(x,y)\,dx \quad (y \in \mathbb{R}).$$

If X and Y are independent, then their joint probability density function $\gamma(x,y) = p(x)q(y)$.

The covariance of two random variables is defined as

$$\text{Cov}(X,Y) = E[(X - EX)(Y - EY)].$$

Especially, if $X = Y$,

$$\text{Cov}(X,X) = E[(X - EX)^2] = \text{Var}(X).$$

Therefore, the covariance is a generalization of variance. From the definition of covariance, it follows that

$$\text{Cov}(X,Y) = E[XY] - (E[X])(E[Y]).$$

The covariance has the following properties.

Property. Let X, Y, and Z be random variables and c and d be constants. Then

(i) $\text{Cov}(c,X) = 0$;

(ii) $\text{Cov}(cX + dY, Z) = c\text{Cov}(X,Z) + d\text{Cov}(Y,Z)$;

(iii) if X and Y are independent, then $\text{Cov}(X,Y) = 0$.

Finally, we discuss the transform of joint random variables.

Theorem 5.2. *Let two random variables X and Y have the joint probability density function $p_{X_1 X_2}(x_1, x_2)$. Denote*

$$A = \{(x_1, x_2) \in \mathbb{R}^2 | p_{X_1 X_2}(x_1, x_2) \neq 0\}.$$

Two bivariate differentiable functions $g_1(x_1, x_2)$ and $g_2(x_1, x_2)$ are such that the transform

$$U: \quad y_1 = g_1(x_1, x_2), \quad y_2 = g_2(x_1, x_2)$$

maps the set A in the $x_1 x_2$-plane onto the set B in the $y_1 y_2$-plane one-to-one. Denote the inverse transform

$$U^{-1}: \quad x_1 = h_1(y_1, y_2), \quad x_2 = h_2(y_1, y_2).$$

Then $Y_1 = g_1(X_1, X_2)$ and $Y_2 = g_2(X_1, X_2)$ have the joint probability density function

$$p_{Y_1 Y_2}(y_1, y_2) = \begin{cases} p_{X_1 X_2}(h_1(y_1, y_2), h_2(y_1, y_2))|J(y_1, y_2)|, & (y_1, y_2) \in B, \\ 0, & otherwise, \end{cases}$$

(5.5)

where $J(y_1, y_2)$ is the Jacobian determinant and

$$J(y_1, y_2) = \begin{vmatrix} \frac{\partial x_1}{\partial y_1} & \frac{\partial x_1}{\partial y_2} \\ \frac{\partial x_2}{\partial y_1} & \frac{\partial x_2}{\partial y_2} \end{vmatrix}.$$

Proof. Let $\widetilde{A}$ be a subset of A, and let $\widetilde{B}$ be the one-to-one mapping of $\widetilde{A}$ under transform U. From $Y_1 = g_1(X_1, X_2)$ and $Y_2 = g_2(X_1, X_2)$, it follows that $(X_1, X_2) \in \widetilde{A}$ is equivalent to $(Y_1, Y_2) \in \widetilde{B}$. Therefore,

$$P((Y_1, Y_2) \in \widetilde{B}) = P((X_1, X_2) \in \widetilde{A}) = \int \int_{\widetilde{A}} p_{X_1 X_2}(x_1, x_2)\, dx_1\, dx_2.$$

It has been proved in calculus that under the transform U^{-1}: $x_1 = h_1(y_1, y_2)$, $x_2 = h_2(y_1, y_2)$,

$$\int \int_{\widetilde{A}} p_{X_1 X_2}(x_1, x_2)\, dx_1\, dx_2 =$$
$$\int \int_{\widetilde{B}} p_{X_1 X_2}(h_1(y_1, y_2), h_2(y_1, y_2))|J(y_1, y_2)|\, dy_1\, dy_2.$$

Therefore, for each subset $\widetilde{B} \subset B$,

$$P((y_1, y_2) \in \widetilde{B}) = \int \int_{\widetilde{B}} p_{X_1 X_2}(h_1(y_1, y_2), h_2(y_1, y_2))|J(y_1, y_2)|\, dy_1\, dy_2.$$

This implies (5.5). □

The *joint distribution* of N random variables $X_1, X_2, \ldots, X_N$ is defined as

$$F(x_1, x_2, \ldots, x_N) = P(X_1 \le x_1, X_2 \le x_2, \ldots, X_N \le x_N).$$

If each X_k is discrete, then their *joint probability mass function* is defined as

$$p(k_1 \cdots k_N) = P(X_1 = k_1, X_2 = k_2, \ldots, X_N = k_N).$$

If each X_k is continuous and there exists an N-variate function $p(x_1, \ldots, x_N)$ such that

$$F(x_1, \ldots, x_N) = \int_{-\infty}^{x_1} \cdots \int_{-\infty}^{x_N} p(x_1, \ldots, x_N)\, dx_1 \cdots dx_N,$$

then $p(x_1, \ldots, x_N)$ is called the *joint probability density function* of $X_1, X_2, \ldots, X_N$.

Suppose that $X_1, X_2, \ldots, X_N$ have the joint probability density function $\gamma(x_1, \ldots, x_N)$. Then the *conditional joint probability density function*

$$\gamma(x_1, \ldots, x_m | x_{m+1}, \ldots, x_N)$$

of $X_1, X_2, \ldots, X_m$ given $X_{m+1}, \ldots, X_N$ is defined as

$$\gamma(x_1, \ldots, x_m | x_{m+1}, \ldots, x_N) = \frac{\gamma(x_1, \ldots, x_N)}{\gamma(x_{m+1}, \ldots, x_N)},$$

where $\gamma(x_{m+1}, \ldots, x_N)$ is the joint probability density function of $X_{m+1}, \ldots, X_N$. Take the transforms

$$Y_1 = g_1(X_1, \ldots, X_n), \quad Y_2 = g_2(X_1, \ldots, X_n), \quad \ldots \quad, Y_n = g_n(X_1, \ldots, X_n).$$

The inverse transforms are

$$X_1 = h_1(Y_1, \ldots, Y_n), \quad X_2 = h_2(Y_1, \ldots, Y_n), \quad \ldots \quad, X_n = h_n(Y_1, \ldots, Y_n).$$

The joint probability density function of $Y_1, \ldots, Y_n$ is

$$p_Y(y_1, \ldots, y_n) = p_X(h_1(y_1, \ldots, y_n), \ldots, h_n(y_1, \ldots, y_n)) |J(y_1, \ldots, y_n)|,$$

where

$$J(y_1, \ldots, y_n) = \begin{vmatrix} \frac{\partial h_1}{\partial y_1} & \cdots & \frac{\partial h_1}{\partial y_n} \\ \vdots & \vdots & \vdots \\ \frac{\partial h_n}{\partial y_1} & \cdots & \frac{\partial h_n}{\partial y_n} \end{vmatrix}$$

is the Jacobian determinant.

5.3   CENTRAL LIMIT THEOREM AND LAW OF LARGE NUMBERS

Limit theorems are the fundamental theory in probability and statistics. The central limit theorem shows that the distribution of a sum of independent and identically distributed random variables converges toward a normal distribution as the number of random variables increases. So it establishes the dominant role of the normal distribution. The law of large numbers shows that the average of a large number of samples is a good estimator for the expectation.

For a sequence of random variables, there are various definitions of convergence.

Definition 5.1. Let $\{X_n\}_{n \in \mathbb{Z}_+}$ be a sequence of random variables and X be a random variable.

(i) $\{X_n\}_{n \in \mathbb{Z}_+}$ converges to X in probability if $\lim_{n \to \infty} P(|X_n - X| \geq \epsilon) = 0$ for $\epsilon > 0$. Denote $X_n \xrightarrow{p} X$.

(ii) $\{X_n\}_{n \in \mathbb{Z}_+}$ converges to X in the mean square sense if $E[X_n^2] < \infty (n \in \mathbb{Z}_+)$ and $\lim_{n \to \infty} E[(X_n - X)^2] = 0$. Denote $X_n \xrightarrow{\text{m.s.}} X$.

(iii) $\{X_n\}_{n\in\mathbb{Z}_+}$ converges to X in distribution if $\lim_{n\to\infty} F_{X_n}(x) = F_X(x)$ at all $x \in \mathbb{R}$. Denote $X_n \xrightarrow{d} X$.

Proposition 5.4. *If $X_n \xrightarrow{m.s.} X$, then $X_n \xrightarrow{p} X$. If $X_n \xrightarrow{p} X$, then $X_n \xrightarrow{d} X$.*

Proposition 5.5. *A sequence $\{X_n\}_{n\in\mathbb{Z}_+}$ of random variables converges to a random variable X in distribution if and only if their characteristic functions satisfy*

$$\Phi_{X_n}(t) \to \Phi_X(t) \quad (t \in \mathbb{R}) \quad as\ n \to \infty.$$

This proposition states that convergence in distribution is equivalent to pointwise convergence of characteristic functions.

Central Limit Theorem. *Suppose that $\{X_n\}_{n\in\mathbb{Z}_+}$ is a sequence of independent and identically distributed random variables and each X_n has the expectation μ and variance σ^2. Let $S_n = \sum_1^n X_k$. Then the sequence of random variables*

$$\left\{ \frac{S_n - n\mu}{\sqrt{n}} \right\}$$

converges to a Gaussian random variable $X \sim N(0, \sigma^2)$ in distribution.

Proof. Without loss of generality, we assume $\mu = 0$. Consider the characteristic function of $S_n = \sum_1^n X_k$.

Since $\{X_n\}_{n\in\mathbb{Z}_+}$ are independent, by Proposition 5.1(i), we have

$$\Phi_{S_n}(\omega) = \Phi_{X_1}(\omega) \cdots \Phi_{X_n}(\omega).$$

By the definition of characteristic functions,

$$\Phi_{X_k}(\omega) = \int_{\mathbb{R}} p_k(x)e^{i\omega x}\, dx \quad (k \in \mathbb{Z}),$$

where p_k is the probability density function of X_k. From this and the assumption $p_k(x) = p_1(x)(k \in \mathbb{Z})$, we get

$$\Phi_{X_k}(\omega) = \int_{\mathbb{R}} p_1(x)e^{i\omega x}\, dx = \Phi_{X_1}(\omega) \quad (k \in \mathbb{Z}), \tag{5.6}$$

and so $\Phi_{S_n}(\omega) = \big(\Phi_{X_1}(\omega)\big)^n$.

Let $Z_n = S_n/\sqrt{n}$. Then, by Proposition 5.1(iii),

$$\Phi_{Z_n}(\omega) = \Phi_{S_n}\left(\frac{\omega}{\sqrt{n}}\right) = \left(\Phi_{X_1}\left(\frac{\omega}{\sqrt{n}}\right)\right)^n.$$

By use of Taylor's theorem $f(\omega) \sim f(0) + f'(0)\omega + f''(0)\frac{\omega^2}{2}$, it follows that

$$\Phi_{X_1}\left(\frac{\omega}{\sqrt{n}}\right) \sim \Phi_{X_1}(0) + \Phi'_{X_1}(0)\frac{\omega}{\sqrt{n}} + \Phi''_{X_1}(0)\frac{\omega^2}{2n}. \tag{5.7}$$

It is clear from (5.6) that

$$\Phi_{X_1}(0) = \int_{\mathbb{R}} p_1(x)\, dx = 1.$$

Differentiating both sides of (5.6), we get

$$\Phi'_{X_1}(\omega) = \int_{\mathbb{R}} ix e^{i\omega x} p_1(x)\, dx,$$

$$\Phi''_{X_1}(\omega) = \int_{\mathbb{R}} (ix)^2 e^{i\omega x} p_1(x)\, dx.$$

By the assumption, $E[X] = 0$ and $\sigma^2 = \text{Var}(X_1) = E[X_1^2] - (EX)^2 = E[X^2]$. Let $\omega = 0$ in these two equalities. Then

$$\Phi'_{X_1}(0) = \int_{\mathbb{R}} ix p_1(x)\, dx = iE[X] = 0,$$

$$\Phi''_{X_1}(0) = -\int_{\mathbb{R}} x^2 p_1(x)\, dx = -E[X_1^2] = -\sigma^2.$$

From this and (5.7), it follows that

$$\Phi_{X_1}\left(\frac{\omega}{\sqrt{n}}\right) \sim 1 - \frac{\omega^2}{2n}\sigma^2,$$

and so

$$\Phi_{Z_n}(\omega) = \left(\Phi_{X_1}\left(\frac{\omega}{\sqrt{n}}\right)\right)^n \sim \left(1 - \frac{\omega^2}{2n}\sigma^2\right)^n.$$

By $(1 - \frac{x}{n})^n \to e^{-x}$,

$$\Phi_{Z_n}(\omega) \sim \left(1 - \frac{1}{n}\left(\frac{\sigma^2}{2}\omega^2\right)\right)^n \to e^{-\frac{\sigma^2}{2}\omega^2} \quad \text{as } n \to \infty. \tag{5.8}$$

By (5.4), the characteristic function of a random variable $X \sim N(0, \sigma^2)$ is $e^{-\frac{\sigma^2}{2}\omega^2}$. From this and (5.8), it follows by properties of the covariance that the sequence $\{S_n\}_{n\in\mathbb{Z}_+}$ converges to a Gaussian random variable $X \sim N(0, 1)$ in distribution. $\quad\square$

Law of Large Numbers. *Suppose that $\{X_n\}_{n\in\mathbb{Z}_+}$ is a sequence of independent random variables and each X_k has the same expectation μ. Let $S_n = \sum_1^n X_k (n \in \mathbb{Z}_+)$. If $\text{Var}(X_k) \le c$ for all k and some constant c, then the sequence $\{S_n/n\}_{n\in\mathbb{Z}_+}$ converges to μ in the mean square sense.*

Proof. Since $E[X_n] = \mu (n \in \mathbb{Z}_+)$,

$$E\left[\frac{S_n}{n}\right] = \frac{1}{n}E[X_1 + X_2 + \cdots + X_n] = \frac{1}{n}(E[X_1] + E[X_2] + \cdots + E[X_n]) = \mu.$$

By the definition of variance,

$$E\left[\left(\frac{S_n}{n} - \mu\right)^2\right] = \text{Var}\left(\frac{S_n}{n}\right) = \frac{1}{n^2}\text{Var}(S_n).$$

Since $X_n (n \in \mathbb{Z}_+)$ are independent and $\text{Var} X_k \leq c$,

$$\text{Var} S_n = \text{Var}(X_1 + X_2 + \cdots + X_n) = \text{Var} X_1 + \text{Var} X_2 + \cdots + \text{Var} X_n \leq nc,$$

and so

$$E\left[\left(\frac{S_n}{n} - \mu\right)^2\right] \leq \frac{c}{n}.$$

So $\lim_{n \to \infty} E\left[\left(\frac{S_n}{n} - \mu\right)^2\right] = 0$, i.e., $\frac{S_n}{n} \overset{\text{m.s.}}{\to} \mu$. $\qquad\qquad\square$

Under conditions of this theorem, it follows immediately by Proposition 5.4 that

$$\frac{S_n}{n} \overset{\text{p}}{\to} \mu, \quad \frac{S_n}{n} \overset{\text{d}}{\to} \mu.$$

5.4 MINIMUM MEAN SQUARE ERROR

Now we introduce the approximation theory of random variables. For a random variable X and a constant b, the mean square error $E[(X - b)^2]$ is

$$E[(X - b)^2] = E[(X - EX + EX - b)^2]$$
$$= E[(X - EX)^2] + 2E[(X - EX)(EX - b)] + E[(EX - b)^2].$$

Note that $EX - b$ is constant. Then

$$E[(X - b)^2] = E[(X - EX)^2] + (E[X] - b)^2 = \text{Var} X + (EX - b)^2.$$

This equality shows that the mean square error $E[(X - b)^2]$ attains the minimal value if and only if $b = EX$. Therefore, the minimum mean square error is $\text{Var} X$.

For two random variables X, Y, if $E[XY] = 0$, then X, Y are called *orthogonal*. Denote this by $X \perp Y$. From $E[(X - EX)b] = bE[X - EX] = 0$, it follows that $(X - EX) \perp b$ for all constants b. This result can be generalized to the following orthogonality principle.

Let $Z_1, \ldots, Z_N$ be N random variables with $E[Z_k^2] < \infty (k = 1, \ldots, N)$. Denote by τ_N all linear combinations of $Z_1, \ldots, Z_N$, i.e.,

$$\tau_N := \{c_1 Z_1 + c_2 Z_2 + \cdots + c_N Z_N, \quad \text{where each } c_k \text{ is a constant}\}.$$

Let X be a random variable and $X \notin \tau_N$. We will choose a $Z^* \in \tau_N$ such that $E[(X - Z^*)^2]$ is minimal, i.e.,

$$E[(X - Z^*)^2] \leq E[(X - Z)^2] \quad \text{for all } Z \in \tau_N,$$

where Z^* is called the *best approximation* of X in τ_N.

Theorem 5.3. *Let X be a random variable. Then $Z^* \in \tau_N$ is the best approximation of X if and only if $(X - Z^*) \perp Z$ for $Z \in \tau_N$, and the minimum mean square error is*

$$E[(X - Z^*)^2] = E[X^2] - E[(Z^*)^2].$$

Proof. Suppose that $Z^* \in \tau_N$ is such that $(X - Z^*) \perp Z$ for all $Z \in \tau_N$. Then, for $Z \in \tau_N$,

$$Z^* - Z \in \tau_N, \quad (X - Z^*) \perp (Z^* - Z),$$

i.e., $E[(X - Z^*)(Z^* - Z)] = 0$. So

$$
\begin{aligned}
E[(X - Z)^2] &= E[(X - Z^* + Z^* - Z)^2] \\
&= E[(X - Z^*)^2 + 2(X - Z^*)(Z^* - Z) + (Z^* - Z)^2] \\
&= E[(X - Z^*)^2] + E[(Z^* - Z)^2].
\end{aligned}
$$

From $E[(Z^* - Z)^2] \geq 0$, it follows that

$$E[(X - Z^*)^2] \leq E[(X - Z)^2].$$

Therefore, Z^* is the best approximation of X in τ_N. From this and $E[(X - Z^*)Z^*] = 0$, it follows that

$$E[(X - Z^*)^2] = E[X^2] - E[(Z^*)^2].$$

Conversely, suppose that Z^* is the best approximation of X. Then, for all $Z \in \tau_N$,

$$E[(X - Z^*)^2] \leq E[(X - Z)^2].$$

Let $Z \in \tau_N$ and $c \in \mathbb{R}$. Then $Z^* + cZ \in \tau_N$. So

$$E[(X - Z^*)^2] \leq E[(X - (Z^* + cZ))^2].$$

However, the right-hand side is

$$
\begin{aligned}
E[(X - (Z^* + cZ))^2] &= E[((X - Z^*) - cZ)^2] \\
&= E[(X - Z^*)^2] - 2cE[(X - Z^*)Z] + c^2 E[Z^2].
\end{aligned}
$$

Therefore,

$$\gamma(c) := -2cE[(X - Z^*)Z] + c^2 E[Z^2] \geq 0.$$

Since Z^* is the best approximation of X, $\gamma(c)$ attains the minimal value 0 at $c = 0$. This implies $\gamma'(0) = 0$, and so $E[(X - Z^*)Z] = 0$. Therefore, $(X - Z^*) \perp Z$. $\qquad\square$

Let X and Y be two random variables. We choose constants a and b such that $E[r^2]$ is minimal, where $r = X - (aY + b)$. By Theorem 5.3, we need only to choose a and b such that $r \perp Y$ and $r \perp 1$, i.e., to choose a and b such that $E[r] = 0$ and $E[rY] = 0$. By the formula of covariance

$$\text{Cov}(r, Y) = E[rY] - E[r]E[Y],$$

it follows that $r \perp Y$ and $r \perp 1$ are equivalent to $E[r] = 0$ and $\text{Cov}(r, Y) = 0$. So

$$E[r] = E[X] - aE[Y] - b.$$

Since $E[r] = 0$, we get

$$b = E[X] - aE[Y], \quad r = X - E[X] - a(Y - E[Y]).$$

So

$$\text{Cov}(r, Y) = \text{Cov}(X, Y) - \text{Cov}(EX, Y) - a\text{Cov}(Y, Y) + a\text{Cov}(EY, Y).$$

Since EX and EY are constants, by property (i) of the covariance,

$$\text{Cov}(EX, Y) = \text{Cov}(EY, Y) = 0.$$

From this and $\text{Cov}(Y, Y) = \text{Var}\, Y$, it follows that

$$\text{Cov}(r, Y) = \text{Cov}(X, Y) - a\text{Var}(Y).$$

Since $\text{Cov}(r, Y) = 0$,

$$a = \frac{\text{Cov}(X, Y)}{\text{Var}\, Y}.$$

Denote the best approximation of X in the set $\{aY + b\}$ by $S_X(Y)$. Then $S_X(Y)$ is given by the following proposition

Proposition 5.6. *The best approximation of X in the set $\{aY + b\}$, where a and b are constants, is*

$$S_X(Y) = E[X] + \frac{\text{Cov}(X, Y)}{\text{Var}(Y)}(Y - EY).$$

5.5 χ^2-DISTRIBUTION, t-DISTRIBUTION, AND F-DISTRIBUTION

Starting from this section, we begin to study statistics. A function of samples from one or more random variables that does not depend on any unknown parameter is called a *statistic*. For example,

$$\overline{X} = \frac{1}{N} \sum_{1}^{N} X_k,$$

$$S^2 = \frac{1}{N} \sum_{1}^{N} (X_k - \overline{X})^2,$$

where $X_1, \ldots, X_N$ are N samples from a random variable X. They are used for estimating the mean (expectation) and variance of a random variable X.

Proposition 5.7. *Let $X_1, \ldots, X_N$ be N samples of a random variable X with mean μ and variance σ^2. Then $E[\overline{X}] = \mu$ and $\operatorname{Var} \overline{X} = \frac{\sigma^2}{N}$, and $E[S^2] = (1 - \frac{1}{N})\sigma^2$.*

Proof. It is clear that

$$E[\overline{X}] = \frac{1}{N} \sum_1^N E[X_k] = \frac{1}{N} \sum_1^N \mu = \mu$$

and

$$\operatorname{Var} \overline{X} = E[(\overline{X} - \mu)^2] = E\left[\left(\frac{1}{N} \sum_1^N (X_k - \mu)\right)^2\right].$$

Since $X_1, \ldots, X_N$ are independent,

$$E[(X_k - \mu)(X_l - \mu)] = E[X_k - \mu]E[X_l - \mu] = 0 \quad (k \neq l),$$

$$\operatorname{Var} \overline{X} = \frac{1}{N^2} \sum_1^N \operatorname{Var} X_k = \frac{\sigma^2}{N}.$$

For S^2, we have

$$E[S^2] = E\left[\frac{1}{N} \sum_1^N (X_k - \overline{X})^2\right] = \frac{1}{N} E\left[\sum_1^N X_k^2 - N\overline{X}^2\right]$$

$$= \frac{1}{N} \sum_1^N E[X_k^2] - E[\overline{X}^2] = \frac{1}{N} \sum_1^N (\sigma^2 + \mu^2) - \frac{\sigma^2}{N} - \mu^2 = \left(1 - \frac{1}{N}\right)\sigma^2.$$

$\square$

The following distributions play a key role in statistics:

(i) χ^2-*Distribution.* Let $X_1, X_2, \ldots, X_N$ be N independent normal random variables and $X_k \sim N(0, 1)(k = 1, \ldots, N)$. Then the probability density function of the random variable,

$$\chi^2 = \sum_1^N X_k^2,$$

is called a χ^2-*distribution* with N degrees of freedom. Denote it by $\chi^2(N)$.

Proposition 5.8. *The probability density function of $\chi^2(N)$ is*

$$p(x) = \begin{cases} \dfrac{x^{N/2-1}e^{-\frac{x}{2}}}{\Gamma(\frac{N}{2})2^{\frac{N}{2}}}, & 0 < x < \infty, \\ 0, & otherwise, \end{cases}$$

where the gamma function $\Gamma(\frac{N}{2}) = \int_0^\infty x^{\frac{N}{2}-1} e^{-x} \, dx$.

Proof. We prove Proposition 5.8 only in the case $N = 2$.

Since $X \sim N(0, 1)$, Example 5.1 has shown that the probability density function of X^2 is

$$q(t) = \begin{cases} \frac{1}{\sqrt{2\pi}} t^{-\frac{1}{2}} e^{-\frac{t}{2}}, & t > 0, \\ 0, & t < 0. \end{cases} \tag{5.9}$$

By Proposition 5.1(ii), the probability density function of $\chi^2(2) = X_1^2 + X_2^2$ is

$$p(x) = \int_{\mathbb{R}} q(x - t) q(t) \, dt.$$

By (5.9), the integrand $q(x - t)q(t)$ is not zero only if $0 < t < x$. Therefore,

$$p(x) = \frac{1}{2\pi} \int_0^x (x - t)^{-\frac{1}{2}} e^{-\frac{x-t}{2}} t^{-\frac{1}{2}} e^{-\frac{t}{2}} \, dt = \frac{1}{2\pi} e^{-\frac{x}{2}} \int_0^x x^{-\frac{1}{2}} \left(1 - \frac{t}{x}\right)^{-\frac{1}{2}} t^{-\frac{1}{2}} \, dt$$

$$= \frac{1}{2\pi} e^{-\frac{x}{2}} \int_0^1 (1 - u)^{-\frac{1}{2}} u^{-\frac{1}{2}} \, du = \frac{1}{2} e^{-\frac{x}{2}},$$

i.e., the probability density function of $\chi^2(2)$ is $\frac{1}{2} e^{-\frac{x}{2}}$. □

Property. The χ^2-distribution has the following properties:

(i) If $\chi_1^2 \sim \chi^2(n_1)$ and $\chi_2^2 \sim \chi^2(n_2)$, and χ_1^2, χ_2^2 are independent, then $\chi_1^2 + \chi_2^2 \sim \chi^2(n_1 + n_2)$.

(ii) If $\chi^2 \sim \chi^2(N)$, then the expectation $E[\chi^2] = N$ and the variance $\text{Var } \chi^2 = 2N$.

By $X_k \sim N(0, 1)$, it follows that

$$E[\chi^2] = E\left[\sum_1^N X_k^2\right] = \sum_1^N E[X_k^2] = N,$$

$$\text{Var } \chi^2 = \text{Var}\left(\sum_1^N X_k^2\right) = \sum_1^N \text{Var } X_k^2,$$

$$\text{Var } X_k^2 = E[X_k^4] - 1.$$

Since

$$E[X_k^4] = \frac{1}{\sqrt{2\pi}} \int_{\mathbb{R}} x^4 e^{-\frac{x^2}{2}} \, dx = \frac{2}{\sqrt{2\pi}} \int_0^\infty x^4 e^{-\frac{x^2}{2}} \, dx = 3,$$

we get $\text{Var } \chi^2 = 2N$.

(ii) *t-Distribution.* Define a random variable t by $t = \frac{W}{\sqrt{V/r}}$, where W and V are two independent random variables and $W \sim N(0, 1)$ and $V \sim \chi^2(r)$. The probability density function of the random variable t,

$$p_t(t) = \frac{\Gamma((r+1)/2)}{(1 + t^2/r)^{\frac{r+1}{2}} \sqrt{\pi r} \Gamma(r/2)} \quad (t \in \mathbb{R}),$$

is called a *t-distribution* with r degrees of freedom. Denote it by $t(r)$.

(iii) *F-Distribution*. Let U and V be two independent random variables and $U \sim \chi^2(r_1)$ and $V \sim \chi^2(r_2)$. Define a random variable F by $F = \frac{U}{r_1}/(\frac{V}{r_2})$. The probability density function of the random variable F,

$$p_F(x) = \begin{cases} \dfrac{x^{\frac{r_1}{2}-1} \Gamma\left(\dfrac{r_1 + r_2}{2}\right)(r_1 | r_2)^{\frac{r_1}{2}}}{\left(1 + \dfrac{r_1 x}{r_2}\right)^{\frac{r_1+r_2}{2}} \Gamma\left(\dfrac{r_1}{2}\right)\Gamma\left(\dfrac{r_2}{2}\right)}, & 0 < x < \infty, \\ 0, & \text{otherwise,} \end{cases}$$

is called an *F-distribution* with n_1 and n_2 degrees of freedom. Denote it by $F(r_1, r_2)$.

5.6   PARAMETER ESTIMATION

Suppose that a random variable X has a probability density function which depends on an unknown parameter θ, i.e.,

$$X \sim p(x, \theta) \quad (\theta \in \Omega).$$

We want to determine the parameter θ.

Let $X_1, \ldots, X_N$ be N samples from X. Define a statistic $\widetilde{\theta} = u(X_1, \ldots, X_N)$ such that if $x_1, \ldots, x_N$ are the observed experimental values of $X_1, \ldots, X_N$, then the number $u(x_1, \ldots, x_N)$ will be a good estimate of θ.

For example, let a normal random variable $X \sim N(\theta, 1)$ and $X_1, \ldots, X_N$ be its samples. Since each X_k has probability density function $\frac{1}{\sqrt{2\pi}} e^{-\frac{1}{2}(x_k - \theta)^2}$ and $X_1, \ldots, X_N$ are independent, the joint probability density function of $X_1, \ldots, X_N$ is

$$L(x_1, \ldots, x_N, \theta) = \left(\frac{1}{\sqrt{2\pi}}\right)^N e^{-\frac{1}{2}\sum_1^N (x_k - \theta)^2} \quad (\theta \in \mathbb{Z}).$$

We choose the parameter θ such that the function $L(x_1, \ldots, x_N, \theta)$ takes the maximal value. This maximizing value of θ is a good estimate of θ because it provides the largest probability of these particular samples. Consider

$$\log L(x_1, \ldots, x_N, \theta) = N \log \frac{1}{\sqrt{2\pi}} - \frac{1}{2} \sum_1^N (x_k - \theta)^2.$$

Since $\frac{\mathrm{d}\log L(x_1,\ldots,x_N,\theta)}{\mathrm{d}\theta} = 0$ is equivalent to $\sum_1^N (x_k - \theta) = 0$, i.e.,

$$\theta = \frac{1}{N}\sum_1^N x_k,$$

it is easily checked that L attains the maximal value at $\theta = \frac{1}{N}\sum_1^N x_k$. So $u(x_1,\ldots,x_N) = \frac{1}{N}\sum_1^N x_k$ is a good estimate of θ. The corresponding statistic $\widetilde{\theta}$ is equal to the mean of the samples, i.e.,

$$\widetilde{\theta} = u(X_1,\ldots,X_N) = \frac{1}{N}\sum_1^N X_k =: \overline{X}.$$

In general, the method of maximum likelihood is stated as follows:

(i) Let X be a random variable with the probability density function $p(x,\theta)$. Consider N samples $X_1,\ldots,X_N$. Their joint probability density function is called the *likelihood function* of the samples. Denote it by $L(x_1,\ldots,x_N,\theta)$, i.e.,

$$L(x_1,\ldots,x_N,\theta) = p(x_1,\theta)p(x_2,\theta)\cdots p(x_N,\theta).$$

If, when $\theta = u(x_1,\ldots,x_N)$, the likelihood function L takes the maximal value, then the statistic $u(X_1,\ldots,X_N)$ is called the *maximum likelihood estimator* of parameter θ.

(ii) Let $X, Y, \ldots, Z$ be several random variables with the joint probability density function $g(x, y, \ldots, z; \theta_1, \ldots, \theta_m)$, where $\theta_1, \ldots, \theta_m$ are parameters. If, when $\theta_k = u_k(x, y, \ldots, z)(k = 1, \ldots, m)$, g attains the maximal value, then the statistic $u_k(X, Y, \ldots, Z)(k = 1, \ldots, m)$ is called the *maximum likelihood estimator* of parameters $\theta_1, \ldots, \theta_m$.

Example 5.2. Let X be a random variable and $X \sim N(\theta_1, \theta_2)$, and let $X_1, \ldots, X_N$ be N samples of X. Then their likelihood function

$$L(x_1,\ldots,x_N,\theta_1,\theta_2) = \left(\frac{1}{\sqrt{2\pi\theta_2}}\right)^N e^{-\sum_1^N \frac{(x_k-\theta_1)^2}{2\theta_2}}.$$

Its logarithm is

$$\log L(x_1,\ldots,x_N,\theta_1,\theta_2) = -\frac{N}{2}\log(2\pi\theta_2) - \sum_1^N \frac{(x_k-\theta_1)^2}{2\theta_2},$$

and

$$\frac{\partial \log L(x_1,\ldots,x_N,\theta_1,\theta_2)}{\partial\theta_1} = \sum_1^N \frac{x_k-\theta_1}{\theta_2},$$

$$\frac{\partial \log L(x_1,\ldots,x_N,\theta_1,\theta_2)}{\partial\theta_2} = -\frac{N}{2\theta_2} + \sum_1^N \frac{(x_k-\theta_1)^2}{2\theta_2^2}.$$

Let $\frac{\partial \log L(x_1,\ldots,x_N,\theta_1,\theta_2)}{\partial \theta_1} = 0$ and $\frac{\partial \log L(x_1,\ldots,x_N,\theta_1,\theta_2)}{\partial \theta_2} = 0$. Then

$$\sum_{1}^{N}(x_k - \theta_1) = 0, \quad \sum_{1}^{N}(x_k - \theta_1)^2 = N\theta_2,$$

and so

$$\theta_1 = \frac{1}{N}\sum_{1}^{N} x_k = \overline{x}, \quad \theta_2 = \frac{1}{N}\sum_{1}^{N}(x_k - \overline{x})^2 = s^2.$$

Therefore, the maximum likelihood estimators of θ_1 and θ_2 are $\widetilde{\theta}_1 = \overline{X}$ and $\widetilde{\theta}_2 = S^2$, where

$$\overline{X} = \frac{1}{N}\sum_{1}^{N} X_k, \quad S^2 = \frac{1}{N}\sum_{1}^{N}(X_k - \overline{X})^2,$$

i.e., $\overline{X}$ and S^2 are the maximum likelihood estimators of the expectation θ_1 and the variance θ_2, respectively. It is clear that

$$E[\overline{X}] = \frac{1}{N}\sum_{1}^{N} E[X_k] = \frac{1}{N}\sum_{1}^{N}\theta_1 = \theta_1$$

and

$$E[S^2] = \frac{1}{N}\sum_{1}^{N} E\left[(X_k - \overline{X})^2\right] = \frac{1}{N}\sum_{1}^{N} E\left[(X_k - \theta_1 - \overline{X} + \theta_1)^2\right]. \quad (5.10)$$

Note that
$$E\left[(X_k - \theta_1 - \overline{X} + \theta_1)^2\right] =$$

$$E\left[(X_k - \theta_1)^2\right] - 2E\left[(X_k - \theta_1)(\overline{X} - \theta_1)\right] + E\left[(\overline{X} - \theta_1)^2\right].$$

The first term on the right-hand side is $E[(X_k - \theta_1)^2] = \theta_2$. Since

$$\overline{X} - \theta_1 = \frac{1}{N}\sum_{1}^{N}(X_j - \theta_1), \quad E[(X_k - \theta_1)(X_j - \theta_1)] = 0 \quad (k \neq j),$$

the second term on the right-hand side is

$$2E\left[(X_k - \theta_1)(\overline{X} - \theta_1)\right] = \frac{2}{N}\sum_{j=1}^{N} E\left[(X_j - \theta_1)(X_k - \theta_1)\right]$$

$$= \frac{2}{N}E\left[(X_k - \theta_1)^2\right] = \frac{2\theta_2}{N}$$

and the third term on the right-hand side is

$$E[(\overline{X} - \theta_1)^2] = E\left[\frac{1}{N^2}\sum_1^N(X_j - \theta_1)^2 + \frac{2}{N^2}\sum_{j \neq k}(X_j - \theta_1)(X_k - \theta_1)\right] = \frac{\theta_2}{N}.$$

So $E[(X_k - \theta_1 - \overline{X} + \theta_1)^2] = (1 - \frac{1}{N})\theta_2(k = 1, \ldots, N)$. From this and (5.10), we get

$$E[S^2] = \left(1 - \frac{1}{N}\right)\theta_2.$$

From this, we see that $E[S^2] \neq \theta_2$ but $\widetilde{\theta}_2 \to \theta_2$ in probability.

Any statistic whose expectation is equal to a parameter is called an *unbiased estimator* of this parameter, otherwise it is called a *biased estimator*. If a statistic converges in probability to a parameter, then it is called *a consistent estimator* of this parameter. Therefore, the statistic $\overline{X} = \frac{1}{N}\sum_1^N X_k$ is an unbiased estimator of the parameter θ_1, while the statistic $S^2 = \frac{1}{N}\sum_1^N (X_k - \overline{X})^2$ is a biased and consistent estimator of the parameter θ_2.

Example 5.3. Let X be a random variable with the probability density function

$$p(x) = \begin{cases} \lambda e^{-\lambda x}, & x > 0, \\ 0, & \text{otherwise.} \end{cases}$$

Then the likelihood function of N samples $X_1, \ldots, X_N$ is

$$L(x_1, \ldots, x_N, \lambda) = (\lambda e^{-\lambda x_1}) \cdots (\lambda e^{-\lambda x_N}) = \lambda^N e^{-\lambda \sum_1^N x_k}.$$

Its logarithm is

$$\log L(x_1, \ldots, x_N, \lambda) = N \log \lambda - \lambda \sum_1^N x_k.$$

Let

$$\frac{d \log L(x_1, \ldots, x_N, \lambda)}{d\lambda} = \frac{N}{\lambda} - \sum_1^N x_k = 0.$$

Then $\lambda^{-1} = \frac{1}{N}\sum_1^N x_k$. So the maximum likelihood estimator of λ is $\widetilde{\lambda} = (\overline{X})^{-1}$.

Another method of parameter estimates is based on the moments.

Let X be a random variable. Then $E[X^j]$ is called its *jth moment*. Let $X_1, \ldots, X_N$ be N samples from X. Then $A_j = \frac{1}{N}\sum_1^N X_k^j$ is called the *jth sample moment*. Clearly,

$$E[A_j] = \frac{1}{N}\sum_{k=1}^N E[X_k^j] = \frac{1}{N}\sum_{k=1}^N E[X^j] = E[X^j].$$

If X has the probability density function $p(x, \theta)$, where $\theta = (\theta_1, \ldots, \theta_l)$ are unknown parameters, then

$$E[X^j] = \int_{\mathbb{R}} x^j p(x, \theta) \, dx.$$

The system of equations:

$$\frac{1}{N} \sum_{k=1}^{N} X_k^j = E[X^j] \quad (j = 1, \ldots, l) \tag{5.11}$$

is a system of equations with l unknown parameters $\theta_1, \ldots, \theta_l$. Its solution $\tilde{\theta} = (\tilde{\theta}_1, \ldots, \tilde{\theta}_l)$ is called the *moment estimator*.

Example 5.4. Let X be a uniform random variable on $[a, b]$ and $X_1, \ldots, X_N$ be N samples of X. We want to find the moment estimators of a and b. Since the probability density function of X is

$$p(x, a, b) = \begin{cases} \frac{1}{b-a}, & a \leq x \leq b, \\ 0, & \text{otherwise,} \end{cases}$$

from Section 5.1.2(i), $E[X] = \frac{a+b}{2}$ and $\mathrm{Var}\, X = \frac{(b-a)^2}{12}$, it follows that

$$E[X^2] = \mathrm{Var}\, X + (E[X])^2 = \frac{(b-a)^2}{12} + \frac{(a+b)^2}{4}.$$

Let $j = 1, 2$ in the system of equations (5.11). Then

$$\frac{1}{N} \sum_{1}^{N} X_k = \frac{a+b}{2},$$

$$\frac{1}{N} \sum_{1}^{N} X_k^2 = \frac{(b-a)^2}{12} + \frac{(a+b)^2}{4}.$$

Therefore, the moment estimators of a and b are

$$\tilde{a} = \overline{X} - \sqrt{\frac{3}{N} \sum_{1}^{N} (X_i - \overline{X})^2} = \overline{X} - \sqrt{3S^2},$$

$$\tilde{b} = \overline{X} + \sqrt{\frac{3}{N} \sum_{1}^{N} (X_i - \overline{X})^2} = \overline{X} + \sqrt{3S^2},$$

where $\overline{X} = \frac{1}{N} \sum_{1}^{N} X_k$ and $S^2 = \frac{1}{N} \sum_{1}^{N} (X_k - \overline{X})^2$.

Example 5.5. Let X be a discrete random variable with the probability mass function $p(k) = \frac{n!}{k!(n-k)!} r^k (1-r)^{n-k}$ and $X_1, \ldots, X_N$ be N samples of X. We want to find the moment estimators of parameters n, r. From Section 5.1.1(i), $E[X] = nr$ and $\mathrm{Var}\, X = nr(1-r)$, it follows that

$$E[X^2] = \operatorname{Var} X + (E[X])^2 = nr(1 - r) + n^2 r^2.$$

Let $j = 1, 2$ in the system of equations (5.11). Then

$$\frac{1}{N} \sum_1^N X_k = nr,$$

$$\frac{1}{N} \sum_1^N X_k^2 = nr(1 - r) + n^2 r^2.$$

Therefore, the moment estimators of parameters n, r are

$$\widetilde{n} = \frac{\overline{X}^2}{\overline{X} - S^2}, \quad \widetilde{r} = 1 - \frac{S^2}{\overline{X}},$$

where $\overline{X} = \frac{1}{N} \sum_1^N X_k$ and $S^2 = \frac{1}{N} \sum_1^N (X_k - \overline{X})^2$.

5.7 CONFIDENCE INTERVAL

In statistics, the *confidence interval* is used for evaluating the reliability of the parameter estimate. A confidence interval with confidence level α for a parameter θ is defined as the interval $[a, b]$ satisfying

$$P(a < \theta < b) = \alpha.$$

Let X be a normal random variable with the known variance σ^2 and the unknown mean μ, and let $X_1, \ldots, X_N$ be N samples of X. From Section 5.6, we know that the maximum likelihood estimator is $\widetilde{\mu} = \overline{X}$. Since $\overline{X}$ is $N(\mu, \frac{\sigma^2}{N})$, $\frac{\sqrt{N}(\overline{X} - \mu)}{\sigma}$ is $N(0, 1)$. So there exists a number $Z_{\frac{\alpha}{2}}$ such that

$$P\left(-Z_{\frac{\alpha}{2}} < \frac{\sqrt{N}(\overline{X} - \mu)}{\sigma} < Z_{\frac{\alpha}{2}}\right) = \alpha.$$

This is equivalent to

$$P\left(\overline{X} - \frac{\sigma}{\sqrt{N}} Z_{\frac{\alpha}{2}} < \mu < \overline{X} + \frac{\sigma}{\sqrt{N}} Z_{\frac{\alpha}{2}}\right) = \alpha.$$

This means that the probability of the interval $\left(\overline{X} - \frac{\sigma}{\sqrt{N}} Z_{\frac{\alpha}{2}}, \overline{X} + \frac{\sigma}{\sqrt{N}} Z_{\frac{\alpha}{2}}\right)$ containing μ is α.

Suppose the experiment yields $X_1 = x_1, X_2 = x_2, \ldots, X_N = x_N$. The interval

$$\left(\overline{x} - \frac{\sigma}{\sqrt{N}} Z_{\frac{\alpha}{2}}, \quad \overline{x} + \frac{\sigma}{\sqrt{N}} Z_{\frac{\alpha}{2}}\right)$$

is called a confidence interval with confidence level α for the mean μ, where $\overline{x} = \frac{1}{N} \sum_1^N x_k$.

If we do not know the variance σ^2, how can we find a confidence interval for the mean μ?

Note that $\overline{X} = \frac{1}{N} \sum_1^N X_k$ and $S = \frac{1}{N} \sum_1^N (X_k - \overline{X})^2$. By a known result, the random variable $t = \frac{\sqrt{N-1}(\overline{X}-\mu)}{S}$ has a t-distribution with $N-1$ degrees of freedom. There exists a number $t_{\frac{\alpha}{2}} > 0$ such that $P(|t| < t_{\frac{\alpha}{2}}) = \alpha$. This is equivalent to

$$P\left(\overline{X} - \frac{t_{\frac{\alpha}{2}} S}{\sqrt{N-1}} < \mu < \overline{X} + \frac{t_{\frac{\alpha}{2}} S}{\sqrt{N-1}} \right) = \alpha.$$

Let the experimental value of X_k be $x_k (k = 1, \ldots, N)$ and

$$\overline{x} = \frac{1}{N} \sum_1^N x_k, \quad s^2 = \frac{1}{N} \sum_1^N (x_k - \overline{x})^2.$$

Then the interval $\left(\overline{x} - \frac{s}{\sqrt{N-1}} t_{\frac{\alpha}{2}}, \overline{x} + \frac{s}{\sqrt{N-1}} t_{\frac{\alpha}{2}} \right)$ is a confidence interval with confidence level α for the mean μ.

5.8   TESTS OF STATISTICAL HYPOTHESES

A statistical hypothesis is an assertion about the distribution of one or more random variables. If it determines completely the distribution, then it is called a *simple statistical hypothesis*, otherwise it is called a *composite statistical hypothesis*. For example, for a random variable $X \sim N(\theta, 1)$, the statistical hypothesis $H_0 : \theta = \theta_0$ is a simple statistical hypothesis, while statistical hypotheses $H_1 : \theta \leq \theta_0$ and $H_2 : \theta > \theta_0$ are composite statistical hypotheses.

For given samples, a test of a statistical hypothesis is a rule which leads to acceptance or rejection of the hypothesis. This test is always associated with a subset of the sample space which is called the *critical region*. When samples fall in the critical region, reject the hypothesis, otherwise accept the hypothesis. The probability that the sample falls in the critical region is called the *significance level*. For example, let X be a random variable with a probability density function $p(x, \theta)$, and let $X_1, \ldots, X_N$ be N samples of X. If H_0 is true, then $\theta = \theta_0$. Since $X_1, \ldots, X_N$ are independent, the joint probability density function of $X_1, \ldots, X_N$ is

$$p(x_1, \theta_0) p(x_2, \theta_0) \cdots p(x_N, \theta_0).$$

Take critical region $C \subset \mathbb{R}^N$. Then

$$P((X_1, \ldots, X_N) \in C) = \int_C p(x_1, \theta_0) \cdots p(x_N, \theta_0) \, dx_1 \cdots dx_N =: \tau_0$$

is the significance level. If $(x_1, \ldots, x_N) \in C$, the hypothesis is rejected. If $(x_1, \ldots, x_N) \notin C$, the hypothesis is accepted.

Another method of statistical hypothesis testing is the following likelihood ratio test.

Let $X_1, \ldots, X_N$ be independent random variables with probability density functions $p_k(x_k, \theta_1, \ldots, \theta_m)(k = 1, 2, \ldots, N)$. Let

$$\Omega = \{(\theta_1, \ldots, \theta_m) : \quad \theta_1, \ldots, \theta_m \in \mathbb{R}\}$$

be the total parameter space and ω be its subset. It is desired to test the hypothesis $H_0 : (\theta_1, \ldots, \theta_m) \in \omega$ against all alternative hypotheses. The likelihood function is

$$L(x_1, \ldots, x_N; \theta_1, \ldots, \theta_m) = \prod_1^N p_k(x_k, \theta_1, \ldots, \theta_m).$$

Denote

$$L(\widetilde{\omega}) = \max_{(\theta_1, \ldots, \theta_m) \in \omega} L(x_1, \ldots, x_N; \theta_1, \ldots, \theta_m),$$

$$L(\widetilde{\Omega}) = \max_{(\theta_1, \ldots, \theta_m) \in \Omega} L(x_1, \ldots, x_N, \theta_1, \ldots, \theta_m).$$

The likelihood ratio $\lambda(x_1, \ldots, x_N) = \frac{L(\widetilde{\omega})}{L(\widetilde{\Omega})}$.

Let $0 < \lambda_0 < 1$. The likelihood ratio test principle states that the hypothesis $H_0 : (\theta_1, \ldots, \theta_m) \in \omega$ is rejected if and only if $\lambda(x_1, \ldots, x_N) \leq \lambda_0$. The corresponding statistic is $\lambda(X_1, \ldots, X_N)$ and the significance level

$$\alpha = P(\lambda(X_1, \ldots, X_N) \leq \lambda_0 : H_0 \text{ is true}).$$

5.9 ANALYSIS OF VARIANCE

Let $X_1, \ldots, X_b$ be b independent normal random variables with means $\mu_1, \ldots, \mu_b$ and the same variance σ^2, respectively, where both μ_j and σ^2 are unknown. For each $X_j(j = 1, \ldots, b)$, let $X_{1j}, X_{2j}, \ldots, X_{aj}$ be its samples, i.e.,

$$
\begin{array}{cccc}
X_{11} & X_{12} & \ldots & X_{1b} \\
X_{21} & X_{22} & \ldots & X_{2b} \\
\vdots & \vdots & \ldots & \vdots \\
X_{a1} & X_{a2} & \ldots & X_{ab}
\end{array}
$$

It is desired to test the hypothesis

$$H_0 : \mu_1 = \mu_2 = \cdots = \mu_b = \mu \quad (\mu \text{ is unknown}).$$

We introduce $a + b + 1$ statistics:

$$\overline{X}_{..} = \frac{1}{ab} \sum_{i=1}^{a} \sum_{j=1}^{b} X_{ij},$$

$$\overline{X}_{i\cdot} = \frac{1}{b} \sum_{j=1}^{b} X_{ij} \quad (i = 1, \ldots, a),$$

$$\overline{X}_{\cdot j} = \frac{1}{a} \sum_{i=1}^{a} X_{ij} \quad (j = 1, \ldots, b),$$

where $\overline{X}_{\cdot\cdot}$ is the mean of all ab samples, $\overline{X}_{i\cdot}(i = 1, \ldots, a)$ is the mean of the ith row, and $\overline{X}_{\cdot j}(j = 1, \ldots, b)$ is the mean of the jth column. Denote the total variance as

$$S^2 = \frac{1}{ab} \sum_{i=1}^{a} \sum_{j=1}^{b} (X_{ij} - X_{\cdot\cdot})^2. \tag{5.12}$$

It can be rewritten in the form

$$abS^2 = \sum_{i=1}^{a} \sum_{j=1}^{b} \left((X_{ij} - \overline{X}_{i\cdot}) + (\overline{X}_{i\cdot} - \overline{X}_{\cdot\cdot}) \right)^2 = Q_1 + Q_2 + Q_3,$$

where

$$Q_1 = \sum_{i=1}^{a} \sum_{j=1}^{b} (X_{ij} - \overline{X}_{i\cdot})^2,$$

$$Q_2 = \sum_{i=1}^{a} \sum_{j=1}^{b} (\overline{X}_{i\cdot} - \overline{X}_{\cdot\cdot})^2,$$

$$Q_{12} = 2 \sum_{i=1}^{a} \sum_{j=1}^{b} (X_{ij} - \overline{X}_{i\cdot})(\overline{X}_{i\cdot} - \overline{X}_{\cdot\cdot}).$$

Clearly, $Q_2 = b \sum_{i=1}^{a} (\overline{X}_{i\cdot} - \overline{X}_{\cdot\cdot})^2$. Since $\sum_{j=1}^{b} (X_{ij} - \overline{X}_{i\cdot}) = b\overline{X}_{i\cdot} - b\overline{X}_{i\cdot} = 0$, we have $Q_{12} = 0$. So we get

$$abS^2 = \sum_{i=1}^{a} \sum_{j=1}^{b} (X_{ij} - \overline{X}_{i\cdot})^2 + b \sum_{i=1}^{a} (\overline{X}_{i\cdot} - \overline{X}_{\cdot\cdot})^2 = Q_1 + Q_2.$$

To explain the statistical characteristics of abS^2 and Q_1, Q_2, we need the following two propositions.

Proposition 5.9. *Let* $X_1, \ldots, X_N$ *be* N *independent random variables and each* $X_k \sim N(\mu_k, \sigma^2)(k = 1, \ldots, N)$. *Let* $S^2 = \frac{1}{N} \sum_{1}^{N} (X_k - \overline{X})^2$, *where* $\overline{X} = \frac{1}{N} \sum_{1}^{N} X_k$. *Then* $\frac{NS^2}{\sigma^2} \sim \chi^2(N - 1)$.

Proposition 5.10. *Let* $X_1, \ldots, X_N$ *be* N *independent random variables and each* $X_k \sim N(\mu_k, \sigma^2)(k = 1, \ldots, N)$. *Let* $Q = Q_1 + Q_2$, *where both* Q_1 *and* Q_2

are of quadratic form in $X_1, \ldots, X_N$. *If* $\frac{Q}{\sigma^2} \sim \chi^2(\gamma)$, $\frac{Q_1}{\sigma^2} \sim \chi^2(\beta)$, *and* $Q_2 \geq 0$, *then* Q_1 *and* Q_2 *are independent and* $\frac{Q_2}{\sigma^2} \sim \chi^2(\gamma - \beta)$.

By Proposition 5.9, it follows from (5.12) that $\frac{abS^2}{\sigma^2} \sim \chi^2(ab - 1)$. Similarly,

$$\frac{1}{\sigma^2} \sum_{i=1}^{b} (X_{ij} - \overline{X}_{i\cdot})^2 \sim \chi^2(b - 1).$$

By the additive property of χ^2-distribution,

$$\frac{Q_1}{\sigma^2} \sim \chi^2(a(b - 1)).$$

By Proposition 5.10, Q_1 and Q_2 are independent and

$$\frac{Q_2}{\sigma^2} \sim \chi^2(a - 1).$$

In (5.12), replacing $(X_{ij} - \overline{X}_{\cdot\cdot})$ by $(X_{ij} - \overline{X}_{\cdot j}) + (\overline{X}_{\cdot j} - \overline{X}_{\cdot\cdot})$, we get

$$abS^2 = \sum_{j=1}^{b} \sum_{i=1}^{a} (X_{ij} - \overline{X}_{\cdot j})^2 + a \sum_{j=1}^{b} (\overline{X}_{\cdot j} - \overline{X}_{\cdot\cdot})^2 =: Q_3 + Q_4. \tag{5.13}$$

Similarly, by Propositions 5.9 and 5.10, it follows that Q_3 and Q_4 are independent and

$$\frac{Q_3}{\sigma^2} \sim \chi^2(b(a - 1)), \quad \frac{Q_4}{\sigma^2} \sim \chi^2(b - 1).$$

Therefore,

$$\frac{Q_2/(a - 1)}{Q_1/(a(b - 1))} = \frac{Q_2/(\sigma^2(a - 1))}{Q_1/(\sigma^2 a(b - 1))} \sim F(a - 1, a(b - 1))$$

and

$$\frac{Q_4/(b - 1)}{Q_3/(b(a - 1))} = \frac{Q_4/(\sigma^2(b - 1))}{Q_3/(\sigma^2 b(a - 1))} \sim F(b - 1, b(a - 1)).$$

We know that $X_{1j}, \ldots, X_{aj}$ are samples from X_j with the unknown mean $\mu_j (j = 1, \ldots, b)$ and the unknown but common variance σ^2. It is desired to test the hypothesis

$$H_0 : \mu_1 = \mu_2 = \cdots = \mu_b = \mu,$$

where μ is unspecified. The total parameter space is

$$\Omega = \{(\mu_1, \ldots, \mu_b, \sigma^2) : \mu_1, \ldots, \mu_b \in \mathbb{R}, \quad 0 < \sigma^2 < \infty\}.$$

Let

$$\omega = \{(\mu_1, \ldots, \mu_b, \sigma^2) : \mu_1 = \cdots = \mu_b = \mu, \quad \mu \in \mathbb{R}, \quad 0 < \sigma^2 < \infty\}.$$

The likelihood functions are

$$L(\omega) = \left(\frac{1}{2\pi\sigma^2}\right)^{\frac{ab}{2}} e^{-\frac{1}{2\sigma^2}\sum_{i=1}^{a}\sum_{j=1}^{b}(x_{ij}-\mu)^2},$$

$$L(\Omega) = \left(\frac{1}{2\pi\sigma^2}\right)^{\frac{ab}{2}} e^{-\frac{1}{2\sigma^2}\sum_{i=1}^{a}\sum_{j=1}^{b}(x_{ij}-\mu_j)^2}.$$

When

$$\mu = \frac{1}{ab}\sum_{i=1}^{a}\sum_{j=1}^{b}x_{ij} =: \bar{x}_{..},$$

$$\sigma^2 = \frac{1}{ab}\sum_{i=1}^{a}\sum_{j=1}^{b}(x_{ij} - \bar{x}_{..})^2 =: \sigma_0^2.$$

$L(\omega)$ takes the maximal value. When

$$\mu_j = \frac{1}{a}\sum_{i=1}^{a}x_{ij} =: \bar{x}_{.j} \quad (j = 1, \ldots, b),$$

$$\sigma^2 = \frac{1}{ab}\sum_{i=1}^{a}\sum_{j=1}^{b}(x_{ij} - \bar{x}_{.j})^2 =: \sigma_1^2,$$

$L(\Omega)$ takes the maximal value. A direct calculation shows that the likelihood ratio λ satisfies

$$\lambda^{\frac{2}{ab}} = \frac{\sum_{i=1}^{a}\sum_{j=1}^{b}(x_{ij} - \bar{x}_{.j})^2}{\sum_{i=1}^{a}\sum_{j=1}^{b}(x_{ij} - \bar{x}_{..})^2}. \tag{5.14}$$

Therefore, the maximal likelihood estimators of μ and σ_0^2 are, respectively,

$$\bar{X}_{..} = \frac{1}{ab}\sum_{i=1}^{a}\sum_{j=1}^{b}X_{ij},$$

$$S^2 = \frac{1}{ab}\sum_{i=1}^{a}\sum_{j=1}^{b}(X_{ij} - \bar{X}_{..})^2,$$

and the maximal likelihood estimators of $\mu_j (j = 1, \ldots, b)$ and σ_1^2 are, respectively,

$$\bar{X}_{.j} = \frac{1}{a}\sum_{i=1}^{a}X_{ij} \quad (j = 1, \ldots, b),$$

$$\frac{Q_3}{ab} = \frac{1}{ab} \sum_{i=1}^{a} \sum_{j=1}^{b} (X_{ij} - \overline{X}_{.j})^2.$$

If the hypothesis H_0 is true, the random variables $\{X_{ij}\}_{i=1,...,a;j=1,...,b}$ are samples of b normal random variables with mean μ and variance σ^2. By (5.14) and $abS^2 = Q_3 + Q_4$, the statistic defined by $\lambda^{\frac{2}{ab}}$ may be written

$$\frac{Q_3}{abS^2} = \frac{Q_3}{Q_3 + Q_4} = \frac{1}{1 + \frac{Q_4}{Q_3}}.$$

If $\lambda \leq \lambda_0$, we reject the hypothesis H_0. We need to find λ_0 which corresponds to a desired significance level α, i.e.,

$$\alpha = P\left(\frac{1}{1 + Q_4/Q_3} \leq \lambda_0^{\frac{2}{ab}}\right) = P\left(\frac{Q_4/(b-1)}{Q_3/(b(a-1))} \geq c\right),$$

where $c = \frac{b(a-1)}{b-1}(\lambda_0^{-\frac{2}{ab}} - 1)$. Since $\frac{Q_4/(b-1)}{Q_3/(b(a-1))}$ has an F-distribution with $b-1$ and $b(a-1)$ degrees of freedom, the test of the hypothesis H_0 against all possible alternatives is based on an F statistic.

5.10 LINEAR REGRESSION

Suppose that a random variable Y depends on a variable x, i.e., for a fixed value x, the random variable Y has a determined probability density function, denoted by $p(Y|x)$. The expectation of Y is a function of x. Denote it by $E[Y] = \mu(x)$. The function $\mu(x)$ is called a *regression function*.

Suppose that $x_1, \ldots, x_N$ are N different values of x. Observing the random variable Y for each x_k, we have N pairs of known numbers (x_1, y_1), $(x_2, y_2), \cdots (x_N, y_N)$.

We assume that $Y \sim N(\alpha + \beta x, \sigma^2)$, and we estimate α and β. For each i,

$$Y_i = \alpha + \beta x_i + \epsilon_i,$$

where $\epsilon_i \in N(0, \sigma^2)(i = 1, \ldots, N)$, $Y_1, \ldots, Y_N$ are independent and each Y_i has the probability density function

$$p(Y_i|x) = \frac{1}{\sqrt{2\pi}\sigma} e^{-\frac{1}{2\sigma^2}(y_i - \alpha - \beta x_i)^2} \quad (i = 1, \ldots, N)$$

Therefore, the joint probability density function (likelihood function) of $Y_1, \ldots, Y_N$

$$L(\alpha, \beta, \sigma^2) = \left(\frac{1}{2\pi\sigma^2}\right)^{\frac{N}{2}} e^{-\frac{1}{2\pi\sigma^2}\sum_1^N (y_i - \alpha - \beta x_i)^2},$$

and we get the maximum likelihood estimates of α and β,

$$\tilde{\alpha} = \bar{y} - \tilde{\beta}\bar{x}, \quad \tilde{\beta} = \frac{\sum_1^N (x_i - \bar{x})(y_i - \bar{y})}{\sum_1^N (x_i - \bar{x})^2}, \tag{5.15}$$

where

$$\bar{x} = \frac{1}{N}\sum_1^N x_i, \quad \bar{y} = \frac{1}{N}\sum_1^N y_i.$$

We regard $\tilde{\alpha} + \tilde{\beta}x$ as an estimate of the regression function $\mu(x) = \alpha + \beta x$, i.e., $\tilde{\mu}(x) = \tilde{\alpha} + \tilde{\beta}x$.

Now we estimate σ^2.

Consider the sum $r = \sum_1^N (y_i - \tilde{\alpha} - \tilde{\beta}x_i)^2$. By (5.15),

$$r = \sum_1^N (y_i - \bar{y} - \tilde{\beta}(x_i - \bar{x}))^2 = \sum_1^N (y_i - \bar{y})^2$$

$$-2\tilde{\beta}\sum_1^N (x_i - \bar{x})(y_i - \bar{y}) + \tilde{\beta}^2 \sum_1^N (x_i - \bar{x})^2$$

$$= \sum_1^N (y_i - \bar{y})^2 - \tilde{\beta}\sum_1^N (x_i - \bar{x})(y_i - \bar{y}).$$

The corresponding statistic

$$R = \sum_1^N (Y_i - \bar{Y})^2 - \tilde{\beta}\sum_1^N (x_i - \bar{x})(Y_i - \bar{Y}). \tag{5.16}$$

It can be proved that $\frac{R}{\sigma^2} \sim \chi^2(N-2)$. Therefore,

$$E\left[\frac{R}{\sigma^2}\right] = N - 2, \quad E\left[\frac{R}{N-2}\right] = \sigma^2.$$

From this and (5.16), we get an unbiased estimate of σ^2:

$$\tilde{\sigma}^2 = \frac{R}{N-2} = \frac{1}{N-2}\left(\sum_1^N (Y - \bar{Y})^2 - \tilde{\beta}\sum_1^N (x_i - \bar{x})(Y_i - \bar{Y})\right),$$

where $\tilde{\beta} = \frac{\sum_1^N (x_i - \bar{x})(Y_i - \bar{Y})}{\sum_1^N (x_i - \bar{x})^2}$.

5.11 MANN-KENDALL TREND TEST

The most famous nonparametric trend analysis is the *Mann-Kendall trend test*. The hypothesis H_0 is that data are independent and randomly ordered. The statistic related to the Mann-Kendall trend test is as follows.

Let X be a random variable and $X_1, \ldots, X_n$ be its samples. Denote the statistic S by

$$S = \sum_{i<j} a_{ij}, \quad a_{ij} = \text{sgn}(X_j - X_i), \tag{5.17}$$

where $\text{sgn}\,\lambda = 1(\lambda > 0)$, $\text{sgn}\,\lambda = -1(\lambda < 0)$, $\text{sgn}\,0 = 0$,   and $\sum_{i<j} = \sum_{1 \le i < j \le n}$.

Proposition 5.11. *If $X_1, \ldots, X_n$ are n samples from a random variable X, then the statistic S tends to a normal distribution as $n \to \infty$ and*

$$E[S] = 0, \quad \text{Var}\,S = \frac{n(n-1)(2n+5)}{18}. \tag{5.18}$$

Remark. On the basis of the distribution of the statistic S, we can assess whether there is a significant trend in a climatic time series. This method is called the Mann-Kendall trend test.

Here we give the proof of (5.18). Since X_i and X_j are independent for $i \ne j$,

$$P(X_i > X_j) = P(X_i < X_j) = \frac{1}{2}.$$

From this and $a_{ij} = \text{sgn}(x_j - x_i)$, it follows that for $i \ne j$,

$$P(a_{ij} = 1) = P(a_{ij} = -1) = \frac{1}{2},$$

and so $E[a_{ij}] = 0(i \ne j)$. From this and (5.17), it follows that $E[S] = \sum_{i<j} E[a_{ij}] = 0$.

From $E[S] = 0$ and $\text{Var}\,S = E[S^2] - (E[S])^2$, it follows that

$$\text{Var}\,S = E[S^2] - (E[S])^2 = E[S^2].$$

However,

$$S^2 = \left(\sum_{i<j} a_{ij}\right)^2 = \sum_{i<j,k<l} a_{ij}a_{kl} = P_1 + P_2 + P_3 + P_4 + P_5,$$

where

$$P_1 = \sum_{i<j} a_{ij}^2, \quad P_2 = \sum_{\substack{i<j,i<l \\ j \ne l}} a_{ij}a_{il}, \quad P_3 = \sum_{\substack{i<j,k<j \\ i \ne k}} a_{ij}a_{kj},$$

$$P_4 = 2\sum_{i<j<l} a_{ij}a_{jl}, \quad P_5 = \sum_{\substack{i<j,k<l \\ i,j,k,l \text{ different}}} a_{ij}a_{lk}. \tag{5.19}$$

Therefore,

$$\text{Var}\,S = E[P_1] + E[P_2] + E[P_3] + E[P_4] + E[P_5]. \tag{5.20}$$

Since $a_{ij}^2 = (\text{sgn}(x_j - x_i))^2 = 1(i < j)$, by (5.17),

$$E[P_1] = \sum_{i<j} E[a_{ij}^2] = \sum_{i<j} 1 = \sum_{i=1}^{n-1} \sum_{j=i+1}^{n} 1 = \frac{n(n-1)}{2}.$$

Since $a_{ij}a_{il} = 1$ or $-1 (j \neq l)$, it can be proved that

$$E[a_{ij}a_{il}] = P(a_{ij}a_{il} = 1) - P(a_{ij}a_{il} = -1) = \frac{2}{\pi} \sin^{-1} \rho,$$

where ρ is the correlation coefficient of $Y = x_j - x_i$ and $Z_1 = x_l - x_i$. Denote $E[X] = \mu$ and $\text{Var}(X) = \sigma^2$. Since $x_1, \ldots, x_n$ are independent, for different i, j, l,

$$E[YZ_1] = E[x_j x_l] - E[x_i x_l] - E[x_j x_i] + E[x_i^2] = E[x_i^2] = \sigma^2.$$

Since x_j and x_i are independent,

$$\text{Var } Y = \text{Var}(x_j - x_i) = \text{Var } x_j + \text{Var } x_i = 2\sigma^2.$$

Similarly, $\text{Var } Z_1 = 2\sigma^2$. By $E[Y] = E[Z_1] = 0$ and $\text{Cov}(Y, Z_1) = \sigma^2$, it follows that the correlation coefficient of Y and Z_1 is

$$\rho = \frac{\text{Cov}(Y, Z_1)}{(\text{Var } Y)^{1/2}(\text{Var } Z_1)^{1/2}} = \frac{1}{2}.$$

So $E[a_{ij}a_{il}] = \frac{2}{\pi} \sin^{-1} \frac{1}{2} = \frac{1}{3} (i < j, i < l, j \neq l)$. By (5.19), we get

$$E[P_2] = \sum_{\substack{i<j,i<l \\ j\neq l}} E[a_{ij}a_{il}] = \sum_{i<j,i<l} \frac{1}{3} - \frac{1}{3}E[P_1].$$

The first term of the right-hand side is

$$\sum_{i<j,i<l} \frac{1}{3} = \sum_{i=1}^{n-1} \sum_{j=i+1}^{n} \sum_{l=i+1}^{n} \frac{1}{3} = \frac{1}{3} \sum_{i=1}^{n-1} (n-i)^2 = \frac{n(n-1)(2n-1)}{18}.$$

Therefore,

$$E[P_2] = \frac{n(n-1)(2n-1)}{18} - \frac{n(n-1)}{6} = \frac{n(n-1)(n-2)}{9}.$$

Let $Y = x_j - x_i$ and $Z_2 = x_j - x_k$. The correlation coefficient of Y and Z_2 is also $\rho = 1/2$. So

$$E[a_{ij}a_{kj}] = \frac{1}{3}.$$

By (5.19), we get

$$E[P_3] = \sum_{\substack{i<j,k<j \\ i \neq k}} E[a_{ij}a_{kj}] = \sum_{\substack{i<j,k<j \\ i \neq k}} \frac{1}{3} = \sum_{i=1}^{n-1} \sum_{j=i+1}^{n} \sum_{k=1}^{j-1} \frac{1}{3} - \frac{1}{3}E[P_1]$$

$$= \frac{n(n-1)(n-2)}{9}.$$

Let $Y = x_j - x_i$ and $Z_3 = x_l - x_j$. Then

$$E[YZ_3] = E[-x_j^2] = -E[x_j^2] = -\sigma^2$$

and the correlation coefficient of Y and Z_3 is $\rho = -1/2$. By (5.18), we get

$$E[P_4] = 2 \sum_{i<j<l} E[a_{ij}a_{jl}] = -\frac{2}{3} \sum_{i<j<l} 1 = -\frac{2}{3} \sum_{i=1}^{n-2} \sum_{j=i+1}^{n-1} \sum_{l=j+1}^{n} 1$$

$$= -\frac{2}{3} \sum_{i=1}^{n-2} \sum_{j=i+1}^{n-1} (n-j) = -\frac{n(n-1)(n-2)}{9}.$$

Since $x_1, \ldots, x_n$ are independent, when i, j, k, l are four different suffixes, a_{ij} and a_{kl} are independent. So

$$E[a_{ij}] = E[a_{kl}] = 0, \quad E[a_{ij}a_{kl}] = E[a_{ij}]E[a_{kl}] = 0.$$

From this and (5.19),

$$E[P_5] = \sum_{\substack{i<j,k<l \\ i,l,j,k \text{ different}}} E[a_{ij}a_{kl}] = 0.$$

Combining all the results with (5.20), we get

$$\text{Var } S = \frac{n(n-1)(2n+5)}{18}.$$

PROBLEMS

5.1 Find the cumulative distribution function of local temperatures, summed over 1 week, 1 month, and 1 year.

5.2 Let X_1 and X_2 be independent and $X \sim N(0, \sigma^2), X_2 \sim N(0, \sigma^2)$. Any point $x \in \mathbb{R}$ can be represented in polar coordinates: $x_1 = r\cos\theta, x_2 = r\sin\theta$. Its inverse transform is $r = \sqrt{x_1^2 + x_2^2}, \theta = \tan^{-1}\frac{x_2}{x_1}$. Define $R = \sqrt{X_1^2 + X_2^2}, \Theta = \tan^{-1}\frac{X_2}{X_1}$. Try to find the joint probability density function $p_{R,\Theta}(r, \theta)$ and the marginal probability density function $p_R(r)$ and $p_\Theta(\theta)$.

5.3 Let the joint probability density function of X and Y be

$$p_{X,Y}(x,y) = \frac{1}{\sqrt{3}\pi} e^{-\frac{2}{3}(x^2 - \frac{1}{2}xy + y^2)}.$$

Find the correlation coefficient of X and Y.

5.4 Predict the daily maximum local temperature by using linear regression.

5.5 With the help of the Mann-Kendall trend test, assess the significance of the trend in local precipitation data.

BIBLIOGRAPHY

Abdallah, N.B., Mouhous-Voyneau, N., Denux, T., 2014. Combining statistical and expert evidence using belief functions: application to centennial sea level estimation taking into account climate change. Int. J. Approx. Reason. 55, 341-354.

Ahmed, K.F., Wang, G., Silander, J., Wilson, A.M., Allen, J.M., Horton, R., Anyah, R., 2013. Statistical downscaling and bias correction of climate model outputs for climate change impact assessment in the U.S. northeast. Global Planet. Change 100, 320-332.

Amatulli, G., Camia, A. San-Miguel-Ayanz, J., 2013. Estimating future burned areas under changing climate in the EU-Mediterranean countries. Sci. Total Environ. 450-451, 209-222.

Anderson, T.W., 1984. An Introduction to Multivariate Statistical Analysis, second ed. John Wiley & Sons, New York.

Andreo, B., Jimenez, P., Duran, J.J., Carrasco, F., Vadillo, I., Mangin, A., 2006. Climatic and hydrological variations during the last 117–166 years in the south of the Iberian Peninsula, from spectral and correlation analyses and continuous wavelet analyses. J. Hydrol. 324, 24-39.

Casas-Prat, M., Wang, X.L., Sierra, J.P., 2014. A physical-based statistical method for modeling ocean wave heights. Ocean Model. 73, 59-75

Ghil, M., Allen, M.R., Dettinger, M.D., Ide, K., Kondrashov, D., Mann, M.E., Robertson, A.W., Saunders, A., Tian, Y., Varadi, F., Yiou, P., 2002. Advanced spectral methods for climatic time series. Rev. Geophys. 40, 1003-1043.

Gocic, M., Trajkovic, S., 2013. Analysis of changes in meteorological variables using Mann-Kendall and Sen's slope estimator statistical tests in Serbia. Global Planet. Change 100, 172-182.

Hamed, K.H., 2008. Trend detection in hydrologic data: the Mann-Kendall trend test under the scaling hypothesis. J. Hydrol. 349, 350-363.

Joshi, D., St-Hilaire, A., Daigle, A., Ouarda, T.B.M.J., 2013. Data-based comparison of sparse Bayesian learning and multiple linear regression for statistical downscaling of low flow indices. J. Hydrol. 488, 136-149.

Kendall, M.G., 1955. Rank Correlation Methods. Griffin, London.

Lee, T., Jeong, C., 2014. Nonparametric statistical temporal downscaling of daily precipitation to hourly precipitation and implications for climate change scenarios. J. Hydrol. 510, 182-196.

Patidar, S., Jenkins, D., Banfill, P., Gibson, G., 2014. Simple statistical model for complex probabilistic climate projections: overheating risk and extreme events. Renew. Energy, 61, 23-28.

Storch, H.V., Zwiers, F.W., 1999. Statistical Analysis in Climate Research. Cambridge University Press, Cambridge.

Tabari, H., Kisi, O., Ezani, A., Talaee, P.H., 2012. SVM, ANFIS, regression and climate based models for reference evapotranspiration modeling using limited climatic data in a semi-arid highland environment. J. Hydrol. 444-445, 78-89.

Willems, P., Vrac, M., 2011. Statistical precipitation downscaling for small-scale hydrological impact investigations of climate change, J. Hydrol. 402, 193-205.

Wu, W., Liu, Y., Ge, M., Rostkier-Edelstein, D., Descombes, G., Kunin, P., Warner, T., Swerdlin, S., Givati, A., Hopson, T., Yates, D., 2012. Statistical downscaling of climate forecast system seasonal predictions for the southeastern Mediterranean. Atmos. Res. 118, 346-356.

Chapter 6

Empirical Orthogonal Functions

Empirical orthogonal function (EOF) analyses are often used to study possible spatial patterns of climate variability and how they change with time. One of the important results from EOF analysis is the discovery of several oscillations in the climate system, including the Pacific Decadal Oscillation and the Arctic Oscillation. Similarly to Fourier analysis and wavelet analysis, in EOF analysis, one also projects the original climate data on an orthogonal basis. However, this orthogonal basis is derived by computing the eigenvectors of a spatially weighted anomaly covariance matrix, and the corresponding eigenvalues provide a measure of the percent variance explained by each pattern. Therefore, EOFs of a space-time physical process can represent mutually orthogonal space patterns where the data variance is concentrated, with the first pattern being responsible for the largest part of the variance, the second for the largest part of the remaining variance, and so on. Later, in order to overcome some limitations of classical EOF analysis and make the resulting patterns more physically interpretable, rotated EOFs, Hilbert EOFs, and complex EOFs are developed. In this chapter, we focus on the theory and algorithms of EOF analyses and their generalization. At the same time, we introduce singular spectrum analysis, canonical correlation analysis, and principal oscillation pattern analysis.

6.1 RANDOM VECTOR FIELDS

Suppose that $\mathbf{X} = (X_1, \ldots, X_m)^{\mathrm{T}}$ is an m-dimensional real-valued random vector, where $X_k (k = 1, \ldots, m)$ are in the same probability space and T expresses the transpose of the vector $(X_1, \ldots, X_m)$. The m-dimensional vector $E[\mathbf{X}]$ is called the *mean* of the random vector $\mathbf{X}$ and is defined by

$$E[\mathbf{X}] = (E[X_1], \ldots, E[X_m])^{\mathrm{T}}.$$

The $m \times m$ matrices $E[\mathbf{X}\mathbf{X}^{\mathrm{T}}]$ and $\mathrm{Cov}(\mathbf{X}, \mathbf{X})$,

$$E[\mathbf{X}\mathbf{Y}^{\mathrm{T}}] = (E[X_i Y_j])_{i,j=1,\ldots,m},$$

$$\mathrm{Cov}(\mathbf{X}, \mathbf{Y}) = (\mathrm{Cov}(X_i, Y_j))_{i,j=1,\ldots,m},$$

are called the *correlation matrix* and the *covariance matrix* of the random vector $\mathbf{X}$, respectively, and $E[\mathbf{X}\mathbf{X}^{\mathrm{T}}]$ and $\mathrm{Cov}(\mathbf{X}, \mathbf{X})$ are both real symmetric matrices.

Mathematical and Physical Fundamentals of Climate Change

Suppose that $\mathbf{X}$ is a complex-valued random vector. The correlation matrix is a complex-valued $m \times m$ matrix $E[\mathbf{X}\overline{\mathbf{X}^{\mathrm{T}}}]$ and the covariance matrix is a complex-valued $m \times m$ matrix $\mathrm{Cov}(\mathbf{X}, \mathbf{X})$, and they are both Hermite matrices, i.e., conjugate symmetric matrices.

Suppose that $\mathbf{Y} = (Y_1, \ldots, Y_n)^{\mathrm{T}}$ is an n-dimensional real-valued random vector in the same probability space as $\mathbf{X}$. The *cross-correlation matrix* of the random vectors $\mathbf{X}$ and $\mathbf{Y}$ is a $m \times n$ matrix $E[\mathbf{X}\mathbf{Y}^{\mathrm{T}}]$. The *cross-covariance matrix* of $\mathbf{X}$ and $\mathbf{Y}$, denoted by $\mathrm{Cov}(\mathbf{X}, \mathbf{Y})$, is a $m \times n$ matrix:

$$\mathrm{Cov}(\mathbf{X}, \mathbf{Y}) = (\mathrm{Cov}(X_i, Y_j))_{i=1,\ldots,m; j=1,\ldots,n}.$$

If $E[\mathbf{X}] = 0$ and $E[\mathbf{Y}] = 0$, then

$$\mathrm{Cov}(\mathbf{X}, \mathbf{Y}) = E[\mathbf{X}\mathbf{Y}].$$

Suppose that $\mathbf{X}$ and $\mathbf{Y}$ are both complex-valued random vectors. Then the cross-correlation matrix of $\mathbf{X}$ and $\mathbf{Y}$ is the complex-valued matrix $E[\mathbf{X}\overline{\mathbf{Y}^{\mathrm{T}}}]$. If $E[\mathbf{X}] = E[\mathbf{Y}] = 0$, then

$$\mathrm{Cov}(\mathbf{X}, \mathbf{Y}) = E\left[\mathbf{X}\overline{\mathbf{Y}}\right].$$

Property. Let $\mathbf{X}$, $\mathbf{Y}$, $\mathbf{Z}$, and $\mathbf{W}$ be random vectors, and let A and B be deterministic matrices and $\mathbf{b}$ and $\mathbf{d}$ be deterministic vectors. Then

(i) $E[A\mathbf{X} + \mathbf{b}] = AE[\mathbf{X}] + \mathbf{b}$;
(ii) $E[(A\mathbf{X})(B\mathbf{Y})^{\mathrm{T}}] = AE[\mathbf{X}\mathbf{Y}^{\mathrm{T}}]B^{\mathrm{T}}$;
(iii) $\mathrm{Cov}(\mathbf{X}, \mathbf{Y}) = E[\mathbf{X}\mathbf{Y}^{\mathrm{T}}] - (E\mathbf{X})(E\mathbf{Y})^{\mathrm{T}}$;
(iv) $\mathrm{Cov}(A\mathbf{X} + \mathbf{b}, B\mathbf{Y} + \mathbf{d}) = A\mathrm{Cov}(\mathbf{X}, \mathbf{Y})B^{\mathrm{T}}$;
(v) $\mathrm{Cov}(\mathbf{W} + \mathbf{X}, \mathbf{Y} + \mathbf{Z}) = \mathrm{Cov}(\mathbf{W}, \mathbf{Y}) + \mathrm{Cov}(\mathbf{W}, \mathbf{Z}) + \mathrm{Cov}(\mathbf{X}, \mathbf{Y}) + \mathrm{Cov}(\mathbf{X}, \mathbf{Z})$.

These properties are similar to those of ordinary scalar random variables.

Proposition 6.1. *The correlation matrix* $E[\mathbf{X}\mathbf{X}^{\mathrm{T}}]$ *and the covariance matrix* $\mathrm{Cov}(\mathbf{X}, \mathbf{X})$ *are positive semidefinite.*

Proof. Let $K = E[\mathbf{X}\mathbf{X}^{\mathrm{T}}]$. For any vector $\mathbf{b}$,

$$\mathbf{b}^{\mathrm{T}} K\mathbf{b} = \mathbf{b}^{\mathrm{T}} E[\mathbf{X}\mathbf{X}^{\mathrm{T}}]\mathbf{b} = E[\mathbf{b}^{\mathrm{T}}\mathbf{X}\mathbf{X}^{\mathrm{T}}\mathbf{b}].$$

Since $\mathbf{X}^{\mathrm{T}}\mathbf{b}$ is a scalar random variable,

$$\mathbf{X}^{\mathrm{T}}\mathbf{b} = (\mathbf{X}^{\mathrm{T}}\mathbf{b})^{\mathrm{T}} = \mathbf{b}^{\mathrm{T}}\mathbf{X}.$$

So, for any vector $\mathbf{b}$,

$$E[\mathbf{b}^{\mathrm{T}}\mathbf{X}\mathbf{X}^{\mathrm{T}}\mathbf{b}] = E[(\mathbf{b}^{\mathrm{T}}\mathbf{X})^2] \geq 0.$$

Therefore, $\mathbf{b}^{\mathrm{T}} K \mathbf{b} \geq 0$, i.e., $K = E[\mathbf{X}\mathbf{X}^{\mathrm{T}}]$ is a positive semidefinite matrix. Let $K = \mathrm{Cov}(\mathbf{X}, \mathbf{X})$. Then, for any vector $\mathbf{b}$,

$$\mathbf{b}^{\mathrm{T}} K = \mathbf{b}^{\mathrm{T}} \mathrm{Cov}(\mathbf{X}, \mathbf{X})\mathbf{b} = \mathrm{Cov}(\mathbf{b}^{\mathrm{T}}\mathbf{X}, \mathbf{b}^{\mathrm{T}}\mathbf{X}).$$

Since $\mathbf{b}^{\mathrm{T}}\mathbf{X}$ is a scalar random variable, for any vector $\mathbf{b}$,

$$\mathrm{Cov}(\mathbf{b}^{\mathrm{T}}\mathbf{X}, \mathbf{b}^{\mathrm{T}}\mathbf{X}) = \mathrm{Var}(\mathbf{b}^{\mathrm{T}}\mathbf{X}) \geq 0.$$

Therefore, $\mathbf{b}^{\mathrm{T}} K \mathbf{b} \geq 0$, i.e., $K = \mathrm{Cov}(\mathbf{X}, \mathbf{X})$ is a positive semidefinite matrix. $\square$

6.2 CLASSICAL EOFs

Let $\mathbf{X}$ be an m-dimensional random vector with mean 0. We will give EOFs of $\mathbf{X}$.

First, we find one pattern $\mathbf{e}_1 \in \mathbb{R}^m$ with $\|\mathbf{e}_1\|_2 = 1$ such that $\epsilon_1 = E\left[\|\mathbf{X} - (\mathbf{X}, \mathbf{e}_1)\mathbf{e}_1\|_2^2\right]$ is minimized.

Notice that

$$\|\mathbf{X} - (\mathbf{X}, \mathbf{e}_1)\mathbf{e}_1\|_2^2 = (\mathbf{X} - (\mathbf{X}, \mathbf{e}_1)\mathbf{e}_1, \mathbf{X} - (\mathbf{X}, \mathbf{e}_1)\mathbf{e}_1)$$
$$= \|\mathbf{X}\|_2^2 - (\mathbf{X}, \mathbf{e}_1)^2.$$

Then

$$\epsilon_1 = E[\|\mathbf{X}\|_2^2] - E[(\mathbf{X}, \mathbf{e}_1)^2]. \qquad (6.1)$$

Let $\mathbf{X} = (X_1, \ldots, X_m)^{\mathrm{T}}$. Notice that $\|\mathbf{X}\|_2^2 = \sum_1^m |X_k|^2$. Since each X_k is a random variable and $E[X_k] = 0$, it follows that

$$E[\|\mathbf{X}\|_2^2] = E\left[\sum_1^m |X_k|^2\right] = \sum_1^m E[|X_k|^2]$$
$$= \sum_1^m \mathrm{Var}(X_k) + (E[X_k])^2$$
$$= \sum_1^m \mathrm{Var}(X_k).$$

On the other hand, the variance of $\mathbf{X}$ is defined as the sum of the variances of all components X_j:

$$\mathrm{Var}\,\mathbf{X} = \sum_1^m \mathrm{Var}\,X_k.$$

Thus, the first term of (6.1) is

$$E[\|\mathbf{X}\|_2^2] = \mathrm{Var}\,\mathbf{X}.$$

Let $\mathbf{e}_1 = (e_{11}, \ldots, e_{1m})^{\mathrm{T}}$. Notice that

$$(\mathbf{X}, \mathbf{e}_1)^2 = \left(\sum_{1}^{m} X_k e_{1k} \right)^2 = \sum_{k=1}^{m} \sum_{l=1}^{m} X_k X_l e_{1k} e_{1l}.$$

Thus, the second term of (6.1) is

$$E[(\mathbf{X}, \mathbf{e}_1)^2] = \sum_{k=1}^{m} \sum_{l=1}^{m} E[X_k X_l] e_{1k} e_{1l}.$$

Therefore, by (6.1), it follows that

$$\epsilon_1 = \operatorname{Var} \mathbf{X} - \sum_{k=1}^{m} \sum_{l=1}^{m} E[X_k X_l] e_{1k} e_{1l}. \tag{6.2}$$

This shows that ϵ_1 is a function of $e_{11}, \ldots, e_{1m}$.

We find the minimal value of ϵ_1 under the constraint $\|\mathbf{e}_1\|_2^2 = \sum_{1}^{m} e_{1k}^2 = 1$. By using the Lagrange multiplier method, let

$$F(e_{11}, \ldots, e_{1m}) = \epsilon_1 + \lambda \left(\sum_{1}^{m} e_{1k}^2 - 1 \right),$$

where λ is the Lagrange multiplier. By (6.2),

$$F(e_{11}, \ldots, e_{1m}) = \operatorname{Var}(\mathbf{X}) - \sum_{k=1}^{m} \sum_{l=1}^{m} E[X_k X_l] e_{1k} e_{1l} + \lambda \left(\sum_{1}^{m} e_{1k}^2 - 1 \right).$$

Let the partial derivatives of F be zero,

$$\frac{\partial F}{\partial e_{1k}} = -2 \sum_{l=1}^{m} E[X_k X_l] e_{1l} + 2\lambda e_{1k} = 0 \quad (k = 1, \ldots, m).$$

Then

$$\sum_{l=1}^{m} E[X_k X_l] e_{1l} = \lambda e_{1k} \quad (k = 1, \ldots, m). \tag{6.3}$$

Write it in a matrix form:

$$\Sigma_{\mathbf{XX}} \mathbf{e}_1 = \lambda \mathbf{e}_1, \tag{6.4}$$

where $\Sigma_{\mathbf{XX}} = (E[X_k X_l])_{k,l=1,\ldots,m}$ is the covariance matrix of $\mathbf{X}$, i.e.,

$$\Sigma_{\mathbf{XX}} = (E[X_k X_l])_{k,l=1,\ldots,m} = (\operatorname{Cov}(X_k, X_l))_{k,l=1,\ldots,m}.$$

Therefore, λ and $\mathbf{e}_1$ are the *eigenvalue* and the *eigenvector* of Σ_{XX}, respectively. From (6.2) and (6.3), and the constraint $\|\mathbf{e}_1\|_2^2 = \sum_1^m e_{1k}^2 = 1$,

$$\epsilon_1 = \operatorname{Var}\mathbf{X} - \sum_{k=1}^m \left(\sum_{l=1}^m E[X_k X_l] e_{1l} \right) e_{1k}$$

$$= \operatorname{Var}\mathbf{X} - \lambda \sum_1^m e_{1k}^2$$

$$= \operatorname{Var}\mathbf{X} - \lambda. \tag{6.5}$$

Let $\lambda = \lambda_1$ be the largest eigenvalue. Then ϵ_1 is minimal. So

$$\epsilon_1 = \operatorname{Var}\mathbf{X} - \lambda_1,$$

and $\mathbf{e}_1$ is the first EOF.

Second, we choose the $\mathbf{e}_2 \in \mathbb{R}^m$ satisfying the conditions $\|\mathbf{e}_2\|_2 = 1$ and $\mathbf{e}_2 \perp \mathbf{e}_1$ such that the error

$$\epsilon_2 = E\left[\|\mathbf{X} - (\mathbf{X}, \mathbf{e}_1)\mathbf{e}_1 - (\mathbf{X}, \mathbf{e}_2)\mathbf{e}_2\|_2^2 \right]$$

is minimal. Similarly to the argument of (6.1), it follows by (6.5) that

$$\epsilon_2 = E\left[\|\mathbf{X} - (\mathbf{X}, \mathbf{e}_1)\mathbf{e}_1\|_2^2 \right] - E\left[(\mathbf{X}, \mathbf{e}_2)^2 \right]$$

$$= \epsilon_1 - E[(\mathbf{X}, \mathbf{e}_2)^2]$$

$$= \operatorname{Var}\mathbf{X} - \lambda_1 - E[(\mathbf{X}, \mathbf{e}_2)^2]. \tag{6.6}$$

We find the minimal value of ϵ_2 under the constraint $\|\mathbf{e}_2\|_2 = 1$ and $\mathbf{e}_1 \perp \mathbf{e}_2$.

Let $\mathbf{e}_2 = (e_{21}, \ldots, e_{2m})^{\mathrm{T}}$, and let

$$F_2(e_{21}, \ldots, e_{2m}) = \epsilon_2 + \lambda(\|\mathbf{e}_2\|_2^2 - 1) + \mu(\mathbf{e}_1, \mathbf{e}_2),$$

where λ and μ are the Lagrange multipliers. Let $\frac{\partial F_2}{\partial e_{2k}} = 0 (k = 1, \ldots, m)$. Similarly to (6.4), this implies that

$$-2\Sigma_{XX}\mathbf{e}_2 + 2\lambda\mathbf{e}_2 + \mu\mathbf{e}_1 = 0. \tag{6.7}$$

Multiplying both sides of (6.7) by $\mathbf{e}_1^{\mathrm{T}}$, we get

$$-2\mathbf{e}_1^{\mathrm{T}}\Sigma_{XX}\mathbf{e}_2 + 2\lambda\mathbf{e}_1^{\mathrm{T}}\mathbf{e}_2 + \mu\mathbf{e}_1^{\mathrm{T}}\mathbf{e}_1 = 0.$$

By (6.4) and $\lambda = \lambda_1$,

$$\mathbf{e}_1^{\mathrm{T}}\Sigma_{XX} = (\Sigma_{XX}\mathbf{e}_1)^{\mathrm{T}} = \lambda_1\mathbf{e}_1^{\mathrm{T}},$$

and so

$$-2\lambda_1\mathbf{e}_1^{\mathrm{T}}\mathbf{e}_2 + 2\lambda\mathbf{e}_1^{\mathrm{T}}\mathbf{e}_2 + \mu\mathbf{e}_1^{\mathrm{T}}\mathbf{e}_1 = 0.$$

From this and $\mathbf{e}_1^{\mathrm{T}}\mathbf{e}_2 = (\mathbf{e}_1, \mathbf{e}_2) = 0$ and $\mathbf{e}_1^{\mathrm{T}}\mathbf{e}_1 = \|\mathbf{e}_1\|_2^2 = 1$, it follows that

$$\mu = 0.$$

Combining this with (6.7), we get

$$\Sigma_{\mathbf{XX}} \mathbf{e}_2 = \lambda \mathbf{e}_2. \tag{6.8}$$

This shows that $\mathbf{e}_2$ and λ are the eigenvector and the eigenvalue of $\Sigma_{\mathbf{XX}}$, respectively. From this and $\|\mathbf{e}_2\|_2 = 1$ and $E[(\mathbf{X}, \mathbf{e}_2)^2] = \mathbf{e}_2^{\mathsf{T}} \Sigma_{\mathbf{XX}} \mathbf{e}_2$, it follows by (6.6) and (6.8) that

$$\epsilon_2 = \operatorname{Var} \mathbf{X} - \lambda_1 - \mathbf{e}_2^{\mathsf{T}} \Sigma_{\mathbf{XX}} \mathbf{e}_2$$

$$= \operatorname{Var} \mathbf{X} - \lambda_1 - \lambda.$$

Let $\lambda = \lambda_2$ be the second largest eigenvalue. Then ϵ_2 is minimal, and so

$$\mathbf{e}_2 = \operatorname{Var} \mathbf{X} - \lambda_1 - \lambda_2,$$

and $\mathbf{e}_2$ is the second EOF.

Continuing this procedure, we have the following.

Definition 6.1. Let $\mathbf{X} = (X_1, \ldots, X_m)$ be an m-dimensional random vector with mean 0 and covariance matrix $\Sigma_{\mathbf{XX}}$, where

$$\Sigma_{\mathbf{XX}} = (\operatorname{Cov}(X_k, X_l))_{k,l=1,\ldots,m},$$

and let $\lambda_1 \geq \lambda_2 \geq \cdots \geq \lambda_m$ be eigenvalues of $\Sigma_{\mathbf{XX}}$. Then the corresponding orthogonal eigenvectors $\mathbf{e}_1, \ldots, \mathbf{e}_m$ with unit length are called EOFs of $\mathbf{X}$.

Theorem 6.1. *Let $\mathbf{e}_1, \ldots, \mathbf{e}_m$ be EOFs of an m-dimensional random vector $\mathbf{X}$ with mean 0 and covariance matrix $\Sigma_{\mathbf{XX}}$ with eigenvalues $\lambda_1 > \lambda_2 > \cdots > \lambda_m$. Then*

(i) *for any $1 \leq k \leq m$, the first k EOFs $\mathbf{e}_1, \ldots, \mathbf{e}_k$ minimize the mean square error:*

$$\epsilon_k = E\left[\left\| \mathbf{X} - \sum_1^k (\mathbf{X}, \mathbf{e}_j) \mathbf{e}_j \right\|_2^2 \right].$$

Precisely, any linear combination $\sum_1^k c_j \mathbf{g}_j$ of m-dimensional vectors $\mathbf{g}_1, \ldots, \mathbf{g}_k$ satisfies

$$E\left[\left\| \mathbf{X} - \sum_1^k c_j \mathbf{g}_j \right\|_2^2 \right] \geq \epsilon_k.$$

(ii) *the minimal mean square error*

$$\epsilon_k = \operatorname{Var} \mathbf{X} - \sum_1^k \lambda_j,$$

where $\lambda_j (j = 1, \ldots, k)$ are the corresponding first k eigenvalues of $\Sigma_{\mathbf{XX}}$.

Since the covariance matrix Σ_{XX} of an m-dimensional random vector X is real symmetric, the eigenvalues are non-negative and the eigenvectors are orthogonal, and the EOF is an orthonormal basis of $\mathbb{R}^m$. So

$$X = \sum_1^m \alpha_j e_j \quad (\alpha_j = (X, e_j)), \tag{6.9}$$

where α_j is a one-dimensional random variable and $E[\alpha_j] = 0 (j = 1, \ldots, m)$.

Definition 6.2. Let $\{e_1, \ldots, e_m\}$ be EOFs of X. Then $\alpha_j = (X, e_j)$ $(j = 1, \ldots, m)$ are called EOF coefficients.

Proposition 6.2. *Let* $\alpha_1, \ldots, \alpha_m$ *be EOF coefficients of* X. *Then*

$$\text{Var}\,\alpha_j = \lambda_j \quad (j = 1, \ldots, m),$$
$$\text{Cov}(\alpha_j, \alpha_l) = 0 \quad (j \neq l, \; j, l = 1, \ldots, m),$$

where λ_j *is the jth eigenvalue corresponding to the jth EOF.*

Proof. Let $X = (X_1, \ldots, X_m)^T$ and $e_j = (e_{j1}, \ldots, e_{jm})^T$. Since $E[\alpha_j] = 0$ $(j = 1, \ldots, m)$,

$$\text{Cov}(\alpha_j, \alpha_l) = E[(X, e_j)(X, e_l)] \quad (j \neq l, \; j, l = 1, \ldots, m).$$

However,

$$(X, e_j)(X, e_l) = \left(\sum_{k=1}^m X_k e_{jk}\right)\left(\sum_{n=1}^m X_n e_{ln}\right)$$
$$= \sum_{k=1}^m \sum_{n=1}^m X_k X_n e_{jk} e_{ln}.$$

Therefore,

$$\text{Cov}(\alpha_j, \alpha_l) = \sum_{k=1}^m \sum_{n=1}^m E[X_k X_n] e_{jk} e_{ln}.$$

Since e_l is an eigenvector of Σ_{XX} and λ_l is the eigenvalue corresponding to e_l,

$$\Sigma_{XX} e_l = \lambda_l e_l$$

or

$$\sum_{n=1}^m E[X_k X_n] e_{ln} = \lambda_l e_{lk} \quad (k = 1, \ldots, m).$$

So

$$\text{Cov}(\alpha_j, \alpha_l) = \lambda_l \sum_{k=1}^m e_{jk} e_{lk} = \lambda_l(e_j, e_l) \quad (j \neq l, \; j, l = 1, \ldots, m).$$

By $e_j \perp e_l$ and $\|e_j\|_2 = \|e_l\|_2 = 1$, it follows that $(e_j, e_l) = \delta_{jl}$. So

$$\text{Cov}(\alpha_j, \alpha_l) = \delta_{jl}\lambda_l \quad (j, l = 1, \ldots, m).$$

$\square$

Since EOFs $e_1, \ldots, e_m$ are an orthonormal basis of $\mathbb{R}^m$,

$$\left\| X - \sum_1^m (X, e_j)e_j \right\|_2^2 = 0.$$

By Theorem 6.1, it follows that

$$\text{Var } X = \sum_1^m \lambda_j.$$

By Proposition 6.2, the variance of each term $\alpha_j e_j$ of (6.9) is

$$\text{Var}(\alpha_j e_l) = \sum_{k=1}^m \text{Var}(\alpha_j e_{lk}) = \text{Var } \alpha_j \sum_{k=1}^m e_{lk}^2 = \lambda_j.$$

This shows that the variance contribution of the jth component e_j to the total variance is just λ_j. The bulk of the variance of X can often be represented by the first few EOFs. This leads to a significant reduction of the data.

On the other hand, $\alpha_k = (X, e_k)$ $(k = 1, \ldots, m)$ is equivalent to

$$\alpha_1 = e_{11}X_1 + e_{12}X_2 + \cdots + e_{1m}X_m,$$

$$\alpha_2 = e_{21}X_1 + e_{22}X_2 + \cdots + e_{2m}X_m,$$

$$\vdots$$

$$\alpha_m = e_{m1}X_1 + e_{m2}X_2 + \cdots + e_{mm}X_m.$$

This means that the EOF coefficients $\alpha_1, \ldots, \alpha_m$ are linear combinations of random variables $X_1, \ldots, X_m$. The relativity among $X_1, \ldots, X_m$ may be complicated but their linear combinations $\alpha_1, \ldots, \alpha_m$ are unrelated and the variances

$$\text{Var } \alpha_j = \lambda_j \quad (j = 1, \ldots, m),$$

where $\lambda_1 > \lambda_2 > \cdots > \lambda_m$. Sometimes, the EOF coefficients $\alpha_1, \ldots, \alpha_m$ are also called *principal components*.

The inverse theorem of Proposition 6.2 is as follows.

Proposition 6.3. *If X is an m-dimensional random vector with mean 0 and $X = \sum_1^m \beta_j p_j$, where $p_j (j = 1, \ldots, m)$ are an orthonormal basis of $\mathbb{R}^m$ and $\beta_j (j = 1, \ldots, m)$ are uncorrelated, $\text{Cov}(\beta_j, \beta_l) = 0 (j \neq l)$, then $p_j (j = 1, \ldots, m)$ are the EOFs of the random vector X.*

Proof. We may assume that each β_j has mean 0. Then covariance matrix Σ_{XX} of X

$$\Sigma_{XX} = E\left[\left(\sum_1^m \beta_j p_j \right) \left(\sum_1^m \beta_l p_l \right)^T \right]$$

$$= E \left[\sum_{l=1}^{m} \sum_{j=1}^{m} \beta_j \beta_l \mathbf{p}_j \mathbf{p}_l^{\mathrm{T}} \right]$$

$$= \sum_{l=1}^{m} \left(\sum_{j=1}^{m} E[\beta_j \beta_l] \mathbf{p}_j \right) \mathbf{p}_l^{\mathrm{T}}$$

$$= \sum_{1}^{m} \mathrm{Var}(\beta_l) \mathbf{p}_l \mathbf{p}_l^{\mathrm{T}}.$$

Since $\mathbf{p}_l (l = 1, \ldots, m)$ are an orthonormal basis,

$$\mathbf{p}_l^{\mathrm{T}} \mathbf{p}_j = \delta_{lj},$$

and so, for $j = 1, \ldots, m$,

$$\Sigma_{\mathbf{XX}} \mathbf{p}_j = \left(\sum_{1}^{m} \mathrm{Var}(\beta_l) \mathbf{p}_l \mathbf{p}_l^{\mathrm{T}} \right) \mathbf{p}_j$$

$$= \sum_{1}^{m} \mathrm{Var}(\beta_l) \mathbf{p}_l (\mathbf{p}_l^{\mathrm{T}} \mathbf{p}_j)$$

$$= \mathrm{Var}(\beta_j) \mathbf{p}_j.$$

Therefore, $\mathbf{p}_j (j = 1, \ldots, m)$ are the eigenvectors of $\Sigma_{\mathbf{XX}}$, i.e., they are the EOFs of the random vector $\mathbf{X}$. $\qquad\qquad\qquad\qquad\qquad\qquad\qquad\qquad \square$

For an orthonormal basis $\mathbf{p}_j (j = 1, \ldots, m)$, we choose a set of uncorrelated random variables $\beta_j (j = 1, \ldots, m)$ with mean 0, then $\mathbf{X} = \sum_1^m \beta_j \mathbf{p}_j$ is the desired random vector with EOFs $\mathbf{p}_j (j = 1, \ldots, m)$.

For the random vector $\mathbf{X} = (X_1, \ldots, X_m)^{\mathrm{T}}$, one can often choose a few EOFs such that

$$\mathbf{X} \approx \sum_{1}^{k} \alpha_j \mathbf{e}_j,$$

where $\alpha_j = (\mathbf{X}, \mathbf{e}_j)$ and $\mathbf{e}_j = (e_{j1}, \ldots, e_{jm})$. In practice, it is also often replaced by

$$\mathbf{X} \approx \sum_{1}^{k} \alpha_j^+ \mathbf{e}_j^+,$$

where $\alpha_j^+ = \frac{\alpha_j}{\sqrt{\lambda_j}}$ and $\mathbf{e}_j^+ = \sqrt{\lambda_j} \mathbf{e}_j$. So

$$\mathrm{Var}\, \alpha_j^+ = \frac{\mathrm{Var}\, \alpha_j}{\lambda_j} = 1,$$

$$\alpha_j^+ = \frac{(\mathbf{X}, \mathbf{e}_j)}{\sqrt{\lambda_j}} = \frac{1}{\lambda_j}(\mathbf{X}, \mathbf{e}_j^+).$$

Example 6.1. Let $\mathbf{X} = (X_1, X_2)$, where X_1 and X_2 are normal random variables and $X_1 \sim N(0, \sigma_1^2)$, $X_2 \sim N(0, \sigma_2^2)$, and $E[X_1 X_2] = \sigma_1 \sigma_2 \rho$. The covariance matrix of $\mathbf{X}$ is

$$\Sigma_{\mathbf{XX}} = \begin{pmatrix} \sigma_1^2 & \sigma_1 \sigma_2 \rho \\ \sigma_1 \sigma_2 \rho & \sigma_2^2 \end{pmatrix}.$$

Let the determinant of $\Sigma_{\mathbf{XX}} - \lambda I$ be equal to zero:

$$\begin{vmatrix} \sigma_1^2 - \lambda & \sigma_1 \sigma_2 \rho \\ \sigma_1 \sigma_2 \rho & \sigma_2^2 - \lambda \end{vmatrix} = 0.$$

Then the eigenvalues λ_1 and λ_2 are

$$\frac{1}{2}(\sigma_1^2 + \sigma_2^2) \pm \frac{1}{2}\sqrt{(\sigma_1^2 - \sigma_2^2)^2 + 4\sigma_1^2 \sigma_2^2 \rho^2}.$$

Let e_1 and e_2 be the corresponding eigenvectors with the unit length satisfying

$$\begin{pmatrix} \sigma_1^2 - \lambda_j & \sigma_1 \sigma_2 \rho \\ \sigma_1 \sigma_2 \rho & \sigma_2^2 - \lambda_j \end{pmatrix} \begin{pmatrix} e_{j1} \\ e_{j2} \end{pmatrix} = \lambda_j \begin{pmatrix} e_{j1} \\ e_{j2} \end{pmatrix} \quad (j = 1, 2).$$

Expand $\mathbf{X}$ into a linear combination of $\mathbf{e}_1$ and $\mathbf{e}_2$:

$$\mathbf{X} = \alpha_1 \mathbf{e}_1 + \alpha_2 \mathbf{e}_2,$$

where α_1 and α_2 are EOF coefficients and

$$\alpha_1 = (\mathbf{X}, \mathbf{e}_1),$$
$$\alpha_2 = (\mathbf{X}, \mathbf{e}_2).$$

Since X_1 and X_2 are both normal random variables with mean 0, the EOF coefficients α_1 and α_2 are also normal random variables with mean 0. By Proposition 6.2,

$$\text{Var}\,\alpha_1 = \lambda_1,$$
$$\text{Var}\,\alpha_2 = \lambda_2,$$
$$\text{Cov}(\alpha_1, \alpha_2) = 0.$$

Therefore, α_1 and α_2 have probability density functions,

$$p_{\alpha_1}(x_1) = \frac{1}{\sqrt{2\pi \lambda_1}} e^{-\frac{x_1^2}{2\lambda_1}},$$

$$p_{\alpha_2}(x_2) = \frac{1}{\sqrt{2\pi \lambda_2}} e^{-\frac{x_2^2}{2\lambda_2}},$$

respectively, and α_1 and α_2 are independent. So the joint probability density function of EOF coefficients α_1 and α_2 is

$$p(x_1, x_2) = p_{\alpha_1}(x_1) p_{\alpha_2}(x_2) = \frac{1}{2\pi \sqrt{\lambda_1 \lambda_2}} e^{-\frac{x_1^2}{2\lambda_1} - \frac{x_2^2}{2\lambda_2}}.$$

In EOFs analysis, X has two components $\alpha_1 e_1$ and $\alpha_2 e_2$. The variance contribution of the jth component to the variance is $\lambda_j (j = 1, 2)$.

Proposition 6.4. *Let X be a random vector and L be an orthogonal transform. Denote $Z = LX$. If X has EOFs $e_k (k = 1, \ldots, m)$, then Z has EOFs $Le_k (k = 1, \ldots, m)$ and the EOF coefficients are invariable.*

Proof. Let X be an m-dimensional random vector. Then

$$X = \sum_1^m \alpha_j e_j.$$

Let the orthogonal transform L be from $\mathbb{R}^m$ to $\mathbb{R}^m$. Then

$$LX = \sum_1^m \alpha_j L e_j.$$

The EOFs $e_j (j = 1, \ldots, m)$ are an orthonormal basis of $\mathbb{R}^m$, so are $Le_j (j = 1, \ldots, m)$. By Proposition 6.2,

$$\text{Cov}(\alpha_j, \alpha_l) = 0 \quad (j \neq l).$$

From this and Proposition 6.3, Z has EOFs $Le_k (k = 1, \ldots, m)$ and the EOF coefficients are invariable. $\qquad \square$

Let X be a random vector. If there are several eigenvectors corresponding to the same eigenvalue in the EOFs of X, then the EOFs are said to be *degenerate*. When there exist k eigenvectors corresponding to the same eigenvalue, the k eigenvectors are not uniquely determined but the space spanned by k eigenvectors is uniquely determined.

Suppose that m components of random vector X are not correlated and their variances are all 1. Then its covariance matrix Σ_{XX} is the unit matrix of order m and it has only an eigenvalue $\lambda_1 = \cdots = \lambda_m = 1$, and any orthonormal basis of $\mathbb{R}^m$ is a set of EOFs of X. This is a degenerate set of EOFs. In this case, there is no preferred direction, and the variances are uniformly in all directions.

6.3 ESTIMATION OF EOFs

In practice, for an m-dimensional random vector $X = (X_1, \ldots, X_m)^T$ with mean 0, its covariance matrix, eigenvalues, and eigenvectors are unknown. Therefore, they must be estimated from a finite sampling $x_1, \ldots, x_n$, where $x_j = (x_{j1}, x_{j2}, \ldots, x_{jm})^T (j = 1, \ldots, n)$ are the m-dimensional vectors. Since $\mu = E[X] = 0$, its estimate is

$$\hat{\mu} = \frac{1}{n} \sum_1^n \mathbf{x}_j \approx 0.$$

Denote $\hat{\mu} = (\hat{\mu}_1, \ldots, \hat{\mu}_m)$,

$$\hat{\mu}_i = \frac{1}{n} \sum_{j=1}^n x_{ji} \approx 0 \quad (i = 1, \ldots, m).$$

Since $\mathrm{Var}\, X_i = E[X_i^2]$ and $\mathrm{Var}\, \mathbf{X} = \sum_1^m \mathrm{Var}\, X_i$, the estimates of $\mathrm{Var}\, X_i$ and $\mathrm{Var}\, \mathbf{X}$ are, respectively,

$$\widehat{\mathrm{Var}}\, X_i = \frac{1}{n} \sum_{j=1}^n x_{ji}^2,$$

$$\widehat{\mathrm{Var}}\, \mathbf{X} = \frac{1}{n} \sum_{j=1}^n \sum_{i=1}^m x_{ji}^2. \tag{6.10}$$

Since the estimate of $\mathrm{Cov}(X_i, X_j)$ is

$$\widehat{\mathrm{Cov}}(X_i, X_j) = \frac{1}{n} \sum_{k=1}^n x_{ki} x_{kj},$$

the covariance matrix $\Sigma_{\mathbf{XX}} = (\mathrm{Cov}(X_i, X_j))_{i,j=1,\ldots,m}$ of $\mathbf{X}$ has an estimation

$$\hat{\Sigma}_{\mathbf{XX}} = \left(\frac{1}{n} \sum_{k=1}^n x_{ki} x_{kj} \right)_{i,j=1,\ldots,m}. \tag{6.11}$$

From this and (6.10), the estimate $\widehat{\mathrm{Var}}\, \mathbf{X}$ is the sum of the diagonal elements of $\hat{\Sigma}_{\mathbf{XX}}$, i.e.,

$$\widehat{\mathrm{Var}}\, \mathbf{X} = \mathrm{tr}\hat{\Sigma}_{\mathbf{XX}}. \tag{6.12}$$

Theorem 6.2. *Let $\hat{\lambda}_1 > \hat{\lambda}_2 > \cdots > \hat{\lambda}_m$ be the eigenvalues of $\hat{\Sigma}_{\mathbf{XX}}$ and $\hat{\mathbf{e}}_1, \hat{\mathbf{e}}_2, \ldots, \hat{\mathbf{e}}_m$ be corresponding eigenvectors with unit length. Then, for $1 \le k \le m$, the first k eigenvectors minimize*

$$\hat{\epsilon}_k = \frac{1}{n} \sum_{j=1}^n \left\| \mathbf{x}_j - \sum_{i=1}^k (\mathbf{x}_j, \hat{\mathbf{e}}_i) \hat{\mathbf{e}}_i \right\|_2^2$$

and $\hat{\epsilon}_k = \mathrm{tr}\hat{\Sigma}_{\mathbf{XX}} - \sum_1^k \hat{\lambda}_j$.

Proof. For the case $k = 1$, from $\| \mathbf{x}_j - (\mathbf{x}_j, \hat{\mathbf{e}}_1)\hat{\mathbf{e}}_1 \|_2^2 = \|\mathbf{x}_j\|_2^2 - (\mathbf{x}_j, \hat{\mathbf{e}}_1)^2$, it follows that

$$\hat{\epsilon}_1 = \frac{1}{n} \sum_1^n \left(\|\mathbf{x}_j\|_2^2 - (\mathbf{x}_j, \hat{\mathbf{e}}_1)^2 \right)$$

$$= \frac{1}{n} \sum_1^n \|\mathbf{x}_j\|_2^2 - \frac{1}{n} \sum_1^n (\mathbf{x}_j, \hat{\mathbf{e}}_1)^2$$

$$= J_1 - J_2.$$

By (6.10) and (6.12),

$$J_1 = \frac{1}{n} \sum_1^n \|\mathbf{x}_j\|_2^2 = \frac{1}{n} \sum_{j=1}^n \sum_{i=1}^m x_{ji}^2 = \widehat{\mathrm{Var}}\, \mathbf{X} = \mathrm{tr}\hat{\Sigma}_{\mathbf{XX}}.$$

Notice that

$$(\mathbf{x}_j, \hat{\mathbf{e}}_1)^2 = \left(\sum_{i=1}^m x_{ji} \hat{e}_{1i} \right)^2 = \sum_{i=1}^m \sum_{l=1}^m x_{jl} x_{ji} \hat{e}_{1l} \hat{e}_{1i}.$$

Then

$$J_2 = \frac{1}{n} \sum_1^n (\mathbf{x}_j, \hat{\mathbf{e}}_1)^2 = \sum_{i=1}^m \sum_{l=1}^m \left(\frac{1}{n} \sum_{j=1}^n x_{ji} x_{jl} \right) \hat{e}_{1i} \hat{e}_{1l}.$$

Under the constraint condition $\|\hat{\mathbf{e}}_1\|_2^2 = \sum_1^m \hat{e}_{1i}^2 = 1$, similarly to the argument of Theorem 6.1, when $\hat{\mathbf{e}}_1$ is the eigenvector of $\hat{\Sigma}_{\mathbf{XX}}$, J_2 takes extreme $\hat{\lambda}_1$, i.e.,

$$J_2 = \hat{\lambda}_1.$$

Therefore,

$$\hat{\epsilon}_1 = \mathrm{tr}\hat{\Sigma}_{\mathbf{XX}} - \hat{\lambda}_1.$$

More generally, $\hat{\mathbf{e}}_1, \ldots, \hat{\mathbf{e}}_k$ are such that $\hat{\epsilon}_k$ is minimal and

$$\hat{\epsilon}_k = \mathrm{tr}\hat{\Sigma}_{\mathbf{XX}} - \sum_1^k \hat{\lambda}_j.$$

$\square$

6.4 ROTATION OF EOFs

Let $\mathbf{X} = (X_1, \ldots, X_m)^{\mathrm{T}}$ be an m-dimensional random vector and $\mathbf{e}_1, \ldots, \mathbf{e}_k$ be its first k EOFs. Then

$$\mathbf{X} = \sum_1^k \alpha_i \mathbf{e}_i + \epsilon,$$

where $\alpha_i = (\mathbf{X}, \mathbf{e}_i)$ and $\sum_1^k \alpha_i \mathbf{e}_i$ contains a substantial fraction of the total variance of $\mathbf{X}$. Its matrix form is

$$\mathbf{X} = (\mathbf{e}_1 | \cdots | \mathbf{e}_k)\alpha + \epsilon, \tag{6.13}$$

where $\alpha = (\alpha_1, \ldots, \alpha_k)^{\mathrm{T}}$ and $\epsilon = (\epsilon_1, \ldots, \epsilon_k)^{\mathrm{T}}$. Take an orthogonal matrix $G = (g_{ij})_{i,j=1,\ldots,k}$ of order k. Let

$$F = (\mathbf{e}_1 | \cdots | \mathbf{e}_k)G.$$

Then

$$F = (\mathbf{f}_1 | \cdots | \mathbf{f}_k),$$

where $\mathbf{f}_i = \sum_1^k g_{ij}\mathbf{e}_j (i = 1, \ldots, k)$. The vectors $\mathbf{f}_1, \ldots, \mathbf{f}_k$ are called *rotated EOFs*. This implies that

$$(\mathbf{e}_1 | \cdots | \mathbf{e}_k) = FG^{-1} = FG^{\mathrm{T}}.$$

From this and (6.13),

$$\mathbf{X} = F\beta + \epsilon = \sum_1^k \beta_i \mathbf{f}_i + \epsilon, \tag{6.14}$$

where $\beta = (\beta_1, \ldots, \beta_k)^{\mathrm{T}}$ satisfies $\beta = G^{\mathrm{T}}\alpha$. The set $\beta_1, \ldots, \beta_k$ are called *rotated EOF coefficients*.

Since $\mathbf{e}_1, \ldots, \mathbf{e}_k$ are orthonormal,

$$(\mathbf{e}_1 | \cdots | \mathbf{e}_k)^{\mathrm{T}}(\mathbf{e}_1 | \cdots | \mathbf{e}_k) = I_k,$$

where I_k is the unit matrix of order k. Since G is an orthogonal matrix of order k,

$$G^{\mathrm{T}}G = I_k,$$

and so

$$F^{\mathrm{T}}F = G^{\mathrm{T}}(\mathbf{e}_1 | \cdots | \mathbf{e}_k)^{\mathrm{T}}(\mathbf{e}_1 | \cdots | \mathbf{e}_k)G = G^{\mathrm{T}}I_k G = G^{\mathrm{T}}G = I_k,$$

i.e., F is an orthogonal matrix of order k and its k column vectors are orthonormal.

Since $E[\alpha] = \mathbf{0}$,

$$E[\beta] = G^{\mathrm{T}}E[\alpha] = \mathbf{0}$$

and the covariance matrix of β is

$$\mathrm{Cov}(\beta, \beta) = E[\beta\beta^{\mathrm{T}}] = E[G^{\mathrm{T}}\alpha\alpha^{\mathrm{T}}G] = G^{\mathrm{T}}E[\alpha\alpha^{\mathrm{T}}]G = G^{\mathrm{T}}\mathrm{Cov}(\alpha, \alpha)G.$$

By Proposition 6.2,

$$\mathrm{Cov}(\alpha_i, \alpha_j) = \lambda_i \delta_{ij} \quad (i, j = 1, \ldots, k),$$

where λ_i is the ith eigenvalue of the covariance matrix of $\mathbf{X}$ in decreasing order. So

$$\text{Cov}(\alpha, \alpha) = \text{diag}(\lambda_1, \ldots, \lambda_k),$$
$$\text{Cov}(\beta, \beta) = G^T \text{diag}(\lambda_1, \ldots, \lambda_k)G.$$

In general, $\text{Cov}(\beta, \beta)$ is not a diagonal matrix, so

$$\text{Cov}(\beta_i, \beta_j) \neq 0 \quad (i \neq j),$$

i.e., the rotated EOF coefficients β_i and β_j $(i \neq j)$ are correlated.

How do we choose the orthogonal matrix G in the EOF "rotation" procedure? Denote rotated EOFs by

$$\mathbf{f}_j = (f_{j1}, \ldots, f_{jm})^T \quad (j = 1, \ldots, k).$$

Define the *normalized square variance* of $\mathbf{f}_j$ as

$$V(\mathbf{f}_j) = \frac{1}{m} \sum_1^m \left(\frac{f_{ji}^2}{h_i^2} \right)^2 - \left(\frac{1}{m} \sum_1^k \frac{f_{ji}^2}{h_i^2} \right)^2, \tag{6.15}$$

where

$$h_i^2 = \sum_1^k e_{ji}^2.$$

In application, one often chooses G such that $\sum_1^k V(\mathbf{f}_j)$ attains the maximal value. This method is called the *varimax method*.

Example 6.2. Consider the case $k = 2$. Let $\mathbf{X}$ be an m-dimensional random vector with mean 0, and let $\mathbf{e}_1$ and $\mathbf{e}_2$ be the first two eigenvectors. If the orthogonal matrix G of order 2 is taken as

$$G = \frac{1}{2} \begin{pmatrix} \sqrt{3} & -1 \\ 1 & \sqrt{3} \end{pmatrix},$$

then the rotated EOFs $\mathbf{f}_1$ and $\mathbf{f}_2$ satisfy

$$(\mathbf{f}_1 | \mathbf{f}_2) = \frac{1}{2}(\mathbf{e}_1 | \mathbf{e}_2) \begin{pmatrix} \sqrt{3} & -1 \\ 1 & \sqrt{3} \end{pmatrix} = \frac{1}{2} \left(\sqrt{3}\mathbf{e}_1 + \mathbf{e}_2 | -\mathbf{e}_1 + \sqrt{3}\mathbf{e}_2 \right)$$

or

$$\mathbf{f}_1 = \frac{\sqrt{3}}{2}\mathbf{e}_1 + \frac{1}{2}\mathbf{e}_2,$$
$$\mathbf{f}_2 = -\frac{1}{2}\mathbf{e}_1 + \frac{\sqrt{3}}{2}\mathbf{e}_2.$$

The rotated EOF coefficients β_1 and β_2 satisfy

$$\begin{pmatrix} \beta_1 \\ \beta_2 \end{pmatrix} = G^T \begin{pmatrix} \alpha_1 \\ \alpha_2 \end{pmatrix} = \frac{1}{2} \begin{pmatrix} \sqrt{3} & 1 \\ -1 & \sqrt{3} \end{pmatrix} \begin{pmatrix} \alpha_1 \\ \alpha_2 \end{pmatrix}$$

or

$$\beta_1 = \frac{\sqrt{3}}{2}\alpha_1 + \frac{1}{2}\alpha_2,$$

$$\beta_2 = -\frac{1}{2}\alpha_1 + \frac{\sqrt{3}}{2}\alpha_2.$$

The eigenvectors $\mathbf{e}_1$ and $\mathbf{e}_2$ are orthonormal, so are $\mathbf{f}_1$ and $\mathbf{f}_2$. The EOF coefficients α_1 and α_2 are uncorrelated, while the rotated EOF coefficients β_1 and β_2 are correlated since

$$\text{Cov}(\beta_1, \beta_2) = E[\beta_1\beta_2] = \frac{\sqrt{3}}{4}(-E[\alpha_1^2] + E[\alpha_2^2])$$

$$= \frac{\sqrt{3}}{4}(-\text{Cov}(\alpha_1, \alpha_1) + \text{Cov}(\alpha_2, \alpha_2))$$

$$= \frac{\sqrt{3}}{4}(\lambda_2 - \lambda_1),$$

where λ_1 and λ_2 are the first two eigenvalues.

The EOF decomposition of $\mathbf{X}$ is

$$\mathbf{X} = \alpha_1\mathbf{e}_1 + \alpha_2\mathbf{e}_2 + \epsilon,$$

where $\alpha_i = (\mathbf{X}, \mathbf{e}_i)$ $(i = 1, 2)$, and

$$\text{Var}(\alpha_1\mathbf{e}_1) = \lambda_1,$$
$$\text{Var}(\alpha_2\mathbf{e}_2) = \lambda_2,$$
$$\text{Var}(\alpha_1\mathbf{e}_1 + \alpha_2\mathbf{e}_2) = \lambda_1 + \lambda_2.$$

The rotated EOF decomposition of $\mathbf{X}$ is

$$\mathbf{X} = \beta_1\mathbf{f}_1 + \beta_2\mathbf{f}_2 + \epsilon,$$

where $\beta_i = (\mathbf{X}, \mathbf{f}_i)$ $(i = 1, 2)$. Notice that

$$\text{Var}(\beta_1\mathbf{f}_1) = \text{Var }\beta_1 = \text{Var}\left(\frac{\sqrt{3}}{2}\alpha_1 + \frac{1}{2}\alpha_2\right)$$

$$= E\left[\frac{3}{4}\alpha_1^2 + \frac{\sqrt{3}}{2}\alpha_1\alpha_2 + \frac{1}{4}\alpha_2^2\right] = \frac{3}{4}\lambda_1 + \frac{1}{4}\lambda_2$$

and

$$\text{Var}(\beta_2\mathbf{f}_2) = \text{Var }\beta_2 = \text{Var}\left(-\frac{1}{2}\alpha_1 + \frac{\sqrt{3}}{2}\alpha_2\right)$$

$$= E\left[\frac{1}{4}\alpha_1^2 - \frac{\sqrt{3}}{2}\alpha_1\alpha_2 + \frac{3}{4}\alpha_2^2\right] = \frac{1}{4}\lambda_1 + \frac{3}{4}\lambda_2.$$

So

$$\text{Var}(\beta_1 \mathbf{f}_1 + \beta_2 \mathbf{f}_2) = \lambda_1 + \lambda_2,$$

but

$$\text{Var}(\alpha_1 \mathbf{e}_1) \neq \text{Var}(\beta_1 \mathbf{f}_1),$$
$$\text{Var}(\alpha_2 \mathbf{e}_2) \neq \text{Var}(\beta_2 \mathbf{f}_2).$$

Let $\alpha_i^+ = \frac{\alpha_i}{\sqrt{\lambda_i}}$ and $\mathbf{e}_i^+ = \sqrt{\lambda_i} \mathbf{e}_i$ ($i = 1, 2$). Then the resulting rotated EOFs are

$$\mathbf{f}_i^+ = \sqrt{\lambda_i} \mathbf{f}_i \quad (i = 1, 2).$$

Since $\|\mathbf{f}_i^+\|_2 = \sqrt{\lambda_i}$, the rotated EOFs $\mathbf{f}_1^+$ and $\mathbf{f}_2^+$ are not orthonormal. Let

$$\beta_1^+ = \frac{\sqrt{3}}{2}\alpha_1^+ + \frac{1}{2}\alpha_2^+,$$
$$\beta_2^+ = -\frac{1}{2}\alpha_1^+ + \frac{\sqrt{3}}{2}\alpha_2^+.$$

Then the resulting rotated EOF coefficients β_1^+ and β_2^+ are uncorrelated since

$$\text{Cov}(\beta_1^+, \beta_2^+) = E[\beta_1^+ \beta_2^+] = \frac{\sqrt{3}}{4}(E[(\alpha_2^+)^2] - E[(\alpha_1^+)^2])$$
$$= \frac{\sqrt{3}}{4}\left(\frac{E[\alpha_2^2]}{\lambda_2} - \frac{E[\alpha_1^2]}{\lambda_1}\right)$$
$$= \frac{\sqrt{3}}{4}\left(\frac{\text{Var}\,\alpha_2}{\lambda_2} - \frac{\text{Var}\,\alpha_1}{\lambda_1}\right) = 0.$$

Example 6.3. Let $\mathbf{X}$ be an m-dimensional random vector with mean 0, and let $\mathbf{e}_1, \ldots, \mathbf{e}_k$ be the first k EOFs. Take an orthogonal matrix of order k:

$$G_{ls} = (g_{ij})_{i,j=1,\ldots,k} \quad (l < s),$$

where

$$g_{ij} = \begin{cases} 1, & i = j \text{ and } i \neq l, s, \\ \cos\varphi, & i, j = l \text{ or } i, j = s, \\ -\sin\varphi, & i = l, j = s, \\ \sin\varphi, & i = s, j = l, \\ 0, & \text{otherwise.} \end{cases}$$

The rotated EOFs $\mathbf{f}_1, \ldots, \mathbf{f}_k$ satisfy

$$(\mathbf{f}_1 | \cdots | \mathbf{f}_k) = (\mathbf{e}_1 | \cdots | \mathbf{e}_k)(g_{ij})_{i,j=1,\ldots,k},$$

where $\mathbf{f}_k = \mathbf{e}_k (k \neq l, s)$. Denote

$$\mathbf{f}_l = (f_{1l}, \ldots, f_{kl})^{\mathrm{T}},$$

$$\mathbf{f}_s = (f_{1s}, \ldots, f_{ks})^\mathrm{T},$$
$$\mathbf{e}_l = (e_{1l}, \ldots, e_{kl})^\mathrm{T},$$
$$\mathbf{e}_s = (e_{1s}, \ldots, e_{ks})^\mathrm{T}.$$

Then

$$\begin{pmatrix} f_{1l} & f_{1s} \\ \vdots & \vdots \\ f_{kl} & f_{ks} \end{pmatrix} = \begin{pmatrix} e_{1l} & e_{1s} \\ \vdots & \vdots \\ e_{kl} & e_{ks} \end{pmatrix} \begin{pmatrix} \cos\varphi & -\sin\varphi \\ \sin\varphi & \cos\varphi \end{pmatrix},$$

i.e.,

$$f_{il} = e_{il}\cos\varphi + e_{is}\sin\varphi,$$
$$f_{is} = -e_{il}\sin\varphi + e_{is}\cos\varphi \quad (i = 1, \ldots, k). \tag{6.16}$$

Choose φ such that $\sum_1^k V(\mathbf{f}_j)$ attains the maximal value. Since $\mathbf{f}_j = \mathbf{e}_j (j \neq l, s)$,

$$V(\mathbf{f}_j) = V(\mathbf{e}_j) \quad (j \neq l, s)$$

are independent of φ. So we need only to choose φ such that $V(\mathbf{f}_l) + V(\mathbf{f}_s)$ attains the maximal value. The extremal point φ can be found from the equation $\frac{\partial}{\partial \varphi}(V(\mathbf{f}_l) + V(\mathbf{f}_s)) = 0$.

6.5 COMPLEX EOFs AND HILBERT EOFs

The EOF analysis of complex-valued random vectors is called a *complex EOF analysis*. In climate change research, for a real-valued random vector, one often adds its Hilbert transforms as the imaginary part to form a complex-valued random vector, and then does a complex EOF analysis. Such a complex EOF analysis is called a *Hilbert EOF analysis*.

Let $\mathbf{Z} = (Z_1, \ldots, Z_m)$ be an m-dimensional complex-valued vector with mean 0 and

$$\Sigma_{\mathbf{ZZ}} = (E[Z_j \overline{Z}_k])_{j,k=1,\ldots,m}$$

be the covariance matrix of $\mathbf{Z}$. Clearly, $\Sigma_{\mathbf{ZZ}}$ is a conjugate symmetric matrix and its eigenvalues are non-negative real numbers. Suppose that these eigenvalues are in decreasing order $\lambda_1 > \cdots > \lambda_m$. Denote the corresponding eigenvectors with unit length by $\mathbf{e}_1, \ldots, \mathbf{e}_m$. Then

$$\Sigma_{\mathbf{ZZ}}\mathbf{e}_k = \lambda_k \mathbf{e}_k,$$
$$\|\mathbf{e}_k\|_2 = 1 \quad (k = 1, \ldots, m),$$
$$\mathbf{Z} = \sum_1^m \alpha_k \mathbf{e}_k, \quad \text{where } \alpha_k = (\mathbf{Z}, \mathbf{e}_k). \tag{6.17}$$

Let $e_k = (e_{1k}, \ldots, e_{mk})^T$. Then

$$\alpha_k = \sum_{j=1}^{m} Z_j e_{jk}.$$

The vectors $\mathbf{e}_1, \ldots, \mathbf{e}_m$ are called *complex EOFs*. Similarly to the argument of Theorem 6.1, for each $n(n \leq m)$, the expected error is

$$\epsilon_n = E\left[\left\|\mathbf{Z} - \sum_1^n \alpha_k \mathbf{e}_k\right\|_2^2\right] = \mathrm{Var}\,\mathbf{Z} - \sum_1^n \lambda_k,$$

which is smaller for the EOFs than for any other basis. By (6.17), for any φ_k,

$$\Sigma_{ZZ} e^{\mathrm{i}\varphi_k} \mathbf{e}_k = \lambda_k e^{\mathrm{i}\varphi_k} \mathbf{e}_k,$$

$$\|e^{\mathrm{i}\varphi_k} \mathbf{e}_k\|_2 = \|\mathbf{e}_k\|_2 = 1 \quad (k = 1, \ldots, m),$$

i.e., $e^{\mathrm{i}\varphi_k} \mathbf{e}_k (k = 1, \ldots, m)$ are also EOFs of $\mathbf{Z}$. Denote

$$\widetilde{\mathbf{e}}_k = e^{\mathrm{i}\varphi_k} \mathbf{e}_k.$$

We may choose φ_k such that $\mathrm{Re}\widetilde{\mathbf{e}}_k$ and $\mathrm{Im}\widetilde{\mathbf{e}}_k$ are orthogonal, i.e.,

$$(\mathrm{Re}\widetilde{\mathbf{e}}_k, \mathrm{Im}\widetilde{\mathbf{e}}_k) = 0.$$

In fact,

$$\widetilde{\mathbf{e}}_k = e^{\mathrm{i}\varphi_k} \mathbf{e}_k = (\cos\varphi_k + \mathrm{i}\sin\varphi_k)(\mathrm{Re}\mathbf{e}_k + \mathrm{i}\mathrm{Im}\mathbf{e}_k).$$

Taking the real part and imaginary part on both sides, we get

$$\mathrm{Re}\widetilde{\mathbf{e}}_k = (\mathrm{Re}\mathbf{e}_k)\cos\varphi_k - (\mathrm{Im}\mathbf{e}_k)\sin\varphi_k,$$

$$\mathrm{Im}\widetilde{\mathbf{e}}_k = (\mathrm{Re}\mathbf{e}_k)\sin\varphi_k + (\mathrm{Im}\mathbf{e}_k)\cos\varphi_k.$$

The inner product of two vectors $\mathrm{Re}\widetilde{\mathbf{e}}_k$ and $\mathrm{Im}\widetilde{\mathbf{e}}_k$ is

$$(\mathrm{Re}\widetilde{\mathbf{e}}_k, \mathrm{Im}\widetilde{\mathbf{e}}_k) = (\mathrm{Re}\mathbf{e}_k, \mathrm{Re}\mathbf{e}_k)\cos\varphi_k\sin\varphi_k - (\mathrm{Re}\mathbf{e}_k, \mathrm{Im}\mathbf{e}_k)\sin^2\varphi_k$$

$$+ (\mathrm{Re}\mathbf{e}_k, \mathrm{Im}\mathbf{e}_k)\cos^2\varphi_k - (\mathrm{Im}\mathbf{e}_k, \mathrm{Im}\mathbf{e}_k)\sin\varphi_k\cos\varphi_k$$

$$= \frac{1}{2}\left(\|\mathrm{Re}\mathbf{e}_k\|_2^2 - \|\mathrm{Im}\mathbf{e}_k\|_2^2\right)\sin(2\varphi_k) + (\mathrm{Re}\mathbf{e}_k, \mathrm{Im}\mathbf{e}_k)\cos(2\varphi_k).$$

So $(\mathrm{Re}\widetilde{\mathbf{e}}_k, \mathrm{Im}\widetilde{\mathbf{e}}_k) = 0$ if and only if

$$\tan(2\varphi_k) = -\frac{2(\mathrm{Re}\mathbf{e}_k, \mathrm{Im}\mathbf{e}_k)}{\|\mathrm{Re}\mathbf{e}_k\|_2^2 - \|\mathrm{Im}\mathbf{e}_k\|_2^2}, \tag{6.18}$$

i.e., $\mathrm{Re}\widetilde{\mathbf{e}}_k$ and $\mathrm{Im}\widetilde{\mathbf{e}}_k$ are orthogonal if and only if φ_k satisfies (6.18).

Similarly, we may also choose φ_k such that $\widetilde{\alpha}_k = (\mathbf{Z}, \widetilde{\mathbf{e}}_k)$ satisfies $\mathrm{Cov}(\mathrm{Re}\widetilde{\alpha}_k, \mathrm{Im}\widetilde{\alpha}_k) = 0$.

Let $\mathbf{X}(t) = (X_1(t), \ldots, X_m(t))$ be an m-dimensional vector consisting of real-valued random processes. Its *Hilbert transform* is defined as $\widetilde{\mathbf{X}}(t) = (\widetilde{X}_1(t), \ldots, \widetilde{X}_m(t))$, where $\widetilde{X}_k(t)$ is the Hilbert transform of the component $X_k(t)$. The random process $\mathbf{Y}(t) = \mathbf{X}(t) + i\widetilde{\mathbf{X}}(t)$ is an m-dimensional complex-valued vector. The EOFs of $\mathbf{Y}(t)$ are called *Hilbert EOFs*.

We know from Section 2.4 the following two points:

(1) The Hilbert transform of $f \in L^2_{2\pi}$ is

$$\widetilde{f}(t) = -\frac{1}{2\pi} \lim_{\epsilon \to 0} \int_\epsilon^\pi \frac{f(t+\tau) - f(t-\tau)}{\tan \frac{\tau}{2}} d\tau.$$

Especially, if $f(t) = \cos t$, then $\widetilde{f}(t) = \sin t$; if $f(t) = \sin t$, then $\widetilde{f}(t) = -\cos t$. Moreover, the Fourier coefficients of f and $\widetilde{f}$ satisfy

$$c_n(\widetilde{f}) = -ic_n(f) \quad (n \in \mathbb{Z}).$$

(2) The Hilbert transform of $f \in L^2(\mathbb{R})$ is

$$\widetilde{f}(t) = \frac{1}{\pi} \lim_{\epsilon \to 0} \int_{|t-\tau|>\epsilon} \frac{f(\tau)}{t-\tau} d\tau =: \text{p.v.} \frac{1}{\pi} \int_\mathbb{R} \frac{f(\tau)}{t-\tau} d\tau,$$

where p.v. is the principal value. Write it in a convolution form:

$$\widetilde{f}(t) = \frac{1}{\pi} \left(f * \frac{1}{\tau} \right)(t).$$

Moreover, the Fourier transforms of f and $\widetilde{f}$ satisfy

$$\widehat{\widetilde{f}}(\omega) = -i\widehat{f}(\omega)\text{sgn}\omega.$$

On the basis of the theory of continuous Hilbert transforms, one defines the Hilbert transform of a finite time series with mean 0 by the following methods:

(i) *A simple method.* Suppose that $\{x_n\}_{n=0,\ldots,N-1}$ is a finite time series. The discrete Fourier transforms of $x_n (n = 0, \ldots, N-1)$ are

$$X_k = \frac{1}{N} \sum_{n=0}^{N-1} x_n e^{-ik\frac{2\pi n}{N}} \quad (k = 0, \ldots, N-1).$$

For $m = 0, \ldots, N-1$, let

$$W_m = \begin{cases} 2X_m, & 0 < m < \frac{N}{2}, \\ 0, & \text{otherwise.} \end{cases}$$

The inverse discrete Fourier transforms of $W_m (m = 0, \ldots, N - 1)$ are

$$\xi_n = \sum_{m=0}^{N-1} W_m e^{in\frac{2\pi m}{N}} \quad (n = 0, \ldots, N - 1).$$

Define the *Hilbert transform* of x_n as $\tilde{x}_n = \text{Im}\xi_n$.

(ii) *Convolution method.* Suppose that $\{x_n\}_{n \in \mathbb{Z}}$ is a time series. Let

$$h_l = \begin{cases} \frac{2}{\pi l} & \text{if } l \text{ is odd,} \\ 0 & \text{if } l \text{ is even.} \end{cases} \tag{6.19}$$

Define the *Hilbert transforms* of $x_n (n \in \mathbb{Z})$ as

$$\tilde{x}_n = \sum_l h_l x_{n-l} \quad (n \in \mathbb{Z}).$$

Their frequency response

$$\sum_n \tilde{x}_n e^{-in\omega} = -\text{isgn}\omega \sum_n x_n e^{-in\omega}.$$

(iii) *Fourier series method.* Let the Fourier expansions of $x_n (n = 1, \ldots, N)$ be

$$x_n = \sum_{k=1}^{N} \left(a_k \cos \frac{2\pi kn}{N} + b_k \sin \frac{2\pi kn}{N} \right),$$

where

$$a_k = \frac{1}{N} \sum_{n=1}^{N} x_n \cos \frac{2\pi kn}{N},$$

$$b_k = \frac{1}{N} \sum_{n=1}^{N} x_n \sin \frac{2\pi kn}{N}.$$

Define the *Hilbert transform* of x_n as

$$\tilde{x}_n = \sum_{k=1}^{N} \left(b_k \cos \frac{2\pi kn}{N} - a_k \sin \frac{2\pi kn}{N} \right).$$

Proposition 6.5. *Let $X(n)$ $(n \in \mathbb{Z})$ be a stationary time series with $E[X(n)] = 0$ and $\tilde{X}(n)$ be the Hilbert transform of $X(n)$. Then $\tilde{X}(n)$ is also a stationary time series with $E[\tilde{X}(n)] = 0$ and the autorelation functions satisfy $R_{XX}(\tau) = R_{\tilde{X}\tilde{X}}(\tau)$ and the cross-correlation functions satisfy $R_{\tilde{X}X}(\tau) = -R_{X\tilde{X}}(\tau)$.*

Let $X(t)$ be a stationary random process with mean 0. Consider a complexified process:

$$Y(t) = X(t) + i\widetilde{X}(t).$$

The covariance matrix Σ_{YY} of $Y(t)$ is

$$
\begin{aligned}
\Sigma_{YY} &= E\left[Y(t)\overline{Y}(t)^{\mathrm{T}}\right] \\
&= E\left[\left(X(t) + i\widetilde{X}(t)\right)\left(X(t) - i\widetilde{X}(t)\right)^{\mathrm{T}}\right] \\
&= E\left[X(t)X(t)^{\mathrm{T}}\right] + E\left[\widetilde{X}(t)\widetilde{X}(t)^{\mathrm{T}}\right] \\
&\quad + i\left(E\left[\widetilde{X}(t)X(t)^{\mathrm{T}}\right] - E\left[X(t)\widetilde{X}(t)^{\mathrm{T}}\right]\right).
\end{aligned}
$$

By Proposition 6.5,

$$
\begin{aligned}
\Sigma_{YY} &= 2E\left[X(t)X(t)^{\mathrm{T}}\right] + 2iE\left[\widetilde{X}(t)X(t)^{\mathrm{T}}\right] \\
&= 2(\Sigma_{XX} + \Sigma_{\widetilde{X}X}),
\end{aligned}
$$

where Σ_{YY} is a Hermitian matrix with non-negative eigenvalues λ_k. The eigenvectors $\mathbf{e}_k$ corresponding to λ_k are called *Hilbert EOFs* of $X(t)$. Expand $Y(t)$ into a series:

$$Y(t) = \sum_k \alpha_k(t)\mathbf{e}_k,$$

where $\alpha_k(t) = (Y(t), \mathbf{e}_k)$. The coefficients $\alpha_k(t)$ are called *Hilbert EOF coefficients* of $Y(t)$.

Proposition 6.6. *Let $X(t)$ be an m-dimensional real-valued random process, $Y(t) = X(t) + i\widetilde{X}(t)$, $\{\mathbf{e}_k\}$ be Hilbert EOFs of $X(t)$, and $\alpha_k(t)$ be Hilbert EOF coefficients. Then*

(i) $\widetilde{\mathrm{Re}}(\alpha_k(t)) = \mathrm{Im}(\alpha_k(t))$ *and* $\widetilde{\mathrm{Im}}(\alpha_k(t)) = -\mathrm{Re}(\alpha_k(t))$;

(ii) $\mathrm{Var}(\mathrm{Re}(\alpha_k(t))) = \mathrm{Var}(\widetilde{\mathrm{Im}}(\alpha_k(t)))$;

(iii) $\widetilde{\mathrm{Re}}(\alpha_k(t)\mathbf{e}_k) = \mathrm{Im}(\alpha_k(t)\mathbf{e}_k)$.

6.6 SINGULAR VALUE DECOMPOSITION

Let C be an $m \times n$ real matrix with rank r. Then there exists an $m \times m$ orthogonal matrix A and an $n \times n$ orthogonal matrix B such that

$$C = A \begin{pmatrix} D & 0 \\ 0 & 0 \end{pmatrix} B^{\mathrm{T}}, \tag{6.20}$$

where $D = \mathrm{diag}(\sigma_1, \ldots, \sigma_r)$ $(r \leq \min\{m, n\}$ and $\sigma_1 \geq \sigma_2 \geq \cdots \sigma_r > 0)$, i.e.,

$$C = \sum_{1}^{r} \sigma_k \mathbf{A}_k \mathbf{B}_k^{\mathrm{T}},$$

where $\mathbf{A}_k, \mathbf{B}_k$ are the kth column vectors of the matrices A, B, respectively. Equality (6.20) is called the *singular value decomposition* of C, and each σ_k is called a *singular value*. The column vectors $\mathbf{A}_k$ of the matrix A are called the *left singular vectors* of C. The column vectors $\mathbf{B}_k$ of the matrix B are called the *right singular vectors* of C.

Proposition 6.7. *The following three points hold:*

(i) $C^{\mathrm{T}}\mathbf{A}_k = \sigma_k \mathbf{B}_k$ *and* $C\mathbf{B}_k = \sigma_k \mathbf{A}_k$.

(ii) $\sigma_k^2 (k = 1, \ldots, r)$ *is the kth non-zero eigenvalue of the matrices $C^{\mathrm{T}}C$ and CC^{T}. Other eigenvalues of the matrices $C^{\mathrm{T}}C$ and CC^{T} are both zero.*

(iii) $\mathbf{A}_k$ *is the eigenvector of the matrix CC^{T} and $\mathbf{B}_k$ is the eigenvector of the matrix $C^{\mathrm{T}}C$.*

If C is a real symmetric square matrix, then the left and right singular vectors of C are both ordinary eigenvectors and the singular values are both ordinary eigenvalues.

Let

$$\mathbf{X} = (X_1, \ldots, X_m)^{\mathrm{T}},$$

$$\mathbf{Y} = (Y_1, \ldots, Y_n)^{\mathrm{T}}$$

be two random vectors with mean $\mathbf{0}$. Choose a unit vector $\mathbf{P} \in \mathbb{R}^m$ and a unit vector $\mathbf{Q} \in \mathbb{R}^n$ such that the covariance $\mathrm{Cov}(\alpha, \beta)$ attains the maximal value, where

$$\alpha = (\mathbf{X}, \mathbf{P}),$$

$$\beta = (\mathbf{Y}, \mathbf{Q}).$$

From $\alpha\beta = \mathbf{P}^{\mathrm{T}}\mathbf{X}\mathbf{Y}^{\mathrm{T}}\mathbf{Q}$, it follows that

$$\mathrm{Cov}(\alpha, \beta) = E[\alpha\beta] = \mathbf{P}^{\mathrm{T}}E[\mathbf{X}\mathbf{Y}^{\mathrm{T}}]\mathbf{Q}$$

$$= \mathbf{P}^{\mathrm{T}}\mathrm{Cov}(\mathbf{X}, \mathbf{Y})\mathbf{Q}.$$

Now we find the maximal value of $\mathbf{P}^{\mathrm{T}}\mathrm{Cov}(\mathbf{X}, \mathbf{Y})\mathbf{Q}$ under the conditions $\mathbf{P}^{\mathrm{T}}\mathbf{P} = 1$ and $\mathbf{Q}^{\mathrm{T}}\mathbf{Q} = 1$.

By the Lagrange multiplier method, the vector form of a Lagrange function is

$$L(\mathbf{P}, \mathbf{Q}) = \mathbf{P}^{\mathrm{T}}\mathrm{Cov}(\mathbf{X}, \mathbf{Y})\mathbf{Q} - \mu_1(\mathbf{P}^{\mathrm{T}}\mathbf{P} - 1) - \mu_2(\mathbf{Q}^{\mathrm{T}}\mathbf{Q} - 1). \qquad (6.21)$$

Denote

$$\mathbf{P} = (p_1, \ldots, p_m)^{\mathrm{T}},$$

$$\mathbf{Q} = (q_1, \ldots, q_n)^{\mathrm{T}}.$$

Then

$$L(\mathbf{P}, \mathbf{Q}) = \sum_{k=1}^{m} \sum_{l=1}^{n} \mathrm{Cov}(X_k, Y_l) p_k q_l - \mu_1 \left(\sum_{1}^{m} p_k^2 - 1 \right) - \mu_2 \left(\sum_{1}^{n} q_l^2 - 1 \right).$$

Let $\frac{\partial L}{\partial p_k} = 0$ and $\frac{\partial L}{\partial q_l} = 0$. Then

$$\sum_{l=1}^{n} \mathrm{Cov}(X_k, Y_l) q_l = 2\mu_1 p_k,$$

$$\sum_{k=1}^{m} \mathrm{Cov}(X_k, Y_l) p_k = 2\mu_2 q_l. \tag{6.22}$$

Since $\mathbf{P}^\mathrm{T}\mathbf{P} = \|\mathbf{P}\|^2 = 1$ and $\mathbf{Q}^\mathrm{T}\mathbf{Q} = \|\mathbf{Q}\| = 1$,

$$\mathbf{P}^\mathrm{T}\Sigma_{\mathbf{XY}}\mathbf{Q} = 2\mu_1,$$

$$\mathbf{Q}^\mathrm{T}\Sigma_{\mathbf{XY}}^\mathrm{T}\mathbf{P} = 2\mu_2,$$

where $\Sigma_{\mathbf{XY}} = (\mathrm{Cov}(X_k, Y_l))_{k=1,\dots,m; l=1,\dots,n}$. Since $\mathbf{P}^\mathrm{T}\Sigma_{\mathbf{XY}}\mathbf{Q}$ is a constant,

$$\mathbf{P}^\mathrm{T}\Sigma_{\mathbf{XY}}\mathbf{Q} = \mathbf{Q}^\mathrm{T}\Sigma_{\mathbf{XY}}^\mathrm{T}\mathbf{P}.$$

So $2\mu_1 = 2\mu_2 =: \sigma$, and so

$$\mathbf{P}^\mathrm{T}\Sigma_{\mathbf{XY}}\mathbf{Q} = \sigma,$$

$$\mathbf{Q}^\mathrm{T}\Sigma_{\mathbf{XY}}^\mathrm{T}\mathbf{P} = \sigma. \tag{6.23}$$

By (6.22), it follows that

$$\Sigma_{\mathbf{XY}}\mathbf{Q} = \sigma\mathbf{P},$$

$$\Sigma_{\mathbf{XY}}^\mathrm{T}\mathbf{P} = \sigma\mathbf{Q}.$$

So

$$\Sigma_{\mathbf{XY}}\Sigma_{\mathbf{XY}}^\mathrm{T}\mathbf{P} = \sigma^2\mathbf{P},$$

$$\Sigma_{\mathbf{XY}}^\mathrm{T}\Sigma_{\mathbf{XY}}\mathbf{Q} = \sigma^2\mathbf{Q}.$$

Corresponding to same eigenvalue σ^2, the vectors $\mathbf{P}$ and $\mathbf{Q}$ are the eigenvectors of $\Sigma_{\mathbf{XY}}\Sigma_{\mathbf{XY}}^\mathrm{T}$ and $\Sigma_{\mathbf{XY}}^\mathrm{T}\Sigma_{\mathbf{XY}}$, respectively. The matrix $\Sigma_{\mathbf{XY}}\Sigma_{\mathbf{XY}}^\mathrm{T}$ is an $m \times m$ real symmetric matrix, while $\Sigma_{\mathbf{XY}}^\mathrm{T}\Sigma_{\mathbf{XY}}$ is an $n \times n$ real symmetric matrix. Therefore, their eigenvalues are both non-negative real numbers. Assume that $\Sigma_{\mathbf{XY}}$ has the rank r. Then both $\Sigma_{\mathbf{XY}}\Sigma_{\mathbf{XY}}^\mathrm{T}$ and $\Sigma_{\mathbf{XY}}^\mathrm{T}\Sigma_{\mathbf{XY}}$ have r non-zero same eigenvalues. Denote

$$\sigma_1^2 \geq \sigma_2^2 \geq \cdots \geq \sigma_r^2 > 0$$

and denote by $\mathbf{P}_1, \ldots, \mathbf{P}_r$ the corresponding r orthonormal eigenvectors of $\Sigma_{\mathbf{XY}} \Sigma_{\mathbf{XY}}^{\mathrm{T}}$, and denote by $\mathbf{Q}_1, \ldots, \mathbf{Q}_r$ the corresponding r orthonormal eigenvectors of $\Sigma_{\mathbf{XY}}^{\mathrm{T}} \Sigma_{\mathbf{XY}}$.

By (6.23), the first pair of patterns, $\mathbf{P}_1$ and $\mathbf{Q}_1$, is such that the correlation ρ_1 of α_1 and β_1,

$$\rho_1 = \mathrm{Cov}(\alpha_1, \beta_1) = \mathbf{P}_1^{\mathrm{T}} \mathrm{Cov}(\mathbf{X}, \mathbf{Y}) \mathbf{Q}_1 = \sigma_1,$$

is the maximal value, where

$$\alpha_1 = (\mathbf{X}, \mathbf{P}_1),$$

$$\beta_1 = (\mathbf{Y}, \mathbf{Q}_1).$$

The second pair of patterns, $\mathbf{P}_2$ and $\mathbf{Q}_2$, is such that the correlation ρ_2 of α_2 and β_2,

$$\rho_2 = \mathrm{Cov}(\alpha_2, \beta_2) = \mathbf{P}_2^{\mathrm{T}} \mathrm{Cov}(\mathbf{X}, \mathbf{Y}) \mathbf{Q}_2 = \sigma_2,$$

is the second maximal value, where

$$\alpha_2 = (\mathbf{X}, \mathbf{P}_2),$$

$$\beta_2 = (\mathbf{Y}, \mathbf{Q}_2).$$

Continuing this procedure, we find the rth pair of patterns, $\mathbf{P}_r$ and $\mathbf{Q}_r$, is such that the correlation ρ_r of α_r and β_r,

$$\rho_r = \mathrm{Cov}(\alpha_r, \beta_r) = \mathbf{P}_r^{\mathrm{T}} \mathrm{Cov}(\mathbf{X}, \mathbf{Y}) \mathbf{Q}_r = \sigma_r,$$

is the rth maximal value, where

$$\alpha_r = (\mathbf{X}, \mathbf{P}_r),$$

$$\beta_r = (\mathbf{Y}, \mathbf{Q}_r).$$

Proposition 6.8. *Let $\mathbf{X}$ be an m-dimensional vector and $\mathbf{Y}$ be an n-dimensional vector. Let the cross-covariance matrix $\Sigma_{\mathbf{XY}}$ have the rank r and the singular values $\sigma_k (k = 1, \ldots, r)$ satisfying $\sigma_1 \geq \cdots \geq \sigma_r > 0$. Let $\mathbf{P}_k (k = 1, \ldots, r)$ and $\mathbf{Q}_k (k = 1, \ldots, r)$ be the left and right singular vectors of $\Sigma_{\mathbf{XY}}$ corresponding to $\sigma_k (k = 1, \ldots, r)$, respectively. Then, for each $k (k = 1, \ldots, r)$, the pair $\mathbf{P}_k, \mathbf{Q}_k$ is such that the cross-covariance matrix $\mathrm{Cov}(\alpha_k, \beta_k)$ attains the local maximal value σ_k, where $\alpha_k = (\mathbf{X}, \mathbf{P}_k)$ and $\beta_k = (\mathbf{Y}, \mathbf{Q}_k)$.*

6.7 CANONICAL CORRELATION ANALYSIS

To discuss canonical correlation analysis, we first state the concept of the square root of a real symmetric matrix.

Let $C = (c_{ij})_{i,j=1,\ldots,m}$ be a real symmetric matrix. Then there exists an $m \times m$ orthogonal matrix G such that

$$C = G \operatorname{diag}(\lambda_1, \ldots, \lambda_m) G^{\mathrm{T}},$$

where $\lambda_1 \geq \lambda_2 \geq \cdots \geq \lambda_m > 0$. Define the *square root* of the real symmetric matrix C as

$$C^{1/2} = G \operatorname{diag}(\sqrt{\lambda_1}, \ldots, \sqrt{\lambda_m}) G^{\mathrm{T}}.$$

Then $C^{1/2} C^{1/2} = C$.

$C^{1/2}$ is a real symmetric matrix and $G^{\mathrm{T}} = G^{-1}$, so

$$
\begin{aligned}
C^{1/2} C^{1/2} &= C^{1/2} (C^{1/2})^{\mathrm{T}} \\
&= G \operatorname{diag}(\sqrt{\lambda_1}, \ldots, \sqrt{\lambda_m}) G^{\mathrm{T}} G \operatorname{diag}(\sqrt{\lambda_1}, \ldots, \sqrt{\lambda_m}) G^{\mathrm{T}} \\
&= G \operatorname{diag}(\sqrt{\lambda_1}, \ldots, \sqrt{\lambda_m}) \operatorname{diag}(\sqrt{\lambda_1}, \ldots, \sqrt{\lambda_m}) G^{\mathrm{T}} \\
&= G \operatorname{diag}(\lambda_1, \ldots, \lambda_m) G^{\mathrm{T}} = C.
\end{aligned}
$$

Canonical correlation analysis discusses the correlativity of two random vectors.

Let $\mathbf{X} = (X_1, \ldots, X_m)^{\mathrm{T}}$ be an m-dimensional real-valued random vector with mean 0 and let $\mathbf{Y} = (Y_1, \ldots, Y_n)^{\mathrm{T}}$ be an n-dimensional real-valued random vector with mean 0. Choose an m-dimensional vector $\mathbf{f}_1 = (f_{11}, \ldots, f_{1m})$ and an n-dimensional vector $\mathbf{g}_1 = (g_{11}, \ldots, g_{1n})$ such that

$$\alpha_1 = (\mathbf{X}, \mathbf{f}_1),$$

$$\beta_1 = (\mathbf{Y}, \mathbf{g}_1),$$

where α_1, β_1 satisfy the following two conditions:

(i) $\operatorname{Var} \alpha_1 = \operatorname{Var} \beta_1 = 1$;

(ii) the covariance $\operatorname{Cov}(\alpha_1, \beta_1)$ attains the maximal value.

From $\alpha_1 = \sum_1^m X_k f_{1k}$ and $\beta_1 = \sum_1^n Y_l g_{1l}$, it follows that

$$E[\alpha_1 \beta_1] = \sum_{k=1}^m \sum_{l=1}^n E[X_k Y_l] f_{1k} g_{1l}.$$

From $E[\mathbf{X}] = 0$ and $E[\mathbf{Y}] = 0$, it follows that

$$E[X_k] = E[Y_l] = 0,$$
$$E[\alpha_1] = E[\beta_1] = 0.$$

This implies that

$$\text{Cov}(\alpha_1, \beta_1) = \sum_{k=1}^{m} \sum_{l=1}^{n} \text{Cov}(X_k, Y_l) f_{1k} g_{1l} = \mathbf{f}_1^T \Sigma_{\mathbf{XY}} \mathbf{g}_1. \tag{6.24}$$

The constraint condition (i) can be written in the form

$$\mathbf{f}_1^T \Sigma_{\mathbf{XX}} \mathbf{f}_1 = \sum_{k=1}^{m} \sum_{l=1}^{m} \text{Cov}(X_k, X_l) f_{1k} f_{1l} = 1,$$

$$\mathbf{g}_1^T \Sigma_{\mathbf{YY}} \mathbf{g}_1 = \sum_{k=1}^{n} \sum_{l=1}^{n} \text{Cov}(Y_k, Y_l) g_{1k} g_{1l} = 1.$$

Let

$$C_1 = \Sigma_{\mathbf{XX}}^{1/2} \mathbf{f}_1,$$

$$D_1 = \Sigma_{\mathbf{YY}}^{1/2} \mathbf{g}_1.$$

Notice that

$$\Sigma_{\mathbf{XX}} = \Sigma_{\mathbf{XX}}^{1/2} \Sigma_{\mathbf{XX}}^{1/2},$$

$$\Sigma_{\mathbf{YY}} = \Sigma_{\mathbf{YY}}^{1/2} \Sigma_{\mathbf{YY}}^{1/2},$$

$$\left(\Sigma_{\mathbf{XX}}^{1/2}\right)^T = \Sigma_{\mathbf{XX}}^{1/2},$$

$$\left(\Sigma_{\mathbf{YY}}^{1/2}\right)^T = \Sigma_{\mathbf{YY}}^{1/2}.$$

It follows that

$$\mathbf{f}_1^T \Sigma_{\mathbf{XX}} \mathbf{f}_1 = \mathbf{f}_1^T \Sigma_{\mathbf{XX}}^{1/2} \Sigma_{\mathbf{XX}}^{1/2} \mathbf{f}_1 = C_1^T C_1,$$

$$\mathbf{g}_1^T \Sigma_{\mathbf{YY}} \mathbf{g}_1 = \mathbf{g}_1^T \Sigma_{\mathbf{YY}}^{1/2} \Sigma_{\mathbf{YY}}^{1/2} \mathbf{g}_1 = D_1^T D_1,$$

Therefore, $C_1^T C_1 = 1$ and $D_1^T D_1 = 1$.

Let $\tau_1 = \mathbf{f}_1^T \Sigma_{\mathbf{XY}} \mathbf{g}_1$. From

$$\mathbf{f}_1 = \Sigma_{\mathbf{XX}}^{-(1/2)} C_1,$$

$$\mathbf{g}_1 = \Sigma_{\mathbf{YY}}^{-(1/2)} D_1,$$

it follows that

$$\tau_1 = C_1^T S D_1, \tag{6.25}$$

where $S = \Sigma_{\mathbf{XX}}^{-(1/2)} \Sigma_{\mathbf{XY}} \Sigma_{\mathbf{YY}}^{-(1/2)}$. Therefore, the problem of finding the maximal value of the covariance $\text{Cov}(\alpha_1, \beta_1)$ reduces to that of finding the maximal value of τ_1 under the constraint condition $C_1^T C_1 = 1$ and $D_1^T D_1 = 1$.

By use of the Lagrange multiplier method, an argument similar to that in Section 6.6 shows that there exists a λ_1 such that

$$SS^T C_1 = \lambda_1 C_1,$$

$$S^T S D_1 = \lambda_1 D_1,$$

i.e., if C_1 and D_1 are the eigenvectors of SS^T and $S^T S$ corresponding to same eigenvalue λ_1, then τ_1 attains the maximal value and

$$\tau_1^2 = C_1^T S D_1 (C_1^T S D_1)^T$$
$$= \lambda_1,$$

where λ_1 is the maximal eigenvalue of SS^T. So $\tau_1 = \sqrt{\lambda_1}$.

In general, we have the following proposition.

Proposition 6.9. *Let* $\mathbf{X}$ *be an m-dimensional vector with mean* $\mathbf{0}$, $\mathbf{Y}$ *be an n-dimensional vector with mean* $\mathbf{0}$, *and the cross-covariance matrix* $\Sigma_{\mathbf{XY}}$ *have the rank r. Let the matrix*

$$S = \Sigma_{\mathbf{XX}}^{-(1/2)} \Sigma_{\mathbf{XY}} \Sigma_{\mathbf{YY}}^{-(1/2)}$$

and S have the singular values $\lambda_1 \geq \cdots \geq \lambda_r > 0$ *and* $\mathbf{C}_k, \mathbf{D}_k (k = 1, \ldots, r)$ *be the left and right singular vectors of the matrix S corresponding to* $\lambda_k (k = 1, \ldots, r)$. *Then, for each* $k(k = 1, \ldots, r)$,

$$\mathbf{f}_k = \Sigma_{\mathbf{XX}}^{-(1/2)} \mathbf{C}_k,$$

$$\mathbf{g}_k = \Sigma_{\mathbf{YY}}^{-(1/2)} \mathbf{D}_k$$

are such that

$$\alpha_k = (\mathbf{X}, \mathbf{f}_k),$$

$$\beta_k = (\mathbf{Y}, \mathbf{g}_k),$$

satisfy the following:

(i) *Var* $\alpha_k = $ *Var* $\beta_k = 1$;
(ii) *the cross-covariance* $\text{Cov}(\alpha_k, \beta_k)$ *attains the kth maximal value* $\sqrt{\lambda_k}$, *i.e.,*

$$\tau_k := \text{Cov}(\alpha_k, \beta_k) = \sqrt{\lambda_k} \quad (k = 1, \ldots, r).$$

Proposition 6.10. *Let* $\mathbf{f}_k, \mathbf{g}_k (k = 1, \ldots, r)$ *be stated as in Proposition 6.9. Then*

$$(\mathbf{f}_k, \mathbf{f}_l) = (\mathbf{g}_k, \mathbf{g}_l) = 0 \quad (k \neq l).$$

$$(\mathbf{f}_k, \mathbf{g}_l) = \delta_{k,l} \quad (k, l = 1, \ldots, r).$$

Proof. Notice that

$$(\mathbf{f}_k, \mathbf{g}_l) = \mathbf{f}_k^T \Sigma_{\mathbf{XY}} \mathbf{g}_l = \mathbf{C}_k^T \Sigma_{\mathbf{XX}}^{-(1/2)} \Sigma_{\mathbf{XY}} \Sigma_{\mathbf{YY}}^{-(1/2)} \mathbf{D}_l = \mathbf{C}_k^T SD_l.$$

Since $\mathbf{D}_l$ is the lth right singular vector of S corresponding to the singular value λ_l, by Proposition 6.7(i),

$$SD_l = \lambda_l \mathbf{C}_l,$$

where $\mathbf{C}_l$ is the lth left singular vector. So

$$(\mathbf{f}_k, \mathbf{g}_l) = \mathbf{C}_k^T SD_l = \lambda_l \mathbf{C}_k^T \mathbf{C}_l \quad (k \neq l).$$

By Proposition 6.7(iii), $\mathbf{C}_l$ is the eigenvector of SS^T. So $(\mathbf{f}_k, \mathbf{g}_l) = \delta_{k,l}$. Similarly, $(\mathbf{f}_k, \mathbf{f}_l) = (\mathbf{g}_k, \mathbf{g}_l) = 0 (k \neq l)$. □

6.8 SINGULAR SPECTRUM ANALYSIS

Singular spectrum analysis is a time-series analysis techniques to identify recurrent patterns in univariate time series.

Suppose that $\mathbf{X}_t$ is a wide-sense stationary (WSS) time series with $E[\mathbf{X}_t] = 0$ and $\mathbf{Y}_t = (\mathbf{X}_{t+1}, \ldots, \mathbf{X}_{t+m})^T$. Let $\lambda_k (k = 1, \ldots, m)$ be eigenvalues of the covariance matrix $\Sigma_{\mathbf{Y}_t \mathbf{Y}_t}$ and $\mathbf{e}_k (k = 1, \ldots, m)$ be the eigenvectors corresponding to λ_k. Since $\mathbf{X}_t$ is WSS,

$$\text{Var} \, \mathbf{X}_{t+k} = \text{Var} \, \mathbf{X}_t \quad (k = 1, \ldots, m),$$

and so the total variance of $\mathbf{Y}_t$

$$\text{Var} \, Y_t = \sum_1^m \text{Var} \, \mathbf{X}_{t+k} = m \text{Var} \, \mathbf{X}_t.$$

From this and $\text{Var} \, \mathbf{Y}_t = \sum_1^m \lambda_k$, it follows that

$$\text{Var} \, \mathbf{Y}_t = \sum_1^m \lambda_k = m \text{Var} \, \mathbf{X}_t.$$

Notice that $\mathbf{e}_k (k = 1, \ldots, m)$ are EOFs of $\mathbf{Y}_t$. Let $\alpha_k(t)$ $(k = 1, \ldots, m)$ be the EOF coefficients. Then

$$\mathbf{Y}_t = \sum_1^m \alpha_k(t) \mathbf{e}_k, \tag{6.26}$$

where $\alpha_k(t) = (\mathbf{Y}_t, \mathbf{e}_k) = \sum_1^m \mathbf{X}_{t+j} e_{kj}$ $(\mathbf{e}_k = (e_{k1}, \ldots, e_{km}))$. Clearly, $\alpha_k(t) \in$ WSS. Denote

$$\alpha(t) = (\alpha_1(t), \ldots, \alpha_m(t))^T.$$

$$P = (e_{kj})_{k,j=1,\ldots,m}.$$

Then $\alpha(t) = P\mathbf{Y}_t$.

The τ lag covariance matrix of $\mathbf{Y}_t$ is

$$\Sigma_{\mathbf{Y}_t\mathbf{Y}_t}(\tau) = (\mathrm{Cov}(\mathbf{X}_{t+j}, \mathbf{X}_{t+k+\tau}))_{j,k=1,\dots,m}$$

$$= (R_{\mathbf{X}_t\mathbf{X}_t}(|j-k-\tau|))_{j,k=1,\dots,m},$$

where $R_{\mathbf{X}_t\mathbf{X}_t}$ is the autocorrelation function of $\mathbf{X}_t$. From this and (6.26), the τ lag covariance matrix of $\alpha(t)$ is

$$\Sigma_{\alpha\alpha}(\tau) = E[\alpha(t)(\alpha(t+\tau))^T]$$

$$= P\Sigma_{\mathbf{Y}_t\mathbf{Y}_t}(\tau)P^T$$

$$= P(R_{\mathbf{X}_t\mathbf{X}_t}(|j-k-\tau|))_{j,k}P^T.$$

Especially, the zero lag covariance matrix of α is

$$\Sigma_{\alpha\alpha}(0) = P\Sigma_{\mathbf{Y}_t\mathbf{Y}_t}(0)P^T$$

$$= P(R_{\mathbf{X}_t\mathbf{X}_t}(|j-k|))_{j,k}P^T$$

$$= \mathrm{diag}(\lambda_1,\dots,\lambda_m).$$

From (6.26), it follows that

$$\mathbf{X}_{t+k} = \sum_{j=1}^{m} \alpha_j(t)e_{jk}$$

or

$$\mathbf{X}_t = \sum_{j=1}^{m} \alpha_j(t-k)e_{jk} \quad (k=1,\dots,m).$$

This means that $\mathbf{X}_t$ has m equivalent expansions. Since α_k are EOF coefficients of $\mathbf{Y}_t$,

$$\mathrm{Var}\,\alpha_k = \lambda_k \quad (k=1,\dots,m).$$

Since

$$\alpha_k(t) \in \mathrm{WSS},$$

$$\mathrm{Cov}(\alpha_k(t), \alpha_l(t)) = 0 \quad (k \neq l),$$

the variance of $\mathbf{X}_t$ is

$$\mathrm{Var}\,\mathbf{X}_t = \sum_{j=1}^{m} \mathrm{Var}\,\alpha_j(t-k)e_{jk}^2 = \sum_{j=1}^{m} \lambda_j e_{jk}^2.$$

6.9 PRINCIPAL OSCILLATION PATTERNS

The principal oscillation pattern (POP) analysis is a technique used to simultaneously infer the characteristic patterns and timescales of a vector time series.

6.9.1 Normal Modes

Let $\mathbf{X}(t) = (X_1(t), \ldots, X_m(t))$ be an m-dimensional real-valued random vector with time parameter t. Assume that $\mathbf{X}(t)$ satisfies the equation

$$\mathbf{X}(t+1) = D\mathbf{X}(t) \quad (t \in \mathbb{R}), \tag{6.27}$$

where $D = (d_{jk})_{j,k=1,\ldots,m}$ is a real-valued matrix. We say $\mathbf{X}(t)$ is an AR(1) process. If (6.27) holds, then

$$E[\mathbf{X}(t+1)\mathbf{X}(t)^{\mathrm{T}}] = E[D\mathbf{X}(t)\mathbf{X}(t)^{\mathrm{T}}] = DE[\mathbf{X}(t)\mathbf{X}(t)^{\mathrm{T}}].$$

Without loss of generality, assume that $E[\mathbf{X}(t)] = 0$. Then

$$\mathrm{Cov}(\mathbf{X}(t+1), \mathbf{X}(t)) = D\,\mathrm{Cov}(\mathbf{X}(t), \mathbf{X}(t))$$

or

$$D = \mathrm{Cov}(\mathbf{X}(t+1), \mathbf{X}(t))(\mathrm{Cov}(\mathbf{X}(t), \mathbf{X}(t)))^{-1}.$$

If the matrix D has m different non-zero eigenvalues $\lambda_k (k = 1, \ldots, m)$, then the corresponding m eigenvectors $\mathbf{P}_k (k = 1, \ldots, m)$ form a basis and

$$D\mathbf{P}_k = \lambda_k \mathbf{P}_k \quad (k = 1, \ldots, m).$$

Since $\mathbf{P}_k (k = 1, \ldots, m)$ are a basis, the random vector $\mathbf{X}(t)$ can be represented in the form

$$\mathbf{X}(t) = \sum_{1}^{m} \alpha_k(t)\mathbf{P}_k \quad (t \in \mathbb{R}) \tag{6.28}$$

for any time t. Substituting (6.28) in (6.27), we obtain

$$\sum_{1}^{m} \alpha_k(t+1)\mathbf{P}_k = D\sum_{1}^{m} \alpha_k(t)\mathbf{P}_k = \sum_{1}^{m} \alpha_k(t)D\mathbf{P}_k = \sum_{1}^{m} \alpha_k(t)\lambda_k \mathbf{P}_k.$$

Comparing coefficients of both sides, for any time t, we obtain

$$\alpha_k(t+1) = \alpha_k(t)\lambda_k \quad (k = 1, \ldots, m).$$

This implies that for any positive integer n,

$$\alpha_k(n) = \alpha_k(n-1)\lambda_k = \cdots = \alpha_k(0)\lambda_k^n.$$

Let $\alpha_k(0) = r_k e^{i\theta_k}$ and denote $\lambda_k = \rho_k e^{i\eta_k}$. Then

$$\begin{aligned}
\alpha_k(n)\mathbf{P}_k &= \alpha_k(0)\lambda_k^n \mathbf{P}_k = r_k \rho_k^n e^{i(\eta_k n + \theta_k)}\mathbf{P}_k \\
&= r_k \rho_k^n (\cos(\eta_k n + \theta_k) + i\sin(\eta_k n + \theta_k))\mathbf{P}_k.
\end{aligned}$$

Since $\mathbf{X}$ is real-valued, by (6.28), we obtain that for any integer n,

$$\mathbf{X}(n) = \sum_1^m \mathrm{Re}(\alpha_k(n)\mathbf{P}_k).$$

Denote $\mathbf{P}_k = \mathbf{Q}_k + i\mathbf{R}_k$. So

$$\mathrm{Re}(\alpha_k(n)\mathbf{P}_k) = r_k \rho_k^n (\mathbf{Q}_k \cos(\eta_k n + \theta_k) - \mathbf{R}_k \sin(\eta_k n + \theta_k)),$$

and so

$$\mathbf{X}(n) = \sum_1^m r_k \rho_k^n (\mathbf{Q}_k \cos(\eta_k n + \theta_k) - \mathbf{R}_k \sin(\eta_k n + \theta_k)).$$

Let

$$\mathbf{Q}_k = (Q_{k1}, \ldots, Q_{km})^{\mathrm{T}},$$
$$\mathbf{R}_k = (R_{k1}, \ldots, R_{km})^{\mathrm{T}}.$$

Notice that $\mathbf{X}(t) = (X_1(t), \ldots, X_m(t))$. Then

$$\mathbf{X}_l(n) = \sum_1^m r_k \rho_k^n (Q_{kl} \cos(\eta_k n + \theta_k) - R_{kl} \sin(\eta_k n + \theta_k)) \quad (l = 1, \ldots, m).$$

Denote

$$\tau_{kl} = \sqrt{Q_{kl}^2 + R_{kl}^2},$$
$$\tan\varphi_{kl} = \frac{R_{kl}}{Q_{kl}}.$$

Then the lth component of $\mathbf{X}(n)$ is

$$\begin{aligned}
\mathbf{X}_l(n) &= \sum_1^m r_k \rho_k^n \tau_{kl} (\cos\varphi_{kl} \cos(\eta_k n + \theta_k) - \sin\varphi_{kl} \sin(\eta_k n + \theta_k)) \\
&= \sum_1^m r_k \rho_k^n \tau_{kl} \cos(\eta_k n + \theta_k - \varphi_{kl}) \quad (l = 1, \ldots, m).
\end{aligned}$$

Example 6.4. Let $\mathbf{X}(t) = (X_1(t), X_2(t))$ and let $\mathbf{X}(t)$ satisfy (6.27), where D is a 2×2 matrix:

$$D = \frac{1}{\sqrt{2}} \begin{pmatrix} 1 & 1 \\ -1 & 1 \end{pmatrix}.$$

Denote by λ_1, λ_2 the eigenvalues of D. Then

$$\lambda_1 = e^{i\frac{\pi}{4}},$$
$$\lambda_2 = e^{-i\frac{\pi}{4}}.$$

Denote by $\mathbf{P}_1, \mathbf{P}_2$ the eigenvectors corresponding to λ_1, λ_2. Then

$$\mathbf{P}_1 = (1, i)^T,$$
$$\mathbf{P}_2 = (1, -i)^T$$

and

$$D\mathbf{P}_1 = \lambda_1 \mathbf{P}_1,$$
$$D\mathbf{P}_2 = \lambda_2 \mathbf{P}_2.$$

Since $\mathbf{P}_1$ and $\mathbf{P}_2$ are a basis of $\mathbb{R}^2$,

$$\mathbf{X}(t) = \alpha_1(t)\mathbf{P}_1 + \alpha_2(t)\mathbf{P}_2. \tag{6.29}$$

From this and (6.27), it follows that

$$\alpha_1(t+1)\mathbf{P}_1 + \alpha_2(t+1)\mathbf{P}_2 = D\alpha_1(t)\mathbf{P}_1 + D\alpha_2(t)\mathbf{P}_2$$
$$= \alpha_1(t)D\mathbf{P}_1 + \alpha_2(t)D\mathbf{P}_2$$
$$= \alpha_1(t)\lambda_1\mathbf{P}_1 + \alpha_2(t)\lambda_2\mathbf{P}_2.$$

So

$$\alpha_1(t+1) = \alpha_1(t)\lambda_1,$$
$$\alpha_2(t+1) = \alpha_2(t)\lambda_2.$$

Using repeatedly this formula, we obtain that for any integer n,

$$\alpha_1(n) = \alpha_1(0)\lambda_1^n = \alpha_1(0)e^{i\frac{\pi}{4}n},$$
$$\alpha_2(n) = \alpha_2(0)\lambda_2^n = \alpha_2(0)e^{-i\frac{\pi}{4}n}. \tag{6.30}$$

From $\mathbf{P}_1 = (1, i)^T$ and $\mathbf{P}_2 = (1, -i)^T$, and (6.29), it follows that

$$X_1(n) = \alpha_1(n) + \alpha_2(n),$$
$$X_2(n) = i(\alpha_1(n) - \alpha_2(n)).$$

So

$$\alpha_1(0) = \frac{1}{2}(X_1(0) - iX_2(0)),$$

$$\alpha_2(0) = \frac{1}{2}(X_1(0) + iX_2(0)),$$

i.e., $\alpha_1(0) = \overline{\alpha}_2(0)$.

Denote $\alpha_1(0) = re^{i\theta}$. Then $\alpha_2(0) = re^{-i\theta}$. From this and (6.30), the components of $\mathbf{X}(n)$ are, respectively,

$$
\begin{aligned}
X_1(n) &= \alpha_1(n) + \alpha_2(n) \\
&= re^{i(\frac{\pi}{4}n+\theta)} + re^{-i(\frac{\pi}{4}n+\theta)} \\
&= 2r\cos\left(\frac{\pi}{4}n + \theta\right),
\end{aligned}
$$

$$
\begin{aligned}
X_2(n) &= i(\alpha_1(n) - \alpha_2(n)) \\
&= i\left(re^{i(\frac{\pi}{4}n+\theta)} - re^{-i(\frac{\pi}{4}n+\theta)}\right) \\
&= -2r\sin\left(\frac{\pi}{4}n + \theta\right).
\end{aligned}
$$

6.9.2 Estimates of Principal Oscillation Patterns

Let $\mathbf{x}(0), \ldots, \mathbf{x}(n-1)$ be n sampling vectors of the random vector $\mathbf{X}(t) = (X_1(t), \ldots, X_m(t))^{\mathrm{T}}$ satisfying

$$\mathbf{X}(t+1) = D\mathbf{X}(t) \quad (t = 0, \ldots, n-1),$$

where D is an $m \times m$ matrix. Let

$$\mathbf{x}(t) = (x_1(t), \ldots, x_m(t))^{\mathrm{T}} \quad (t = 0, \ldots, n-1).$$

Then the covariance matrix and the cross-covariance matrix,

$$\mathrm{Cov}(\mathbf{X}(t), \mathbf{X}(t)) = (E[X_k(t)X_l(t)])_{k,l=1,\ldots,m},$$
$$\mathrm{Cov}(\mathbf{X}(t+1), \mathbf{X}(t)) = (E[X_k(t+1)X_l(t)])_{k,l=1,\ldots,m},$$

are estimated, respectively, by

$$\widehat{\mathrm{Cov}}(\mathbf{X}(t), \mathbf{X}(t)) = \left(\frac{1}{n}\sum_{t=0}^{n-1} x_k(t)x_l(t)\right)_{k,l=1,\ldots,m},$$

$$\widehat{\mathrm{Cov}}(\mathbf{X}(t+1), \mathbf{X}(t)) = \left(\frac{1}{n}\sum_{t=0}^{n-1} x_k(t+1)x_l(t)\right)_{k,l=1,\ldots,m}.$$

By $\mathbf{X}(t+1) = D\mathbf{X}(t)$, it follows that

$$\mathrm{Cov}(\mathbf{X}(t+1), \mathbf{X}(t)) = \mathrm{Cov}(D\mathbf{X}(t), \mathbf{X}(t)) = D\,\mathrm{Cov}(\mathbf{X}(t), \mathbf{X}(t)),$$

and so

$$D = \mathrm{Cov}(\mathbf{X}(t+1), \mathbf{X}(t))(\mathrm{Cov}(\mathbf{X}(t), \mathbf{X}(t)))^{-1}.$$

Therefore, D is estimated by

$$\hat{D} = \widehat{\mathrm{Cov}}(\mathbf{X}(t+1), \mathbf{X}(t))(\widehat{\mathrm{Cov}}(\mathbf{X}(t), \mathbf{X}(t)))^{-1}.$$

Finally, we can compute the eigenvalues $\hat{\lambda}_1, \ldots, \hat{\lambda}_m$ and eigenvectors $\hat{\mathbf{P}}_1, \ldots, \hat{\mathbf{P}}_m$ of $\hat{D}$.

The eigenvectors $\hat{\mathbf{P}}_1, \ldots, \hat{\mathbf{P}}_m$ are called the *estimated principal oscillation patterns*.

PROBLEMS

6.1 Calculate the EOFs and rotated EOFs of global sea surface temperatures.

6.2 Let $\mathbf{X}$ be an m-dimensional random vector, $e_1, \ldots, e_m$ be the EOFs, and $\alpha_1, \ldots, \alpha_m$ be the EOF coefficients. Denote $e_i^+ = \sqrt{\lambda_i} e_i$ and $\alpha_i^+ = \frac{\alpha_i}{\sqrt{\lambda_i}}$ ($i = 1, \ldots, m$). Let G be an orthogonal matrix of order m. Define rotated EOFs $f_1, \ldots, f_m$ by

$$(f_1 | \cdots | f_m) = (e_1^+ | \cdots | e_m^+)G,$$

and rotated EOF coefficients $\beta_1, \ldots, \beta_m$ by

$$\begin{pmatrix} \beta_1 \\ \vdots \\ \beta_m \end{pmatrix} = G^T \begin{pmatrix} \alpha_1 \\ \vdots \\ \alpha_m \end{pmatrix}.$$

Try to prove that $f_1, \ldots, f_m$ are not orthonormal unless $\lambda_1 = \lambda_2 = \cdots = \lambda_m$.

6.3 Let $\mathbf{X}$ be an m-dimensional random vector with mean 0, e_1 and e_2 be the first two EOFs, and α_1 and α_2 be the first two EOF coefficients. Take an orthogonal matrix G of order 2 as follows:

$$G = \frac{\sqrt{2}}{2} \begin{pmatrix} 1 & -1 \\ 1 & 1 \end{pmatrix}.$$

Try to find rotated EOFs f_1 and f_2, EOF coefficients β_1 and β_2, and $\mathrm{Cov}(\beta_i, \beta_j)$ ($i = 1, 2$)?

6.4 Compare classical EOFs with rotated EOFs and Hilbert EOFs.

6.5 Perform singular spectrum analysis for winter air temperature data.

6.6 Pick one paper employing EOF/singular spectrum analysis from the Bibliography section to read and list the strengths and weaknesses of the particular use of EOF/singular spectrum analysis made in this case study.

BIBLIOGRAPHY

Barnston, A.G., Livezey, R.E., 1987. Classification, seasonality and persistence of low-frequency atmospheric circulation patterns. Mon. Weather Rev. 115, 1083-1126.

Barnston, A.G., Ropelewski, C.F., 1992. Prediction of ENSO episodes using canonical correlation analysis. J. Clim. 5, 1316-1345.

Barnston, A.G., Smith, T.M., 1996. Specification and prediction of global surface temperature and precipitation from global SST using CCA. J. Clim. 9, 2660-2697.

Bayr, T., Dommenget, D., 2014. Comparing the spatial structure of variability in two datasets against each other on the basis of EOF-modes. Clim. Dyn. 42, 1631-1648.

Biggerstaff, M.I., Seo, E.K., Hristova-Veleva, S.M., Kim, K.Y., 2006. Impact of cloud model microphysics on passive microwave retrievals of cloud properties. Part I: Model comparison using EOF analyses. J. Appl. Meteorol. Climatol. 45, 930-954.

Cheng, X., Nitsche, G., Wallace, J.M., 1995. Robustness of low-frequency circulation patterns derived from EOF and rotated EOF analysis. J. Clim. 8, 1709C1713.

Cherry, S., 1996. Singular value analysis and canonical correlation analysis. J. Clim. 9, 2003C2009.

Ding, Q.H., Wang, B., 2007. Intraseasonal teleconnection between the summer Eurasian wave train and the Indian monsoon. J. Clim. 20, 3751-3767.

Dommenget, D., 2007. Evaluating EOF modes against a stochastic null hypothesis. Clim. Dyn. 28, 517-531.

Fan, L., Fu, C., Chen, D., 2011. Long-term trend of temperature derived by statistical downscaling based on EOF analysis. Acta Meteorol. Sin. 25, 327-339.

Finnigan, J.J., Shaw, R.H., 2000. A wind-tunnel study of airflow in waving wheat: an EOF analysis of the structure of the large-eddy motion. Bound.-Layer Meteorol. 96, 211-255.

Gehne, M., Kleeman, R., Trenberth, K.E., 2014. Irregularity and decadal variation in ENSO: a simplified model based on principal oscillation patterns. Clim. Dyn. in press.

Grezio, A., Pinardi, N., Sparnocchia, S., Zavatarelli, M., 2003. The study of seasonal variability in the Adriatic Sea with the use of EOF analysis. Elsevier Oceanogr. Ser. 69, 222-225.

Haddad, M., Hassani, H., Taibi, H., 2013. Sea level in the Mediterranean Sea: seasonal adjustment and trend extraction within the framework of SSA. Earth Sci. Inform. 6, 99-111.

Horel, J.D., 1984. Complex principal component analysis: theory and examples. J. Clim. Appl. Meteorol. 23, 1660-1673.

Huth, R., 2007. Arctic or north Atlantic oscillation? Arguments based on the principal component analysis methodology. Theor. Appl. Climatol. 89, 1-8.

Jawson, S.D., Niemann, J.D., 2007. Spatial patterns from EOF analysis of soil moisture at a large scale and their dependence on soil, land-use, and topographic properties. Adv. Water Resour. 30, 366-381.

Kutzbach, J.E., 1967. Empirical eigenvectors of sea-level pressure, surface pressure, and precipitation complexes. J. Appl. Meteorol. 6, 791-802.

Lorenz, E.N., 1956. Empirical orthogonal functions and statistical weather prediction. Statistical Forecasting Project, MIT, Cambridge.

Monahan, A.H., Fyfe, J.C., Ambaum, M.H.P., Stephenson, D.B., North, G.R., 2009. Empirical orthogonal functions: the medium is the message. J. Clim. 22, 6501-6514.

Peh, Z., Mileusni, M., Miko, S., 2011. Canonical correlation analysis as a tool in the provenance study of overbank sediments from the small mountainous watersheds. Environ. Earth Sci. 64, 1139-1155.

Perry, M.A., Niemann, J.D., 2007. Analysis and estimation of soil moisture at the catchment scale using EOFs. J. Hydrol. 334, 388-404.

Ryu, J.-H., Jenkins, G.S., 2005. Lightning-tropospheric ozone connections: EOF analysis of TCO and lightning data. Atmos. Environ. 39, 5799-5805.

Singh, A., Mohanty, U.C., Mishra, G., 2014. Long-lead prediction skill of Indian summer monsoon rainfall using outgoing longwave radiation (OLR): an application of canonical correlation analysis. Pure Appl. Geophys. 171, 1519-1530.

Taibi, H., Kahlouche, S., Haddad, M., Rami, A., 2013. Trends in global and regional sea level from satellite altimetry within the framework of auto-SSA. Arab. J. Geosci. 6, 4575-4584.

Thompson, D.W.J., Wallace, J.M., 1998. The Arctic oscillation signature in the wintertime geopotential height and temperature fields. Geophys. Res. Lett. 25, 1297-1300.

Tolkova, E., 2010. EOF analysis of a time series with application to tsunami detection. Dyn. Atmos. Oceans 50, 35-54.

Wang, G., Dommenget, D., Frauen, C., 2014. An evaluation of the CMIP3 and CMIP5 simulations in their skill of simulating the spatial structure of SST variability. Clim. Dyn. in press.

Zhang, Q., Wang, B., He, B., Peng, Y., Ren, M., 2011. Singular spectrum analysis and ARIMA hybrid model for annual runoff forecasting. Water Resour. Manage. 25, 2683-2703.

Chapter 7

Random Processes and Power Spectra

Climatic data and theoretical considerations suggest that a large part of climatic variability has a random nature and can be analyzed using the theory of random processes. In this chapter, we will describe the random approach to the study of changes in the climate system, including stationary random processes, calculus of random processes, power spectra, spectrum estimation, and Wiener filter. Moreover, in order to extract intrinsic frequency features from climatic time series, we will show how to do statistical significance tests by Fourier analysis and wavelet analysis.

7.1 STATIONARY AND NON-STATIONARY RANDOM PROCESSES

A random process $X(\xi, t)$ is a function of time t and outcome ξ. For a fixed time t, it is a *random variable*. For a fixed outcome ξ, it is a *simple path*, denoted simply by $X(t)$. If $t \in \mathbb{Z}$ or $t \in \mathbb{Z}_+$, then $X(t)$ is called a *discrete-time random process*. If $t \in \mathbb{R}$, then $X(t)$ is called a *continuous-time random process*.

The following three examples are useful and simple:

(i) Suppose that w_n ($n \in \mathbb{Z}_+$) are independent random variables with

$$P(w_k = 1) = p,$$

$$P(w_k = -1) = 1 - p \quad (k \in \mathbb{Z}_+, \ 0 < p < 1).$$

Take

$$X(n) = w_1 + w_2 + \cdots + w_n \quad (n \in \mathbb{Z}_+).$$

Then $\mathbf{X}(n)$ ($n \in \mathbb{Z}_+$) is a discrete random process and is called a *random walk* with parameter p.

(ii) Suppose that

$$X(t) = \sum_{1}^{N} a_n \cos(\tau_n t + \theta_n) \quad (t \in \mathbb{Z}),$$

where a_n and τ_n are both N constants and θ_n are N uniform random variables with probability density function

$$p(x) = \begin{cases} \frac{1}{2\pi}, & (0, 2\pi), \\ 0, & \text{otherwise} \end{cases}$$

and $\theta_1, \ldots, \theta_N$ are independent. Then $X(t)$ ($t \in \mathbb{Z}$) is a continuous-time random process.

(iii) Let

$$X(t) = A + Bt + t^2 \quad (t \in \mathbb{R}),$$

where A and B are independent Gaussian random variables $N(0, 1)$. This is a continuous-time random process. Let $A = a$ and $B = b$, where a and b are real numbers. Then

$$X(t) = a + bt + t^2 \quad (t \in \mathbb{R})$$

is a parabolic path.

A random process satisfying the condition $E[X^2(t)] < \infty$ ($t \in T$) is called a *second-order random process*. Hereafter, we always assume that $X(t)$ ($t \in T$) is a second-order random process. The expectation of $X(t)$ is called the *mean function*, denoted by $\mu_X(t)$. The expectation of $X(s)X(t)$ is called the *correlation function*, denoted by $R_X(s, t)$. The covariance of $X(s)$ and $X(t)$ is called the *covariance function*, denoted by $C_X(s, t)$. Namely,

$$\mu_X(t) = E[X(t)],$$

$$R_X(s, t) = E[X(s)\overline{X}(t)],$$

$$C_X(s, t) = \text{Cov}(X(s), X(t)).$$

For the random process $X(t) = A + Bt + t^2$ ($t \in \mathbb{R}$), where A and B are independent Gaussian random variables $N(0, 1)$, the mean function

$$\mu_X(t) = E[X(t)] = E[A + Bt + t^2] = E[A] + tE[B] + t^2 = t^2,$$

and the covariance function

$$C_X(s, t) = E[(X(s) - \mu_X(s))(X(t) - \mu_X(t))]$$
$$= E[A^2] + tE[AB] + sE[AB] + stE[B^2] = 1 + st.$$

A random process $X(t)$ is called *stationary* if, for any $t_1, \ldots, t_N$ and s, two random vectors $(X(t_1), \ldots, X(t_N))$ and $(X(t_1 + s), \ldots, X(t_N + s))$ have the same joint distribution:

$$P(X(t_1) \leq x_1, \ldots, X(t_N) \leq x_N)$$
$$= P(X(t_1 + s) \leq x_1, \ldots, X(t_N + s) \leq x_N) \quad (x_1, \ldots, x_N \in \mathbb{R}),$$

i.e., the statistical properties of the random process are unaffected by a shift in time. Otherwise, this random process is called *non-stationary*.

Proposition 7.1. *Suppose that X is a stationary random process. Then, for* $t, s, t_1, t_2 \in \mathbb{R}$,

$$\mu_X(t) = \mu_{\mathbf{X}}(t + s),$$
$$R_X(t_1, t_2) = R_{\mathbf{X}}(t_1 + s, t_2 + s).$$

Proof. Since X is stationary, two random vectors $(X(t_1), \ldots, X(t_N))$ and $(X(t_1 + s), \ldots, X(t_N + s))$ have the same probability density function. Take $N = 1$ and let $X(t)$ and $X(t + s)$ have the same probability density function $p(x)$. Then

$$\mu_X(t) = \int_{\mathbb{R}} x p(x)\, \mathrm{d}x = \mu_X(t + s).$$

Take $N = 2$ and let $(X(t_1), X(t_2))$ and $(X(t_1 + s), X(t_2 + s))$ have the same joint probability density function $p(x, y)$. Then

$$R_X(t_1, t_2) = \int\int_{\mathbb{R}^2} xy p(x, y)\, \mathrm{d}x\, \mathrm{d}y = R_X(t_1 + s, t_2 + s).$$

$\square$

If a random process $X(t)$ satisfies that for any t, s, t_1, t_2,

$$\mu_X(t) = \mu_X(t + s),$$
$$R_X(t_1, t_2) = R_X(t_1 + s, t_2 + s),$$

then the random process is called *wide-sense stationary* (WSS). Denote $X \in$ WSS.

The first condition $\mu_X(t) = \mu_X(t + s)$ shows that $\mu_{\mathbf{X}}(t)$ is independent of t. So $\mu_X(t)$ can be denoted by μ_X. The second condition $R_X(t_1, t_2) = R_X(t_1 + s, t_2 + s)$ shows that $R_X(t_1, t_2)$ depends only on the difference $\tau = t_1 - t_2$. So $R_X(t_1, t_2)$ can be denoted by $R_X(t_1 - t_2)$ or $R_X(\tau)$. Moreover, the covariance function

$$C_X(t_1, t_2) = R_X(t_1, t_2) - \mu_X(t_1)\mu_X(t_2) = R_X(t_1 - t_2) - \mu_X^2.$$

This shows that $C_X(t_1, t_2)$ depends only on $\tau = t_1 - t_2$. Therefore, $C_X(t_1, t_2)$ can be denoted by $C_X(\tau)$, and

$$C_X(\tau) = R_X(\tau) - \mu_X^2.$$

For example, let $X(k)(k \in \mathbb{Z})$ be a random process, where $X(k)(k \in \mathbb{Z})$ are independent with $E[X(k)] = 0$ and $\mathrm{Var}\, X(k) = 1$ $(k \in \mathbb{Z})$. Since

$$\mu_X(k) = E[X(k)] = 0 \quad (k \in \mathbb{Z}),$$
$$R_X(k_1, k_2) = E[X(k_1)X(k_2)] = \begin{cases} 1, & k_1 = k_2, \\ 0, & k_1 \neq k_2, \end{cases}$$

the process $X(k)$ $(k \in \mathbb{Z})$ is a WSS random process.

Proposition 7.2. *For $X(t) \in$ WSS, if $E[X] = 0$, then the correlation function $R_X(\tau)$ satisfies*

$$R_X(0) \geq 0,$$

$$R_X(-\tau) = R_X(\tau) \quad (\tau \in \mathbb{Z}),$$

$$|R_X(\tau)| \leq R_X(0).$$

Proof. It is clear that

$$R_X(0) = E[X^2(t)] \geq 0,$$
$$R_X(-\tau) = E[X(t - \tau)X(t)] = E[X(t)X(t - \tau)] = R_X(\tau).$$

By the Cauchy-Schwarz inequality,

$$|R_X(\tau)| = |E[X(t + \tau)X(t)]| \leq (E[X^2(t + \tau)]E[X^2(t)])^{1/2} = R_X(0).$$

$\square$

For example, let $X = \{X_n\}_{n\in\mathbb{Z}}$ and $X_n = Z_n + \frac{1}{2}Z_{n-1}$, where $\{Z_n\}_{n\in\mathbb{Z}}$ are independent random variables for the different n and $E[Z_n] = 0$, and $\operatorname{Var} Z_n = 1$ $(n \in \mathbb{Z})$. Then

$$E[X_n] = 0.$$

A direct computation shows that the correlation functions satisfies

$$R_X(0) = \frac{5}{4},$$

$$R_X(-1) = R_X(1) = \frac{1}{2},$$

$$R_X(n) = 0 \quad (n \neq 0, \pm 1).$$

So

$$R_X(\tau) = E[R_X(n + \tau)R_X(\tau)]$$

is independent of n, i.e., $\{X_n\}_{n\in\mathbb{Z}} \in$ WSS.

Example 7.1. Let

$$X(t) = A\cos(\alpha t + \theta) \quad (t \in \mathbb{R})$$

be a trigonometric polynomial random process, where A and θ are independent random variables and $A > 0$, and α is a constant.

Since A and $\cos(\alpha t + \theta)$ are independent and $\cos(\alpha t + \theta) = \cos(\alpha t)\cos\theta - \sin(\alpha t)\sin\theta$,

$$\mu_X(t) = E[X(t)] = E[A]E[\cos(\alpha t)\cos\theta - \sin(\alpha t)\sin\theta].$$

Using the linear property of expectation gives

$$\mu_X(t) = E[A](\cos(\alpha t)E[\cos\theta] - \sin(\alpha t)E[\sin\theta]).$$

So $\mu_X(t)$ does not depend on t if and only if $E[\cos\theta] = E[\sin\theta] = 0$.

We turn next to correlation function $R_X(s, t)$. It is clear that

$$R_X(s, t) = E[X(s)X(t)] = E[A^2 \cos(\alpha s + \theta) \cos(\alpha t + \theta)]$$

$$= E[A^2] \cdot E[\cos(\alpha s + \theta) \cos(\alpha t + \theta)]$$

$$= E[A^2] \frac{\cos\alpha(s - t) + \cos\alpha(s + t)E[\cos(2\theta)] - \sin\alpha(s + t)E[\sin(2\theta)]}{2}.$$

So $R_X(s, t)$ is a function of $s - t$ alone if and only if $E[\cos(2\theta)] = E[\sin(2\theta)] = 0$.

Therefore, $X \in \text{WSS}$ if and only if $E[\cos\theta] = E[\cos(2\theta)] = E[\sin\theta] = E[\sin(2\theta)] = 0$.

Let $X(t)$ and $Y(t)$ be two random processes. The expectation of $X(s)\overline{Y}(t)$ is called the *cross-correlation function*, denoted by $R_{XY}(s, t)$, i.e.,

$$R_{XY}(s, t) = E[X(s)\overline{Y}(t)] \quad (s, t \in \mathbb{R}).$$

The covariance of $X(s)$ and $Y(t)$ is called the *cross-covariance function*, denoted by $C_{XY}(s, t)$, i.e.,

$$C_{XY}(s, t) = \text{Cov}[X(s), Y(t)] \quad (s, t \in \mathbb{R}).$$

If, for any t, s,

$$\mu_X(t) = \mu_X(t + s),$$
$$\mu_Y(t) = \mu_Y(t + s),$$

and for any t_1, t_2,

$$R_{XY}(t_1, t_2) = R_{XY}(t_1 + s, t_2 + s),$$

then X and Y are called *jointly WSS*.

Similarly, we can define $R_{XY}(\tau)$ and $C_{XY}(\tau)$.

7.2 MARKOV PROCESS AND BROWNIAN MOTION

If a random process $X(t)$ ($t \in T$) is independent for different t, then $X(t)$ is called an *independent random process*. A Markov process is a direct extension of the independent random processes. Roughly, given a value of the random process at t, if the future values $X(s)$ ($s > t$) are unaffected by the past values $X(s)$ ($s < t$), then the random process is called a *Markov process*.

When $T = \{0, 1, \dots\}$, the Markov process is called a *Markov chain*. In detail, let $X(n)$ ($n = 0, 1, \dots$) be a random process. If, for any $k_0 < k_1 < \cdots < k_{n-1} < k < l$ (each $k_j \in \mathbb{Z}_+$) and any $n \geq 1$, the random process satisfies the Markov property

$$P(X(n + 1) = l | X(0) = k_0, \ldots, X(n - 1) = k_{n-1}, X(n) = k)$$
$$= P(X(n + 1) = l | X(n) = k), \tag{7.1}$$

then the random process $X(n)$ $(n = 0, 1, \ldots)$ is called a *Markov chain*.

Let $\{X(n)\}$ be a Markov chain. Then

$$P_{kl}^{n,n+1} = P(X(n + 1) = l | X(n) = k)$$

is called the *transition probability*. A Markov chain is determined by its initial state and transition probabilities. When the transition probability is independent of n, we say the Markov chain has *stable transition probabilities*, denoted by P_{kl}. Ordinarily, P_{kl} are arranged in a matrix form:

$$P = (P_{kl})_{k,l=0,1,\ldots} = \begin{pmatrix} P_{00} & P_{01} & P_{02} & \cdots \\ P_{10} & P_{11} & P_{12} & \cdots \\ \vdots & \vdots & \vdots & \ddots \\ P_{k0} & P_{k1} & P_{k2} & \cdots \\ \vdots & \vdots & \vdots & \vdots \end{pmatrix}.$$

The matrix P is called a *matrix of transition probabilities*.

Example 7.2. A random walk with parameter p,

$$X(n) = w_1 + w_2 + \cdots + w_n \quad (n \in \mathbb{Z}_+),$$

is a Markov chain. Since

$$X(n) = X(n - 1) + w_n,$$

if $X(n - 1) = l$, then $X(n) = l + 1$ or $l - 1$ and

$$P(X(n) = l + 1) = P(X(n) - X(n - 1) = 1) = P(w_n = 1) = p,$$
$$P(X(n) = l - 1) = P(X(n) - X(n - 1) = -1) = P(w_n = -1) = 1 - p.$$

Therefore, it has transition probabilities

$$P_{kl} = \begin{cases} p, & k = l + 1, \\ 1 - p, & k = l - 1, \\ 0, & \text{otherwise.} \end{cases}$$

Proposition 7.3. *Let $\{X(n)\}$ be a Markov chain with stationary transition probabilities $(P_{kl})_{k,l}$. If $P(X(0) = k) = p_k$, then*

$$P(X(0) = k_0, X(1) = k_1, \ldots, X(n) = k_n) = p_{k_0} P_{k_0 k_1} P_{k_1 k_2}, \ldots, P_{k_{n-1} k_n}.$$

Proof. Notice that

$$J := P(X(0) = k_0, X(1) = k_1, \ldots, X(n) = k_n) = I_1 I_2,$$

where

$$I_1 = P(X(0) = k_0, X(1) = k_1, \ldots, X(n-1) = k_{n-1}),$$
$$I_2 = P(X(n) = k_n | X(0) = k_0, X(1) = k_1, \ldots, X(n-1) = k_{n-1}).$$

Since $\{X(n)\}$ is a Markov chain, by (7.1), it follows that

$$I_2 = P(X(n) = k_n | X(n-1) = k_{n-1}) = P_{k_{n-1}k_n},$$

and so

$$J = I_1 P_{k_{n-1}k_n}$$
$$= P(X(0) = k_0, X(1) = k_1, \ldots, X(n-1) = k_{n-1}) P_{k_{n-1}k_n}.$$

If we repeat this procedure, by induction, the desired result is obtained. $\qquad\square$

Let $X(t)$ ($t \in T$) be a random process. If, for any $t_1, t_2, \ldots, t_n \in T$ and $t_1 < t_2 < \cdots < t_n$,

$$X(t_2) - X(t_1),$$
$$X(t_3) - X(t_2),$$
$$\vdots$$
$$X(t_n) - X(t_{n-1})$$

are independent, then $X(t)$ ($t \in T$) is called an *independent increment process*.

If a random process is independent, then it is an independent increment process. The converse is not true: there exist many random processes which are not independent random processes but are independent increment processes.

Example 7.3. Suppose that $Z(n)$ ($n = 0, 1, \ldots$) are independent random processes. Define

$$X(n) = \sum_0^n Z(k).$$

Then the random process $X(n)$ ($n = 0, 1, \ldots$) is an independent increment process. The sum $X(n)$ ($n = 0, 1, \ldots$) is called an *independent sum*.

For any $n_1, n_2, \ldots, n_l$ and $n_1 < n_2 < \cdots < n_l$,

$$X(n_2) - X(n_1) = \sum_{n_1+1}^{n_2} Z(k),$$

$$X(n_3) - X(n_2) = \sum_{n_2+1}^{n_3} Z(k),$$

$$\vdots$$

$$X(n_l) - X(n_{l-1}) = \sum_{n_{l-1}+1}^{n_l} Z(k).$$

Since $Z(n)$ $(n = 0, 1, \ldots)$ are independent, the increments

$$X(n_2) - X(n_1),$$
$$X(n_3) - X(n_2),$$
$$\vdots$$
$$X(n_l) - X(n_{l-1})$$

are independent, i.e., $X(n)$ $(n = 0, 1, \ldots)$ are independent increment processes. If

$$E[Z(k)] = 0,$$
$$\operatorname{Var} Z(k) = \sigma^2 \quad (k = 0, 1, \ldots),$$

then, for $k = 0, 1, \ldots,$

$$\operatorname{Cov}(Z(k), Z(k)) = \sigma^2,$$
$$\operatorname{Cov}(Z(k), Z(l)) = E[Z(k)Z(l)] = E[Z(k)]E[Z(l)] = 0 \quad (k \neq l)$$

and for $n > m \geq 0$,

$$\operatorname{Cov}(X(n), X(m)) = \sum_{0}^{m} \operatorname{Cov}(Z(k), Z(k)) = (m + 1)\sigma^2 \neq 0.$$

Therefore, $X(n)$ $(n = 0, 1, \ldots)$ are not independent.

A random process $W = W(t)$ $(t \geq 0)$ is called a *Brownian motion* if

(i) $W(0) = 0$;
(ii) W has independent increments;
(iii) for any $t \geq s \geq 0$, $W(t) - W(s)$ is a random variable $N(0, \sigma^2(t - s))$;
(iv) the sample path of $W(t)$ is continuous.

Let W be a Brownian motion. Its mean function $\mu_W(t)$, correlation function $R_W(s, t)$, and the covariance function $C_W(s, t)$ are given as follows.

Since the mean $\mu_W(t) = E[W(t)]$, it follows by (i) that for $t \geq 0$,

$$\mu_W(t) = E[W(t) - W(0)].$$

By (iii), $W(t) - W(0)$ is a random variable $N(0, \sigma^2 t)$, and so the mean function of the Brownian motion is

$$\mu_W(t) = E[W(t) - W(0)] = 0.$$

Since the correlation function $R_W(s, t) = E[W(s)W(t)]$, it follows by (i) that for $s \leq t$,

$$R_W(s, t) = E[(W(s) - W(0))(W(t) - W(s))] + E[(W(s) - W(0))^2].$$

By (ii) and $E[W(s) - W(0)] = 0$, it follows that $W(s) - W(0)$ and $W(t) - W(s)$ are independent and

$$E[(W(s) - W(0))(W(t) - W(s))] = E[W(s) - W(0)]E[W(t) - W(s)] = 0.$$

So

$$R_W(s, t) = E[(W(s) - W(0))^2] = \text{Var}(W(s) - W(0)).$$

From this and (iii),

$$R_W(s, t) = \sigma^2 s \quad (s < t),$$

$$R_W(s, t) = \sigma^2 t \quad (s > t).$$

Therefore, the correlation function of the Brownian motion is

$$R_W(s, t) = \sigma \min\{s, t\}.$$

Notice that

$$C_W(s, t) = R_W(s, t) - \mu_W(s)\mu_W(t),$$

$$\mu_W(t) = 0.$$

The covariance function of the Brownian motion is

$$C_W(s, t) = R_W(s, t) = \sigma^2 \min\{s, t\}.$$

7.3   CALCULUS OF RANDOM PROCESSES

Limit is the most fundamental concept in calculus of deterministic functions, so is limit in calculus of random processes.

Consider sequences of random variables. A sequence of random variables $\{X_n\}$ converges to a random variable X if

$$\lim_{n \to \infty} E[(X_n - X)^2] = 0.$$

Write $\lim_{n \to \infty} X_n = X$ or $X_n \to X$ $(n \to \infty)$.

Property. Let $\{X_n\}$ and $\{Y_n\}$ be sequences of random variables, and let X and Y be random variables.

(i) If $X_n \to X$ and $Y_n \to Y$, then $X_n + Y_n \to X + Y$.

(ii) If $X_n \to X$ and $Y_n \to Y$, then $E[X_nY_n] \to E[XY]$.

(iii) If $X_n \to X$, then $E[X_n^2] \to E[X^2]$ and $E[X_n] \to E[X]$.

(iv) If $c_n \to c$ and $X_n \to X$, where $\{c_n\}$ is a sequence of numbers, then $c_n X_n \to cX$.

Proof. Since

$$E[((X_n + Y_n) - (X + Y))^2] \leq E[2(X_n - X)^2 + 2(Y_n - Y)^2]$$
$$= 2E[(X_n - X)^2] + 2E[(Y_n - Y)^2] \to 0.$$

So $X_n + Y_n \to X + Y$, i.e., (i).

Notice that

$$2X_n Y_n = (X_n + Y_n)^2 - X_n^2 - Y_n^2.$$

Taking the expectation on both sides, we get

$$2E[X_n Y_n] = E[(X_n + Y_n)^2] - E[X_n^2] - E[Y_n^2]. \tag{7.2}$$

The second term on the right-hand side is

$$E[X_n^2] = E[(X - (X_n - X))^2]$$
$$= E[X^2 + 2X(X_n - X) + (X_n - X^2)]$$
$$= E[X^2] + 2E[X(X_n - X)] + E[(X_n - X)^2],$$

i.e.,

$$E[X_n^2] - E[X^2] = 2E[X(X_n - X)] + E[(X_n - X)^2].$$

Applying the Schwarz inequality in Chapter 5 gives

$$|E[X(X_n - X)]| \leq (E[X^2])^{1/2}(E[(X - X_n)^2])^{1/2}.$$

Therefore,

$$|E[X_n^2] - E[X^2]| \leq 2(E[X^2])^{1/2}(E[(X - X_n)^2])^{1/2} + E[(X_n - X)^2].$$

From this and the assumption that $E[(X_n - X)^2] \to 0$, it follows that $E[X_n^2] \to E[X^2]$.

Similarly, by $X_n \to X$ and $Y_n \to Y$, it follows that

$$E[Y_n^2] \to E[Y^2],$$

$$E[(X_n + Y_n)^2] \to E[(X + Y)^2].$$

From this and (7.2),

$$2E[X_n Y_n] \to E[(X + Y)^2] - E[X^2] - E[Y^2].$$

However,

$$E[(X + Y)^2] = E[X^2] + 2E[XY] + E[Y^2].$$

Therefore, $E[X_n Y_n] \to E[XY]$, i.e., (ii).

In (ii), take $Y_n = X_n$ and $Y_n = 1$, respectively, in this equality, then

$$E[X_n^2] \to E[X^2],$$

$$E[X_n] \to E[X],$$

i.e., (iii).

It is clear that

$$(c_n X_n - cX)^2 = ((c_n - c)X_n + c(X_n - X))^2 \le 2(c_n - c)^2 X_n^2 + 2c^2(X_n - X)^2.$$

Taking the expectation on both sides of this inequality, we get

$$E[(c_n X_n - cX)^2] \le 2(c_n - c)^2 E[X_n^2] + 2c^2 E[(X_n - X)^2].$$

By (iii), $E[X_n^2] \to E[X]$. From this and the assumption: $c_n \to c$ and $X_n \to X$, it follows that

$$E[(c_n X_n - cX)^2] \to 0,$$

and so $c_n X_n \to cX$, i.e., (iv). $\qquad\square$

Example 7.4. Let X be a random variable and $X_n = e^{\frac{1}{\sqrt{n}}X} - \frac{1}{\sqrt{n}}X$. Then $\lim_{n \to \infty} X_n = 1$.

From the Taylor formula $e^x = 1 + x + \frac{x^2}{2} + \cdots$, it is clear that there exists a constant M such that $|e^x - 1 - x| \le Mx^2$. Take $x = \frac{1}{\sqrt{n}}X$. Then

$$\left| e^{\frac{1}{\sqrt{n}}X} - \frac{1}{\sqrt{n}}X - 1 \right| \le \frac{M}{n}X^2,$$

and so

$$E[(X_n - 1)^2] = E\left[\left(e^{\frac{1}{\sqrt{n}}X} - \frac{1}{\sqrt{n}}X - 1 \right)^2 \right] \le \frac{M^2}{n^2} E[X^4] \to 0 \quad (n \to \infty),$$

i.e., $\lim_{n \to \infty} X_n = 1$.

The concepts of continuous and derivatives in calculus of random processes are similar to those in calculus of deterministic functions but the limits involved are in terms of expectations.

Let $X = X(t)$ $(t \in T)$ be a random process and $t_0 \in T$, and let Y be a random variable. If

$$\lim_{t \to t_0} E[(X(t) - Y)^2] = 0,$$

then the random process $X(t)$ has the limit Y at t_0. Write $\lim_{t \to t_0} X(t) = Y$ or $X(s) \to Y$ $(t \to t_0)$.

If $\lim_{t \to t_0} X(t) = X(t_0)$, i.e.,

$$\lim_{t \to t_0} E[(X(t) - X(t_0))^2] = 0,$$

then we say that the random process $X(t)$ is continuous at t_0. If

$$\lim_{t \to t_0} \frac{X(t) - X(t_0)}{t - t_0} = Y$$

or

$$\frac{X(t) - X(t_0)}{t - t_0} \to Y \quad (t \to t_0),$$

then we say that the random process $X(t)$ is differentiable and has a derivative Y at t_0. Write $X'(t_0) = Y$.

Random processes have the following operation rules.

Rule 1 (mean function). If X is differentiable, then the mean function $\mu_X(t)$ is differentiable and

$$\mu'_X(t) = \mu_{X'}(t).$$

Simply, the derivative of the mean is equal to the mean of the derivative.

Proof. Since X is differentiable, for any fixed t,

$$\frac{X(s) - X(t)}{s - t} \to X'(t) \quad (s \to t).$$

Applying property (iii) gives

$$\frac{\mu_X(s) - \mu_X(t)}{s - t} = \frac{E[X(s)] - E[X(t)]}{s - t} = E\left[\frac{X(s) - X(t)}{s - t} \right] \to E[X'(t)] \ (s \to t).$$

From this and $E[X'(t)] = \mu_{X'}(t)$, it follows that

$$\mu'_X(t) = \mu_{X'}(t).$$

$\square$

Rule 2 (autocorrelation function). If X is differentiable, then the autocorrelation function

$$R_{X'X}(t, t_0) = \frac{\partial R_X(t, t_0)}{\partial t},$$

where X' is the derivative of X.

By the definition and properties of the autocorrelation function, it follows that

$$
\begin{aligned}
R_{X'X}(t, t_0) &= E[X'(t)X(t_0)] \\
&= E\left[\left(\lim_{s \to t} \frac{X(s) - X(t)}{s - t} \right) X(t_0) \right] \\
&= \lim_{s \to t} \frac{E[X(s)X(t_0)] - E[X(t)X(t_0)]}{s - t} \\
&= \lim_{s \to t} \frac{R_X(s, t_0) - R_X(t, t_0)}{s - t} = \frac{\partial R_X(t, t_0)}{\partial t}.
\end{aligned}
$$

Rule 3 (derivative). Let X be a differentiable random process and f be a deterministic differentiable function. Then $(Xf)' = X'f + Xf'$.

Let $s \to t$ in the identity:

$$\frac{X(s)f(s) - X(t)f(t)}{s - t} = \frac{X(s) - X(t)}{s - t}f(s) + X(t)\frac{f(s) - f(t)}{s - t} \quad (s \neq t).$$

Then $(Xf)' = X'f + Xf'$.

Example 7.5. Let $W(t)$ $(t \geq 0)$ be a Brownian motion. For $t \geq s \geq 0$, from

$$W(t) - W(s) = N(0, \sigma^2(t - s)),$$

it follows that

$$E[(W(t) - W(s))^2] = \text{Var}(W(t) - W(s)) = \sigma^2|t - s|,$$

and so

$$\lim_{s \to t} E[(W(t) - W(s))^2] = 0,$$

$$\lim_{s \to t} E\left[\left(\frac{W(s) - W(t)}{s - t}\right)^2\right] = \lim_{s \to t} \frac{\sigma^2}{|s - t|} = +\infty,$$

i.e., W is continuous but is not differentiable.

Let $X = X(t)$ $(a \leq t \leq b)$ be a random process. Given a partition of a closed interval $[a, b]$,

$$a = t_0 < t_1 < \cdots < t_n = b.$$

Arbitrarily take $\zeta_k \in [t_{k-1}, t_k]$ $(k = 1, \ldots, n)$. If the limit

$$\lim_{\delta \to 0} \sum_{1}^{n} X(\zeta_k)(t_k - t_{k-1}) = I$$

exists, where $\delta = \max_k |t_k - t_{k-1}|$, then the random process $X(t)$ is *integrable* over $[a, b]$. Write $\int_a^b X(t)\, dt = I$.

Proposition 7.4. *If a random process $X(t)$ is integrable over $[a, b]$, then*

$$E\left[\int_a^b X(t)dt\right] = \int_a^b E[X(t)]dt.$$

Simply, the expectation and integral are interchangeable.

Proof. From the definition of the integral and the properties of the expectation, it follows that

$$E\left[\int_a^b X(t)\,dt\right] = E\left[\lim_{\delta \to 0} \sum_1^n X(\zeta_k)(t_k - t_{k-1})\right] = \lim_{\delta \to 0} \sum_1^n (E[X(\zeta_k)])(t_k - t_{k-1}).$$

Applying the definition of integrals in calculus gives

$$\lim_{\delta \to 0} \sum_1^n (E[X(\zeta_k)])(t_k - t_{k-1}) = \int_a^b E[X(t)]\,dt.$$

Therefore,

$$E\left[\int_a^b X(t)\,dt\right] = \int_a^b E[X(t)]\,dt.$$

$\square$

Proposition 7.5. *Let random processes $X(t)$ and $Y(t)$ both be integrable over $[a, b]$. Then*

(i) $E\left[\int_a^b X(t)\,dt\right] = \int_a^b \mu_X(t)\,dt;$

(ii) $E\left[\left(\int_a^b X(t)\,dt\right)^2\right] = \int_a^b \int_a^b R_X(s, t)\,ds\,dt;$

(iii) $\mathrm{Var}\left(\int_a^b X(t)\,dt\right) = \int_a^b \int_a^b C_X(s, t)\,ds\,dt.$

Proof. By Proposition 7.4 and $\mu_X(t) = E[X(t)]$, it follows immediately that

$$E\left[\int_a^b X(t)\,dt\right] = \int_a^b \mu_X(t)\,dt.$$

Since

$$\left(\int_a^b X(t)\,dt\right)^2 = \int_a^b X(t)\,dt \int_a^b X(s)\,ds = \int_a^b \int_a^b X(s)X(t)\,ds\,dt, \qquad (7.3)$$

by Proposition 7.4 and $R_X(s, t) = E[X(s)X(t)]$, it follows that

$$E\left[\left(\int_a^b X(t)\,dt\right)^2\right] = \int_a^b \int_a^b E[X(s)X(t)]\,ds\,dt = \int_a^b \int_a^b R_X(s, t)\,ds\,dt.$$

Let $Y = \int_a^b X(t)\,dt$ in the formula $\mathrm{Var}(Y) = E[Y^2] - (E[Y])^2$, and then by (i) and (ii), it follows that

$$\mathrm{Var}\left(\int_a^b X(t)\,dt\right) = E\left[\left(\int_a^b X(t)\,dt\right)^2\right] - \left(E\left[\int_a^b X(t)\,dt\right]\right)^2$$

$$= \int_a^b \int_a^b R_X(s,t)\,\mathrm{d}s\,\mathrm{d}t - \left(\int_a^b \mu_X(t)\,\mathrm{d}t \right)^2 .$$

Similarly to (7.3), the last term is

$$\left(\int_a^b \mu_X(t)\,\mathrm{d}t \right)^2 = \int_a^b \int_a^b \mu_X(s)\mu_X(t)\,\mathrm{d}s\,\mathrm{d}t.$$

Therefore,

$$\mathrm{Var}\left(\int_a^b X(t)\,\mathrm{d}t \right) = \int_a^b \int_a^b (R_X(s,t) - \mu_X(s)\mu_X(t))\,\mathrm{d}s\,\mathrm{d}t$$

$$= \int_a^b \int_a^b C_X(s,t)\,\mathrm{d}s\,\mathrm{d}t;$$

the last equality is because $C_X(s,t) = R_X(s,t) - \mu_X(s)\mu_X(t)$. $\qquad\square$

Theorem 7.1. *Let X be a continuously differentiable random process. Then, for $a < b$,*

$$\int_a^b X'(t)\,\mathrm{d}t = X(b) - X(a).$$

Proof. Let

$$B = X(b) - X(a) - \int_a^b X'(t)\,\mathrm{d}t.$$

Then

$$E[B^2] = E\left[B\left(X(b) - X(a) - \int_a^b X'(t)\,\mathrm{d}t \right) \right]$$

$$= E[BX(b)] - E[BX(a)] - E\left[B \int_a^b X'(t)\,\mathrm{d}t \right].$$

Denote $\phi(t) = E[BX(t)]$. Then

$$E[B^2] = \phi(b) - \phi(a) - E\left[B \int_a^b X'(t)\,\mathrm{d}t \right]. \qquad (7.4)$$

By the definition of the integral and the properties of the expectation, it follows that

$$E\left[B \int_a^b X'(t)\,\mathrm{d}t \right] = E\left[\lim_{\delta \to 0} \sum_{k=1}^n BX'(\zeta_k)(t_k - t_{k-1}) \right]$$

$$= \lim_{\delta \to 0} \sum_{k=1}^n E[BX'(\zeta_k)](t_k - t_{k-1}).$$

From $\phi(t) = E[BX(t)]$ and $\lim_{s \to t} \frac{X(s) - X(t)}{s - t} = X'(t)$, it follows that

$$\phi'(t) = \lim_{s \to t} \frac{\phi(s) - \phi(t)}{s - t}$$

$$= \lim_{s \to t} \frac{E[BX(s)] - E[BX(t)]}{s - t}$$

$$= E\left[B \left(\lim_{s \to t} \frac{X(s) - X(t)}{s - t} \right) \right] = E[BX'(t)].$$

Therefore,

$$E\left[B \int_a^b X'(t) \, dt \right] = \lim_{\delta \to 0} \sum_{k=1}^n \phi'(\zeta_k)(t_k - t_{k-1}) = \int_a^b \phi'(t) \, dt.$$

From this and (7.4),

$$E[B^2] = \phi(b) - \phi(a) - \int_a^b \phi'(t) \, dt.$$

The fundamental theorem in calculus gives $\phi(b) - \phi(a) = \int_a^b \phi'(t) \, dt$. So $E[B^2] = 0$. Notice that $B^2 \geq 0$. Then $B = 0$, i.e.,

$$\int_a^b X'(t) \, dt = X(b) - X(a).$$

$\square$

7.4 SPECTRAL ANALYSIS

Now we extend the linear time-invariant system given in Section 3.1 to WSS processes and then give the relationship between power spectral densities of input processes and output processes.

7.4.1 Linear Time-Invariant System for WSS Processes

Let T be a linear time-variant system with filter h, which is stated in Section 3.1. Assume that the input signal X and the output signal Y are both random processes. If T is continuous, then

$$Y(s) = (h * X)(s) = \int_{\mathbb{R}} h(s - t)X(t) \, dt \quad (s \in \mathbb{R}),$$

where $X(t)$ $(t \in \mathbb{R})$ and $Y(s)$ $(s \in \mathbb{R})$ are both continuous random processes. If T is discrete, then

$$Y(k) = (h * X)(k) = \sum_n h(k - n)X(n) \quad (k \in \mathbb{Z}),$$

where $X(n)$ $(n \in \mathbb{Z})$ and $Y(k)$ $(k \in \mathbb{Z})$ are both discrete random processes.

Assume that $X(t)$ $(t \in \mathbb{R})$ is a continuous WSS random process. The mean function of the output signal

$$\mu_Y(s) = \int_{\mathbb{R}} h(s - t)\mu_X \, dt = \mu_X \int_{\mathbb{R}} h(t) \, dt.$$

Therefore, the mean function $\mu_Y(s)$ does not depend on s. The cross-correlation function of X and Y

$$R_{YX}(s, \tau) = E[Y(s)X(\tau)] = E\left[X(\tau) \int_{\mathbb{R}} h(s - t)X(t) \, dt\right]$$

$$= \int_{\mathbb{R}} h(s - t)E[X(\tau)X(t)] \, dt.$$

From this and $E[X(\tau)X(t)] = R_X(t - \tau)$, it follows that

$$R_{YX}(s, \tau) = \int_{\mathbb{R}} h(s - t)R_X(t - \tau) \, dt$$

$$= \int_{\mathbb{R}} h(s - \tau - t)R_X(t) \, dt = (h * R_X)(s - \tau).$$

Therefore, the cross-correlation function $R_{YX}(s, t)$ is a function of $s - \tau$ alone. The correlation function of Y

$$R_Y(s, u) = E[Y(s)Y(u)] = E\left[Y(s) \int_{\mathbb{R}} h(u - \tau)X(\tau) \, d\tau\right]$$

$$= \int_{\mathbb{R}} h(u - \tau)E[Y(s)X(\tau) \, d\tau] = \int_{\mathbb{R}} h(u - \tau)R_{YX}(s, \tau) \, d\tau.$$

Define a new function $\widetilde{h}$ by $\widetilde{h}(v) = h(-v)$. Then

$$R_Y(s, u) = \int_{\mathbb{R}} \widetilde{h}(\tau - u)(h * R_X)(s - \tau) \, d\tau$$

$$= \int_{\mathbb{R}} \widetilde{h}(s - u - v)(h * R_X)(v) \, dv = (\widetilde{h} * h * R_X)(s - u).$$

Therefore, the correlation function $R_Y(s, u)$ is also a function of $s - u$ alone.

Similarly, we discuss the cross-covariance of Y and X and the covariance of Y.

Proposition 7.6. *For a continuous linear time-invariant system with filter* $h(t)$ $(t \in \mathbb{R})$, *if an input random process* $X(t) \in$ WSS, *then*

(i) *the output random process* $Y(t) \in$ WSS $(t \in \mathbb{R})$;

(ii) $X(t)$ *and* $Y(t)$ *are jointly WSS;*

(iii) *the mean function* μ_Y, *the cross-correlation function* R_{YX}, *the correlation function* R_Y, *the cross-covariance function* C_{YX}, *and the covariance function* C_Y *satisfy*

$$\mu_Y = \mu_X \int_{\mathbb{R}} h(t) \, dt,$$

$$R_{YX}(\tau) = (h * R_X)(\tau) = \int_{\mathbb{R}} h(\tau - s)R_X(s)\,ds,$$

$$R_Y(\tau) = (\tilde{h} * h * R_X)(\tau) = \int_{\mathbb{R}} \tilde{h}(\tau - u)\left(\int_{\mathbb{R}} h(u - s)R_X(s)\,ds\right)du,$$

$$C_{YX}(\tau) = (h * C_X)(\tau) = \int_{\mathbb{R}} h(\tau - s)C_X(s)\,ds,$$

$$C_Y(\tau) = (\tilde{h} * h * C_X)(\tau) = \int_{\mathbb{R}} \tilde{h}(\tau - u)\left(\int_{\mathbb{R}} h(u - s)C_X(s)\,ds\right)du,$$

where $\tilde{h}(t) = \overline{h}(-t)$ $(t \in \mathbb{R})$.

Assume that $X(n)$ $(n \in \mathbb{Z})$ is a discrete WSS random process. Then Proposition 7.6 still holds if $X(t)$, $Y(t)$, $h(t)$ are replaced by $X(n)$, $Y(n)$, $h(n)$ and the convolution is considered as

$$(f * g)(k) = \sum_k f(n - k)g(k).$$

For example,

$$R_{YX}(k) = (h * R_X)(\tau) = \sum_n h(k - n)R_X(n),$$

$$R_Y(k) = (\tilde{h} * h * R_X)(k) = \sum_n \tilde{h}(k - n)\left(\sum_l h(n - l)R_X(l)\right),$$

where $\tilde{h}(n) = \overline{h}(-n)$ $(n \in \mathbb{Z})$.

7.4.2 Power Spectral Density

Let $X(t)$ $(t \in \mathbb{R})$ be a WSS random process. The Fourier transform of the correlation function $R_X(\tau)$ is called the *power spectral density function* of the WSS random process X, denoted by $S_X(\omega)$, i.e.,

$$S_X(\omega) = \widehat{R}_X(\omega) = \int_{\mathbb{R}} R_X(\tau)e^{-i\omega\tau}\,d\tau.$$

Let $X(t)$ $(t \in \mathbb{R})$ and $Y(t)$ $(t \in \mathbb{R})$ be jointly WSS random processes. The Fourier transform of the cross-correlation function $R_{YX}(\tau)$ of Y and X is called the *cross-power spectral density function* of the jointly WSS random processes Y and X, denoted by S_{YX}, i.e.,

$$S_{YX}(\omega) = \widehat{R}_{YX}(\omega) = \int_{\mathbb{R}} R_{YX}(\tau)e^{-i\omega\tau}\,d\tau.$$

For a linear time-invariant system $Y = h * X$ with filter h, if $X \in$ WSS, then, by Proposition 7.6,

$$R_{YX}(\tau) = (h * R_X)(\tau),$$

$$R_Y(\tau) = (\tilde{h} * h * R_X)(\tau).$$

Taking Fourier transforms on both sides, by the property $\widehat{\tilde{h}}(\omega) = \overline{\widehat{h}(\omega)}$, we get

$$S_{YX}(\omega) = \widehat{h}(\omega)S_X(\omega),$$

$$S_Y(\omega) = \widehat{\tilde{h}}(\omega)\widehat{h}(\omega)\widehat{S}_X(\omega) = |\widehat{h}(\omega)|^2 S_X(\omega). \qquad (7.5)$$

Since $S_X(\omega)$ is the *power spectral density function* of X, the inverse Fourier transform of $S_X(\omega)$ yields

$$R_X(\tau) = \frac{1}{2\pi} \int_{\mathbb{R}} S_X(\omega)\, e^{i\omega\tau}\, d\omega.$$

In particular,

$$R_X(0) = \frac{1}{2\pi} \int_{\mathbb{R}} S_X(\omega)\, d\omega.$$

Since $E[|X(t)|^2]$ is the total power of X and $R_X(0) = E[|X(t)|^2]$, the total power of X is

$$E[|X(t)|^2] = \frac{1}{2\pi} \int_{\mathbb{R}} S_X(\omega)\, d\omega. \qquad (7.6)$$

Example 7.6. Suppose that X is a complex-valued WSS random process with mean function $\mu_X = 0$ and Y is a moving average of X:

$$Y(s) = \frac{1}{T} \int_{s-T}^{s} X(t)\, dt \quad (T > 0).$$

This is equivalent to a linear time-invariant system:

$$Y(s) = \int_{\mathbb{R}} h(s-t)X(t)\, dt = (h * X)(s),$$

where

$$h(\tau) = \frac{1}{T} \quad (0 \le \tau \le T),$$
$$h(\tau) = 0 \quad (\tau < 0, \tau > T).$$

By $\tilde{h}(v) = h(-v)$, a direct calculation gives

$$(\tilde{h} * h)(\tau) = \int_{\mathbb{R}} h(s)h(s-\tau)\, ds = \begin{cases} \frac{1}{T}\left(1 - \frac{|\tau|}{T}\right), & |\tau| \le T, \\ 0, & |\tau| \ge T. \end{cases}$$

From this and $R_Y(s) = (\tilde{h} * h * R_X)(s)$, the correlation function is

$$R_Y(s) = \int_{\mathbb{R}} (\tilde{h} * h)(s-\tau)R_X(\tau)\, d\tau = \frac{1}{T} \int_{s-T}^{s+T} \left(1 - \frac{|s-\tau|}{T}\right) R_X(\tau)\, d\tau.$$

By $\mu_X = 0$,

$$E[Y(s)] = \frac{1}{T} \int_{s-T}^{T} E[X(t)] \, dt = 0,$$

and so $\mu_Y = 0$. This implies that

$$C_Y(s) = R_Y(s) = \frac{1}{T} \int_{s-T}^{s+T} \left(1 - \frac{|s - \tau|}{T} \right) R_X(\tau) \, d\tau.$$

From this and $\mathrm{Var}(Y(t)) = C_Y(0)$, it follows that

$$\mathrm{Var}(Y(t)) = \frac{1}{T} \int_{-T}^{T} \left(1 - \frac{|\tau|}{T} \right) R_X(\tau) \, d\tau.$$

Note that the Fourier transform of h

$$\widehat{h}(\omega) = \frac{1}{T} \int_{0}^{T} e^{-it\omega} \, dt = \frac{1}{i\omega} \left(1 - e^{-iT\omega} \right) = \frac{\sin(T\omega/2)}{\omega/2} e^{-iT\omega/2}.$$

By (7.5), the power spectral density function of Y is

$$S_Y(\omega) = |\widehat{h}(\omega)|^2 S_X(\omega) = \left(\frac{\sin(T\omega/2)}{\omega/2} \right)^2 S_X(\omega).$$

Consider two linear time-invariant system, one has input $X(t)$, filter $h(t)$, and output $U(t) = (h * X)(t)$, and the other has input $Y(t)$, filter $k(t)$, and output $V(t) = (k * Y)(t)$. Suppose that $X(t)$ and $Y(t)$ are jointly WSS. The correlation function of U and V

$$R_{UV}(t, \tau) = E[U(t)\overline{V}(\tau)] = E \left[\int_{\mathbb{R}} \int_{\mathbb{R}} h(t - s)\overline{k}(\tau - v)X(s)\overline{Y}(v) \, ds \, dv \right].$$

Combining this with $E[X(s)\overline{Y}(v)] = R_{XY}(s - v)$, we get

$$R_{UV}(t, \tau) = \int_{\mathbb{R}} \int_{\mathbb{R}} h(t - s)\overline{k}(\tau - v)R_{XY}(s - v) \, ds \, dv.$$

However,

$$\int_{\mathbb{R}} h(t - s)R_{XY}(s - v) \, ds = \int_{\mathbb{R}} h(t - v - u)R_{XY}(u) \, du = (h * R_{XY})(t - v).$$

Therefore,

$$R_{UV}(t, \tau) = \int_{\mathbb{R}} \overline{k}(\tau - v)(h * R_{XY})(t - v) \, dv = \int_{\mathbb{R}} \overline{k}(\tau - t + s)(h * R_{XY})(s) \, ds$$

$$= \int_{\mathbb{R}} \widetilde{k}(t - \tau - s)(h * R_{XY})(s) \, ds = (\widetilde{k} * h * R_{XY})(t - \tau),$$

where $\widetilde{k}(v) = \overline{k}(-v)$. This shows that $R_{UV}(t, \tau)$ is a function of $t - \tau$ alone. From this, we see that if U and V are individually WSS, then they are jointly

WSS, and the correlation function

$$R_{UV} = \tilde{k} * h * R_{XY},$$

and the cross-power spectral density function

$$S_{UV} = \overline{K}HS_{XY},$$

where $K = \widehat{k}$, $H = \widehat{h}$, and S_{XY} is the cross-power spectral density of X and Y.

7.4.3 Shannon Sampling Theorem for Random Processes

Let X be a continuous WSS random process with power spectral density function S_X. If

$$S_X(\omega) = 0 \quad (|\omega| \geq \omega_0),$$

the process X is called a *bandlimited random process*.

Now we extend the Shannon sampling theorem in Chapter 1 from deterministic functions to random processes.

Shannon Sampling Theorem for Random Processes. *Suppose that X is a complex-valued WSS bandlimited random process with $S_X(\omega) = 0$ ($|\omega| \geq \omega_0$). Let $T = \frac{\pi}{\omega_0}$. Then, for each $t \in \mathbb{R}$,*

$$X(t) = \sum_n X(nT) \frac{\sin((\pi(t - nT))/T)}{(\pi(t - nT))/T}.$$

Let B be the process of samples defined by $B_n = X(nT)$. Then the power spectral density functions of B and X satisfy

$$S_B(\omega) = \frac{1}{T} S_X\left(\frac{\omega}{T}\right) \quad (|\omega| \leq \pi).$$

Proof. Let

$$J_N(t) = \left| X(t) - \sum_{-N}^{N} X(nT) \frac{\sin((\pi(t - nT))/T)}{(\pi(t - nT))/T} \right|^2.$$

Notice that

$$J_N(t) = |X(t)|^2 - 2 \sum_{-N}^{N} X(t)\overline{X}(nT) \frac{\sin((\pi(t - nT))/T)}{(\pi(t - nT))/T}$$

$$+ \sum_{n,m=-N}^{N} X(nT)\overline{X}(mT) \frac{\sin((\pi(t - nT))/T)}{(\pi(t - nT))/T} \frac{\sin((\pi(t - nT))/T)}{(\pi(t - mT))/T}.$$

Taking expectation on both sides, we get

$$E[J_N(t)] = E[|X(t)|^2] - 2 \sum_{n=-N}^{N} E[X(t)\overline{X}(nt)] \frac{\sin((\pi(t-nT))/T)}{(\pi(t-nT))/T}$$

$$+ \sum_{n,m=-N}^{N} E[X(nt)\overline{X}(mt)] \frac{\sin((\pi(t-nT))/T)}{(\pi(t-nT))/T} \frac{\sin((\pi(t-mT))/T)}{(\pi(t-mT))/T}.$$

By the assumption $S_X(\omega) = 0$ ($|\omega| \geq \omega_0$), it follows that

$$E[X(a)\overline{X}(b)] = R_X(a-b) = \frac{1}{2\pi} \int_{\mathbb{R}} S_X(\omega) e^{i\omega(a-b)} \, d\omega$$

$$= \frac{1}{2\pi} \int_{-\omega_0}^{\omega_0} S_X(\omega) e^{i\omega(a-b)} \, d\omega.$$

Especially,

$$E[|X(t)|^2] = E[X(t)\overline{X}(t)] = \frac{1}{2\pi} \int_{-\omega_0}^{\omega_0} S_X(\omega) \, d\omega,$$

$$E[X(t)\overline{X}(nt)] = \frac{1}{2\pi} \int_{-\omega_0}^{\omega_0} S_X(\omega) e^{i\omega t} e^{-i\omega nt} \, d\omega,$$

$$E[X(nt)\overline{X}(mt)] = \frac{1}{2\pi} \int_{-\omega_0}^{\omega_0} S_X(\omega) e^{i\omega nt} e^{-i\omega mt} \, d\omega.$$

Therefore,

$$E[J_N(t)] = \frac{1}{2\pi} \int_{-\omega_0}^{\omega_0} S_X(\omega) \left| e^{i\omega t} - \sum_{-N}^{N} e^{i\omega nT} \frac{\sin((\pi(t-nT))/T)}{(\pi(t-nT))/T} \right|^2 \, d\omega.$$

Notice that

$$\left| e^{i\omega t} - \sum_{-N}^{N} e^{i\omega nT} \frac{\sin((\pi(t-nT))/T)}{(\pi(t-nT))/T} \right|^2 \to 0 \quad (N \to \infty)$$

uniformly on $|\omega| \leq \omega_0 - \epsilon$ ($\omega_0 = \frac{\pi}{T}$). Applying the dominated convergence theorem, we get

$$E[J_N(t)] \to 0 \quad (N \to \infty),$$

i.e.,

$$X(t) = \sum_{n} X(nT) \frac{\sin((\pi(t-nT))/T)}{(\pi(t-nT))/T}.$$

Since X is WSS, it is clear that $B_n = X(nT)$ is a WSS discrete-time random process with $\mu_B = \mu_X$. Since $S_X(\omega) = 0$ ($|\omega| \geq \omega_0$),

$$R_B(n) = R_X(nT) = \frac{1}{2\pi} \int_{\mathbb{R}} S_X(\omega) e^{inT\omega} \, d\omega = \frac{1}{2\pi} \int_{-\omega_0}^{\omega_0} S_X(\omega) e^{inT\omega} \, d\omega.$$

Taking a substitution of variable $\nu = T\omega$, by the assumption $T = \frac{\pi}{\omega_0}$, we get for the correlation of B

$$R_B(n) = \frac{1}{2\pi T} \int_{-\pi}^{\pi} S_X\left(\frac{\nu}{T}\right) e^{in\nu} \, d\nu,$$

The theory of Fourier series in Chapter 1 shows that

$$\sum_n R_B(n) e^{-in\omega} = \frac{1}{T} S_X\left(\frac{\omega}{T}\right).$$

Combining this with $S_B(\omega) = \sum_n R_B(n) e^{-in\omega}$, we find the power spectral density functions of B and X satisfy

$$S_B(\omega) = \frac{1}{T} S_X\left(\frac{\omega}{T}\right) \quad (|\omega| \leq \pi).$$

$\square$

7.5 WIENER FILTERING

Let $X(t)$ be a random process. Consider an approximation problem of the random process $X(t)$ at some fixed t by the observation $Y(s)$ ($a \leq s \leq b$).

Define v as the closure of the linear combinations $c_1 Y(s_1) + c_2 Y(s_2) + \cdots + c_n Y(s_n)$, where $c_1, c_2, \ldots, c_n \in \mathbb{R}$ and $a \leq s_1 \leq \cdots \leq s_n \leq b$. If $\hat{X}(t) \in v$ and the mean square error $E[|X(t) - \hat{X}(t)|^2]$ is minimal, then $\hat{X}(t)$ is called the *best linear approximation* of $X(t)$ based on the observation $Y(s)$ ($a \leq s \leq b$). In Section 5.4, we chose the best approximation in linear combinations. However, now we choose the best approximation in the closure of linear combinations.

Proposition 7.7. *The integral $\int_a^b h(t-s)Y(s) \, ds$ is the best linear approximation of $X(t)$ based on the observation $Y(s)$ ($a \leq s \leq b$) if and only if*

$$\left(X(t) - \int_a^b h(t-s)Y(s) \, ds\right) \perp Y(u) \quad (u \in [a,b]).$$

Proof. By the definition of integrals,

$$\int_a^b h(t-s)Y(s) \, ds = \lim_{\delta \to 0} \sum_1^n h(t-s_k)Y(s_k),$$

where $\delta = \min(s_k - s_{k-1})$ $(k = 2, \ldots, n)$. So

$$\int_a^b h(t - s)Y(s)\, ds \in v.$$

However, Theorem 5.3 shows that the random variable $W(t) \in v$ is the best linear approximation of $X(t)$ if and only if $(X(t) - W(t)) \perp Y(u)$ $(a \leq u \leq b)$. Therefore, Proposition 7.7 holds. $\qquad\qquad\square$

If X and Y are the jointly WSS, we want to find an optimal estimator of the form

$$\hat{X}(t) = \int_{\mathbb{R}} h(t - s)Y(s)\, ds.$$

It is known that the optimality condition is

$$\left(X(t) - \int_{\mathbb{R}} h(t - s)Y(s)\, ds\right) \perp Y(u) \quad (u \in \mathbb{R}).$$

This is equivalent to

$$E[X(t)Y(u)] - \int_{\mathbb{R}} h(t - s)E[Y(s)Y(u)]\, ds = 0.$$

Since X and Y are the jointly WSS,

$$E[X(t)Y(u)] = R_{XY}(t - u),$$
$$E[Y(s)Y(u)] = R_Y(s - u),$$

and so

$$R_{XY}(t - u) = \int_{\mathbb{R}} h(t - s)R_Y(s - u)\, ds.$$

Perform the change of variable $t - u = v$,

$$R_{XY}(v) = \int_{\mathbb{R}} h(v - (s - u))R_Y(s - u)\, ds = \int_{\mathbb{R}} h(v - s)R_Y(s)\, ds = (h * R_Y)(v).$$

Taking the Fourier transform on both sides, by the definitions

$$\widehat{R}_{XY}(\omega) = S_{XY}(\omega),$$
$$\widehat{R}_Y(\omega) = S_Y(\omega),$$

we find the optimality condition in the frequency domain is equivalent to

$$S_{XY}(\omega) = \widehat{h}(\omega)S_Y(\omega).$$

Therefore, the optimal filter h^o satisfies

$$\widehat{h^o}(\omega) = \frac{S_{XY}(\omega)}{S_Y(\omega)}, \qquad\qquad (7.7)$$

and $\hat{x}(t)$ is the optimal estimator if and only if

$$\hat{x}(t) = X^\circ(t) = \int_{\mathbb{R}} h^\circ(t - s)Y(s)\,\mathrm{d}s = (h^\circ * Y)(t).$$

By (7.5) and (7.7),

$$S_{X^\circ}(\omega) = |\widehat{h^\circ}(\omega)|^2 S_Y(\omega) = \frac{S_{XY}^2(\omega)}{S_Y(\omega)}.$$

It follows from (7.6) that

$$E[|X^\circ(t)|^2] = \frac{1}{2\pi} \int_{\mathbb{R}} S_{X^\circ}(\omega)\,\mathrm{d}\omega = \frac{1}{2\pi} \int_{\mathbb{R}} \frac{S_{XY}^2(\omega)}{S_Y(\omega)}\,\mathrm{d}\omega,$$

$$E[|X(t)|^2] = \frac{1}{2\pi} \int_{\mathbb{R}} S_X(\omega)\,\mathrm{d}\omega.$$

Therefore, by Theorem 5.3, the optimal mean square error is

$$E[|X(t) - X^\circ(t)|^2] = E[|X(t)|^2] - E[|X^\circ(t)|^2]$$

$$= \frac{1}{2\pi} \int_{\mathbb{R}} \left(S_X(\omega) - \frac{|S_{XY}(\omega)|^2}{S_Y(\omega)} \right) \mathrm{d}\omega. \qquad (7.8)$$

Example 7.7. Suppose that the observation Y is a random process X plus noise N, i.e., $Y = X + N$. Suppose further that X and N are both WSS with mean zero and X and N are jointly WSS. If $R_{XN} = 0$, then

$$\mu_Y(t) = E[Y(t)] = E[X(t)] + E[N(t)] = 0,$$

$$R_Y(t, s) = E[X(t)\overline{X}(s)] + E[X(t)\overline{N}(s)] + E[N(t)\overline{X}(s)] + E[N(t)\overline{N}(s)]$$

$$= R_X(t - s) + R_N(t - s).$$

Notice that $R_Y(t, s)$ depends only on $t - s$. Thus, Y is WSS and

$$R_Y(\tau) = R_X(\tau) + R_N(\tau).$$

Taking Fourier transform on both sides,

$$S_Y(\omega) = S_X(\omega) + S_N(\omega).$$

Similarly, $R_{XY}(t, s)$ depends only on $t - s$. Thus X and Y are jointly WSS and

$$R_{XY}(\tau) = R_X(\tau).$$

Taking Fourier transform on both sides, we get

$$S_{XY}(\omega) = S_X(\omega).$$

By (7.7), the optimal filter

$$\hat{h}^0(\omega) = \frac{S_{XY}(\omega)}{S_Y(\omega)} = \frac{S_X(\omega)}{S_X(\omega) + S_N(\omega)}.$$

By (7.8), the optimal mean square error is

$$E[|X(t) - X^0(t)|^2] = \frac{1}{2\pi} \int_{\mathbb{R}} \frac{S_X(\omega)S_N(\omega)}{S_X(\omega) + S_N(\omega)} \, d\omega.$$

7.6 SPECTRUM ESTIMATION

Suppose that $X(t)$ is a WSS process with mean 0. Denote

$$X_n = X(n\Delta t) \quad (n \in \mathbb{Z}).$$

Then X_n ($n \in \mathbb{Z}$) is a discrete WSS with mean 0. Its correlation sequence

$$\gamma_k = E[X_{n+k}X_n] \quad (n \in \mathbb{Z})$$

is an even sequence. Denote x_n ($n \in \mathbb{Z}$) is a sample of X_n ($n \in \mathbb{Z}$). If $X(t)$ is an ergodic random process, then the correlation sequence is

$$\gamma_k = \lim_{N \to \infty} \frac{1}{2N+1} \sum_{n=-N}^{N} x_{n+k}x_n \quad (k \in \mathbb{Z}).$$

The Fourier transform of correlation sequence γ_k ($k \in \mathbb{Z}$) is called the *power spectrum* of X_n, denoted by $S_X(\omega)$, i.e.,

$$S_X(\omega) = \sum_k \gamma_k e^{-ik\omega} = \gamma_0 + 2 \sum_1^{\infty} \gamma_k \cos(k\omega).$$

In application, one only knows finitely many x_n, say, $x_0, x_1, \ldots, x_{N-1}$, and the correlation sequence γ_k satisfies $\gamma_k = 0$ ($k \geq N$) and $\gamma_k = \gamma_{-k}$. The correlation sequence γ_k is estimated by

$$\hat{\gamma}_k = \frac{1}{N} \sum_{n=0}^{N-1} x_{n+k}x_n$$

or

$$\hat{\gamma}_k = \frac{1}{N} \sum_{n=0}^{N-k-1} x_{n+k}x_n \quad (k = 0, \ldots, N-1). \tag{7.9}$$

Let $\hat{\gamma}_k = 0$ ($k \geq N$) and $\hat{\gamma}_{-k} = \hat{\gamma}_k$ ($k \in \mathbb{Z}$). Then the corresponding power spectrum is estimated by

$$\hat{S}_X(\omega) = \sum_k \hat{\gamma}_k e^{-ik\omega}$$

or

$$\hat{S}_X(\omega) = \sum_{-N+1}^{N-1} \hat{\gamma}_k e^{-ik\omega} = \hat{\gamma}_0 + 2 \sum_0^{N-1} \hat{\gamma}_k \cos(k\omega). \tag{7.10}$$

Denote

$$x(n) = \begin{cases} x_n, & 0 \le n < N - 1, \\ 0, & \text{otherwise.} \end{cases}$$

Notice that $\hat{\gamma}_k = \hat{\gamma}_{-k}$. By (7.9), it follows that

$$\hat{\gamma}_k = \frac{1}{N} \sum_{n=0}^{N-1} x_{n-k} x_n = \frac{1}{N} \sum_n x(n-k)x(n) = \frac{1}{N} \sum_n x(n+k)x(n),$$

and so

$$\hat{S}_X(\omega) = \sum_k \hat{\gamma}_k e^{-ik\omega} = \frac{1}{N} \sum_k \left(\sum_n x(n+k)x(n) \right) e^{-ik\omega}$$

$$= \frac{1}{N} \sum_n x(n) \left(\sum_k x(n+k)e^{-ik\omega} \right)$$

$$= \frac{1}{N} \left(\sum_n x(n)e^{in\omega} \right) \left(\sum_m x(m)e^{-im\omega} \right)$$

$$= \frac{1}{N} \left| \sum_0^{N-1} x_n e^{-in\omega} \right|^2.$$

Since $\gamma_k = E[X_{n+k}X_n]$ $(n \in \mathbb{Z})$, by (7.9), it follows that

$$E[\hat{\gamma}_k] = \frac{1}{N} \sum_{n=0}^{N-1-k} E[X_{n+k}X_n] = \frac{N-k}{N} \gamma_k = \left(1 - \frac{k}{N} \right) \gamma_k.$$

i.e., (7.9) is a biased estimate of the correlation. From this and (7.10), by $\gamma_k = 0$ $(k \ge N)$ and $\gamma_k = \gamma_{-k}$, the expectation of the estimate of the power spectrum is

$$E[\hat{S}_X(\omega)] = \sum_{-N+1}^{N-1} E[\hat{\gamma}_k] e^{-ik\omega}$$

$$= \sum_{-N+1}^{N-1} \left(1 - \frac{|k|}{N} \right) \gamma_k e^{-ik\omega}.$$

By the integral $\frac{1}{2\pi} \int_{-\pi}^{\pi} e^{i(k-l)\theta} \, d\theta = \delta_{kl}$ and $S_X(\omega) = \sum_l \gamma_l e^{-il\theta}$, it follows that

$$E[\hat{S}_X(\omega)] = \sum_{k=-N+1}^{N-1} \left(1 - \frac{|k|}{N}\right) e^{-ik\omega} \sum_l \gamma_l \left(\frac{1}{2\pi} \int_{-\pi}^{\pi} e^{i(k-l)\theta} \, d\theta\right)$$

$$= \frac{1}{2\pi} \int_{-\pi}^{\pi} \left(\sum_{k=-N+1}^{N-1} \left(1 - \frac{|k|}{N}\right) e^{-ik(\omega-\theta)}\right) \left(\sum_l \gamma_l e^{-il\theta}\right) d\theta$$

$$= \frac{1}{2\pi} \int_{-\pi}^{\pi} \left(\sum_{k=-N+1}^{N-1} \left(1 - \frac{|k|}{N}\right) e^{-ik(\omega-\theta)}\right) S_X(\omega) \, d\theta. \qquad (7.11)$$

Notice that

$$\sum_{k=-N+1}^{N-1} \left(1 - \frac{|k|}{N}\right) e^{-iku} = 1 + 2 \sum_{k=1}^{N-1} \left(1 - \frac{k}{N}\right) \cos(ku) = \frac{1}{N} \left(\frac{\sin \frac{N}{2}u}{\sin \frac{1}{2}u}\right)^2.$$

Then

$$E[\hat{S}_X(\omega)] = \frac{1}{2\pi N} \int_{-\pi}^{\pi} \left(\frac{\sin \frac{N}{2}(\omega - \theta)}{\sin \frac{1}{2}(\omega - \theta)}\right)^2 S_X(\theta) \, d\theta.$$

So $E[\hat{S}_X(\omega)]$ is Fejér's sum of the Fourier series of the power spectrum $S_X(\omega)$, i.e., the arithmetic mean of the partial sums of the Fourier series of $S_X(\omega)$. According to Fejér's theorem of Fourier series, we get

$$\lim_{N \to \infty} E[\hat{S}_X(\omega)] = S_X(\omega).$$

This states that the estimate of the power spectrum is an asymptotic unbiased estimate of the power spectrum.

Let $W = \{W_n\}_{n \in \mathbb{Z}}$ be a window $W_n = 0$ $(n \neq 0, \ldots, N-1)$ and $X = \{X_n\}_{n \in \mathbb{Z}}$ be an ergodic WSS sequence with samples $x_0, \ldots, x_{N-1}$. Denote

$$V_N = \frac{1}{N} \sum_0^{N-1} |W_n|^2.$$

Then the estimate

$$\hat{S}_{X,N}^W(\omega) = \frac{1}{NV_N} \left|\sum_n x_n W_n e^{-in\omega}\right|^2$$

is called the *periodogram* with window W.

A direct computation shows that

$$E\left[\hat{S}_{X,N}^W(\omega)\right] = \frac{1}{NV_N} \sum_m \sum_n E[x_n x_m] W_n W_m e^{i(m-n)\omega}$$

$$= \frac{1}{NV_N} \sum_m \sum_n \gamma_{m-n} W_n W_m e^{i(m-n)\omega}$$

$$= \frac{1}{NV_N} \sum_m \sum_k \gamma_k W_{m-k} W_m e^{ik\omega} = \frac{1}{NV_N} \sum_k \gamma_k W_k^* e^{-ik\omega},$$

where $W_k^* = \sum_m W_{m-k} W_m$ and $\gamma_k = \gamma_{-k}$.
An argument similar to (7.11) shows that

$$E\left[\hat{S}_{X,N}^W(\omega)\right] = \frac{1}{2\pi NV_N} \int_{-\pi}^{\pi} \left(\sum_k W_k^* e^{-ik(\omega-\theta)} \right) S_X(\theta)\, d\theta.$$

Notice that

$$\sum_k W_k^* e^{-iku} = \sum_k \left(\sum_m W_{m-k} W_m \right) e^{iku} = \sum_n \left(\sum_m W_n W_m \right) e^{i(m-n)u}$$

$$= \left(\sum_n W_n e^{-inu} \right) \left(\sum_m W_m e^{imu} \right)$$

$$= \left| \sum_n W_n e^{-inu} \right|^2 = \left| \sum_0^{N-1} W_n e^{-inu} \right|^2.$$

Denote

$$Q_N(u) = \left| \sum_0^{N-1} W_n e^{-inu} \right|^2.$$

Then

$$E\left[\hat{S}_{X,N}^W(\omega)\right] = \frac{1}{2\pi NV_N} \int_{-\pi}^{\pi} Q_N(\omega-\theta) S_X(\theta)\, d\theta = \frac{1}{2\pi NV_N} (Q_N * S_X)(\omega).$$

$$(7.12)$$

Using Parseval's identity of Fourier series, by $V_N = \frac{1}{N} \sum_0^{N-1} |W_n|^2$, it follows that

$$\frac{1}{2\pi NV_N} \int_{-\pi}^{\pi} Q_N(\omega-\theta)\, d\theta = \frac{1}{2\pi NV_N} \int_{-\pi}^{\pi} Q_N(\theta)\, d\theta$$

$$= \frac{1}{2\pi N V_N} \int_{-\pi}^{\pi} \left| \sum_{0}^{N-1} W_n e^{-in\theta} \right|^2 d\theta$$

$$= \frac{1}{N V_N} \sum_{0}^{N-1} |W_k|^2 = 1.$$

Therefore, the mean $E[\hat{S}_{X,N}^W(\omega)]$ is a weighted average of the power spectrum $S_X(\omega)$. By (7.12) and

$$\frac{1}{2\pi N V_N} \int_{-\pi}^{\pi} Q_N(\theta) \, d\theta = 1,$$

it follows that

$$E\left[\hat{S}_{X,N}^W(\omega) \right] - S_X(\omega)$$

$$= \frac{1}{2\pi N V_N} \int_{-\pi}^{\pi} Q_N(\omega - \theta) S_X(\omega) \, d\theta$$

$$- \frac{1}{2\pi N V_N} \int_{-\pi}^{\pi} Q_N(\theta) S_X(\omega) \, d\theta$$

$$= \frac{1}{2\pi N V_N} \int_{-\pi}^{\pi} Q_N(\theta)(S_X(\theta + \omega) - S_X(\omega)) \, d\theta.$$

The theory of the classical Fourier series summation shows that under the mild condition,

$$\lim_{N \to \infty} E\left[\hat{S}_{X,N}^W(\omega) \right] = S_X(\omega).$$

This states that the estimate of the power spectrum, $\hat{S}_{X,N}^W(\omega)$ is an asymptotically unbiased estimate of the power spectrum $S_X(\omega)$.

Let X_n ($n \in \mathbb{Z}$) be an ergodic WSS sequence. For $j = 1, \ldots, l$, denote x_{jn} ($n = 0, \ldots, N - 1$) as the jth uncorrelated sample. Then the corresponding periodogram obtained is

$$S_X^{(j)}(\omega) = \frac{1}{N} \left| \sum_{n=0}^{N-1} x_{jn} e^{-in\omega} \right|^2 \qquad (j = 1, \ldots, l).$$

Taking the average of these periodograms, we get

$$MS_X(\omega) = \frac{1}{l} \sum_{1}^{l} S_X^{(j)}(\omega).$$

It is an asymptotically unbiased estimate of the power spectrum $S_X(\omega)$.

7.7 SIGNIFICANCE TESTS OF CLIMATIC TIME SERIES

In climatic time series analysis, significance tests are used for extracting statistically significant features from randomness. To determine significance levels for Fourier and wavelet power spectra, one needs to choose an appropriate background spectrum. For many climatic signals, an appropriate background spectrum is either white noise or red noise.

7.7.1 Fourier Power Spectra

We say $\{x_n\}$ is a discrete-time *white noise* if, for each n, x_n is a random variable with mean 0 and variance σ^2 and for $n \neq m$, x_n and x_m are independent, i.e.,

$$E[x_n] = 0,$$
$$\text{Var}(x_n) = \sigma^2,$$
$$E[x_n x_m] = 0 \quad (n \neq m).$$

We say $\{x_n\}$ is a *Gaussian white noise* if it is a white noise and is a Gaussian random variable for each n.

A simple model for red noise is the univariate lag-1 autoregressive AR(1) process if a time series (discrete random process) $\{x_n\}_{n=0,1,\ldots}$ satisfies

$$x_0 = 0,$$
$$x_n = \alpha x_{n-1} + z_n \quad (n = 1, 2, \ldots), \tag{7.13}$$

where $0 < \alpha < 1$ and z_n is the Gaussian white noise.

For an AR(1) process $\{x_k\}_{k=0,\ldots,N-1}$, its parameters α and σ^2 can be estimated by

$$\alpha = \frac{(1/N - 1) \sum_0^{N-1} (x_i - \overline{x})(x_{i+1} - \overline{x})}{(1/N) \sum_0^{N-1} (x_i - \overline{x})}$$

and

$$\sigma^2 = \frac{1 - \alpha^2}{N} \sum_0^{N-1} (x_i - \overline{x})^2 = (1 - \alpha^2)\widetilde{\sigma}^2,$$

where $\widetilde{\sigma}^2$ is the variance of the sequence $\{x_k\}_{k=0,\ldots,N-1}$, and $\overline{x} = \frac{1}{N-1} \sum_0^{N-1} x_i$.

For a discrete random process $x = \{x_n\}_{n=0,\ldots,N-1}$, its discrete Fourier transform is

$$X_k = \frac{1}{N} \sum_{n=0}^{N-1} x_n \, e^{-ik\frac{2\pi n}{N}} \quad (k = 0, \ldots, N-1), \tag{7.14}$$

and the Fourier power spectrum of x is defined as $|X_k|^2$.

Torrence and Compo (1998) provided an empirical formula for the distribution of the Fourier power spectrum of an AR(1) process. Zhang and Moore

(2011b) proved this empirical formula in a rigorous statistical framework, and applied it to significance tests of temperatures in central England.

Theorem 7.2. *Let $\{X_k\}_{k=0,\ldots,N-1}$ be the discrete Fourier transform of an AR(1) process $x = \{x_k\}_{k=0,\ldots,N-1}$ with parameters α and σ^2. Then the Fourier power spectrum of x satisfies the following condition:*

$$\frac{N|X_k|^2}{\widetilde{\sigma}^2} \quad \text{is distributed as} \quad \frac{1-\alpha^2}{2\left(1 - 2\alpha \cos\frac{2\pi k}{N} + \alpha^2\right)}\chi_2^2,$$

where $\widetilde{\sigma}^2 = \frac{\sigma^2}{1-\alpha}$ is the variance of the sequence $\{x_k\}_{k=0,\ldots,N-1}$ and χ_2^2 is the chi-squared distribution with two degrees of freedom.

Proof. From (7.13), we have

$$x_k = \sum_{l=1}^{k} \alpha^{k-l} z_l \quad (k = 1, 2, 3, \ldots).$$

Consider the first N terms of this recurrence formula. From this and (7.14), the discrete Fourier transform

$$X_k = \sum_{j=0}^{N-1} c_{kj} z_j,$$

where $N c_{kj} = \sum_{l=j}^{N-1} \epsilon_k^l \alpha^{l-j} = \frac{\epsilon_k^j - \alpha^{N-j}}{1 - \epsilon_k \alpha}$ and $\epsilon_k = e^{-i\frac{2k\pi}{N}}$. So

$$N(1 - \epsilon_k \alpha)X_k = \sum_{j=0}^{N-1} (\epsilon_k^j - \alpha^{N-j}) z_j.$$

Let $\zeta_{1k} = \mathrm{Re}\,(\sqrt{N}(1 - \epsilon_k \alpha)X_k)$. Then

$$\zeta_{1k} = \frac{1}{\sqrt{N}} \sum_{j=1}^{N-1} z_j \mathrm{Re}\{\epsilon_k^j - \alpha^{N-j}\} = \frac{1}{\sqrt{N}} \sum_{j=1}^{N-1} \left(\cos\frac{2\pi kj}{N} - \alpha^{N-j}\right) z_j.$$

Since $E(z_j) = 0$,

$$E(\zeta_{1k}) = \frac{1}{\sqrt{N}} \sum_{j=1}^{N-1} \mathrm{Re}\{\epsilon_k^j - \alpha^{N-j}\} E(z_j) = 0.$$

Since $\{z_j\}$ is the Gaussian white noise with variance σ^2,

$$\mathrm{Var}(\zeta_{1k}) = E(\zeta_{1k}^2) = \frac{\sigma^2}{N} \sum_{j=1}^{N-1} \left(\cos\frac{2\pi kj}{N} - \alpha^{N-j}\right)^2,$$

and so

$$\text{Var}(\zeta_{1k}) = \frac{\sigma^2}{N} \sum_{j=1}^{N-1} \cos^2 \frac{2\pi kj}{N} - \frac{2\sigma^2}{N} \sum_{j=1}^{N-1} \left(\cos \frac{2\pi kj}{N} \right) \alpha^{N-j} + \frac{\sigma^2}{N} \sum_{j=1}^{N-1} \alpha^{2N-2j}.$$

Since N is large enough and $|\alpha| < 1$,

$$\text{Var}(\zeta_{1k}) = \frac{1}{2}\sigma^2.$$

Let $\zeta_{2k} = \text{Im} \left(\sqrt{N}(1 - \epsilon_k \alpha) X_k \right)$. Similarly to ζ_{1k}, it follows that

$$E(\zeta_{2k}) = 0,$$

$$\text{Var}(\zeta_{2k}) = \frac{1}{2}\sigma^2.$$

From

$$E(\zeta_{1k}\zeta_{2k}) = -\frac{\sigma^2}{N} \sum_{j=1}^{N-1} \left(\cos \frac{2\pi kj}{N} - \alpha^{N-j} \right) \sin \frac{2\pi kj}{N},$$

it follows that the correlation of ζ_{1k} and ζ_{2k} is

$$E(\zeta_{1k}\zeta_{2k}) = -\frac{\sigma^2}{N} \sum_{j=1}^{N-1} \cos \frac{2\pi kj}{N} \sin \frac{2\pi kj}{N} + \frac{1}{N} \sum_{j=1}^{N-1} \alpha^{N-j} \sin \frac{2\pi kj}{N} \sigma^2.$$

Since N is large enough and $|\alpha| < 1$,

$$E(\zeta_{1k}\zeta_{2k}) = 0.$$

Since ζ_{1k} and ζ_{2k} are independent Gaussian random variables with mean 0 and variance $\frac{1}{2}\sigma^2$, by $N|1 - \epsilon_k \alpha|^2 |X_k|^2 = \zeta_{1k}^2 + \zeta_{2k}^2$, it follows that

$$N|1 - \epsilon_k \alpha|^2 |X_k|^2 \quad \text{is distributed as} \quad \frac{1}{2}\sigma^2 \chi_2^2.$$

Since $|1 - \epsilon_k \alpha|^2 = |1 - \alpha e^{-\frac{2\pi k}{N}i}|^2 = 1 - 2\alpha \cos \frac{2\pi k}{N} + \alpha^2$,

$$\frac{N|X_k|^2}{\sigma^2} \quad \text{is distributed as} \quad \frac{1}{2\left(1 - 2\alpha \cos \frac{2\pi k}{N} + \alpha^2 \right)} \chi_2^2.$$

Since $\sigma^2 = \frac{\tilde{\sigma}^2}{1 - \alpha^2}$, the desired result is obtained immediately. $\qquad \square$

When one applies the discrete Fourier transform to analyze finite-length time series, discontinuities at the data boundaries will distort its Fourier power spectrum in the high-frequency domain. In order to reduce the boundary effect in the high-frequency domain, Zhang and Moore (2011c) presented the following modified Fourier method: First, they removed a line which connects two endpoints of a time series. Next, they performed odd extension and periodic extension of the residual. Finally, for the periodic time series obtained, they used a Fourier expansion to obtain its frequency information. Zhang and Moore (2011c) used their modified Fourier tests to analyze the Arctic Oscillation (AO) and the Nino3.4 time series. They found significant peaks with periods of about 2.3 and 4.5 years in the AO and about 12 years in Nino3.4 in a test against red noise. These peaks are not significant in traditional Fourier tests.

7.7.2 Wavelet Power Spectra

The continuous wavelet transform possesses the ability to construct a time-frequency representation of a signal that offers very good time and frequency localization, so wavelet transforms can analyze localized intermittent periodicities of potentially great interest in climatic time series. One often performs the significance test on a wavelet power spectrum of climatic time series.

A continuous-time Gaussian white noise $x(t)$ satisfies

$$E[x(t)] = 0,$$

$$E[x(t)x(t')] = \sigma^2 \delta(t - t'), \tag{7.15}$$

where δ is the Dirac function. In this section, we use the Morlet wavelet with parameter ω_0:

$$\psi(t) = \pi^{-(1/4)} e^{i\omega_0 t} e^{-(t^2/2)}.$$

As in Chapter 2, denote

$$\psi_{a,b}(t) = \frac{1}{\sqrt{a}} \psi\left(\frac{t - b}{a}\right).$$

The wavelet transform of $x(t)$ is

$$Wx(a, b) = \int_{\mathbb{R}} x(t) \overline{\psi}_{a,b}(t) \, dt.$$

The *wavelet power spectrum* of $x(t)$ is defined as

$$|Wx(a, b)|^2,$$

i.e., the square of the modulus of the wavelet transform of the signal.

Taking real and imaginary parts on both sides, respectively, we get

$$\mathrm{Re}(Wx(a, b)) = \int_{\mathbb{R}} x(t) \, \mathrm{Re}(\overline{\psi}_{a,b}(t)) \, dt,$$

$$\mathrm{Im}(Wx(a,b)) = \int_{\mathbb{R}} x(t)\,\mathrm{Im}(\overline{\psi}_{a,b}(t))\,dt. \tag{7.16}$$

So $\mathrm{Re}(Wx(a,b))$ and $\mathrm{Im}(Wx(a,b))$ are both Gaussian random variables with mean 0.

The *wavelet power spectrum* of $x(t)$ is defined as

$$|Wx(a,b)|^2 = (\mathrm{Re}(Wx(a,b)))^2 + (\mathrm{Im}(Wx(a,b)))^2.$$

Theorem 7.3 (Zhang and Moore, 2012). *Let $x(t)$ be the continuous-time Gaussian white noise which is stated in (7.15) and let ψ be the Morlet wavelet with parameter ω_0. Then*

(i) *for any $a > 0$ and b, $\mathrm{Re}(Wx(a,b))$ and $\mathrm{Im}(Wx(a,b))$ are independent and*

$$\mathrm{Var}(\mathrm{Re}(Wx(a,b))) = \sigma^2 \int_{\mathbb{R}} (\mathrm{Re}[\overline{\psi}_{a,b}(t)])^2\,dt = \frac{\sigma^2}{2}\left(1 + e^{-\omega_0^2}\right),$$

$$\mathrm{Var}(\mathrm{Im}(Wx(a,b))) = \sigma^2 \int_{\mathbb{R}} (\mathrm{Im}[\overline{\psi}_{a,b}(t)])^2\,dt = \frac{\sigma^2}{2}\left(1 - e^{-\omega_0^2}\right);$$

(ii) *the wavelet power spectrum of $x(t)$ is distributed as follows:*

$$|Wx(a,b)|^2 \Longrightarrow \frac{\sigma^2}{2}\left(1 + e^{-\omega_0^2}\right)X_1^2 + \frac{\sigma^2}{2}\left(1 - e^{-\omega_0^2}\right)X_2^2,$$

where X_1 and X_2 are independent Gaussian random variables with mean 0 and variance 1.

Proof. By (7.16),

$$E[\mathrm{Re}(Wx(a,b))\,\mathrm{Im}(Wx(a,b))] = \iint_{\mathbb{R}^2} E[x(t)x(t')]\mathrm{Re}[\overline{\psi}_{a,b}(t)]\mathrm{Im}[\overline{\psi}_{a,b}(t')]\,dt\,dt'.$$

By (7.15), the right-hand side of this equality is equal to

$$\int_{\mathbb{R}} \mathrm{Im}[\overline{\psi}_{a,b}(t')]\left(\int_{\mathbb{R}} \sigma^2\delta(t-t')\,\mathrm{Re}[\overline{\psi}_{a,b}(t)]\,dt\right)dt'$$

$$= \sigma^2 \int_{\mathbb{R}} \mathrm{Im}[\overline{\psi}_{a,b}(t')]\,\mathrm{Re}[\overline{\psi}_{a,b}(t')]\,dt'$$

$$= \sigma^2 \int_{\mathbb{R}} \mathrm{Im}[\overline{\psi}(t')]\,\mathrm{Re}[\overline{\psi}(t')]\,dt'$$

$$= \pi^{-(1/2)} \int_{\mathbb{R}} e^{-(t')^2} \sin(t'\omega_0)\cos(t'\omega_0)\,dt' = 0.$$

Therefore,

$$E[\mathrm{Re}(Wx(a,b))\,\mathrm{Im}(Wx(a,b))] = 0.$$

From this and

$$E[\text{Re}(Wx(a,b))] = 0,$$
$$E[\text{Im}(Wx(a,b))] = 0,$$

it follows that

$$\text{Cov}(\text{Re}\,[T(a,b)],\ \text{Im}\,[T(a,b)]) = 0.$$

Therefore, $\text{Re}(Wx(a,b))$ and $\text{Im}(Wx(a,b))$ are independent. By (7.16),

$$\text{Var}(\text{Re}(Wx(a,b))) = E[(\text{Re}(Wx(a,b)))^2]$$
$$= \int_{\mathbb{R}^2} E[x(t)x(t')]\,\text{Re}[\overline{\psi}_{a,b}(t)]\,\text{Re}[\overline{\psi}_{a,b}(t')]\,dt\,dt'.$$

By (7.15), the right-hand side is equal to

$$\sigma^2 \int_{\mathbb{R}^2} \delta(t-t')\,\text{Re}[\overline{\psi}_{a,b}(t)]\,\text{Re}[\overline{\psi}_{a,b}(t')]\,dt\,dt'$$

$$= \sigma^2 \int_{\mathbb{R}} \text{Re}[\overline{\psi}_{a,b}(t)]\left(\int_{\mathbb{R}} \delta(t-t')\text{Re}[\overline{\psi}_{a,b}(t')]\,dt'\right) dt$$

$$= \sigma^2 \int_{\mathbb{R}} \left(\text{Re}[\overline{\psi}_{a,b}(t)]\right)^2 dt.$$

Therefore,

$$\text{Var}(\text{Re}(Wx(a,b))) = \sigma^2 \int_{\mathbb{R}} \left(\text{Re}[\overline{\psi}_{a,b}(t)]\right)^2 dt. \tag{7.17}$$

Notice that

$$\text{Re}\,[\overline{\psi}_{a,b}(t)] = \frac{1}{\sqrt{a}}\,\text{Re}\left[\overline{\psi}\left(\frac{t-b}{a}\right)\right] = \frac{\pi^{-(1/4)}}{\sqrt{a}}\cos\frac{\omega_0(t-b)}{a}\,e^{-\frac{(t-b)^2}{2a^2}}.$$

By $\int_{\mathbb{R}} e^{-u^2}\,du = \sqrt{\pi}$ and $\int_{\mathbb{R}} \cos(2\omega_0 t)\,e^{-t^2}\,dt = \sqrt{\pi}\,e^{-\omega_0^2}$, the integral on the right-hand side of (7.17) is

$$\int_{\mathbb{R}} \left(\text{Re}\,[\overline{\psi}_{a,b}(t)]\right)^2 dt$$

$$= \frac{1}{a\sqrt{\pi}} \int_{\mathbb{R}} \cos^2\frac{\omega_0(t-b)}{a}\,e^{-\frac{(t-b)^2}{a^2}}\,dt$$

$$= \pi^{-(1/2)} \int_{\mathbb{R}} \cos^2(\omega_0 u)\,e^{-u^2}\,du = \frac{1}{2\sqrt{\pi}} \int_{\mathbb{R}} (1 + \cos(2\omega_0 u))\,e^{-u^2}\,du$$

$$= \frac{1}{2} + \frac{1}{2\sqrt{\pi}} \int_{\mathbb{R}} \cos(2\omega_0 u)\,e^{-u^2}\,du = \frac{1}{2}\left(1 + e^{-\omega_0^2}\right).$$

From this and (7.17),

$$\mathrm{Var}(\mathrm{Re}(Wx(a,b))) = \frac{\sigma^2}{2}\left(1 + \mathrm{e}^{-\omega_0^2}\right),$$

and Similarly,

$$\mathrm{Var}(\mathrm{Im}(Wx(a,b))) = \frac{\sigma^2}{2}\left(1 - \mathrm{e}^{-\omega_0^2}\right),$$

i.e., (i).

Since $\mathrm{Re}(Wx(a,b))$ is a Gaussian random variable with mean 0, it is distributed as follows:

$$\mathrm{Re}(Wx(a,b)) \Longrightarrow \left(\frac{\sigma^2}{2}\left(1 - \mathrm{e}^{-\omega_0^2}\right)\right)^{1/2} X_1,$$

where X_1 is a Gaussian random variable with mean 0 and variance 1. Since $\mathrm{Im}(Wx(a,b))$ is also a Gaussian random variable with mean 0, it is distributed as follows:

$$\mathrm{Im}(Wx(a,b)) \Longrightarrow \left(\frac{\sigma^2}{2}\left(1 + \mathrm{e}^{-\omega_0^2}\right)\right)^{1/2} X_2,$$

where X_2 is a Gaussian random variable with mean 0 and variance 1. By (i) and noticing that the wavelet power spectrum of $x(t)$ is

$$|Wx(a,b)|^2 = |\mathrm{Re}(Wx(a,b))|^2 + |\mathrm{Im}(Wx(a,b))|^2,$$

i.e., (ii). □

For large ω_0 (e.g., $\omega_0 = 6$), $\mathrm{e}^{-\omega_0^2} \approx 0$. By Theorem 7.3,

$$|Wx(a,b)|^2 \Rightarrow \frac{\sigma^2}{2}\left(X_1^2 + X_2^2\right)$$

or

$$|Wx(a,b)|^2 \Longrightarrow \frac{\sigma^2}{2} \chi_2^2,$$

where χ_2^2 is the chi-squared distribution with two degrees of freedom.

By using the Monte Carlo method, Torrence and Compo (1998) gave an empirical formula for the Morlet wavelet power spectra of an AR(1) process as follows:

$$\frac{|W_n(s)|^2}{\widetilde{\sigma}^2} \quad \text{is distributed as} \quad \frac{1 - \alpha^2}{2\left(1 - 2\alpha\cos\frac{2\pi k}{N} + \alpha^2\right)} \chi_2^2,$$

where the Fourier frequency k corresponds to the wavelet scale s. If a peak in the wavelet power spectrum is significantly above this background spectrum, then it can be assumed to be a true feature of climatic time series with a certain percent confidence.

Let $W_n^X(s)$ and $W_n^Y(s)$ be wavelet transforms of two time series X and Y. Then the *cross-wavelet spectrum* is defined as

$$W_n^{XY}(s) = W_n^X(s)\overline{W_n^Y(s)},$$

and the *cross-wavelet power* is defined as

$$\left| W_n^{XY}(s) \right|.$$

Grinsted et al. (2004) discussed the application of the cross-wavelet transform in examining relationships in time-frequency space between two time series. Since the AO index and the Baltic maximum sea ice extent record are in antiphase across all wavelet scales, they concluded that there is a stronger link between the AO and the Baltic maximum sea ice extent. Later, Moore et al. (2007) used wavelet lag coherence to analyze the relation between two climatic time series.

PROBLEMS

7.1 Let $X(t) = \cos(t + \theta)$, where θ is a random variable. What additional assumptions are needed on θ to guarantee that $X(t)$ is WSS?

7.2 Consider a random process $X(t) = A + Bt^2$, where A and B are independent Gaussian random variables with mean 0 and variance 1. Find its mean $\mu_X(t)$, correlation function $R_X(s, t)$, and covariance function $C_X(s, t)$.

7.3 Let the random process $X(t) = A\cos(\omega t) + B\sin(\omega t)$ ($t \in \mathbb{R}$), where A and B are two independent random variables with mean 0 and variance 1. Try to prove $X(t) \in$ WSS and find its correlation function.

7.4 Give a spectral estimate for the Nino3.4 index.

7.5 Perform the wavelet-based significance test for the AO index.

BIBLIOGRAPHY

Adamowski, J., Adamowski, K., Prokoph, A., 2013. Quantifying the spatial temporal variability of annual streamflow and meteorological changes in eastern Ontario and southwestern Quebec using wavelet analysis and GIS. J. Hydrol. 499, 27-40.

Debret, M., Sebag, D., Crosta, X., Massei, N., Petit, J.-R., Chapron, E., Bout-Roumazeilles, V., 2009. Evidence from wavelet analysis for a mid-Holocene transition in global climate forcing. Quat. Sci. Rev. 28, 2675-2688.

Falamarzi, Y., Palizdan, N., Huang, Y.F., Lee, T.S., 2014. Estimating evapotranspiration from temperature and wind speed data using artificial and wavelet neural networks (WNNs). Agric. Water Manage. 140, 26-36.

Meyer, S.D., Kelly, B.G., O'Brien, J.J., 1993. An introduction to wavelet analysis in oceanography and meteorology: with application to the dispersion of Yanai waves. Mon. Weather Rev. 121, 2858-2866.

Moore, J.C., Grinsted, A., Jevrejeva, S., 2009. Wavelet-lag regression analysis of Atlantic tropical cyclone dependence on ENSO and Atlantic thermohaline variability. In: Elsner, J.B., Jagger, T. (Eds.), Proceedings of the 1st International Summit on Hurricanes and Climate Change. Springer, Berlin.

Galloway, J.M., Wigston, A., Patterson, R.T., Swindles, G.T., Reinhardt, E., Roe, H.M., 2013. Climate change and decadal to centennial-scale periodicities recorded in a late Holocene NE Pacific marine record: examining the role of solar forcing. Palaeogeogr. Palaeoclimatol. Palaeoecol. 386, 669-689.

Ghil, M., Allen, M.R., Dettinger, M.D., Ide, K., Kondrashov, D., Mann, M.E., Robertson, A.W., Saunders, A., Tian, Y., Varadi, F., Yiou, P., 2002. Advanced spectral methods for climatic time series. Rev. Geophys. 40, 1003-1043.

Gilman, D.L., Fuglister, F.J., Mitchell, J.M., 1962. On the power spectrum of red noise. J. Atmos. Sci. 20, 182-185.

Grinsted, A., Moore, J.C., Jevrejeva, S., 2004. Application of the cross wavelet transform and wavelet coherence to geophysical time series. Nonlinear Process. Geophys. 11, 561-566.

Mann, M.E., Lees, J.M., 1996. Robust estimation of background noise and signal detection in climatic time series. Clim. Change 33, 409-445.

Moore, J.C., Grinsted, A., Jevrejeva, S., 2007. Evidence from wavelet lag coherence for negligible solar forcing of climate at multi-year and decadal periods. In: Tsonis, A., Elsner, J.B. (Eds.), Nonlinear Dynamics in Geosciences. Springer, New York.

Moosavi, V., Malekinezhad, H., Shirmohammadi, B., 2014. Fractional snow cover mapping from MODIS data using wavelet-artificial intelligence hybrid models. J. Hydrol. 511, 160-170.

Mullon, L., Chang, N.-B., Yang, Y.J., Weiss, J., 2013. Integrated remote sensing and wavelet analyses for screening short-term teleconnection patterns in northeast America. J. Hydrol. 499, 247-264.

Nakken, M., 1999. Wavelet analysis of rainfall-runoff variability isolating climatic from anthropogenic patterns. Environ. Modell. Softw. 14, 283-295.

Nalley, D., Adamowski, J., Khalil, B., Ozga-Zielinski, B., 2013. Trend detection in surface air temperature in Ontario and Quebec, Canada during 1967C2006 using the discrete wavelet transform. Atmos. Res. 132-133, 375-398.

Rossi, A., Massei, N., Laignel, B., Sebag, D., Copard, Y., 2009. The response of the Mississippi river to climate fluctuations and reservoir construction as indicated by wavelet analysis of streamflow and suspended-sediment load, 1950-1975. J. Hydrol. 377, 237-244.

Rossi, A., Massei, N., Laignel, B., 2011. A synthesis of the time-scale variability of commonly used climate indices using continuous wavelet transform. Glob. Planet. Change 78, 1-13.

Soon, W., Herrera, V.M.V., Selvaraj, K., Traversi, R., Usoskin, I., Chen, C.-T.A., Lou, J.-Y., Kao, S.-J., Carter, R.M., Pipin, V., Severi, M., Becagli, S., 2014. A review of Holocene solar-linked climatic variation on centennial to millennial timescales: physical processes, interpretative frameworks and a new multiple cross-wavelet transform algorithm. Earth Sci. Rev. 134, 1-15.

Thompson, D.W.J., Wallace, J.M., 1998. The Arctic oscillation signature in the winter geopotential height and temperature fields. Geophys. Res. Lett. 25, 1297-1300.

Torrence, C., Compo, G.P., 1998. A practical guide to wavelet analysis. Bull. Am. Meteorol. Soc. 79, 61-78.

Vadrevu, K.P., Choi, Y., 2011. Wavelet analysis of airborne CO_2 measurements and related meteorological parameters over heterogeneous landscapes. Atmos. Res. 102, 77-90.

Zhang, Z., Moore, J.C., 2011a. Intrinsic feature extraction in the COI of wavelet power spectra of climatic signals. In: Proc. IEEE 4th International Conference on Image and Signal Processing, pp. 2380-2382.

Zhang, Z., Moore, J.C., 2011b. Distribution of Fourier power spectrum of climatic background noise. In: Proc. IEEE International Conference on Computational Intelligence and Software Engineering.

Zhang, Z., Moore, J.C., 2011c. New significance test methods for Fourier analysis of geophysical time series, Nonli. Processes Geophys., 18, 643-652.

Zhang, Z., Moore, J.C., 2011d. Improved significance testing of wavelet power spectrum near data boundaries as applied to polar research. Adv. Polar Sci. 22, 192-198.

Zhang, Z., Moore, J.C., 2012. Comments on Z. Ge's paper: significance tests for the wavelet power and the wavelet power spectrum. Ann. Geophys. 30, 1743-1750.

Zhang, Z., Moore, J.C., Grinsted, A., 2014. Haar wavelet analysis of climatic time series. Int. J. Wavelets Multiresolut. Inf. Process. 12, 1450020.

Chapter 8

Autoregressive Moving Average Models

Autoregressive moving average (ARMA) models play a key role in the modeling of time series. The linear structure of ARMA processes also leads to a substantial simplification of linear prediction. Compared with the pure autoregressive (AR) or moving average (MA) models, ARMA models provide the most effective linear model of stationary time series since they are capable of modeling the unknown process with the minimum number of parameters. In this chapter, for ARMA models, we study covariance structure, parameter estimation, asymptotic normality, and power spectral density, and introduce Yule-Walker equations and the Durbin-Levinson prediction algorithm. In addition, we also introduce autoregressive integrated moving average (ARIMA) models and multivariate ARMA.

8.1 ARMA PROCESSES

Let $\{Z_n\}_{n \in \mathbb{Z}}$ be a time series. If it satisfies

$$Z_n \sim N(0, \sigma^2) \quad (n \in \mathbb{Z}),$$
$$\mathrm{Cov}(Z_n, Z_m) = \sigma^2 \delta_{nm} \quad (n, m \in \mathbb{Z}),$$

then it is called a *white noise*, denoted by $\{Z_n\}_{n \in \mathbb{Z}} \sim \mathrm{WN}(0, \sigma^2)$.

For a time series $\{Z_n\}_{n \in \mathbb{Z}}$, if $\{Z_n\}_{n \in \mathbb{Z}} \sim \mathrm{WN}(0, \sigma^2)$ and $\{Z_n\}_{n \in \mathbb{Z}}$ has an independent, identical distribution, then denote it by $\{Z_n\}_{n \in \mathbb{Z}} \sim \mathrm{IID}(0, \sigma^2)$.

Let $\{Y_n\}_{n \in \mathbb{Z}}$ be a stationary time series satisfying a linear difference equation:

$$Y_n - \sum_{1}^{p} \varphi_j Y_{n-j} = Z_n + \sum_{1}^{q} \theta_j Z_{n-j} \quad (n \in \mathbb{Z}), \tag{8.1}$$

where $\{Z_n\}_{n \in \mathbb{Z}} \sim \mathrm{WN}(0, \sigma^2)$ and φ_j, θ_j are constants. Then the time series $\{Y_n\}_{n \in \mathbb{Z}}$ is called an *ARMA* (p, q) process, where (p, q) is the order, and (8.1) is called an *ARMA equation*.

If $\theta_j = 0$ $(j = 1, \ldots, q)$, then the ARMA equation becomes

$$Y_n - \sum_{1}^{p} \varphi_j Y_{n-j} = Z_n \quad (n \in \mathbb{Z}).$$

Mathematical and Physical Fundamentals of Climate Change

This equation is called an AR(p) equation, and the time series $\{Y_n\}_{n \in \mathbb{Z}}$ is called an AR(p) process, where p is the order.

If $\varphi_j = 0$ ($j = 1, \ldots, p$), then the ARMA equation becomes

$$Y_n = Z_n + \sum_1^q \theta_j Z_{n-j} \quad (n \in \mathbb{Z}).$$

This equation is called an MA(q) equation, and the time series $\{Y_n\}_{n \in \mathbb{Z}}$ is called an MA(q) process, where q is the order.

8.1.1 AR(p) Processes

The AR(1) process is the simplest case of AR(p) processes.

Let $\{Y_n\}_{n \in \mathbb{Z}}$ be an AR(1) process satisfying

$$Y_n = Z_n + \varphi Y_{n-1} \quad (n \in \mathbb{Z}),$$
$$\{Z_n\}_{n \in \mathbb{Z}} \sim \text{WN}(0, \sigma^2),$$

where φ is a constant. We find the solution of the AR(1) equation.

For $|\varphi| < 1$, it follows from the AR(1) equation that

$$Y_n = \sum_0^k \varphi^j Z_{n-j} + \varphi^{k+1} Y_{n-(k+1)}.$$

Since $\{Y_n\}_{n \in \mathbb{Z}}$ is stationary,

$$E[Y_{n-(k+1)}^2] = E[Y_n^2].$$

Since $|\varphi| < 1$,

$$E\left[\left(Y_n - \sum_0^k \varphi^j Z_{n-j}\right)^2\right] = \varphi^{2k+2} E\left[Y_{n-(k+1)}^2\right] \to 0 \quad (k \to \infty),$$

and so the unique solution of the AR(1) equation is

$$Y_n \overset{\text{m.s.}}{=} \sum_0^\infty \varphi^j Z_{n-j}.$$

By $\{Z_n\}_{n \in \mathbb{Z}} \sim \text{WN}(0, \sigma^2)$, the expectation of the AR(1) process $\{Y_n\}_{n \in \mathbb{Z}}$ is

$$E[Y_n] = \sum_0^\infty \varphi^j E[Z_{n-j}] = 0,$$

and the covariance function of the AR(1) process $\{Y_n\}_{n \in \mathbb{Z}}$ is

$$\gamma_k = E[Y_{n+k} Y_n] = \sum_{j=0}^\infty \sum_{l=0}^\infty \varphi^{j+l} E[Z_{n+k-j} Z_{n-l}]$$

$$= \sigma^2 \sum_{j=0}^{\infty} \sum_{l=0}^{\infty} \varphi^{j+l} \delta_{j-k,l}$$

$$= \sigma^2 \sum_{j=0}^{\infty} \varphi^{2j-k} = \frac{\sigma^2}{(1-\varphi^2)\varphi^k} \quad (k \in \mathbb{Z}).$$

For $|\varphi| > 1$, the series $\sum_0^{\infty} \varphi^j Z_{n-j}$ does not converge in the mean square sense. But the AR(1) equation can be written in the form

$$Y_n = -\frac{Z_{n+1}}{\varphi} + \frac{Y_{n+1}}{\varphi} \quad (n \in \mathbb{Z}),$$

and so a stationary solution of the AR(1) equation is

$$Y_n = -\sum_1^{\infty} \frac{Z_{n+j}}{\varphi^j} \quad (n \in \mathbb{Z}),$$

but this solution is not causal.

For $|\varphi| = 1$, $E[Y_n] = 0$ but $E[Y_n Y_{n-k}]$ depends on n, so an AR(1) process is not stationary.

8.1.2 MA(q) Processes

Let a time series $\{Y_n\}_{n \in \mathbb{Z}}$ be an MR(q) process satisfying

$$Y_n = Z_n + \sum_1^{q} \theta_j Z_{n-j} \quad (n \in \mathbb{Z}),$$

$$\{Z_n\}_{n \in \mathbb{Z}} \sim \mathrm{WN}(0, \sigma^2).$$

The expectation of the MR(q) process $\{Y_n\}_{n \in \mathbb{Z}}$ is

$$E[Y_n] = E[Z_n] + \sum_1^{q} \theta_j E[Z_{n-j}] = 0.$$

The covariance function of the MR(q) process $\{Y_n\}_{n \in \mathbb{Z}}$ is

$$\gamma_\tau = \mathrm{Cov}(Y_{n+\tau}, Y_n)$$

$$= E\left[\left(\sum_{k=0}^{q} \theta_k Z_{n+\tau-k} \right) \left(\sum_{l=0}^{q} \theta_l Z_{n-l} \right) \right]$$

$$= \sum_{k=0}^{q} \sum_{l=0}^{q} \theta_k \theta_l E[Z_{n-l} Z_{n+\tau-k}]$$

$$= \sigma^2 \sum_{k=0}^{q} \sum_{l=0}^{q} \theta_k \theta_l \delta_{k-\tau,l} \quad (k \in \mathbb{Z}),$$

where $\theta_0 = 1$. A direct computation shows that

$$\gamma_k = \begin{cases} 0, & |k| > q, \\ \sigma^2 \sum_{j=0}^{q-|k|} \theta_j \theta_{j+|k|}, & |k| \leq q. \end{cases} \tag{8.2}$$

We use the following notation of projection and spanning space in probability space $\mathcal{H} = L^2(\Omega, F, P)$.

Let $X, Y \in \mathcal{H}$. Then the inner product and the norm are, respectively,

$$(X, Y) = E[XY],$$
$$\|X\| = (E[X^2])^{1/2}.$$

Let $Y \in \mathcal{H}$ and M be a subspace of $\mathcal{H}$. Then there exists a unique decomposition:

$$Y = X + Z, \quad \text{where } X \in M, Z \perp M.$$

The component X is called the projection of Y in the subspace M. Write

$$X = \text{Proj}_M Y.$$

Let $\{Y_n\}_{n \in \mathbb{Z}_+} \subset \mathcal{H}$. Denote by $Q_N = \text{span}\{Y_1, \ldots, Y_N\}$ the linear combination of $Y_1, \ldots, Y_N$. Denote by $\overline{\text{span}}\{Y_n, n \in \mathbb{Z}_+\}$ the closed closure of all linear combination of $\{Y_n\}_{n \in \mathbb{Z}}$.

The following theorem shows that for a zero-mean stationary process, if its covariances γ_k ($k \in \mathbb{Z}_+$) have only finitely many nonzeros, then it must be an MA process of finite order.

Theorem 8.1. *If $\{Y_n\}_{n \in \mathbb{Z}}$ is a zero-mean stationary process with covariances γ_k ($k \in \mathbb{Z}_+$) such that $\gamma_k = 0$ ($k > q$) and $\gamma_q \neq 0$, then $\{Y_n\}_{n \in \mathbb{Z}}$ is an MA(q) process.*

Proof. Denote

$$M_n = \overline{\text{span}}\{Y_k, -\infty < k \leq n\},$$
$$P_n = Y_n - \text{Proj}_{M_{n-1}} Y_n \quad (n \in \mathbb{Z}). \tag{8.3}$$

Then $P_n \perp M_{n-1}$. From $P_s \in M_s \subset M_{n-1}$ ($s < n$), it follows that $P_s \perp P_n$ ($s < n$), i.e.,

$$E[P_s P_n] = 0.$$

Denote $W_{n,k} = \text{span}\{Y_s, s = n - k, \ldots, n - 1\}$. Then

$$\|P_{n+1}\|^2 = \lim_{k \to \infty} \left\| Y_{n+1} - \text{Proj}_{W_{n+1,k}} Y_{n+1} \right\|^2$$
$$= \lim_{k \to \infty} E\left[\left(Y_{n+1} - \text{Proj}_{W_{n+1,k}} Y_{n+1} \right)^2 \right].$$

From stationarity of $\{Y_n\}_{n \in \mathbb{Z}}$,

$$\|P_{n+1}\| = \|P_n\| \quad (n \in \mathbb{Z}).$$

Denote $\|P_n\|^2 = \sigma^2$ $(n \in \mathbb{Z})$. Then

$$\text{Cov}(P_n, P_m) = \sigma^2 \delta_{n,m} \quad (n \in \mathbb{Z}).$$

Therefore, P_n $(n \in \mathbb{Z})$ is a white noise, i.e., $\{P_n\}_{n \in \mathbb{Z}} \sim \text{WN}(0, \sigma^2)$.

By (8.3), the space M_{n-1} can be decomposed into the two orthogonal subspaces M_{n-q-1} and $V_{n,q}$, where $V_{n,q} = \text{span}\{P_{n-q}, \ldots, P_{n-1}\}$. Therefore,

$$\text{Proj}_{M_{n-1}} Y_n = \text{Proj}_{M_{n-q-1}} Y_n + \text{Proj}_{V_{n,q}} Y_n.$$

Since $U_n := \text{Proj}_{V_{n,q}} Y_n \in V_{n,q}$, it is a linear combination of $P_{n-q}, \ldots, P_{n-1}$, i.e.,

$$U_n = \sum_1^q c_l P_{n-l}.$$

So

$$(U_n, P_{n-k}) = \sum_{l=1}^q c_l(P_{n-l}, P_{n-k}) = \sum_{l=1}^q c_l \delta_{l,k} \sigma^2 = c_k \sigma^2 \quad (k = 1, \ldots, q).$$

Since U_n is a projection operator, $(Y_n - U_n) \perp V_{n,q}$, and so

$$(Y_n - U_n) \perp P_{n-k} \quad (k = 1, \ldots, q).$$

This implies that $c_k = \frac{1}{\sigma^2}(Y_n, P_{n-k})$. Therefore,

$$U_n = \frac{1}{\sigma^2} \sum_{l=1}^q (Y_n, P_{n-l}) P_{n-l},$$

$$\text{Proj}_{M_{n-1}} Y_n = \text{Proj}_{M_{n-q-1}} Y_n + \frac{1}{\sigma^2} \sum_{l=1}^q (Y_n, P_{n-l}) P_{n-l}.$$

By the assumption $\gamma_k = 0$ $(k > q)$, it follows that $E[Y_n Y_{n-k}] = 0$ $(k > q)$, i.e.,

$$Y_n \perp Y_{n-k} \quad (k = q+1, \ldots).$$

Thus, Y_n is orthogonal to subspace M_{n-q-1}. This implies that

$$\text{Proj}_{M_{n-q-1}} Y_n = 0.$$

Let $\theta_k = \sigma^{-2} E[Y_n P_{n-k}]$. Then

$$\text{Proj}_{M_{n-1}} Y_n = \sum_1^q \theta_k P_{n-k}.$$

From this and (8.3), $\{Y_n\}_{n \in \mathbb{Z}}$ satisfies

$$Y_n = P_n + \sum_1^q \theta_k P_{n-k},$$

$$\{P_n\}_{n\in\mathbb{Z}} \sim \text{WN}(0,\sigma^2).$$

Therefore, $\{Y_n\}_{n\in\mathbb{Z}}$ is an MA(q) process. $\qquad\qquad\qquad\square$

8.1.3 Shift Operator

Let $\{X_n\}_{n\in\mathbb{Z}}$ be a stationary time series. Define a time shift operator:

$$T : TX_n = X_{n-1}.$$

The time shift operator T is a linear operator. Since $\{X_n\}$ is stationary,

$$\|TX_n\|^2 = \|X_{n-1}\|^2 = E[X_{n-1}^2] = E[X_n^2] = \|X_n\|^2.$$

Therefore, the norm of the shift operator $\|T\| = 1$. It is clear that

$$T^l X_n = X_{n-l} \quad (l \in \mathbb{Z}_+),$$
$$\|T^l\| = 1.$$

Define $T^0 = I$, where I is the identity operator. Then $T^0 X_n = X_n$.

Let $P(z)$ be a polynomial and $P(z) = a_0 + a_1 z + \cdots + a_m z^m$. Then

$$P(T)X_n = (a_0 I + a_1 T + \cdots + a_m T^m)X_n = \sum_{0}^{m} a_k X_{n-k}.$$

Similarly to $(1 - \alpha z)^{-1} = \sum_n \alpha^n z^n (|\alpha| < 1, |z| = 1)$, the inverse operator

$$(1 - \alpha T)^{-1} = \sum_n \alpha^n T^n \quad \text{in the norm sense.}$$

More generally, if $F(z)$ can be expanded into a Laurent series $F(z) = \sum_l b_l z^l$ ($|z| = 1$), then

$$F(T) = \sum_l b_l T^l \quad \text{in the norm sense,}$$

$$F(T)X_n = \sum_l b_l (T^l X_n) = \sum_l b_l X_{n-l},$$

where the series $\sum_l b_l X_{n-l}$ converges in mean square sense.

Let

$$\varphi(z) = 1 - \sum_1^p \varphi_j z^j,$$

$$\theta(z) = 1 + \sum_1^q \theta_j z^j. \qquad (8.4)$$

Then the ARMA equation (8.1) can be written in the form

$$\varphi(T)Y_n = \theta(T)Z_n \quad (n \in \mathbb{Z}). \qquad (8.5)$$

8.1.4 ARMA(p, q) Processes

We discuss the existence and uniqueness of the solution of the ARMA equation, and then discuss causal and invertible ARMA processes.

Theorem 8.2. *Suppose that the polynomial $\varphi(z) \neq 0$ ($|z| = 1$). Then*

(i) *the ARMA equation (8.6) has a unique solution*

$$Y_n = \sum_j \psi_j Z_{n-j} \quad (n \in \mathbb{Z})$$

in the mean square sense, where the coefficients ψ_j ($j \in \mathbb{Z}$) are determined by

$$\psi(z) := \frac{\theta(z)}{\varphi(z)} = \sum_j \psi_j z^j \quad (|z| = 1)$$

and $\theta(z)$ and $\varphi(z)$ are stated as in (8.4);

(ii) *$\{Y_n\}_{n \in \mathbb{Z}}$ is stationary and its covariance function is*

$$\gamma_k = \sigma^2 \sum_j \psi_{j+|k|} \psi_j \quad (k \in \mathbb{Z}).$$

Proof. Since $\theta(z)$ and $\varphi(z)$ are both polynomials and $\varphi(z) \neq 0$ ($|z| = 1$), $\frac{\theta(z)}{\varphi(z)}$ can be expanded in a Laurent series:

$$\frac{\theta(z)}{\varphi(z)} = \sum_j \psi_j z^j \quad (|z| = 1).$$

So $\sum_j |\psi_j| < \infty$. By the argument in Section 8.1.3, the equality

$$\frac{\theta(T)}{\varphi(T)} = \sum_j \psi_j(T)$$

holds in the norm sense. From this and (8.5), it follows that

$$Y_n = \frac{\theta(T)}{\varphi(T)} Z_n = \sum_j \psi_j T^j Z_n = \sum_j \psi_j Z_{n-j} \quad (n \in \mathbb{Z}).$$

For $n \in \mathbb{Z}$, since

$$E[Y_n] = \sum_j \psi_j E[Z_{n-j}] = 0,$$

$$\text{Cov}(Y_{n+k}, Y_n) = E[Y_{n+k}\overline{Y}_n] = \sum_j \sum_l \psi_j \psi_l E[Z_{n+k-j}\overline{Z}_{n-l}]$$

$$= \sigma^2 \sum_j \sum_l \psi_j \overline{\psi}_l \delta_{j-k,l} = \sigma^2 \sum_j \psi_{j+|k|}\overline{\psi}_j$$

is independent of n, $\{Y_n\}_{n \in \mathbb{Z}}$ is stationary, and the covariance function

$$\gamma_k = \sigma^2 \sum_j \psi_{j+|k|} \overline{\psi}_j \quad (k \in \mathbb{Z}).$$

$\square$

If $Y_n = \sum_0^\infty \psi_j Z_{n-j}$ $(n \in \mathbb{Z})$, where $\sum_0^\infty |\psi_j| < \infty$, is a solution of the ARMA equation (8.1), then $\{Y_n\}_{n \in \mathbb{Z}}$ is called a *causal ARMA process*.

Theorem 8.3. *If two polynomials $\varphi(z)$ and $\theta(z)$ have no common zeros in $|z| < 1$, then the ARMA equation is a causal process if and only if $\varphi(z) \neq 0$ ($|z| \leq 1$). The solution of the causal process is*

$$Y_n = \sum_0^\infty \psi_j Z_{n-j} \quad (n \in \mathbb{Z})$$

in the mean square sense, where coefficients ψ_j are determined by

$$\psi(z) = \frac{\theta(z)}{\varphi(z)} = \sum_j \psi_j z^j \quad (|z| = 1).$$

Proof. If $\varphi(z) \neq 0$ ($|z| \leq 1$), then $\frac{\theta(z)}{\varphi(z)}$ can be expanded in a power series:

$$\frac{\theta(z)}{\varphi(z)} = \sum_0^\infty \psi_j z^j \quad (|z| = 1).$$

So

$$Y_n = \sum_0^\infty \psi_j T^j Z_n = \sum_0^\infty \psi_j Z_{n-j} \quad (n \in \mathbb{Z})$$

in the mean square sense, i.e., the ARMA equation is a causal process.

Conversely, if $\{Y_n\}_{n \in \mathbb{Z}}$ is a causal solution, then there is a sequence $\{\psi_j\}_{j=0,1,\ldots}$ such that $Y_n = \sum_0^\infty \psi_j Z_{n-j}$ $(n \in \mathbb{Z})$ satisfies the ARMA equation (8.5), where $\sum_0^\infty |\psi_j| < \infty$. So

$$Y_n = \sum_0^\infty \psi_j T^j Z_n = \psi(T) Z_n \quad (n \in \mathbb{Z}),$$

where

$$\psi(z) = \sum_0^\infty \psi_j z^j \quad (|z| = 1),$$

$$\theta(T) Z_n = \varphi(T) Y_n = \varphi(T) \psi(T) Z_n \quad (n \in \mathbb{Z}).$$

Let $\eta(z) = \varphi(z) \psi(z)$. Since $\sum_0^\infty |\psi_j| < \infty$ and $\varphi(z)$ is a polynomial, $\eta(z)$ can be expanded in a power series,

$$\eta(z) = \sum_0^\infty \eta_j z^j \quad (|z| \leq 1), \tag{8.6}$$

and so

$$\theta(T)Z_n = \varphi(T)\psi(T)Z_n = \sum_0^\infty \eta_j Z_{n-j} \quad (n \in \mathbb{Z}).$$

From this and $\theta(T)Z_n = \sum_0^q \theta_j Z_{n-j}$,

$$\sum_0^q \theta_j Z_{n-j} = \sum_0^\infty \eta_j Z_{n-j} \quad (n \in \mathbb{Z}),$$

where $\theta_0 = 1$. This implies that

$$\sum_0^q \theta_j E[Z_{n-j} Z_{n-k}] = \sum_0^\infty \eta_j E[Z_{n-j} Z_{n-k}] \quad (n \in \mathbb{Z}).$$

Since $E[Z_{n-j} Z_{n-k}] = \sigma^2 \delta_{j,k}$,

$$\eta_k = \theta_k \quad (k = 0, 1, \ldots, q),$$
$$\eta_k = 0 \quad (k > q).$$

By (8.4) and (8.6),

$$\theta(z) = \eta(z) = \varphi(z)\psi(z) \quad (|z| \le 1).$$

If there is a point z^* such that $\varphi(z^*) = 0$ $(|z^*| < 1)$, then $\theta(z^*) = 0$. This contradicts the assumption that $\varphi(z)$ and $\theta(z)$ have no common zeros in $|z| < 1$. So

$$\varphi(z) \ne 0 \quad (|z| \le 1). \qquad \square$$

If $Z_n = \sum_0^\infty \zeta_j Y_{n-j}$ $(n \in \mathbb{Z})$ satisfies the ARMA equation, where $\sum_0^\infty |\zeta_j| < \infty$, then $\{Y_n\}_{n \in \mathbb{Z}}$ is called an *invertible ARMA process*. If $\varphi(z)$ and $\theta(z)$ have no common zeros in $|z| < 1$, then the ARMA process is invertible if and only if $\theta(z) \ne 0$ $(|z| \le 1)$. Its proof is similar to the proof of Theorem 8.3.

Theorem 8.4. *Suppose that $\{Y_n\}_{n \in \mathbb{Z}}$ is an ARMA(p, q) process satisfying*

$$\varphi(T)Y_n = \theta(T)Z_n \quad (n \in \mathbb{Z}),$$
$$\{Z_n\}_{n \in \mathbb{Z}} \sim WN(0, \sigma^2),$$

and $\varphi(z) \ne 0$, $\theta(z) \ne 0$ $(|z| = 1)$, and let α_j $(j = \mu, \ldots, p)$ and β_j $(j = \nu, \ldots, q)$ be zeros of $\varphi(z)$ and $\theta(z)$ in $0 < |z| < 1$, respectively. Then $\{Y_n\}_{n \in \mathbb{Z}}$ must be a causal invertible ARMA(p, q) process satisfying the equation $\widetilde{\varphi}(T)Y_n = \widetilde{\theta}(T)Z_n^*$, where

$$\widetilde{\varphi}(z) = \varphi(z) \prod_\mu^p \frac{1 - \alpha_j z}{1 - \alpha_j^{-1} z},$$

$$\widetilde{\theta}(z) = \theta(z) \prod_\nu^q \frac{1 - \beta_j z}{1 - \beta_j^{-1} z},$$

and $\{Z_n^*\}_{n \in \mathbb{Z}} \sim WN(0, \widetilde{\sigma}^2)$, where $\widetilde{\sigma}^2 = \sigma^2 \prod_\mu^p |\alpha_j|^2 \prod_\nu^q |\beta_j|^2$.

Proof. Since the numerator and denominator of $\widetilde{\varphi}(z)$ have the same zeros with the same order, $\widetilde{\varphi}(z) \neq 0$ $(|z| < 1)$. From this and

$$|\widetilde{\varphi}(z)| = |\varphi(z)| \prod_{\mu}^{p} \left| \frac{1 - \alpha_j z}{1 - \alpha_j^{-1} z} \right| \neq 0 \quad (|z| = 1),$$

it follows that $\widetilde{\varphi}(z)$ is a polynomial of degree p and $\widetilde{\varphi}(z) \neq 0$ $(|z| \leq 1)$.

Similarly,

$$\widetilde{\theta}(z) = \theta(z) \prod_{\nu}^{q} \frac{1 - \beta_j z}{1 - \beta_j^{-1} z}$$

is a polynomial of degree q and $\theta(z) \neq 0$ $(|z| \leq 1)$.

From this, it follows that $\widetilde{\varphi}(T) Y_n = \widetilde{\theta}(T) Z_n^*$ $(n \in \mathbb{Z})$, where

$$Z_n^* := \frac{\widetilde{\varphi}(T)}{\widetilde{\theta}(T)} Y_n = \psi^*(T) Z_n,$$

$$\psi^*(z) = \prod_{\mu}^{p} \frac{1 - \alpha_j z}{1 - \alpha_j^{-1} z} \left(\prod_{\nu}^{q} \frac{1 - \beta_j z}{1 - \beta_j^{-1} z} \right)^{-1}.$$

Furthermore, we get

$$\{Z_n^*\}_{n \in \mathbb{Z}} \sim \mathrm{WN}(0, \widetilde{\sigma}^2),$$

$$\widetilde{\sigma}^2 = \sigma^2 \prod_{\mu}^{p} |\alpha_j|^2 \prod_{\nu}^{q} |\beta_j|^2.$$

$\square$

8.2 YULE-WALKER EQUATION AND SPECTRAL DENSITY

A zero-mean causal ARMA(p, q) process $\varphi(T) Y_n = \theta(T) Z_n$ $(n \in \mathbb{Z})$ has the solution

$$Y_n = \sum_{0}^{\infty} \psi_j Z_{n-j} \quad (n \in \mathbb{Z}),$$

where the coefficients ψ_j $(j = 0, 1, \ldots)$ are determined by

$$\psi(z) := \frac{\theta(z)}{\varphi(z)} = \sum_{0}^{\infty} \psi_j z^j \quad (|z| \leq 1).$$

Now we compute the covariance function of Y_n $(n \in \mathbb{Z})$.

Comparing coefficients of z^j on both sides of $\psi(z)\varphi(z) = \theta(z)$, by (8.5), we find that

$$\psi_0 = \theta_0 = 1,$$
$$\psi_1 = \theta_1 + \varphi_1,$$
$$\psi_2 = \theta_2 + \varphi_2 + \theta_1\varphi_1 + \varphi_1^2,$$
$$\vdots$$

In general,

$$\psi_l - \sum_{k=1}^{l} \varphi_k \psi_{l-k} = \theta_l \quad (l \in \mathbb{Z}_+), \tag{8.7}$$

where $\theta_l = 0$ $(l > q)$ and $\varphi_l = 0$ $(l > p)$.

Starting from (8.1), we compute covariance function γ_k. Notice that

$$E[Y_n Y_{n-k}] - \sum_{l=1}^{p} \varphi_l E[Y_{n-l} Y_{n-k}] = \sum_{l=0}^{q} \theta_l E[Z_{n-l} Y_{n-k}] \quad (\theta_0 = 1).$$

By $Y_n = \sum_0^\infty \psi_j Z_{n-j}$, the right-hand side is

$$\sum_{l=0}^{q} \theta_l E[Z_{n-l} Y_{n-k}] = \sum_{l=0}^{q} \sum_{j=0}^{\infty} \theta_l \psi_j E[Z_{n-l} Z_{n-k-j}]$$

$$= \sigma^2 \sum_{l=k}^{q} \sum_{j=0}^{\infty} \theta_l \psi_j \delta_{l,k+j} = \sigma^2 \sum_{l=k}^{q} \sum_{j=k}^{q} \theta_l \psi_{j-k} \delta_{l,j}$$

$$= \sigma^2 \sum_{j=k}^{q} \theta_j \psi_{j-k} \quad (0 \le k \le q)$$

and

$$\sum_{l=0}^{q} \theta_j E[Z_{n-l} Y_{n-k}] = 0 \quad (k > q).$$

From this and $\gamma_{k-l} = E[Y_{n-l} Y_{n-k}]$, it follows that

$$\gamma_k - \sum_{1}^{p} \varphi_l \gamma_{k-l} = \begin{cases} \sigma^2 \sum_{j=k}^{q} \theta_j \psi_{j-k}, & 0 \le k < s, \\ 0, & k \ge s, \end{cases} \tag{8.8}$$

where $s = \max\{p, q+1\}$.

Let $q = 0$. Then the ARMA(p, q) process reduces to an AR(p) process. Equation (8.8) reduces to

$$\sum_{1}^{p} \varphi_l \gamma_{k-l} = \gamma_k \quad (k = 1, \dots, p),$$

$$\sigma^2 = \gamma_0 - \sum_1^p \varphi_l \gamma_l. \tag{8.9}$$

Equations (8.9) are called the *Yule-Walker equations*.

If $\{Y_n\}_{n \in \mathbb{Z}}$ is a stationary process with covariance function γ_k ($k \in \mathbb{Z}$) and the series $\sum_k \gamma_k z^k$ is convergent in some annulus $1 - \epsilon < |z| < 1 + \epsilon$ ($\epsilon > 0$), then the sum $G(z) = \sum_k \gamma_k z^k$ is called a *covariance-generating function* of $\{Y_n\}_{n \in \mathbb{Z}}$.

Suppose that $\{Y_n\}_{n \in \mathbb{Z}}$ is an ARMA(p, q) process with covariance function γ_k ($k \in \mathbb{Z}$):

$$Y_n = \sum_j \psi_j Z_{n-j} \quad (n \in \mathbb{Z}),$$

$$\{Z_n\}_{n \in \mathbb{Z}} \sim \mathrm{WN}(0, \sigma^2),$$

where $\sum_j |\psi_j| < \infty$. If

$$\psi(z) = \sum_j \psi_j z^j \quad (1 - \epsilon < |z| < 1 + \epsilon),$$

by Theorem 8.2(ii), the generating function

$$G(z) = \sigma^2 \sum_k \sum_j \overline{\psi}_j \psi_{j+|k|} z^k$$

$$= \sigma^2 \sum_k \psi_k z^{-k} \sum_j \overline{\psi}_j z^j = \sigma^2 \overline{\psi}(z) \psi(z^{-1}). \tag{8.10}$$

Since $\psi(z) = \frac{\theta(z)}{\varphi(z)}$,

$$G(z) = \sigma^2 \frac{\overline{\theta}(z) \theta(z^{-1})}{\overline{\varphi}(z) \varphi(z^{-1})} \quad (1 - \epsilon < |z| < 1 + \epsilon).$$

From this, it is easy to compute the covariance function by means of the generating function.

Spectral Density of ARMA(p, q). *If $\varphi(z)$ and $\theta(z)$ have no common zeros and $\varphi(z) \neq 0$ on $|z| = 1$, then the ARMA(p, q) process $\{Y_n\}_{n \in \mathbb{Z}}$ has the spectral density*

$$g_Y(\tau) = \sigma^2 \frac{|\theta(e^{-i\tau})|^2}{|\varphi(e^{-i\tau})|^2} \quad (|\tau| \leq \pi).$$

Proof. By Theorem 8.2,

$$Y_n = \sum_k \psi_k Z_{n-k} \quad (n \in \mathbb{Z}),$$

and the covariance function

$$\gamma_\alpha = \sigma^2 \sum_k \psi_k \psi_{k+|\alpha|}.$$

Therefore, the spectral density g_Y is

$$g_Y(\tau) = \sum_\alpha \gamma_\alpha e^{-i\alpha\tau} = \sigma^2 \sum_k \sum_\alpha \psi_k \psi_{k+|\alpha|} e^{-i\alpha\tau}$$

$$= \sigma^2 \left(\sum_k \psi_k e^{ik\tau} \right) \left(\sum_\alpha \psi_{k+\alpha} e^{-i(k+\alpha)\tau} \right)$$

$$= \sigma^2 \left| \sum_k \psi_k e^{-ik\tau} \right|^2 .$$

Since $\psi(z) = \frac{\theta(z)}{\varphi(z)} = \sum_k \psi_k z^k$,

$$g_Y(\tau) = \sigma^2 |\psi(e^{-i\tau})|^2 = \sigma^2 \frac{|\theta(e^{-i\tau})|^2}{|\varphi(e^{-i\tau})|^2} \quad (|\tau| \le \pi).$$ $\square$

For MA(q) process $Y_n = Z_n + \sum_1^q \theta_k Z_{n-k}$ $(n \in \mathbb{Z})$, the spectral density

$$g_Y(\tau) = \sigma^2 \left| 1 + \sum_1^q \theta_k e^{-i\tau} \right|^2 .$$

In the case $q = 1$, the spectral density

$$g_Y(\tau) = \sigma^2 |1 + \theta_1 e^{-i\tau}|^2 = \sigma^2 (1 + 2\theta_1 \cos\tau + \theta_1^2).$$

For AR(p) process $Y_n - \sum_1^p \varphi_k Y_{n-k} = Z_n$ $(n \in \mathbb{Z})$, the spectral density

$$g_Y(\tau) = \sigma^2 \left| 1 - \sum_1^p \varphi_k e^{-i\tau} \right|^{-2} .$$

In the case $p = 1$, the spectral density

$$g_Y(\tau) = \frac{\sigma^2}{|1 - \varphi_1 e^{-i\tau}|^2} = \frac{\sigma^2}{1 - 2\varphi_1 \cos\tau + \varphi_1^2}.$$

8.3 PREDICTION ALGORITHMS

We introduce the theory and method of the prediction of time series, in particular, the prediction of ARMA processes. For a time series $\{Y_n\}_{n\in\mathbb{Z}}$, we use observations $Y_1, \ldots, Y_n$ to predict the values of $Y_{n+1}, Y_{n+2}, \ldots$. For a time series $\{Y_n\}_{n\in\mathbb{Z}}$, let the space L_n be a set of linear combinations of $Y_1, \ldots, Y_n$. Denote

$$L_n = \text{span}\{Y_1, \ldots, Y_n\}.$$

$\text{Proj}_{L_n} Y_{n+1}$ is regarded as the best linear prediction of Y_{n+1} from $Y_1, \ldots, Y_n$, i.e., the best linear prediction

$$Y_{n+1}^o = \text{Proj}_{L_n} Y_{n+1}.$$

By the orthogonality principle, $(Y_{n+1} - Y_{n+1}^0) \perp L_n$ and the mean square error

$$\|Y_{n+1} - Y_{n+1}^0\|^2 = \|Y_{n+1}\|^2 - \|Y_{n+1}^0\|^2,$$

where $\|X\| = (E[X^2])^{1/2}$.

8.3.1 Innovation Algorithm

The innovation algorithm is used for computing the best linear prediction. This algorithm allows $\{Y_n\}_{n \in \mathbb{Z}_+}$ to be not a stationary random process.

The One-step Innovation Algorithm. *Let $\{Y_n\}_{n \in \mathbb{Z}_+}$ be a random process with mean 0 and*

$$E[Y_\mu Y_\nu] = h_{\mu,\nu},$$

where the covariance matrix $(h_{\mu,\nu})_{\mu,\nu=1,\ldots,n}$ is nonsingular. Then the best linear prediction Y_{n+1}^0 is

$$Y_{n+1}^0 = \sum_1^n \alpha_{nj}(Y_{n+1-j} - Y_{n+1-j}^0) \quad (n \in \mathbb{Z}_+), \tag{8.11}$$

where $Y_1^0 = 0$ and the coefficients α_{nj} and the mean square errors $\lambda_n = \|Y_{n+1} - Y_{n+1}^0\|^2$ satisfy

(i) $\lambda_0 = h_{1,1}$;

(ii) $\alpha_{n,n} = \frac{1}{\lambda_0} h_{n+1,1}$ *and* $\alpha_{n,n-k} = \frac{1}{\lambda_k} h_{n+1,k+1} - \frac{1}{\lambda_k} \sum_{j=0}^{k-1} \alpha_{k,k-j} \alpha_{n,n-j} \lambda_j$ *($k = 1, \ldots, n-1$; $n \in \mathbb{Z}_+$);*

(iii) $\lambda_n = h_{n+1,n+1} - \sum_{j=0}^{n-1} \alpha_{n,n-j}^2 \lambda_j$ *($n \in \mathbb{Z}_+$).*

Proof. Clearly, $\lambda_0 = \|Y_1\|^2 = h_{1,1}$, i.e., (i).

Denote $L_j = \text{span}\{Y_1, \ldots, Y_j\}$. By the orthogonality principle,

$$(Y_j - Y_j^0) \perp L_{j-1} \quad (j \in \mathbb{Z}_+).$$

Let $Y_1^0 = 0$. Then $\{Y_1 - Y_1^0, Y_2 - Y_2^0, \ldots, Y_n - Y_n^0\}$ is an orthogonal basis of the space L_n. So the best linear prediction Y_{n+1}^0 can be expanded in an orthogonal series as in (8.11). By orthogonality, the coefficients

$$\alpha_{n,n-k} = \frac{1}{\lambda_k}(Y_{n+1}^0, Y_{k+1} - Y_{k+1}^0),$$

where $\lambda_k = \|Y_{k+1} - Y_{k+1}^o\|$. Notice that $Y_{n+1} - Y_{n+1}^0 \perp L_n$. Then

$$\alpha_{n,n-k} = \frac{1}{\lambda_k}(Y_{n+1}, Y_{k+1} - Y_{k+1}^o).$$

Combining this with (8.11), we get $\alpha_{n,n} = \frac{1}{\lambda_0} h_{n+1,1}$ ($n \in \mathbb{Z}$) and

$$\alpha_{n,n-k} = \frac{1}{\lambda_k}(Y_{n+1}, Y_{k+1}) - \frac{1}{\lambda_k}(Y_{n+1}, Y_{k+1}^o)$$

$$= \frac{1}{\lambda_k}h_{n+1,k+1} - \frac{1}{\lambda_k}\sum_{j=0}^{k-1}\alpha_{k,k-j}(Y_{n+1}, Y_{j+1} - Y_{j+1}^o)$$

$$= \frac{1}{\lambda_k}h_{n+1,k+1} - \frac{1}{\lambda_k}\sum_{j=0}^{k-1}\alpha_{k,k-j}\alpha_{n,n-j}\lambda_j \quad (k = 1, \ldots, n-1;\ n \in \mathbb{Z}_+),$$

i.e., (ii).

The mean square errors are

$$\lambda_n = \|Y_{n+1} - Y_{n+1}^o\|^2 = \|Y_{n+1}\|^2 - \|Y_{n+1}^o\|^2 = h_{n+1,n+1} - \|Y_{n+1}^o\|^2.$$

By using Parseval's identity, it follows from (8.11) that

$$\|Y_{n+1}^o\|^2 = \sum_{1}^{n}\alpha_{n,j}^2\|Y_{n+1-j} - Y_{n+1-j}^o\|^2 = \sum_{1}^{n}\alpha_{n,j}^2\lambda_{n-j} = \sum_{0}^{n-1}\alpha_{n,n-j}^2\lambda_j,$$

and so

$$\lambda_n = h_{n+1,n+1} - \sum_{0}^{n-1}\alpha_{n,n-j}^2\lambda_j,$$

i.e., (iii). $\qquad\qquad\qquad\qquad\qquad\qquad\qquad\qquad\qquad\qquad\qquad\qquad\Box$

Remark. In the one-step innovation algorithm, the use of the recursive formula is in the order

$$\lambda_0, \alpha_{1,1}, \lambda_1, \alpha_{2,2}, \alpha_{2,1}, \lambda_2, \alpha_{3,3}, \alpha_{3,2}, \alpha_{3,1}, \lambda_3, \ldots.$$

In detail, start from $\lambda_0 = h_{11}$. By (ii), $\alpha_{11} = \frac{h_{21}}{h_{11}}$. Using the values of λ_0 and α_{11}, by (iii), we find $\lambda_1 = h_{22} - \alpha_{11}^2\lambda_0$. After that, by (ii), $\alpha_{22} = \frac{h_{31}}{h_{11}}$. Using the values of λ_1 and α_{22}, and λ_0, by (ii) and (iii), we find

$$\alpha_{21} = \frac{1}{\lambda_1}(h_{32} - \alpha_{11}\alpha_{22}h_{11}),$$

$$\lambda_2 = h_{33} - \alpha_{22}^2\lambda_0 - \alpha_{21}^2\lambda_1,$$

and so on. In the computational process, $\alpha_{k,l}, \lambda_j$ are determined by covariance $h_{\mu,\nu}$ and are computed by (ii) and (iii) recursively.

On the basis of the innovation algorithm, the recursive calculation of the τ-step predictors is given easily. The best linear prediction of $Y_{n+\tau}$ in $L_n = \text{span}\{Y_1, \ldots, Y_n\}$ is

$$\text{Proj}_{L_n}Y_{n+\tau} = \sum_{k=\tau}^{n+\tau-1}\alpha_{n+\tau-1,k}(Y_{n+\tau-k} - Y_{n+\tau-k}^o).$$

The mean squared error is

$$\|Y_{n+\tau} - \mathrm{Proj}_{H_n} Y_{n+\tau}\|^2 = h_{n+\tau,n+\tau} - \sum_{k=\tau}^{n+\tau-1} \alpha_{n+\tau-1}^2 \lambda_{n+\tau-k-1},$$

where the coefficients $\alpha_{n,k}$ are stated as in the one-step innovation algorithm.

For ARMA(p, q) processes, the innovation algorithm is reduced to a simple algorithm as follows.

Let $\{Y_n\}_{n \in \mathbb{Z}_+}$ be an MA(1) process satisfying

$$Y_n = Z_n + \theta Z_{n-1} \quad (n \in \mathbb{Z}),$$
$$\{Z_n\}_{n \in \mathbb{Z}_+} \sim \mathrm{WN}(0, 1).$$

Then $E[Y_n] = 0$ $(n \in \mathbb{Z})$, and for $n \in \mathbb{Z}$,

$$E[Y_{n+\tau} Y_n] = \begin{cases} 1 + \theta^2, & \tau = 0, \\ \theta, & \tau = 1, \\ 0, & \tau \geq 2, \end{cases}$$

i.e.,

$$h_{\mu,\nu} = 0(|\mu - \nu| > 1),$$
$$h_{\mu,\mu} = 1 + \theta^2,$$
$$h_{\mu,\mu+1} = \theta.$$

From this, it follows by the innovation algorithm that

$$\alpha_{n,1} = \frac{1}{\lambda_{n-1}} \theta,$$
$$\alpha_{n,j} = 0 \quad (2 \leq j \leq n),$$
$$\lambda_0 = 1 + \theta^2,$$
$$\lambda_n = 1 + \theta^2 - \frac{1}{\lambda_{n-1}} \theta^2,$$

and so the best linear prediction is

$$Y_{n+1}^{\mathrm{o}} = \frac{\theta}{\lambda_{n-1}} (Y_n - Y_n^{\mathrm{o}}).$$

Suppose that $\{Y_n\}_{n \in \mathbb{Z}_+}$ is a causal ARMA process satisfying

$$\varphi(T) Y_n = \theta(T) Z_n \quad (n \in \mathbb{Z}),$$
$$\{Z_n\}_{n \in \mathbb{Z}_+} \sim \mathrm{WN}(0, \sigma^2).$$

Let

$$S_n = \begin{cases} \frac{1}{\sigma} Y_n, & 1 \leq n \leq \max\{p, q\}, \\ \frac{1}{\sigma} \varphi(T) Y_n, & n > \max\{p, q\}. \end{cases}$$

Then

$$
E[S_\mu S_\nu] =
\begin{cases}
\frac{1}{\sigma^2}\gamma_{\mu-\nu}, & 1 \le \mu \le \nu \le m, \\[2mm]
\frac{1}{\sigma^2}\left(\gamma_{\mu-\nu} - \sum_{k=1}^{p}\varphi_k\gamma_{k-|\mu-\nu|}\right), & \mu \le m \le \nu \le 2m, \\[2mm]
\sum_{k=0}^{q}\theta_k\theta_{k+|\mu-\nu|}, & \nu \ge \mu \ge m, \\[2mm]
0, & \text{otherwise},
\end{cases}
\tag{8.12}
$$

where $m = \max\{p,q\}$ and $\theta_\nu = 0$ $(\nu > q)$.

Replacing $h_{\mu,\nu}$ by $E[S_\mu S_\nu]$, we get α_{nj} and λ_n. Finally, the recursive prediction of the ARMA process is as follows:

$$
Y^o_{n+1} = \sum_{1}^{n}\alpha_{n,j}(Y_{n+1-j} - Y^o_{n+1-j}) \quad (1 \le n < m),
$$

$$
Y^o_{n+1} = \varphi_1 Y_n + \cdots + \varphi_p Y_{n+1-p} + \sum_{1}^{q}\alpha_{n,j}(Y_{n+1-j} - Y^o_{n+1-j}) \quad (n \ge m),
$$

$$
E[(Y_{n+1} - Y^o_{n+1})^2] = \sigma^2\lambda_n.
$$

This algorithm is simple and for $n > m$, it requires only p past observations Y_{n+1-k} $(k = 1,\ldots,p)$ and q past innovations $Y_{n+1-k} - Y^o_{n+1-k}$ $(k = 1,\ldots,q)$. In particular, for an AR(p) process, the best linear prediction is

$$
Y^o_{n+1} = \varphi_1 Y_n + \cdots + \varphi_p Y_{n+1-p} \quad (n \ge p),
$$

and for an MA(q) process with $\varphi_1 = 0$, the best linear prediction is

$$
Y^o_{n+1} = \sum_{k=1}^{\min\{n,q\}}\alpha_{n,k}(Y_{n+1-k} - Y^o_{n+1-k}) \quad (n \ge 1).
$$

Especially, for $\{Y_n\}_{n\in\mathbb{Z}}$ satisfying

$$
Y_n - \frac{1}{2}Y_{n-1} = Z_n + Z_{n-1} \quad (n \in \mathbb{Z}),
$$

$$
\{Z_n\}_{n\in\mathbb{Z}} \sim WN(0,\sigma^2).
$$

the prediction formula is

$$
Y^o_{n+1} = \frac{1}{2}Y_n + \alpha_{n,1}(Y_n - Y^o_n) \quad (n \ge 1).
$$

By (8.12),

$$
h_{\mu,\nu} =
\begin{cases}
\frac{9}{4}, & \mu = \nu = 1, \\[1mm]
2, & \mu = \nu \ge 2, \\[1mm]
1, & |\mu - \nu| = 1, \\[1mm]
0, & \text{otherwise}.
\end{cases}
$$

For an ARMA process, assume $n > m$; then the τ-step prediction is

$$Y_{n+\tau} - \text{Proj}_{L_n} Y_{n+\tau} - \sum_{k=1}^{p} \varphi_k (Y_{n+\tau-k} - \text{Proj}_{L_n} Y_{n+\tau-k})$$

$$= \sum_{k=0}^{\tau-1} \alpha_{n+\tau-1,k} (Y_{n+\tau-k} - Y_{n+\tau-k}^{\text{o}})$$

and the errors

$$E[(Y_{n+\tau} - \text{Proj}_{L_n} Y_{n+\tau})^2] = \sum_{k=0}^{\tau-1} \left(\sum_{l=0}^{k} x_l \alpha_{n+\tau-l-1,k-l} \right)^2 \lambda_{n+\tau-k-1},$$

where the coefficients x_l satisfy

$$\sum_{0}^{\infty} x_l z^l = \frac{1}{1 - \varphi_1 z - \cdots - \varphi_p z^p} \quad (|z| \le 1).$$

The following theorem gives the prediction of a causal invertible ARMA process.

Theorem 8.5. *Let $\{Y_n\}_{n \in \mathbb{Z}}$ be a causal invertible ARMA process satisfying*

$$\varphi(T) Y_n = \theta(T) Z_n \quad (n \in \mathbb{Z}),$$
$$\{Z_n\}_{n \in \mathbb{Z}} \sim \text{WN}(0, \sigma^2).$$

Denote

$$Y_{n+\tau}^{\text{o}} = \text{Proj}_{M_n} Y_{n+\tau} \quad (\tau \in \mathbb{Z}_+),$$

where $M_n = \overline{\text{span}}\{Y_k, -\infty < k \le n\}$. Then

$$Y_{n+\tau}^{\text{o}} = - \sum_{k \in \mathbb{Z}_+} \eta_k Y_{n+\tau-k}^{\text{o}} \quad (\tau \in \mathbb{Z}_+),$$

where $\frac{\varphi(z)}{\theta(z)} = \sum_{0}^{\infty} \eta_k z^k$ ($|z| \le 1$), and the mean square error

$$E[(Y_{n+\tau} - Y_{n+\tau}^{\text{o}})^2] = \sigma^2 \sum_{0}^{\tau-1} \psi_k^2 \quad (\tau \in \mathbb{Z}_+),$$

where $\frac{\theta(z)}{\varphi(z)} = \sum_{0}^{\infty} \psi_k z^k$ ($|z| \le 1$) and $\psi_1 = 1$.
Proof. Since $\{Y_n\}_{n \in \mathbb{Z}}$ is causal invertible,

$$\varphi(z) \ne 0 \quad (|z| \le 1),$$
$$\theta(z) \ne 0 \quad (|z| \le 1).$$

From the ARMA equation, it follows that

$$Y_l = \sum_{0}^{\infty} \psi_k Z_{l-k}. \tag{8.13}$$

So the best linear prediction

$$Y^o_{n+\tau} = \text{Proj}_{M_n} Y_{n+\tau} = \sum_{k=0}^{\infty} \psi_k \text{Proj}_{M_n} Z_{n+\tau-k}.$$

This implies that

$$\text{Proj}_{M_n} Z_{n+\tau-k} = \begin{cases} 0, & k < \tau, \\ Z_{n+\tau-k}, & k \geq \tau, \end{cases}$$

and so

$$Y^o_{n+\tau} = \sum_{k=\tau}^{\infty} \psi_k Z_{n+\tau-k}. \tag{8.14}$$

Since $\{Y_n\}_{n\in\mathbb{Z}}$ is invertible,

$$Z_{n+\tau} = Y_{n+\tau} + \sum_{k\in\mathbb{Z}_+} \eta_k Y_{n+1-k}.$$

From this and $\text{Proj}_{M_n} Z_{n+\tau} = 0$, it follows that

$$Y^o_{n+\tau} = -\sum_{k\in\mathbb{Z}_+} \eta_k Y^o_{n+\tau-k} \quad (\tau \in \mathbb{Z}_+).$$

The combination of (8.13) and (8.14) gives

$$Y_{n+\tau} - Y^o_{n+\tau} = \sum_{k=0}^{\tau-1} \psi_k Z_{n+\tau-k},$$

and so

$$\|Y_{n+\tau} - Y^o_{n+\tau}\|^2 = \sum_{k=0}^{\tau-1} \psi_k^2 \|Z_{n+\tau-k}\|^2 = \sigma^2 \sum_0^{\tau} \psi_k^2.$$

$\square$

8.3.2 Durbin-Lovinson Algorithm

The Durbin-Lovinson algorithm is used in the stationary processes.

Proposition 8.1. *Let $\{Y_n\}_{n\in\mathbb{Z}_+}$ be a stationary random process with mean 0 and correlation function γ_k, and let the matrix $\Sigma_n = (\gamma_{k-j})_{k,j=1,...,n}$ be nonsingular and the space*

$$L_n = \text{span}\{Y_1, \dots, Y_n\}.$$

Then

(i) *the best linear prediction Y^o_{n+1} of Y_{n+1} in L_n is*

$$Y^o_{n+1} = \sum_{k=1}^{n} \xi_{n,k} Y_{n+1-k}, \tag{8.15}$$

where $\xi_{n,k}$ $(k = 1, \ldots, n)$ *satisfy* $(\xi_{n,1}, \ldots, \xi_{n,n})^{\mathrm{T}} = \Sigma_n^{-1}(\gamma_1, \ldots, \gamma_n)^{\mathrm{T}}$;

(ii) *the mean square error*

$$\alpha_n = \|Y_{n+1} - Y_{n+1}^{\mathrm{o}}\|^2 = \gamma_0 - (\gamma_1, \ldots, \gamma_n)\Sigma_n^{-1}(\gamma_1, \ldots, \gamma_n)^{\mathrm{T}}.$$

Proof. Let Y_{n+1}^{o} be the best linear prediction of Y_{n+1} in L_n. Then there are coefficients ξ_{nk} $(k = 1, \ldots, n)$ such that

$$Y_{n+1}^{\mathrm{o}} = \sum_1^n \xi_{nj}Y_{n+1-j}.$$

By the orthogonality principle,

$$E[(Y_{n+1} - Y_{n+1}^{\mathrm{o}})Y_{n+1-k}] = 0 \quad (k = 1, \ldots, n),$$

and so

$$E[Y_{n+1}^{\mathrm{o}}Y_{n+1-k}] = E[Y_{n+1}Y_{n+1-k}] = \gamma_k \quad (k = 1, \ldots, n).$$

This implies that

$$\gamma_k = E[Y_{n+1}^{\mathrm{o}}Y_{n+1-k}] = \sum_{j=1}^n \xi_{nj}E[Y_{n+1-j}Y_{n+1-k}] = \sum_{j=1}^n \xi_{nj}\gamma_{k-j}.$$

Since the coefficient matrix $\Sigma_n = (\gamma_{k-j})_{k,j=1,\ldots,n}$ is nonsingular, (i) follows. The mean square error

$$\alpha_n = E[Y_{n+1}^2] - E[(Y_{n+1}^{\mathrm{o}})^2].$$

It is clear that

$$E[Y_{n+1}^2] = \gamma_0,$$

$$E[(Y_{n+1}^{\mathrm{o}})^2] = \sum_{k=1}^n \sum_{j=1}^n \xi_{n,k}\xi_{nj}E[Y_{n+1-k}Y_{n+1-j}]$$

$$= \sum_{k=1}^n \sum_{j=1}^n \xi_{n,k}\xi_{nj}\gamma_{k-j} = (\xi_{n1}, \ldots, \xi_{nn})\Sigma_n(\xi_{n1}, \ldots, \xi_{nn})^{\mathrm{T}}$$

$$= (\gamma_1, \ldots, \gamma_n)(\Sigma_n^{-1}\Sigma_n\Sigma_n^{-1})(\gamma_1, \ldots, \gamma_n)^{\mathrm{T}}$$

$$= (\gamma_1, \ldots, \gamma_n)\Sigma_n^{-1}(\gamma_1, \ldots, \gamma_n),$$

and (ii) follows. $\qquad\square$

Remark. If $\gamma_0 > 0$ and $\gamma_k \to 0$ $(k \to \infty)$, then the covariance matrix $\Sigma_n = (\gamma_{k-j})_{k,j\in\mathbb{Z}}$ is nonsingular.

Similarly, we have the h-step predictors as follows.

Let $\{Y_n\}_{n\in\mathbb{Z}_+}$ be stated as in Proposition 8.1. For any $\tau \geq 1$, the best linear prediction $Y^o_{n+\tau}$ can be found in the exactly same manner as Y^o_{n+1}:

$$Y^o_{n+\tau} = \sum_{k=1}^{n} \xi^\tau_{n,k} Y_{n+1-k}.$$

If the matrix

$$\Sigma^\tau_n = (\gamma_{k-j+\tau-1})_{k,j=1,\ldots,n}$$

is nonsingular, then $\xi^\tau_{n,k}$ $(k = 1,\ldots,n)$ satisfy

$$(\xi^\tau_{n,1},\ldots,\xi^\tau_{n,n})^\mathrm{T} = (\Sigma^\tau_n)^{-1}(\gamma_\tau,\ldots,\gamma_{\tau+n-1})^\mathrm{T}.$$

To determine the predictors Y^o_{n+1} $(n \in \mathbb{Z}_+)$, a fast algorithm is important.

Theorem 8.6 (Durbin-Lovinson Algorithm). *Under the conditions of Proposition 8.1, if $\gamma_0 > 0$, then the coefficients $\xi_{n,k}$ and the mean square errors α_n satisfy $\xi_{1,1} = \frac{\gamma_1}{\gamma_0}$ and $\alpha_0 = \gamma_0$, and*

(i) $\xi_{n,n} = \frac{1}{\alpha_{n-1}}(\gamma_n - \sum_{j=1}^{n-1} \xi_{n-1,j}\gamma_{n-j})$;
(ii) $\xi_{n,l} = \xi_{n-1,l} - \xi_{n,n}\xi_{n-1,n-l}$ $(l = 1,\ldots,n-1)$;
(iii) $\alpha_n = \alpha_{n-1}(1 - \xi^2_{n,n})$.

Proof. Denote

$$V_n = \mathrm{span}\{Y_2,\ldots,Y_n\},$$
$$L_n = \mathrm{span}\{Y_1,\ldots,Y_n\}.$$

Then

$$\left(Y_1 - \mathrm{Proj}_{V_n} Y_1\right) \perp V_n.$$

Therefore, the best linear prediction of Y_{n+1} in L_n is

$$Y^o_{n+1} = \mathrm{Proj}_{V_n} Y_{n+1} + \beta(Y_1 - \mathrm{Proj}_{V_n} Y_1). \tag{8.16}$$

Since $Y_1 - \mathrm{Proj}_{V_n} Y_1 \in L_n$ and $(Y_{n+1} - Y^o_{n+1}) \perp L_n$,

$$(Y_{n+1}, Y_1 - \mathrm{Proj}_{V_n} Y_1) = \beta\|Y_1 - \mathrm{Proj}_{V_n} Y_1\|^2.$$

From this, it follows that

$$\beta = \frac{(Y_{n+1}, Y_1 - \mathrm{Proj}_{V_n} Y_1)}{\|Y_1 - \mathrm{Proj}_{V_n} Y_1\|^2} = \frac{\gamma_n - (Y_{n+1}, \mathrm{Proj}_{V_n} Y_1)}{\alpha_{n-1}}. \tag{8.17}$$

Since $\{Y_n\}_{n\in\mathbb{Z}_+}$ is a stationary random process, $(Y_1,\ldots,Y_n)$ and $(Y_2,\ldots,Y_{n+1})$ have the same covariance matrix. By Proposition 8.1,

$$Y^o_n = \sum_{k=1}^{n-1} \xi_{n-1,k} Y_{n-k}, \tag{8.18}$$

and so

$$\text{Proj}_{V_n} Y_1 = \sum_{k=1}^{n-1} \xi_{n-1,k} Y_{k+1},$$

$$\text{Proj}_{V_n} Y_{n+1} = \sum_{k=1}^{n-1} \xi_{n-1,k} Y_{n+1-k},$$

and the mean square errors

$$\|Y_1 - \text{Proj}_{V_n} Y_1\|^2 = \|Y_{n+1} - \text{Proj}_{V_n} Y_{n+1}\|^2 = \|Y_n - Y_n^o\|^2 = \alpha_{n-1}.$$

This implies that

$$Y_{n+1}^o = \beta Y_1 + \sum_{k=1}^{n-1} (\xi_{n-1,k} - \beta \xi_{n-1,n-k}) Y_{n+1-k}, \tag{8.19}$$

where

$$\beta = \frac{1}{\alpha_{n-1}} \left(\gamma_n - \sum_{k=1}^{n-1} \xi_{n-1,k} \gamma_{n-k} \right).$$

Comparing coefficients in (8.15) and (8.19), we get (i) and (ii).

From $(Y_1 - \text{Proj}_{V_n} Y_1) \perp V_n$, it follows that

$$(\text{Proj}_{V_n} Y_{n+1}, \beta(Y_1 - \text{Proj}_{V_n} Y_1)) = 0.$$

From this and (8.16), the mean square error

$$\begin{aligned}
\alpha_n &= \|Y_{n+1} - Y_{n+1}^o\|^2 \\
&= \|Y_{n+1} - \text{Proj}_{V_n} Y_{n+1} - \beta(Y_1 - \text{Proj}_{V_n} Y_1)\|^2 \\
&= \|Y_{n+1} - \text{Proj}_{V_n} Y_{n+1}\|^2 + \beta^2 \|Y_1 - \text{Proj}_{V_n} Y_1\|^2 \\
&\quad - 2\beta(Y_{n+1}, Y_1 - \text{Proj}_{V_n} Y_1).
\end{aligned}$$

Since $\{Y_n\}_{n \in \mathbb{Z}}$ is stationary, by (8.17), we get

$$\begin{aligned}
\alpha_n &= \alpha_{n-1} + \beta^2 \alpha_{n-1} - 2\beta(Y_{n+1}, Y_1 - \text{Proj}_{V_n} Y_1) = (1 - \beta^2)\alpha_{n-1} \\
&= (1 - \xi_{n,n}^2)\alpha_{n-1},
\end{aligned}$$

i.e., (iii). $\qquad \square$

In the Durbin-Lovinson algorithm, start from the known ξ_{11} and γ_0. By (iii), α_1 is computed. Then, by (i) and (ii), ξ_{22} and ξ_{21} are computed. Similarly, α_2 and $\xi_{33}, \xi_{32}, \xi_{31}$ are computed. Continuing this procedure, one can compute α_n and ξ_{nk} $(k = 1, \ldots, n)$ fast.

8.3.3 Kolmogorov's Formula

The general prediction error has the following formula.

Kolmogorov's Formula. *The one-step square prediction error* $\tilde{\sigma}^2$ *of the stationary process* $\{Y_n\}_{n \in \mathbb{Z}}$ *is*

$$\tilde{\sigma}^2 = e^{\frac{1}{2\pi} \int_{-\pi}^{\pi} \log g_Y(\tau) \, d\tau},$$

where $g_Y(\tau)$ *is the spectral density.*

For a causal AR(p) process,

$$\varphi(T)Y_n = Z_n \quad (n \in \mathbb{Z}),$$
$$\{Z_n\}_{n \in \mathbb{Z}} \sim WN(0, \sigma^2),$$

where $\varphi(z) = 1 - \sum_1^p \varphi_k z^k \neq 0$ ($|z| \leq 1$), this formula can be checked directly. By Theorem 8.5, the one-step prediction error of $\{Y_n\}_{n \in \mathbb{Z}}$ is

$$E[(Y_{n+1} - Y_{n+1}^o)^2] = \sigma^2 \psi_1 = \sigma^2.$$

Since $\varphi(z) \neq 0$ ($|z| \leq 1$),

$$\varphi(z) = \prod_1^p (1 - \alpha_j z) \quad (|\alpha_j| < 1).$$

So the spectral density of Y_n is $g_Y(\tau) = \sigma^2 |\varphi(e^{-i\tau})|^{-2}$ (see Section 8.2), and

$$\int_{-\pi}^{\pi} \log g_Y(\tau) \, d\tau = \int_{-\pi}^{\pi} \log \sigma^2 \, d\tau - 2 \sum_{j=1}^{p} \int_{-\pi}^{\pi} \log |1 - \alpha_j e^{-i\tau}| \, d\tau = 2\pi \log \sigma^2$$

or

$$e^{\frac{1}{2\pi} \int_{-\pi}^{\pi} \log f(\theta) \, d\theta} = \sigma^2.$$

8.4 ASYMPTOTIC THEORY

Many estimators in time series have asymptotic normality when the number of observation is large enough. The proof needs to use the central limit theorem for dependent random variables in the high-dimensional case.

8.4.1 Gramer-Wold Device

In Section 5.3, we stated various convergences of random variable sequences and showed that if $Y_n \overset{m.s.}{\to} Y$, then $Y_n \overset{p}{\to} Y$; if $Y_n \overset{p}{\to} Y$, then $Y_n \overset{d}{\to} Y$. Now we study further the high-dimensional case.

Suppose that $\{\mathbf{Y}_n\}_{n \in \mathbb{Z}}$ is a sequence of the k-dimensional random vectors with distribution functions $F_{\mathbf{Y}_n}$. If there is a k-dimensional random vector $\mathbf{Y}$ with the distribution function $F_{\mathbf{Y}}$ such that

$$\lim_{n \to \infty} F_{\mathbf{Y}_n}(t) = F_{\mathbf{Y}}(t) \quad (t \in \mathbb{R}^k),$$

then we say that $\{\mathbf{Y}_n\}_{n \in \mathbb{Z}_+}$ converges to $\mathbf{Y}$ in distribution, denoted by $\mathbf{Y}_n \overset{d}{\to} \mathbf{Y}$.

The *eigenfunction* of a k-dimensional random vector $\mathbf{Y}$ is defined as

$$\varphi_{\mathbf{Y}}(t) = \int_{\mathbb{R}^k} e^{it^T x} \rho_{\mathbf{Y}}(x) \, dx \quad (\mathbf{t} \in \mathbb{R}^k),$$

where $\rho_{\mathbf{Y}}(x)$ is the density function of $\mathbf{Y}$ and t^T is the transpose of the k-dimensional vector t. The eigenfunction $\varphi_{\mathbf{Y}}(t)$ is the conjugate of the k-dimensional Fourier transform of the density function $\rho_{\mathbf{Y}}$,

$$\varphi_{\mathbf{Y}}(t) = \overline{\hat{\rho}_{\mathbf{Y}}}(t),$$

and it may be regarded as the expectation of $e^{it^T \mathbf{Y}}$:

$$\varphi_{\mathbf{Y}}(t) = E\left[e^{it^T \mathbf{Y}} \right] \quad (t \in \mathbb{R}^k). \tag{8.20}$$

Characterization of Convergence in Distribution. *If $\{\mathbf{Y}_n\}$ is a sequence of k-dimensional random vectors, then $\mathbf{Y}_n \overset{d}{\to} \mathbf{Y}$ if and only if $\lim_{n \to \infty} \varphi_{\mathbf{Y}_n}(t) = \varphi_{\mathbf{Y}}(t)$ ($t \in \mathbb{R}^k$).*

Gramer-Wold Device. *Let $\{\mathbf{Y}_n\}$ be a sequence of the k-dimensional random vector. Then $\mathbf{Y}_n \overset{d}{\to} \mathbf{Y}$ if and only if $\alpha^T \mathbf{Y}_n \overset{d}{\to} \alpha^T \mathbf{Y}$ for any $\alpha \in \mathbb{R}^k$.*

Proof. Suppose that $\mathbf{Y}_n \overset{d}{\to} \mathbf{Y}$. Then, for $t \in \mathbb{R}$ and $\alpha \in \mathbb{R}^k$, the eigenfunction satisfies

$$\varphi_{\mathbf{Y}_n}(t\alpha) \to \varphi_{\mathbf{Y}}(t\alpha).$$

However,

$$\varphi_{\mathbf{Y}_n}(t\alpha) = E\left[e^{it\alpha^T \mathbf{Y}_n} \right] = \varphi_{\alpha^T \mathbf{Y}_n}(t),$$

$$\varphi_{\mathbf{Y}}(t\alpha) = E\left[e^{it\alpha^T \mathbf{Y}} \right] = \varphi_{\alpha^T \mathbf{Y}}(t).$$

Therefore, for $t \in \mathbb{R}$ and $\alpha \in \mathbb{R}^k$,

$$\varphi_{\alpha^T \mathbf{Y}_n}(t) \to \varphi_{\alpha^T \mathbf{Y}}(t),$$

i.e., $\alpha^T \mathbf{Y}_n \overset{d}{\to} \alpha^T \mathbf{Y} (\alpha \in \mathbb{R}^k)$.

Suppose that $\alpha^T \mathbf{Y}_n \overset{d}{\to} \alpha^T \mathbf{Y}$ ($\alpha \in \mathbb{R}^k$). Then the eigenfunction satisfies $\varphi_{\alpha^T \mathbf{Y}_n}(t) \to \varphi_{\alpha^T \mathbf{Y}}(t)$. Especially,

$$\varphi_{\alpha^T \mathbf{Y}_n}(1) \to \varphi_{\alpha^T \mathbf{Y}}(1).$$

However,

$$\varphi_{\alpha^T \mathbf{Y}_n}(1) = E\left[e^{i\alpha^T \mathbf{Y}_n} \right] = \varphi_{\mathbf{Y}_n}(\alpha),$$

$$\varphi_{\alpha^T \mathbf{Y}}(1) = E\left[e^{i\alpha^T \mathbf{Y}} \right] = \varphi_{\mathbf{Y}}(\alpha).$$

Therefore,

$$\varphi_{\mathbf{Y}_n}(\alpha) \to \varphi_{\mathbf{Y}}(\alpha),$$

i.e., $\mathbf{Y}_n \overset{d}{\to} \mathbf{Y}$ ($\alpha \in \mathbb{R}^k$). □

Let $\mathbf{Y}_n = (Y_{n1}, \ldots, Y_{nk})^{\mathrm{T}}$ and $\mathbf{Y} = (Y_1, \ldots, Y_k)^{\mathrm{T}}$ be random vectors. If $\mathbf{Y}_n \overset{d}{\to} \mathbf{Y}$, then $Y_{nj} \overset{d}{\to} Y_j$ ($j = 1, \ldots, k$). The converse is not true. The Gramer-Wold method gives that $\mathbf{Y}_n \overset{d}{\to} \mathbf{Y}$ if and only if

$$\sum_{j=1}^{k} \alpha_j Y_{nj} \overset{d}{\to} \sum_{1}^{k} \alpha_j Y_j \quad (\alpha_1, \ldots, \alpha_k \in \mathbb{R}).$$

Proposition 8.2. *Let $\{\mathbf{X}_n\}$ and $\{\mathbf{Y}_n\}$ both be sequences of the k-dimensional random vectors. If*

$$\mathbf{X}_n - \mathbf{Y}_n \overset{p}{\to} 0, \quad \mathbf{X}_n \overset{d}{\to} \mathbf{X},$$

then $\mathbf{Y}_n \overset{d}{\to} \mathbf{X}$.

Proof. Since $\mathbf{X}_n \overset{d}{\to} \mathbf{X}$, the characteristic functions of $\mathbf{X}_n$ and $\mathbf{X}$ satisfy

$$\varphi_{\mathbf{X}_n}(u) - \varphi_{\mathbf{X}}(u) \to 0.$$

By (8.20),

$$
\begin{aligned}
|\varphi_{\mathbf{Y}_n}(u) - \varphi_{\mathbf{X}_n}(u)| &= \left| E\left[e^{iu^{\mathrm{T}}\mathbf{Y}_n} - e^{iu^{\mathrm{T}}\mathbf{X}_n} \right] \right| \le E\left[\left| 1 - e^{iu^{\mathrm{T}}(\mathbf{Y}_n - \mathbf{X}_n)} \right| \right] \\
&= E\left[\left| 1 - e^{iu^{\mathrm{T}}(\mathbf{Y}_n - \mathbf{X}_n)} \right| I_{A_1}(X_n, Y_n) \right] \\
&\quad + E\left[\left| 1 - e^{iu^{\mathrm{T}}(\mathbf{Y}_n - \mathbf{X}_n)} \right| I_{A_2}(X_n, Y_n) \right] \\
&= I_1 + I_2,
\end{aligned}
$$

where

$$A_1 = \{(X_n, Y_n) : |X_n - Y_n| < \delta\},$$
$$A_2 = \{(X_n, Y_n) : |X_n - Y_n| \ge \delta\},$$

and

$$I_{A_i}(X_n, Y_n) = \begin{cases} 1, & (X_n, Y_n) \in A_i, \\ 0, & (X_n, Y_n) \notin A_i \end{cases}$$

and $|X_n - Y_n|$ is the norm of $X_n - Y_n$ in the space $\mathbb{R}^k$. For a fixed u and $\epsilon > 0$, there is a $\delta > 0$ such that $|1 - e^{iu^{\mathrm{T}}(\mathbf{y} - \mathbf{x})}| < \epsilon(|\mathbf{x} - \mathbf{y}| < \delta)$. So

$$|I_1| \le \epsilon I_{A_1}(\mathbf{X}_n, \mathbf{Y}_n) \le \epsilon.$$

Since $\mathbf{X}_n - \mathbf{Y}_n \xrightarrow{\text{p}} 0$,

$$|I_2| \leq 2E[I_{A_2}(X_n, Y_n)] = 2P(|\mathbf{Y}_n - \mathbf{X}_n| \geq \delta) \to 0.$$

The combination of these two results for I_1 and I_2 implies $\varphi_{\mathbf{Y}_n}(u) - \varphi_{\mathbf{X}_n}(u) \to 0$.

Therefore, from

$$|\varphi_{\mathbf{Y}_n}(u) - \varphi_{\mathbf{X}}(u)| \leq |\varphi_{\mathbf{Y}_n}(u) - \varphi_{\mathbf{X}_n}(u)| + |\varphi_{\mathbf{X}_n}(u) - \varphi_{\mathbf{X}}(u)| \to 0,$$

it follows that $\mathbf{Y}_n \xrightarrow{\text{d}} \mathbf{X}$. $\qquad\square$

Theorem 8.7 plays a key role in establishing asymptotic normality of the sample mean and covariance functions.

Theorem 8.7. *Let $\mathbf{X}_{l,j}$ be a k-dimensional random vector for each $l, j \in \mathbb{Z}_+$. Suppose that the double sequence $\{\mathbf{X}_{l,j}\}_{l,j \in \mathbb{Z}_+}$ satisfies*

$$\mathbf{X}_{l,j} \xrightarrow{\text{d}} \mathbf{X}_j \quad (l \to \infty) \text{ for each } j,$$

$$\mathbf{X}_j \xrightarrow{\text{d}} \mathbf{X} \quad (j \to \infty).$$

Let $\{\mathbf{Y}_l\}_{l \in \mathbb{Z}_+}$ be a sequence of k-dimensional random vectors. Suppose that $\{\mathbf{Y}_l\}_{l \in \mathbb{Z}_+}$ satisfies

$$\lim_{j \to \infty} \overline{\lim_{l \to \infty}} P(|\mathbf{Y}_l - \mathbf{X}_{l,j}| \geq \delta) = 0 \quad \text{for any } \delta > 0. \tag{8.21}$$

Then $\mathbf{Y}_l \xrightarrow{\text{d}} \mathbf{X}$ $(l \to \infty)$.

Proof. Similarly to the argument of Proposition 8.2, for a fixed u and $\epsilon > 0$, there is a $\delta > 0$ such that

$$|\varphi_{\mathbf{Y}_l}(u) - \varphi_{\mathbf{X}_{l,j}}(u)| \leq \epsilon + E\left[\left|1 - e^{iu^{\mathrm{T}}(\mathbf{Y}_l - \mathbf{X}_{l,j}(u))}\right| I_B(\mathbf{Y}_l, \mathbf{X}_{l,j})\right],$$

where $B = \{(Y_\mu, Y_{\mu,\nu}) : |Y_\mu - Y_{\mu,\nu}| \geq \delta\}$. By (8.21), it follows that

$$\lim_{j \to \infty} \overline{\lim_{l \to \infty}} |\varphi_{\mathbf{Y}_l}(u) - \varphi_{\mathbf{X}_{l,j}}(u)| = 0.$$

By the assumption, it follows that

$$\lim_{j \to \infty} |\varphi_{\mathbf{X}_j}(u) - \varphi_{\mathbf{X}}(u)| = 0,$$

$$\lim_{l \to \infty} |\varphi_{\mathbf{X}_{l,j}}(u) - \varphi_{\mathbf{X}_j}(u)| = 0.$$

The combination of these results gives

$$|\varphi_{\mathbf{Y}_l}(u) - \varphi_{\mathbf{X}}(u)| \leq |\varphi_{\mathbf{Y}_l}(u) - \varphi_{\mathbf{X}_{l,j}}(u)| + |\varphi_{\mathbf{X}_{l,j}}(u) - \varphi_{\mathbf{X}_j}(u)| + |\varphi_{\mathbf{X}_j}(u)$$
$$- \varphi_{\mathbf{X}}(u)| \to 0.$$

So $\mathbf{Y}_l \xrightarrow{\text{d}} \mathbf{X}$ $(l \to \infty)$. $\qquad\square$

8.4.2 Asymptotic Normality

The following law of large numbers in MA processes is derived easily from Theorem 8.7 and the law of large numbers in Chapter 5.

Law of Large Numbers in MA Processes Let $\{Y_n\}_{n\in\mathbb{Z}_+}$ be an MA process:

$$Y_n = \sum_j \psi_j Z_{n-j} \quad (n \in \mathbb{Z}),$$

$$\{Z_n\}_{n\in\mathbb{Z}} \sim \text{IID}(\mu, \sigma^2),$$

where $\sum_j |\psi_j| < \infty$. Then

$$\overline{Y}_n \xrightarrow{\text{p}} \mu \sum_j \psi_j, \quad \text{where } \overline{Y}_n = \frac{1}{n}\sum_1^n Y_k.$$

Definition 8.1. Let $\{X_n\}_{n\in\mathbb{Z}}$ be a sequence of random variables. If

$$\frac{X_n - \mu_n}{\sigma_n} \xrightarrow{\text{d}} Z,$$

where $\sigma_n > 0$ and $Z \sim N(0,1)$, then $\{X_n\}_{n\in\mathbb{Z}}$ is called *asymptotically normal* (AN) with mean μ_n and standard covariance σ_n, say, X_n is $\text{AN}(\mu_n, \sigma_n^2)$.

The central limit theorem in Chapter 5 shows that if

$$\{X_k\}_{k\in\mathbb{Z}_+} \sim \text{IID}(\mu, \sigma^2),$$
$$E[X_k] = \mu,$$
$$\text{Var}\, X_k = \sigma^2,$$

then the sum $S_n = \sum_1^n X_k$ $(n \in \mathbb{Z}_+)$ satisfies

$$\frac{S_n - n\mu}{\sqrt{n}\sigma} \xrightarrow{\text{p}} Y,$$

where $Y \sim N(0,1)$.

Since $\dfrac{S_n - n\mu}{\sqrt{n}\sigma} = \dfrac{\overline{X}_n - \mu}{\sigma/\sqrt{n}}$, the central limit theorem now has the form

Let $\{X_n\}_{n\in\mathbb{Z}_+} \sim \text{IID}(\mu, \sigma^2)$ and $\overline{X}_n = \frac{1}{n}\sum_1^n X_j$. Then $\overline{X}_n$ is $AN(\mu, \frac{\sigma^2}{n})$,

and an extended central limit theorem is as follows:

If X_n is $AN(\mu, \sigma_n^2)$ and $\sigma_n \to 0$, and g is differentiable at μ, then $g(X_n)$ is $AN(g(\mu), (g'(\mu))^2 \sigma_n^2)$.

These results can be generalized to the high-dimensional case.

The sequence $\{\mathbf{X}_n\}_{n\in\mathbb{Z}_+}$ of k-dimensional random vector is AN with mean vector μ_n and covariance matrix Σ_n if, for all sufficient large n,

(i) the matrix Σ_n has no zero diagonal elements;
(ii) $\alpha^T \mathbf{X}_n$ is $\text{AN}(\alpha^T \mu_n, \alpha^T \Sigma_n \alpha)$ for $\alpha \in \mathbb{R}^k$ and $\alpha^T \Sigma_n \alpha > 0$.

Proposition 8.3. *Let* $\mathbf{X}_n$ *be* $AN(\mu, c_n^2\Sigma)$, *where* Σ *is a symmetric non-negative definite matrix. Again let*

$$\mathbf{g}(\mathbf{X}) = (g_1(\mathbf{X}), \ldots, g_m(\mathbf{X}))^\mathsf{T}$$

be a continuous mapping from $\mathbb{R}^k$ *to* $\mathbb{R}^m$, *where* $\mathbf{X} \in \mathbb{R}^k$ *and each* g_i *is continuously differentiable in a neighborhood of* μ. *Denote the* $m \times k$ *Jacobian matrix*

$$D = \left(\frac{\partial g_l}{\partial x_j}(\mu) \right).$$

If $c_n \to 0$ *and* $D\Sigma D^\mathsf{T}$ *has no zero diagonal elements, then*

$$g(\mathbf{X}_n) \quad is \quad AN\left(g(\mu), c_n^2 D\Sigma D^\mathsf{T} \right).$$

The following concept of m-dependent is a generalization of that of independent.

Definition 8.2. A stationary sequence $\{X_n\}_{n \in \mathbb{Z}}$ of random variables is called m-dependent if X_n $(n \leq k)$ and X_n $(n \geq k + m + 1, m \geq 0)$ are independent.

Theorem 8.8. *Let* $\{X_n\}_{n \in \mathbb{Z}_+}$ *be a stationary m-dependent sequence of random variables with mean 0 and covariance function* γ_k $(k \in \mathbb{Z}_+)$. *Denote* $\overline{X}_n = \frac{1}{n} \sum_1^n X_j$. *Let*

$$U_m = \gamma_0 + 2 \sum_{j=1}^m \gamma_j \neq 0.$$

Then $\lim_{n \to \infty} n \operatorname{Var}(\overline{X}_n) = U_m$.

Proof. Notice that $\overline{X}_n = \frac{1}{n} \sum_1^n X_j$. Since $\{X_n\}_{n \in \mathbb{Z}_+}$ is stationary,

$$n \operatorname{Var}(\overline{X}_n) = \frac{1}{n} \sum_{i=1}^n \sum_{j=1}^n E[X_i X_j] = \frac{1}{n} \sum_{i=1}^n \sum_{j=1}^n \gamma_{i-j} = \sum_{|j| \leq n} \left(1 - \frac{|j|}{n} \right) \gamma_j.$$

Since $\{X_n\}_{n \in \mathbb{Z}_+}$ is m-dependent,

$$\gamma_l = E[X_k X_{k+l}] = 0 \quad (l > m),$$

and so

$$n \operatorname{Var}(\overline{X}_n) = \sum_{|j| \leq m} \left(1 - \frac{|j|}{n} \right) \gamma_j \quad (n \geq m). \tag{8.22}$$

In view of $\gamma_j = \gamma_{-j}$, when $n \to \infty$,

$$n \operatorname{Var}(\overline{X}_n) \to \sum_{|j| \leq m} \gamma_j = \gamma_0 + 2 \sum_1^m \gamma_j. \qquad \square$$

Theorem 8.9 (The Central Limit Theorem for Stationary m-Dependent Sequences). *Under the condition of Theorem 8.8,* $\overline{X}_n$ *is* $AN(0, \frac{U_m}{n})$.

This theorem can be derived from Theorem 8.8 and the central limit theorem.

Example 8.1. The MA(q) process satisfying

$$X_n = Z_n + \sum_1^q \theta_j Z_{n-j} \quad (n \in \mathbb{Z}),$$

$$\{Z_n\}_{n\in\mathbb{Z}_+} \sim \text{IID}(0, \sigma^2),$$

where $\sum_0^q \theta_j \neq 0$, is a q-dependent stationary sequence. Its covariance is

$$\gamma_k = 0 \quad (|k| > q),$$

$$\gamma_k = E[X_n X_{n+k}] = \sum_{j=0}^q \sum_{l=0}^q \theta_j \theta_l E[Z_{n-j} Z_{n+k-l}]$$

$$= \sigma^2 \sum_{j=0}^k \theta_j \theta_{k-j} \quad (|k| < q).$$

So

$$U = \sum_{k=-q}^q \gamma_k = \sigma^2 \left(\sum_0^q \theta_j \right)^2.$$

Applying Theorem 8.9, we find $\overline{X}_n$ is AN$(0, \frac{\sigma^2}{n}(\sum_0^q \theta_j)^2)$.

8.5 ESTIMATES OF MEANS AND COVARIANCE FUNCTIONS

First, we study the asymptotic normality of the estimators of means.

Let $\{Y_n\}_{n\in\mathbb{Z}_+}$ be a stationary process with mean μ and covariance function γ_k. An often used estimator of the mean μ is the sample mean:

$$\overline{Y}_n = \frac{Y_1 + \cdots + Y_n}{n}.$$

The mean squared error $E[(\overline{Y}_n - \mu)^2]$ is estimated as follows.

Since $\{Y_n\}_{n\in\mathbb{Z}}$ is a stationary process,

$$E[(\overline{Y}_n - \mu)^2] = \text{Var}(\overline{Y}_n) = \text{Cov}(\overline{Y}_n, \overline{Y}_n)$$

$$= \frac{1}{n^2} \sum_{l=1}^n \sum_{j=1}^n \text{Cov}(Y_l, Y_j) = \frac{1}{n^2} \sum_{l=1}^n \sum_{j=1}^n \gamma_{l-j},$$

and so

$$E[(\overline{Y}_n - \mu)^2] = \frac{1}{n} \sum_{|k| \leq n-1} \left(1 - \frac{|k|}{n}\right) \gamma_k \leq \frac{1}{n} \sum_{|k| \leq n-1} |\gamma_k|.$$

If $\gamma_k \to 0$ $(k \to \infty)$, it is clear that $\frac{1}{n} \sum_0^{n-1} |\gamma_k| \to 0$ $(n \to \infty)$. So

$$E[(\overline{Y}_n - \mu)^2] \to 0 \quad (n \to \infty).$$

If the series $\sum_k |\gamma_k|$ converges, then

$$\lim_{n \to \infty} nE[(\overline{Y}_n - \mu)^2] = \lim_{n \to \infty} \sum_{|k| \leq n-1} \left(1 - \frac{|k|}{n}\right) \gamma_k = \sum_k \gamma_k.$$

Let $\{Y_n\}_{n \in \mathbb{Z}}$ be an ARMA process with mean μ. Then $\sum_n |\psi_n| < \infty$. By Theorem 8.2(ii),

$$\sum_k |\gamma_k| \leq \sigma^2 \sum_j \sum_k |\psi_j \psi_{j+|k|}| \leq 2\sigma^2 \left(\sum_k |\psi_k|\right)^2 < \infty.$$

So the sample mean $\overline{Y}_n \overset{\text{m.s.}}{\to} \mu$ and $nE[(\overline{Y}_n - \mu)^2] \to \sum_k \gamma_k$.

The following theorem discusses the asymptotic normality.

Theorem 8.10. *Suppose that a process $\{Y_n\}_{n \in \mathbb{Z}}$ satisfying*

$$Y_n = \mu + \sum_j \psi_j Z_{n-j} \quad (n \in \mathbb{Z}),$$

$$\{Z_n\}_{n \in \mathbb{Z}} \sim \text{IID}(0, \sigma^2)$$

is a stationary process with covariance function γ_k, where $\sum_j \psi_j$ converges absolutely and $\sum_j \psi_j \neq 0$. Then $\overline{Y}_n$ is $\text{AN}(\mu, n^{-1}U)$, where $U = \sum_k \gamma_k = \sigma^2 (\sum_k \psi_k)^2$.

Proof. Let $S_m = \sum_{-m}^m \psi_j$. Denote

$$Y_{n,m} = \mu + \sum_{j=-m}^m \psi_j Z_{n-j},$$

$$\overline{Y}_{n,m} = \frac{1}{n} \sum_{l=1}^n Y_{l,m}.$$

By Example 8.1, $\overline{Y}_{n,m}$ is $\text{AN}(\mu, \frac{\sigma^2}{n} S_m^2)$, i.e.,

$$\sqrt{n}(\overline{Y}_{n,m} - \mu) \overset{\text{d}}{\to} Y_m^o,$$

where $Y_m^o \sim \text{N}(0, \sigma^2 S_m^2)$, i.e., Y_m^o is a normal random variable with mean 0 and variance $\sigma^2 S_m^2$.

Let $S = \sum_j \psi_j$. Then $Y_m^o \overset{\text{d}}{\to} Y^o$, where $Y^o \sim \text{N}(0, \sigma^2 S^2)$. Since

$$\overline{Y}_n - \overline{Y}_{n,m} = \frac{1}{n} \sum_{l=1}^n \sum_j \psi_j Z_{l-j} - \overline{Y}_{n,m} = \frac{1}{n} \sum_{l=1}^n \sum_{|j|>m} \psi_j Z_{l-j},$$

clearly,

$$\lim_{m\to\infty} \overline{\lim_{n\to\infty}} P\left(\sqrt{n}|\overline{Y}_n - \overline{Y}_{n,m}| > \delta\right) = 0.$$

From this and

$$\sqrt{n}(\overline{Y}_{n,m} - \mu) \xrightarrow{d} Y_m^o,$$

$$Y_m^o \xrightarrow{d} Y^o,$$

by Theorem 8.7, it follows that

$$\sqrt{n}(\overline{Y}_n - \mu) \xrightarrow{d} Y^o.$$

□

Second, we study the asymptotic normality of the estimates of covariance functions.

Define *estimators* of covariance functions γ_τ by the statistics

$$\hat{\gamma}_\tau = \frac{1}{n}\sum_{k=1}^{n-\tau}(Y_k - \overline{Y}_n)(Y_{k+\tau} - \overline{Y}_n) \quad (0 \le \tau \le n-1) \tag{8.23}$$

as estimates of covariance function γ_τ. For each $n \in \mathbb{Z}_+$, the sample covariance matrix $\hat{\Sigma}_n = (\hat{\gamma}_{|l-j|})_{l,j=1,\dots,n}$ is non-negative definite.

Let

$$\gamma_\tau^* = \frac{1}{n}\sum_{1}^{n} Y_k Y_{k+\tau}. \tag{8.24}$$

We study the asymptotic normality of γ_τ^*, and then study that of $\hat{\gamma}_\tau$.

Suppose that $\{Y_n\}_{n\in\mathbb{Z}}$ is a random process satisfying

$$Y_n = \sum_k \psi_k Z_{n-k} \quad (n \in \mathbb{Z}),$$

$$\{Z_n\}_{n\in\mathbb{Z}} \sim \text{IID}(0, \sigma^2),$$

where $E[Z_n^4] = \lambda\sigma^4$, and $\sum_k |\psi_k| < \infty$. From

$$E[\gamma_\tau^*] = \frac{1}{n}\sum_{1}^{n} E[Y_l Y_{l+\tau}] = \frac{1}{n}\sum_{1}^{n} \gamma_\tau = \gamma_\tau,$$

it follows that

$$\text{Cov}(\gamma_\mu^*, \gamma_\nu^*) = E[\gamma_\mu^* \gamma_\nu^*] - E[\gamma_\mu^*]E[\gamma_\nu^*] = \frac{1}{n^2}\sum_{l=1}^{n}\sum_{k=1}^{n} E[Y_l Y_{l+\mu} Y_k Y_{k+\nu}] - \gamma_\mu \gamma_\nu.$$

Since $E[Z_n^4] = \lambda\sigma^4$ and $E[Z_m^2 Z_n^2] = \sigma^4 (m \ne n)$, a direct computation implies

$$\lim_{n\to\infty} n\text{Cov}(\gamma_\mu^*, \gamma_\nu^*) = g_{\mu,\nu}, \tag{8.25}$$

where $g_{\mu,\nu} = (\lambda - 3)\gamma_\mu \gamma_\nu + \sum_k (\gamma_k \gamma_{k-\mu+\nu} + \gamma_{k+\nu}\gamma_{k-p})$.

Proposition 8.4. *Let*

$$Y_n = \sum_{k=-m}^{m} \psi_k Z_{n-k} \quad (n \in \mathbb{Z}),$$

$$\{Z_n\}_{n \in \mathbb{Z}} \sim \text{IID}(0, \sigma^2).$$

If $E[Z_n^4] = \lambda \sigma^2$, *then for* $k \in \mathbb{Z}_+$,

$$(\gamma_0^*, \ldots, \gamma_k^*)^{\mathrm{T}} \quad \text{is AN}\left((\gamma_0, \ldots, \gamma_k)^{\mathrm{T}}, \frac{1}{n} G\right),$$

where $G = (g_{\mu\nu})_{\mu,\nu=0,\ldots,k}$, *and* $g_{\mu\nu}$ *is stated in* (8.25).

Proof. Let $\mathbf{X}_l = (X_{l0}, \ldots, X_{l\tau})^{\mathrm{T}}$ $(l = 1, \ldots, n)$ be the $(\tau + 1)$-dimensional vectors, where $X_{lj} = Y_l Y_{l+j}$ $(j = 0, \ldots, \tau)$. By (8.24), the mean of $\{\mathbf{X}_l\}_{l=1,\ldots,n}$ is

$$\frac{1}{n} \sum_{1}^{n} \mathbf{X}_l = (\gamma_0^*, \ldots, \gamma_\tau^*)^{\mathrm{T}}. \tag{8.26}$$

Notice that

$$\alpha^{\mathrm{T}}\left(\frac{1}{n} \sum_{1}^{n} \mathbf{X}_l\right) = \frac{1}{n} \sum_{1}^{n} \alpha^{\mathrm{T}} \mathbf{X}_l \quad (\alpha \in \mathbb{R}^k).$$

We need prove only that

$$\frac{1}{n} \sum_{1}^{n} \alpha^{\mathrm{T}} \mathbf{X}_l \quad \text{is AN}\left(\alpha^{\mathrm{T}}(\gamma_0, \ldots, \gamma_\tau)^{\mathrm{T}}, \frac{1}{n} \alpha^{\mathrm{T}} G \alpha\right), \tag{8.27}$$

where $\alpha \in \mathbb{R}^{\tau+1}$ and $\alpha^{\mathrm{T}} G \alpha > 0$.

By the assumption, the sequence $\{\alpha^{\mathrm{T}} \mathbf{X}_l\}$ is $(2m + \tau)$-dependent. By (8.26),

$$\text{Var}\left(\frac{1}{n} \sum_{1}^{n} \alpha^{\mathrm{T}} \mathbf{X}_l\right) = \alpha^{\mathrm{T}} \text{Cov}\left(\frac{1}{n} \sum_{1}^{n} \mathbf{X}_l, \frac{1}{n} \sum_{1}^{n} \mathbf{X}_l\right) \alpha$$

$$= \alpha^{\mathrm{T}} \text{Cov}\left((\gamma_0^*, \ldots, \gamma_\tau^*)^{\mathrm{T}}, (\gamma_0^*, \ldots, \gamma_\tau^*)^{\mathrm{T}}\right) \alpha$$

$$= \alpha^{\mathrm{T}} \left(\text{Cov}(\gamma_\mu^*, \gamma_\nu^*)\right)_{\mu,\nu=0,\ldots,\tau} \alpha.$$

By (8.25),

$$\lim_{n \to \infty} n \text{Var}\left(\frac{1}{n} \sum_{1}^{n} \alpha^{\mathrm{T}} \mathbf{X}_l\right) = \alpha^{\mathrm{T}} (g_{\mu\nu})_{\mu,\nu=0,\ldots,\tau} \alpha = \alpha^{\mathrm{T}} G \alpha > 0.$$

From this and Theorem 8.9, (8.27) follows. □

Theorem 8.4 also holds in the case $m = \infty$ by using Theorem 8.7.

The following theorem shows that $(\gamma_0^*, \ldots, \gamma_\tau^*)^{\mathrm{T}}$ and $(\hat{\gamma}_0, \ldots, \hat{\gamma}_\tau)^{\mathrm{T}}$ have same asymptotic normality.

Theorem 8.11. *Let $\{Y_n\}_{n \in \mathbb{Z}_+}$ be a stationary process satisfying*

$$Y_n = \sum_j \psi_j Z_{n-j} \quad (n \in \mathbb{Z}),$$

$$\{Z_n\}_{n \in \mathbb{Z}_+} \sim \text{IID}(0, \sigma^2),$$

where $\sum_j |\psi_j| < \infty$ and the fourth moment $E[Z_n^4] = \lambda \sigma^4$. Then, for any $k \in \mathbb{Z}_+$,

$$(\widehat{\gamma}_0, \dots, \widehat{\gamma}_k)^{\text{T}} \quad \text{is AN}\left((\gamma_0, \dots, \gamma_k)^{\text{T}}, \frac{G}{n}\right),$$

where G is stated in Proposition 8.4.

Proof. A simple calculation gives that

$$\sqrt{n}(\gamma_\mu^* - \widehat{\gamma}_\mu) = \sqrt{n}\overline{Y}_n V_n + \frac{1}{\sqrt{n}} \sum_{l=n-\mu+1}^{n} Y_l Y_{l+\mu} = I_1 + I_2,$$

where

$$V_n = \frac{1}{n} \sum_{l=1}^{n-\mu} Y_{l+\mu} + \frac{1}{n} \sum_{l=1}^{n-\mu} Y_l + \left(1 - \frac{\mu}{n}\right)\overline{Y}_n.$$

Since $\{Y_n\}_{n \in \mathbb{Z}}$ is stationary,

$$E[|Y_l Y_{l+\mu}|] \leq \left(E[Y_l^2]\right)^{1/2} \left(E[Y_{l+\mu}^2]\right)^{1/2} = \gamma_0,$$

and so

$$E[|I_2|] \leq \frac{1}{\sqrt{n}} \sum_{l=n-\mu+1}^{n} E[|Y_l Y_{l+\mu}|] = \frac{1}{\sqrt{n}} \sum_{l=n-\mu+1}^{n} \gamma_0 = \frac{1}{\sqrt{n}}\mu\gamma_0.$$

By the Chebyshev inequality,

$$P(|I_2| \geq \delta) \leq \frac{1}{\delta}E[|I_2|] \leq \frac{1}{\sqrt{n}\delta}\mu\gamma_0.$$

So $I_2 \xrightarrow{\text{p}} 0 \ (n \to \infty)$.

By Theorem 8.10, it follows that

$$\sqrt{n}\,\overline{Y}_n \xrightarrow{\text{d}} Y, \quad \text{where } Y \sim \text{N}(0, \sigma^2 S^2),$$

$$S = \sum_k \psi_k,$$

and so $\sqrt{n}\,\overline{Y}_n$ is p-bounded. According to the law of large numbers in MA processes, $V_n \xrightarrow{\text{p}} 0$, and so $I_1 \xrightarrow{\text{p}} 0$.

Therefore, $\sqrt{n}\,(\gamma_p^* - \hat{\gamma}_p) \xrightarrow{p} 0$. Finally, by Proposition 8.4, we get the desired result. □

Define $\hat{\rho}_\tau = \dfrac{\hat{\gamma}_\tau}{\hat{\gamma}_0}$ as estimates of correlation coefficients $\rho_\tau = \dfrac{\gamma_\tau}{\gamma_0}$, where $\hat{\gamma}_\tau$ is stated in (8.23).

Corollary 8.1. *Let $\{Y_n\}_{n\in\mathbb{Z}_+}$ be a stationary process satisfying*

$$Y_n - \mu = \sum_j \psi_j Z_{n-j} \quad (n \in \mathbb{Z}),$$

$$\{Z_n\}_{n\in\mathbb{Z}_+} \sim \text{IID}(0, \sigma^2),$$

where $\sum_j |\psi_j| < \infty$ and the fourth moment $E[Z_n^4] < \infty$. Then $(\hat{\rho}_1, \ldots, \hat{\rho}_k)^T$ is $\text{AN}((\rho_1, \ldots, \rho_k)^T, \frac{W}{n})$, where the matrix $W = (\omega_{ij})_{i,j\in\mathbb{Z}_+}$, the elements of which ω_{ij} are determined by Barlett's formula:

$$\omega_{ij} = \sum_k \{\rho_{k+i}\rho_{k+j} + \rho_{k-i}\rho_{k+j} + 2\rho_i\rho_j\rho_k^2 - 2\rho_i\rho_k\rho_{k+j} - 2\rho_j\rho_k\rho_{k+i}\}.$$

Proof. Let **g** be a mapping from $\mathbb{R}^{k+1}$ into $\mathbb{R}^k$ defined by

$$\mathbf{g}(x_0, \ldots, x_k) = (g_1(x_0, \ldots, x_k), \ldots, g_k(x_0, \ldots, x_k))^T \quad (x_0 \neq 0),$$

where $g_i(x_0, \ldots, x_k) = \dfrac{x_i}{x_0}$ $(i = 1, \ldots, k)$. Its Jacobian matrix

$$D(x_0, \ldots, x_k) = \left(\frac{\partial g_i}{\partial x_j}\right)_{i=1,\ldots,;j=0,\ldots,k} = \frac{1}{x_0}\begin{pmatrix} -\dfrac{x_1}{x_0} & 1 & 0 & \cdots & 0 \\ -\dfrac{x_2}{x_0} & 0 & 1 & \cdots & 0 \\ \vdots & \vdots & \vdots & \ddots & \vdots \\ -\dfrac{x_k}{x_0} & 0 & 0 & \cdots & 1 \end{pmatrix}.$$

From $\rho_i = \dfrac{\gamma_i}{\gamma_0}$ $(i = 0, \ldots, k)$, it follows that

$$U = D(\gamma_0, \ldots, \gamma_k) = \frac{1}{\gamma_0}\begin{pmatrix} -\rho_1 & 1 & 0 & \cdots & 0 \\ -\rho_2 & 0 & 1 & \cdots & 0 \\ \vdots & \vdots & \vdots & \vdots & \vdots \\ -\rho_h & 0 & 0 & \cdots & 1 \end{pmatrix},$$

and

$$\mathbf{g}(\gamma_0, \ldots, \gamma_k) = (\rho_1, \ldots, \rho_k)^T,$$
$$\mathbf{g}(\hat{\gamma}_0, \ldots, \hat{\gamma}_k) = (\hat{\rho}_1, \ldots, \hat{\rho}_k)^T.$$

Without loss of generality, we assume the mean $\mu = 0$. By Theorem 8.11 and Proposition 8.3, $(\hat{\rho}_1, \ldots, \hat{\rho}_k)^T$ is $\text{AN}((\rho_1, \ldots, \rho_k)^T, \frac{1}{n}UGU^T)$ and it is easy to check that $W = UGU^T$. □

8.6   ESTIMATION FOR ARMA MODELS

We will choose an ARMA(p, q) model,

$$Y_n - \sum_1^p \varphi_k Y_{n-k} = Z_n + \sum_1^q \theta_k Z_{n-k} \quad (n \in \mathbb{Z}),$$

$$\{Z_n\}_{n \in \mathbb{Z}_+} \sim \text{WN}(0, \sigma^2),$$

to express an observed stationary time series, i.e., we will choose the orders p, q, the coefficients $\varphi_k (k = 1, \ldots, p)$ and $\theta_k (k = 1, \ldots, q)$, and the white noise variance σ^2 in the ARMA(p, q) model.

8.6.1   General Linear Model

Given a general linear model

$$\mathbf{X} = Y\theta + \mathbf{Z},$$

where $\mathbf{X} = (X_1, \ldots, X_n)^T$ is of observations, $\theta = (\theta_1, \ldots, \theta_m)^T$ is of parameter values, $Y = (y_{ij})_{i=1,\ldots,n; j=1,\ldots,m}$ is an $n \times m (m < n)$ matrix, and $\mathbf{Z} = (Z_1, \ldots, Z_n)^T$, we will estimate parameters θ.

In the case $\mathbf{Z} = 0$, the linear model $\mathbf{X} = Y\theta$. Let $\mathbf{Y}_l = (y_{1l}, \ldots, y_{nl})^T$ ($l = 1, \ldots, m$) be the lth column of the matrix Y. Then

$$\mathbf{X} = \sum_1^m \theta_k \mathbf{Y}_k.$$

We want to choose parameters $\hat{\theta}_1, \ldots, \hat{\theta}_m$ such that for any $\theta_1, \ldots, \theta_m \in \mathbb{R}$,

$$\left\| \mathbf{X} - \sum_1^m \hat{\theta}_k \mathbf{Y}_k \right\| \leq \left\| \mathbf{X} - \sum_1^m \theta_k \mathbf{Y}_k \right\|,$$

i.e., the $\mathbb{R}^n$-norm of the difference $\mathbf{X} - \sum_1^m \theta_k \mathbf{Y}_k$ attains the minimal value. Denote $M = \text{span}\{\mathbf{Y}_1, \ldots, \mathbf{Y}_m\}$. By the orthogonality principle,

$$\text{Proj}_M \mathbf{X} = \sum_1^m \hat{\theta}_k \mathbf{Y}_k = Y\hat{\theta} \quad (\hat{\theta} = (\hat{\theta}_1, \ldots, \hat{\theta}_m)^T),$$

$$(\mathbf{X} - \text{Proj}_M \mathbf{X}) \perp \mathbf{Y}_l \quad (l = 1, \ldots, m).$$

This implies that

$$(\mathbf{X}, \mathbf{Y}_l) = (\text{Proj}_M \mathbf{X}, \mathbf{Y}_l) = (Y\hat{\theta}, \mathbf{Y}_l) \quad (l = 1, \ldots, m),$$

i.e., $\mathbf{Y}_l^T \mathbf{X} = \mathbf{Y}_l^T Y\hat{\theta}$ ($l = 1, \ldots, m$). So

$$Y^T \mathbf{X} = Y^T Y\hat{\theta}.$$

If $Y^T Y$ is nonsingular, then $\hat{\theta}$ and $\text{Proj}_M \mathbf{X}$ can be represented by the matrix Y and the vector $\mathbf{X}$ as

$$\hat{\theta} = (Y^T Y)^{-1} Y^T \mathbf{X},$$

$$\text{Proj}_M \mathbf{X} = Y(Y^T Y)^{-1} Y^T \mathbf{X}.$$

When $\{\mathbf{Y}_l\}_{l=1,\ldots,m}$ are m orthogonal vectors of $\mathbb{R}^n$, the matrix $Y^T Y$ is an $m \times m$ unit matrix, and so

$$\text{Proj}_M \mathbf{X} = YY^T \mathbf{X} = \sum_1^m (\mathbf{Y}_l^T \mathbf{X}) \mathbf{Y}_l = \sum_1^m (\mathbf{X}, \mathbf{Y}_l) \mathbf{Y}_l.$$

This is a generalization of the well-known orthogonal expansion formula.

In the case $\{Z_k\}_{k=1,\ldots,n} \sim \text{IID}(0, \sigma^2)$, the linear model $\mathbf{X} = Y\theta + \mathbf{Z}$ satisfies

$$X_i = \sum_1^m y_{ij}\theta_j + Z_i \quad (i = 1, \ldots, n).$$

This implies that

$$E[X_i] = \sum_1^m y_{ij}\theta_j,$$

$$\text{Var } X_i = \sigma^2 \quad (i = 1, \ldots, n),$$

$$E[X_i X_j] = 0 \quad (i \neq j),$$

and so $\mathbf{X} \sim N(Y\theta, \sigma^2 I_n)$, where I_n is the n-dimensional unit matrix. The *estimator* of the parameter θ for the linear model is still defined as

$$\hat{\theta} = (Y^T Y)^{-1} Y^T \mathbf{X}.$$

It is clear that

$$E[\hat{\theta}] = (Y^T Y)^{-1} Y^T E[\mathbf{X}] = (Y^T Y)^{-1} (Y^T Y)\theta = \theta,$$

$$E[\hat{\theta}\hat{\theta}^T] = (Y^T Y)^{-1} Y^T E[\mathbf{X}\mathbf{X}^T] Y (Y^T Y)^{-1}.$$

Notice that

$$E[\mathbf{X}\mathbf{X}^T] = E[(Y\theta + \mathbf{Z})(\theta^T Y^T + \mathbf{Z}^T)] = E[Y\theta\theta^T Y^T] + E[\mathbf{Z}\mathbf{Z}^T]$$

$$= Y\theta\theta^T Y^T + \sigma^2 I_n.$$

Then $E[\hat{\theta}\hat{\theta}^T] = \theta\theta^T + \sigma^2 (Y^T Y)^{-1}$. So the covariance matrix of $\hat{\theta}$ is

$$\Sigma = E[\hat{\theta}\hat{\theta}^T] - E[\hat{\theta}](E[\hat{\theta}])^T = \sigma^2 (Y^T Y)^{-1},$$

and so $\hat{\theta} \sim N(\theta, \sigma^2 (Y^T Y)^{-1})$.

The estimator $\hat{\theta}$ is called the *linear regression estimator* of the parameter vector θ.

8.6.2 Estimation for AR(p) Processes

In the Yule-Walker equation (8.9), replacing the covariance γ_j by the sample covariance $\hat{\gamma}_j$, we get

$$
\begin{cases}
\sum_1^p \hat{\varphi}_k \hat{\gamma}_{j-k} = \hat{\gamma}_j & (j = 1, \ldots, p), \\
\hat{\sigma}^2 = \hat{\gamma}_0 - \sum_1^p \hat{\varphi}_k \hat{\gamma}_k.
\end{cases}
$$

This system of equations is called the Yule-Walker equations of estimators, where $\hat{\Phi} = (\hat{\varphi}_1, \ldots, \hat{\varphi}_p)^T$ is the estimator of $\Phi = (\varphi_1, \ldots, \varphi_p)^T$ and $\hat{\sigma}$ is the estimator of σ.

For convenience, we rewrite the AR(p) process in the form

$$
\mathbf{Y} = \widetilde{Y}\Phi + \mathbf{Z}, \tag{8.28}
$$

where $\mathbf{Y} = (Y_1, \ldots, Y_n)^T$, $\widetilde{Y}$ is the $n \times p$ matrix

$$
\widetilde{Y} = \begin{pmatrix}
Y_0 & Y_{-1} & \ldots & Y_{1-p} \\
Y_1 & Y_0 & \ldots & Y_{2-p} \\
\vdots & \vdots & \ldots & \vdots \\
Y_{n-1} & Y_{n-2} & \ldots & Y_{n-p}
\end{pmatrix},
$$

$\Phi = (\varphi_1, \ldots, \varphi_p)^T$, and $\mathbf{Z} = (Z_1, \ldots, Z_n)^T$, where $\{Z_l\}_{l \in \mathbb{Z}} \sim \mathrm{IID}(0, \sigma^2)$.

Similar to the general linear model in subsection 8.6.1 we introduce the linear regression estimate Φ^* of Φ defined by

$$
\Phi^* = (\widetilde{Y}^T \widetilde{Y})^{-1} \widetilde{Y}^T \mathbf{Y}. \tag{8.29}
$$

Since Φ^* depends on the values $Y_{1-p}, Y_{2-p}, \ldots, Y_n$, the vector Φ^* is not an estimator of Φ.

Notice that Φ^* is the product of $(\widetilde{Y}^T \widetilde{Y})^{-1}$ and $\widetilde{Y}^T \mathbf{Y}$ and the ijth element of $n^{-1}(\widetilde{Y}^T \widetilde{Y})$:

$$
\frac{1}{n} \sum_{k=1-i}^{n-i} Y_k Y_{k+i-j} \to \gamma_{i-j}.
$$

Then

$$
\frac{1}{n}(\widetilde{Y}^T \widetilde{Y}) \xrightarrow{p} \Sigma_p (n \to \infty),
$$

where $\Sigma_p = (\gamma_{i-j})_{i,j=1,\ldots,p}$.

Theorem 8.12. Let $\{Y_n\}_{n \in \mathbb{Z}_+}$ be a causal AR(p) process satisfying

$$
Y_n - \sum_1^p \varphi_k Y_{n-k} = Z_n \quad (n \in \mathbb{Z}),
$$

$$
\{Z_n\}_{n \in \mathbb{Z}_+} \sim \mathrm{IID}(0, \sigma^2).
$$

Let $\hat{\Phi}$ be the Yule-Walker estimator of Φ. Then $\sqrt{n}(\hat{\Phi} - \Phi)$ is $AN(0, \sigma^2\Sigma_p^{-1})$, where Σ_p is the covariance matrix $(\gamma_{i-j})_{i,j=1,\ldots,p}$. Let $\tilde{\sigma}^2$ be the Yule-Walker estimator of σ^2. Then $\hat{\sigma}^2 \xrightarrow{\mathrm{p}} \sigma^2$.

The proof of this theorem needs the following several propositions.

Proposition 8.5. *Let $U_k = (Y_{k-1}, \ldots, Y_{k-p})^{\mathrm{T}}Z_k$. Then*

$$E[U_k U_{k+\tau}^{\mathrm{T}}] = \begin{cases} \sigma^2\Sigma_p, & \tau = 0, \\ 0, & \tau \neq 0. \end{cases}$$

Proof. Since $\{Y_n\}$ is a causal AR(p) process,

$$Y_k = \sum_0^\infty \psi_j Z_{k-j} \quad (k \in \mathbb{Z}), \tag{8.30}$$

For $\tau \neq 0$,

$$U_k U_{k+\tau}^{\mathrm{T}} = (\alpha_{ij})_{i,j=1,\ldots,p}, \quad \text{where } \alpha_{ij} = Y_{k-i}Y_{k-j+\tau}Z_kZ_{k+\tau}.$$

Without loss of generality, assume that $\tau > 0$. Then $Y_{k-i}Y_{k-j+\tau}Z_k$ and $Z_{k+\tau}$ are independent. So

$$E[\alpha_{ij}] = E[Y_{k-i}Y_{k-j+\tau}Z_k]E[Z_{k+\tau}] = 0 \quad (i,j = 1,\ldots,p).$$

Therefore, $E[U_k U_{k+\tau}^{\mathrm{T}}] = 0$ $(\tau \neq 0)$.

For $\tau = 0$,

$$U_k U_{k+\tau}^{\mathrm{T}} = (\beta_{ij})_{i,j=1,\ldots,p}, \quad \text{where } \beta_{ij} = Y_{k-i}Y_{k-j}Z_k^2.$$

By (8.30), $Y_{k-i}Y_{k-j}$ and Z_k^2 are independent. This implies that

$$E[\beta_{ij}] = E[Y_{k-i}Y_{k-j}]E[Z_k^2] = \gamma_{|i-j|}\sigma^2 \quad (i,j = 1,\ldots,p).$$

Therefore, $E[U_k U_{k+\tau}^{\mathrm{T}}] = \sigma^2\Sigma_p$. $\qquad\square$

Proposition 8.6. *Let*

$$Y_k^{(m)} = \sum_0^m \psi_j Z_{k-j},$$

$$U_k^{(m)} = (Y_{k-1}^{(m)}, \ldots, Y_{k-p}^{(m)})^{\mathrm{T}} Z_k.$$

Then, for any $\alpha \in \mathbb{R}^p$,

$$\frac{1}{\sqrt{n}} \sum_1^n \alpha^{\mathrm{T}} U_j^{(m)} \xrightarrow{\mathrm{d}} \alpha^{\mathrm{T}} V^{(m)} \quad (n \to \infty),$$

$$V^{(m)} \xrightarrow{\mathrm{d}} V \quad (m \to \infty),$$

where $V \sim N(0, \sigma^2\Sigma_p)$.

Proof. Take $\alpha \in \mathbb{R}^p$. The sequence $\{\alpha^{\mathrm{T}} U_j^{(m)}\}_{j\in\mathbb{Z}}$ is a stationary m-dependent sequence. We compute its covariance function $\gamma_\tau^{(m)}$.

For $\tau = 0$,

$$\gamma_0^{(m)} = \text{Var}(\alpha^{\text{T}} U_0^{(m)}) = \alpha^{\text{T}} E[U_0^{(m)} (U_0^{(m)})^{\text{T}}]\alpha = \sigma^2(\alpha^{\text{T}} \Sigma_p^{(m)} \alpha),$$

where $\Sigma_p^{(m)}$ is the covariance matrix of $Y_1^{(m)}, \ldots, Y_p^{(m)}$.

For $\tau > 0$,

$$\gamma_\tau^{(m)} = \text{Cov}(\alpha^{\text{T}} U_0^{(m)}, \alpha^{\text{T}} U_\tau^{(m)}) = \alpha^{\text{T}} E[U_0^{(m)} (U_\tau^{(m)})^{\text{T}}]\alpha = 0.$$

According to Theorem 8.9, it follows that for any $\alpha \in \mathbb{R}^p$,

$$\frac{1}{n} \sum_1^n \alpha^{\text{T}} U_j^{(m)} \quad \text{is AN} \left(0, \frac{h_m}{n}\right),$$

where $h_m = \gamma_0^{(m)} + 2 \sum_1^m \gamma_j^{(m)} = \sigma^2(\alpha^{\text{T}} \Sigma_p^{(m)} \alpha)$. Therefore,

$$\frac{1}{\sqrt{n}} \sum_1^n \alpha^{\text{T}} U_j^{(m)} \xrightarrow{\text{d}} W^{(m)}(\alpha) \quad (n \to \infty),$$

$$W^{(m)}(\alpha) \sim N(0, \sigma^2(\alpha^{\text{T}} \Sigma_p^{(m)})\alpha). \tag{8.31}$$

Denote p-dimensional unit vectors by $\{\alpha^{(i)}\}_{i=1,\ldots,p}$, where

$$\alpha^{(1)} = (1, 0, \ldots, 0)^{\text{T}},$$
$$\alpha^{(2)} = (0, 1, \ldots, 0)^{\text{T}},$$
$$\vdots$$
$$\alpha^{(p)} = (0, 0, \ldots, 1)^{\text{T}}.$$

Let

$$V_i^{(m)} = W^{(m)}(\alpha^{(i)}) \quad (i = 1, \ldots, p),$$
$$V^{(m)} = (V_1^{(m)}, \ldots, V_p^{(m)})^{\text{T}}.$$

Since $\frac{1}{\sqrt{n}} \sum_{j=1}^n (\alpha^{(i)})^{\text{T}} U_j^{(m)} \xrightarrow{\text{d}} W^{(m)}(\alpha^{(i)})$, it follows that

$$W^{(m)}(\alpha) = \sum_1^p \alpha_i W^{(m)}(\alpha^{(i)}) = \alpha^{\text{T}} V^{(m)}, \tag{8.32}$$

and so the covariance matrix of $W^{(m)}(\alpha)$ is

$$E[W^{(m)}(\alpha)(W^{(m)}(\alpha))^{\text{T}}] = E[\alpha^{\text{T}} V^{(m)} (V^{(m)})^{\text{T}} \alpha] = \alpha^{\text{T}} E[V^{(m)} (V^{(m)})^{\text{T}}]\alpha.$$

Comparing this with $W^{(m)}(\alpha) \sim N(0, \sigma^2(\alpha^{\text{T}} \Sigma_p^{(m)} \alpha))$, for any $\alpha \in \mathbb{R}^p$, we get

$$\alpha^{\text{T}} \text{Cov}(V^{(m)}, V^{(m)})\alpha = \alpha^{\text{T}} \sigma^2 \Sigma_p^{(m)} \alpha,$$

and so

$$\text{Cov}(V^{(m)}, V^{(m)}) = \sigma^2 \Sigma_p^{(m)}.$$

From this and (8.31) and (8.32), it follows that

$$\frac{1}{\sqrt{n}} \sum_1^n \alpha^{\mathrm{T}} U_j^{(m)} \xrightarrow{\mathrm{d}} \alpha^{\mathrm{T}} V^{(m)} \quad (n \to \infty),$$

where $V^{(m)} \sim \mathrm{N}(0, \sigma^2 \Sigma_p^{(m)})$.

Now we prove that $\Sigma_p^{(m)} \to \Sigma_p \ (m \to \infty)$.

From $\sum_1^\infty |\psi_k| < \infty$, it follows that $\sum_1^\infty \psi_k^2 < \infty$. By (8.30),

$$E[|Y_k^{(m)} - Y_k|^2] = E\left[\left(\sum_{j=m+1}^\infty \psi_j Z_{k-j}\right)^2\right] = \sigma^2 \sum_{m+1}^\infty \psi_j^2.$$

So $Y_k^{(m)} \xrightarrow{\mathrm{p}} Y_k \ (m \to \infty)$.

Since

$$|E[Y_i^{(m)} Y_j^{(m)}] - E[Y_i Y_j]| \le |E[(Y_i^{(m)} - Y_i) Y_j^{(m)}]| + |E[Y_i(Y_j^{(m)} - Y_j)]|$$
$$\le (E[(Y_i^{(m)} - Y_i)^2] E[(Y_j^{(m)})^2])^{1/2} + (E[(Y_j^{(m)} - Y_j)^2] E[Y_i^2])^{1/2},$$

from

$$E[(Y_j^{(m)})^2] = \sum_1^m \psi_j^2 \le \sum_1^\infty \psi_j^2,$$

$$E[Y_j^2] = \sum_1^\infty \psi_j^2,$$

it follows that

$$E[Y_i^{(m)} Y_j^{(m)}] \to E[Y_i Y_j] \quad (m \to \infty),$$

i.e., $\Sigma_p^{(m)} \to \Sigma_p \ (m \to \infty)$. From this and $V^{(m)} \sim \mathrm{N}(0, \sigma^2 \Sigma_p^{(m)})$, we get

$$V^{(m)} \xrightarrow{\mathrm{d}} V \quad (m \to \infty),$$
$$V \sim \mathrm{N}(0, \sigma^2 \Sigma_p).$$

$\square$

Proposition 8.7. *Let*

$$F_{n,m} = \frac{1}{\sqrt{n}} \alpha^{\mathrm{T}} \sum_1^n U_j^{(m)},$$

$$F_n = \frac{1}{\sqrt{n}} \alpha^{\mathrm{T}} \sum_1^n U_j,$$

where $\alpha \in \mathbb{R}^p$. Then, for any $\delta > 0$,

$$\lim_{m \to \infty} \overline{\lim_{n \to \infty}} P(|F_{n,m} - F_n| \ge \delta) = 0.$$

Proof. Using the Chebyshev inequality, we get

$$P(|F_{nm} - F_n| \geq \delta) \leq \delta^{-2} E[(F_{n,m} - F_n)^2].$$

A direct computation shows that

$$E[(F_{n,m} - F_n)^2] = \frac{1}{n} \alpha^T \left(\sum_{i=1}^{n} \sum_{j=1}^{n} E[(U_i^{(m)} - U_i)(U_j^{(m)} - U_j)^T] \right) \alpha.$$

By definitions of $U_k^{(m)}$ and U_k,

$$E[U_i^{(m)}(U_j^{(m)})^T] = E[U_i(U_j^{(m)})^T] = E[U_i U_j^T] = 0 \quad (i \neq j).$$

This implies that

$$E[(F_{n,m} - F_n)^2] = \frac{1}{n} \alpha^T \left(\sum_{1}^{n} E[(U_j^{(m)} - U_j)(U_j^{(m)} - U_j)^T] \right) \alpha.$$

By the known results

$$E[U_j^{(m)} U_j^T] = E[U_j (U_j^{(m)})^T] = E[U_j^{(m)} (U_j^{(m)})^T] = \sigma^2 \Sigma_p^{(m)},$$

$$E[U_j U_j^T] = \sigma^2 \Sigma_p,$$

it follows that

$$E[(F_{n,m} - F_n)^2] = \sigma^2 \alpha^T (\Sigma_p - \Sigma_p^{(m)}) \alpha.$$

From this and $\Sigma_p^{(m)} \to \Sigma_p$ $(m \to \infty)$, it follows that $E[(F_{n,m} - F_n)^2] \to 0$ as $m \to \infty$ for any n. By the Chebyshev inequality, Proposition 8.7 is derived. $\square$

Proposition 8.8. $\sqrt{n}(\Phi^* - \Phi)$ *is* $AN(0, \sigma^2 \Sigma_p^{-1})$, *where the linear regression estimate* Φ^* *is stated in* (8.29).

Proof. By (8.28) and (8.29), we get

$$\sqrt{n}(\Phi^* - \Phi) = \sqrt{n}((\tilde{Y}^T \tilde{Y})^{-1} \tilde{Y}^T (\tilde{Y}\Phi + Z) - \Phi) = n(\tilde{Y}^T \tilde{Y})^{-1} \left(\frac{1}{\sqrt{n}} \tilde{Y}^T Z \right).$$

By Propositions 8.6 and 8.7 and Theorem 8.7, for any $\alpha \in \mathbb{R}^p$, we get

$$V \sim N(0, \sigma^2 \Sigma_p),$$

$$\alpha^T \frac{1}{\sqrt{n}} \sum_{1}^{n} U_j \overset{d}{\to} \alpha^T V \quad (n \to \infty).$$

By the definition of U_k and the Gramer-Wold device,

$$\frac{1}{\sqrt{n}} \tilde{Y}^T Z = \frac{1}{\sqrt{n}} \sum_{1}^{n} U_j \overset{d}{\to} V \quad (n \to \infty),$$

$$V \sim N(0, \sigma^2 \Sigma_p).$$

Again, by $\frac{1}{n}(\widetilde{Y}^T\widetilde{Y}) \xrightarrow{p} \Sigma_p$ and $\det\Sigma_p \neq 0$, it follows that

$$\sqrt{n}(\Phi^* - \Phi) = n(\widetilde{Y}^T\widetilde{Y})^{-1}\left(\frac{1}{\sqrt{n}}\widetilde{Y}^T Z\right) \xrightarrow{d} \Sigma_p^{-1} V,$$

$$\Sigma_p^{-1} V \sim N(0, \sigma^2\Sigma_p^{-1}).$$

$\square$

Proof of Theorem 8.12. *We will compute* $\hat{\Phi} - \Phi^*$.
The matrix form of the equation of the Yule-Walker estimator is

$$\hat{\Sigma}_p\hat{\Phi} = \hat{\Gamma}_p, \quad \text{where } \hat{\Gamma}_p = (\hat{\gamma}_1, \ldots, \hat{\gamma}_p)^T.$$

Since the covariance matrix $\hat{\Sigma}_p$ *is nonsingular,* $\hat{\Phi} = \hat{\Sigma}_p^{-1}\hat{\Gamma}_p$. *From this with*
(8.29),

$$\sqrt{n}(\hat{\Phi} - \Phi^*) = \sqrt{n}\left(\hat{\Sigma}_p^{-1}\hat{\Gamma}_p - (\widetilde{Y}^T\widetilde{Y})^{-1}\widetilde{Y}^T\mathbf{Y}\right)$$

$$= \hat{\Sigma}_p^{-1}\sqrt{n}\left(\hat{\Gamma}_p - \frac{1}{n}\widetilde{Y}^T\mathbf{Y}\right) + \sqrt{n}\left(\hat{\Sigma}_p^{-1} - n(\widetilde{Y}^T\widetilde{Y})^{-1}\right)\left(\frac{1}{n}\widetilde{Y}^T\mathbf{Y}\right)$$

$$= \hat{\Sigma}_p^{-1}S_n + Q_n\left(\frac{1}{n}\widetilde{Y}^T\mathbf{Y}\right), \tag{8.33}$$

where

$$S_n = \sqrt{n}\left(\hat{\Gamma}_p - \frac{1}{n}\widetilde{Y}^T\mathbf{Y}\right),$$

$$Q_n = \sqrt{n}\left(\hat{\Sigma}_p^{-1} - n(\widetilde{Y}^T\widetilde{Y})^{-1}\right).$$

Now we prove that $S_n \xrightarrow{p} 0$ *and* $Q_n \xrightarrow{p} 0$.
Let $S_n = (S_{n1}, \ldots, S_{nn})$. *By* (8.23), *the* τ*th component of* $\hat{\Gamma}_p$ *is*

$$\frac{1}{n}\sum_{k=1}^{n-\tau}(Y_k - \overline{Y}_n)(Y_{k+\tau} - \overline{Y}_n).$$

Notice that the τ*th component of* $\widetilde{Y}^T\mathbf{Y}$ *is*

$$\sum_{j=1}^{n} Y_{j-\tau}Y_j = \sum_{k=1-\tau}^{n-\tau} Y_k Y_{k+\tau}.$$

So the τ*th component of* S_n *is*

$$S_{n\tau} = \frac{1}{\sqrt{n}}\left(\sum_{k=1}^{n-\tau}(Y_k - \overline{Y}_n)(Y_{k+\tau} - \overline{Y}_n) - \sum_{k=1-\tau}^{n-\tau} Y_k Y_{k+\tau}\right)$$

$$= \frac{1}{\sqrt{n}}\sum_{k=1}^{n-\tau}(Y_k Y_{k+\tau} + \overline{Y}_n^2 - Y_{k+\tau}\overline{Y}_n - Y_k\overline{Y}_n) - \frac{1}{\sqrt{n}}\sum_{k=1-\tau}^{n-\tau} Y_k Y_{k+\tau}$$

$$= -\frac{1}{\sqrt{n}} \sum_{k=1-\tau}^{0} Y_k Y_{k+\tau} + \frac{1}{\sqrt{n}}(n - \tau)\overline{Y}_n^2 - \frac{1}{\sqrt{n}} \sum_{k=1}^{n-\tau}(Y_k + Y_{k+\tau})\overline{Y}_n$$

$$= S_{n\tau}^{(1)} + S_{n\tau}^{(2)} + S_{n\tau}^{(3)}.$$

By the central limit theorem, it follows that $\overline{Y}_n$ is $AN(0, \frac{1}{n} \sum_j \gamma_j)$, and so

$$S_{n\tau}^{(2)} \overset{P}{\to} 0,$$

$$S_{n\tau}^{(3)} \overset{P}{\to} 0.$$

Clearly, $S_{n\tau}^{(1)} \overset{P}{\to} 0$. Therefore, $S_{n\tau} \overset{P}{\to} 0$ $(\tau = 1, \ldots, p)$, and so $S_n \overset{P}{\to} 0$.

Consider the norm of the matrix Q_n. Define the norm of a $p \times p$ matrix $D = (d_{ij})_{i,j=1,\ldots,p}$ as

$$\|D\| = \left(\sum_{i=1}^{p} \sum_{j=1}^{p} d_{ij}^2 \right)^{1/2}.$$

If $D = AB$, where $A = (\alpha_{ij})_{i,j=1,\ldots,p}$ and $B = (\beta_{ij})_{i,j=1,\ldots,p}$, then application of the Schwarz inequality gives

$$d_{ij}^2 = \left(\sum_{l=1}^{p} \alpha_{il}\beta_{lj} \right)^2 \leq \left(\sum_{l=1}^{p} \alpha_{il}^2 \right) \left(\sum_{k=1}^{p} \beta_{kj}^2 \right),$$

and so

$$\|D\|^2 \leq \sum_{i=1}^{p} \sum_{j=1}^{p} d_{ij}^2 \leq \left(\sum_{i=1}^{p} \sum_{l=1}^{p} \alpha_{il}^2 \right) \left(\sum_{j=1}^{p} \sum_{k=1}^{p} \beta_{kj}^2 \right) = \|A\|^2 \|B\|^2.$$

Let

$$D_n = (d_{ij}^{(n)})_{i,j=1,\ldots,p}$$

be a sequence of p^2-dimensional random variables. Then $D_n \overset{P}{\to} 0$ $(n \to \infty)$ if and only if $\|D_n\| \overset{P}{\to} 0$ $(n \to \infty)$. Therefore, the norm of the matrix Q_n is

$$\sqrt{n}\|\hat{\Sigma}_p^{-1} - n(\widetilde{Y}^{\mathsf{T}}\widetilde{Y})^{-1}\| = \sqrt{n} \left\| \hat{\Sigma}_p^{-1} \left(\frac{1}{n}\widetilde{Y}^{\mathsf{T}}\widetilde{Y} - \hat{\Sigma}_p \right) n(\widetilde{Y}^{\mathsf{T}}\widetilde{Y})^{-1} \right\|$$

$$\leq \sqrt{n}\|\hat{\Sigma}_p^{-1}\| \left\| \frac{1}{n}\widetilde{Y}^{\mathsf{T}}\widetilde{Y} - \hat{\Sigma}_p \right\| \|n(\widetilde{Y}^{\mathsf{T}}\widetilde{Y})^{-1}\|.$$

From $S_n \overset{P}{\to} 0$, it follows that $\sqrt{n}(\hat{\Sigma}_p - \frac{1}{n}\widetilde{Y}^{\mathsf{T}}\widetilde{Y}) \overset{P}{\to} 0$. Notice that

$$\hat{\Sigma}_p^{-1} \overset{P}{\to} \Sigma_p^{-1} \quad (n \to \infty),$$

$$n(\widetilde{Y}^{\mathsf{T}}\widetilde{Y})^{-1} \overset{P}{\to} \Sigma_p^{-1} \quad (n \to \infty).$$

Then $\sqrt{n}\|\hat{\Sigma}_p^{-1} - n(\widetilde{Y}^T\widetilde{Y})^{-1}\| \xrightarrow{p} 0$ $(n \to \infty)$, *i.e.*, $Q_n \xrightarrow{p} 0$.

Finally, by (8.33), $\sqrt{n}(\hat{\Phi} - \Phi^*) \xrightarrow{p} 0$. *Combining this with Proposition 8.8, we have*

$$\sqrt{n}(\hat{\Phi} - \Phi) \quad is \; AN(0, \sigma^2\Sigma_p^{-1}).$$

From $\hat{\gamma}_p \xrightarrow{p} \gamma_p$ *and* $\hat{\Phi} \xrightarrow{p} \Phi$, *it follows by the Yule-Walker equation that*

$$\hat{\sigma}^2 \xrightarrow{p} \gamma_0 - \sum_1^p \varphi_k\gamma_k = \sigma^2 \quad (n \to \infty).$$

8.6.3 Estimation for ARMA(p, q) Processes

We have an MA(s) process

$$Y_n = Z_n + \theta_1 Z_{n-1} + \cdots + \theta_s Z_{n-s} \quad (n \in \mathbb{Z}),$$
$$\{Z_n\}_{n\in\mathbb{Z}} \sim \text{IID}(0, \sigma^2),$$

where $E[Z_n^4] < \infty$, and the samples y_k ($k = 1, \ldots, \nu$) with sample covariance estimators $\hat{\gamma}_k$ ($k = 1, \ldots, \nu$), where ν is the number of samples. The innovation estimates $\hat{\theta}_{s1}, \ldots, \hat{\theta}_{ss}, \hat{d}_s$ of $\theta_1, \ldots, \theta_s, \sigma^2$ satisfy the equation

$$\hat{\theta}_{s,s-l} = \frac{1}{\hat{d}_l}\left(\hat{\gamma}_{s-l} - \sum_{j=0}^{l-1}\hat{\theta}_{l,l-j}\hat{\theta}_{s,s-j}\hat{d}_j\right) \quad (l = 0, \ldots, s-1),$$

where

$$\hat{d}_s = \hat{\gamma}_0 - \sum_{j=0}^{s-1}\hat{\theta}_{s,s-l}\hat{d}_j,$$
$$\hat{d}_0 = \hat{\gamma}_0.$$

Then $\hat{d}_m \xrightarrow{p} \sigma^2$ ($\nu \to \infty$) and for each k,

$$\sqrt{\nu}(\hat{\theta}_{s1} - \theta_1, \hat{\theta}_{s2} - \theta_2, \ldots, \hat{\theta}_{sk} - \theta_k)^T \xrightarrow{d} N(0, \Lambda) \quad (\nu \to \infty),$$

where the matrix Λ is

$$\Lambda = (a_{jl})_{j,l=1,\ldots,k}, \quad a_{jl} = \sum_{\mu=1}^{\min\{j,l\}}\psi_{j-\mu}\psi_{j-l}.$$

Suppose that $\{Y_n\}_{n\in\mathbb{Z}}$ is a zero-mean ARMA(p, q) process satisfying

$$Y_n - \sum_1^p \varphi_k Y_{n-k} = Z_n + \sum_1^q \theta_k Z_{n-k} \quad (n \in \mathbb{Z}),$$
$$\{Z_n\}_{n\in\mathbb{Z}} \sim \text{WN}(0, \sigma^2).$$

If it has a causal solution

$$Y_n = \sum_0^\infty \psi_k Z_{n-k} \quad (n \in \mathbb{Z}),$$

then

$$\psi_0 = 1,$$

$$\psi_k = \theta_k + \sum_{l=1}^{\min\{k,p\}} \varphi_l \psi_{k-l} \quad (k \in \mathbb{Z}_+), \tag{8.34}$$

where $\theta_k = 0$ $(k > q)$ and $\varphi_k = 0$ $(k > p)$. Replacing ψ_k by $\hat{\theta}_{sk}$ in (8.34), we get

$$\hat{\theta}_{sk} = \theta_k + \sum_{l=1}^{\min\{k,p\}} \varphi_l \hat{\theta}_{s,k-l} \quad (k = 1, \ldots, q+p).$$

Solving these equations, we get

$$(\hat{\theta}_{s,q+1}, \ldots, \hat{\theta}_{s,q+p})^{\mathrm{T}} = (\hat{\theta}_{s,q+l-k})_{l,k=1,\ldots,p} (\hat{\varphi}_1, \ldots, \hat{\varphi}_p)^{\mathrm{T}},$$

$$\hat{\theta}_k = \hat{\theta}_{sk} - \sum_{l=1}^{\min(k,p)} \hat{\varphi}_l \hat{\theta}_{s,k-l} \quad (k = 1, \ldots, q).$$

8.7 ARIMA MODELS

We extend the ARMA models to ARIMA models for non-stationary time series. The strength of ARIMA models lies in their ability to reveal complex structures of temporal interdependence in time series. It has also been shown that ARIMA models are highly efficient in short-term forecasting. For $d = 0, 1, \ldots$, let

$$\widetilde{Y}_n = (1 - T)^d Y_n = \sum_{k=0}^d (-1)^k \frac{d!}{k!(d-k)!} Y_{n-k}.$$

If $\{\widetilde{Y}_n\}_{n\in\mathbb{Z}}$ is a causal ARMA(p,q) process, then $\{Y_n\}_{n\in\mathbb{Z}}$ is called an *ARIMA(p,d,q) process*. In detail, if a time series $\{Y_n\}_{n\in\mathbb{Z}}$ satisfies

$$\varphi(T)(1 - T)^d Y_n = \theta(T) Z_n \quad (n \in \mathbb{Z}),$$

$$\{Z_n\}_{n\in\mathbb{Z}} \sim \mathrm{WN}(0, \sigma^2), \tag{8.35}$$

where $\varphi(z), \theta(z)$ are polynomials of degree p, q, respectively, and $\varphi(z) \neq 0$ $(|z| \leq 1)$, then $\{Y_n\}_{n\in\mathbb{Z}}$ is an ARIMA (p, d, q) process. When $d = 0$, $\{Y_n\}_{n\in\mathbb{Z}}$ is a causal ARMA process.

Let $\varphi^*(z) = \varphi(z)(1 - z)^d$. Then (8.35) can be written as

$$\varphi^*(T) Y_n = \theta(T) Z_n \quad (n \in \mathbb{Z}), \quad \text{where } \varphi^*(T) = \varphi(T)(I - T)^d.$$

Since $\varphi^*(z)$ is a polynomial of degree $p + d$ and for $d > 1$, $\varphi^*(z)$ has a zero on $|z| = 1$, Y_n cannot be solved out as in Theorem 8.1. We discuss the solution of (8.35).

Let $\widetilde{Y}_n = (I - T)^d Y_n$ $(n \in \mathbb{Z})$ in (8.35). Then

$$\varphi(T)\widetilde{Y}_n = \theta(T)Z_n \quad (n \in \mathbb{Z}).$$

By Theorem 8.2,

$$\widetilde{Y}_n = \sum_0^\infty \psi_l Z_{n-l}, \tag{8.36}$$

where $\frac{\theta(z)}{\varphi(z)} = \psi(z) = \sum_0^\infty \psi_l z^l$ $(|z| < 1 + \epsilon)$. Now we find Y_n:

(i) In the case $d = 1$, $\widetilde{Y}_n = (I - T)Y_n$ $(n \in \mathbb{Z})$. Notice that $IY_n = Y_n$ and $TY_n = Y_{n-1}$. Then

$$\widetilde{Y}_n = Y_n - Y_{n-1} \quad (n \in \mathbb{Z}),$$

and so $Y_k = Y_{k-1} + \widetilde{Y}_k$ $(k = 1, \ldots, n)$. So

$$Y_n = Y_0 + \sum_1^n \widetilde{Y}_k \quad (n \in \mathbb{Z}_+).$$

(ii) In the case $d = 2$, $\widetilde{Y}_n = (I - T)^2 Y_n$ $(n \in \mathbb{Z})$. This can be decomposed into

$$\widetilde{Y}_n = (I - T)Y_n^*,$$

$$Y_n^* = (I - T)Y_n.$$

From this, imitating the procedure of (i), we get

$$Y_n^* = Y_0^* + \sum_1^n \widetilde{Y}_k,$$

$$Y_n = Y_0 + \sum_1^n Y_k^*,$$

and so

$$Y_n = Y_0 + nY_0^* + \sum_{k=1}^n \sum_{j=1}^k \widetilde{Y}_k.$$

From this, $Y_0^* = Y_0 - Y_{-1}$, and

$$\sum_{j=1}^n \sum_{k=1}^j \widetilde{Y}_k = \sum_{l=1}^n (n - l + 1)\widetilde{Y}_l,$$

it follows that

$$Y_n = (n+1)Y_0 - nY_1 + \sum_{l=1}^{n}(n-l+1)\widetilde{Y}_l,$$

where $\widetilde{Y}_l$ is stated in (8.36).

8.8 MULTIVARIATE ARMA PROCESSES

Consider an m-variate time series $\{\mathbf{Y}_n\}_{n\in\mathbb{Z}}$, where $\mathbf{Y}_n = (Y_{n1}, \ldots, Y_{nm})^{\mathrm{T}}$. The *mean vector* is defined as

$$\mu_n = E[\mathbf{Y}_n] = (\mu_{n1}, \ldots, \mu_{nm})^{\mathrm{T}},$$

where $\mu_{nj} = E[Y_{nj}]$ $(j = 1, \ldots, m)$. Use the notation $\mu := (\mu_1, \ldots, \mu_m)^{\mathrm{T}}$. The *covariance matrix function* is defined as

$$B_{n+\tau,n} = E[(\mathbf{Y}_{n+\tau} - \mu_{n+\tau})(\mathbf{Y}_n - \mu_n)^{\mathrm{T}}] = (\gamma_{kl}(n+\tau, n))_{k,l=1,\ldots,m},$$

where $\gamma_{kl}(n+\tau, n) = E[(Y_{n+\tau,k} - \mu_{n+\tau,k})(Y_{n,l} - \mu_{n,l})]$. Use the notation $B_\tau = (\gamma_{kl}(\tau))_{k,l=1,\ldots,m}$.

The multivariate time series $\{\mathbf{Y}_n\}_{n\in\mathbb{Z}}$ is *stationary* if μ_n and $B_{n+\tau,n}$ are both independent of n.

An m-variate time series $\{\mathbf{Z}_n\}_{n\in\mathbb{Z}}$ is called white noise if

$$E[\mathbf{Z}_n] = (0, \ldots, 0)^{\mathrm{T}},$$
$$E[\mathbf{Z}_n\mathbf{Z}_n^{\mathrm{T}}] = (u_{ij})_{i,j=1,\ldots,m} =: U,$$
$$E[\mathbf{Z}_n\mathbf{Z}_m^{\mathrm{T}}] = O \quad (n \neq m),$$

denoted by $\{\mathbf{Z}_n\}_{n\in\mathbb{Z}} \sim \mathrm{WN}(0, U)$

Suppose that an m-variate stationary time series $\{\mathbf{Y}_n\}_{n\in\mathbb{Z}}$ satisfies

$$\mathbf{Y}_n - \sum_{k=1}^{p}\Phi_k\mathbf{Y}_{n-k} = \mathbf{Z}_n + \sum_{l=1}^{q}\Theta_l\mathbf{Z}_{n-l},$$
$$\{\mathbf{Z}_n\}_{n\in\mathbb{Z}} \sim \mathrm{WN}(\mathbf{0}, U), \tag{8.37}$$

where each Φ_k and Θ_l is a real $m \times m$ matrix. Then $\{\mathbf{Y}_n\}_{n\in\mathbb{Z}}$ is called an *m-variate ARMA process*, and (8.37) is called an *m-variate ARMA equation*. The m-variate ARMA process is an important kind of multivariate stationary process.

Let

$$\Phi(z) = I - \sum_{1}^{p}\Phi_k z^k,$$

$$\Theta(z) = I + \sum_{1}^{q}\Theta_l z^l.$$

Then the m-variate (8.37) is written in the operator form:

$$\Phi(T)\mathbf{Y}_n = \Theta(T)\mathbf{Z}_n,$$

where $T\mathbf{Z}_n = \mathbf{Z}_{n-1}$.

Theorem 8.13. *If* $\det\Phi(z) \neq 0$ ($|z| \leq 1$), *then*

$$\mathbf{Y}_n = \sum_0^\infty \Psi_k \mathbf{Z}_{n-k}, \tag{8.38}$$

where $\Psi(z) = \Phi^{-1}(z)\Theta(z) = \sum_0^\infty \Psi_k z^k$ ($|z| \leq 1$). *If* $\det\Theta(z) \neq 0$ ($|z| \leq 1$), *then*

$$\mathbf{Z}_n = \sum_0^\infty \Lambda_k \mathbf{Y}_{n-k},$$

where $\Lambda(z) = \Theta^{-1}(z)\Phi(z) = \sum_0^\infty \Lambda_k z^k$ ($|z| \leq 1$). *Especially, for a multivariate AR(1) process*

$$\mathbf{Y}_n = \Phi\mathbf{Y}_{n-1} + \mathbf{Z}_n \quad (n \in \mathbb{Z}),$$

if $\det(I - z\Phi) \neq 0$ ($|z| \leq 1$), *then* $\mathbf{Y}_n = \sum_0^\infty \Phi^k \mathbf{Z}_{n-k}$.

From (8.38), the covariance matrix function

$$B_\tau = E[\mathbf{Y}_{n+\tau}\mathbf{Y}_n^{\mathrm{T}}] = \sum_{k=0}^\infty \sum_{l=0}^\infty \Psi_k E[Z_{n+\tau-k}Z_{n-l}^{\mathrm{T}}]\Psi_l^{\mathrm{T}} = \sum_{k=0}^\infty \Psi_{k+\tau} U(\Psi^k)^{\mathrm{T}}.$$

It can be determined by the Yule-Walker equations that

$$B_j - \sum_1^p \Phi_k B_{j-k} = \sum_{k=j}^q \Theta_k U \Psi_{k-j}^{\mathrm{T}} \quad (j = 0, 1, \ldots).$$

The covariance matrix mother function is defined by

$$G(z) = \sum_\tau B_\tau z^\tau = \Psi(z) U \Psi(z^{-1})^{\mathrm{T}}.$$

For a bivariate stationary time series $\{\mathbf{Y}_n\}_{n \in \mathbb{Z}}$, *where* $Y_n = (Y_{n,1}, Y_{n,2})^{\mathrm{T}}$, *with mean* $\mathbf{0}$ *and covariance* $\gamma_{kl}(\tau) = E[Y_{n+\tau+k}Y_{n+l}]$ *satisfying* $\sum_\tau |\gamma_{kl}(\tau)| < \infty (k, l = 1, 2)$, *the function*

$$f_{k,l}(\alpha) = \sum_\tau e^{-i\tau\alpha}\gamma_{kl}(\tau)$$

is called the cross-spectral density of $Y_{n,k}$ *and* $Y_{n,l}$. *The matrix*

$$f(\alpha) = \sum_\tau e^{-i\tau\alpha}B_\tau = \begin{pmatrix} f_{11}(\tau) & f_{12}(\tau) \\ f_{21}(\tau) & f_{22}(\tau) \end{pmatrix}$$

is called the spectral density matrix of $\{\mathbf{Y}_n\}_{n \in \mathbb{Z}}$.

Suppose that $\{\mathbf{Y}_n\}_{n\in\mathbb{Z}}$ is a causal ARMA(p, q) process satisfying

$$\Phi(T)\mathbf{Y}_n = \Theta(T)\mathbf{Z}_n \quad (n \in \mathbb{Z}),$$

$$\{\mathbf{Z}_n\}_{n\in\mathbb{Z}} \sim \text{WN}(\mathbf{0}, U),$$

and

$$\mathbf{Y}_n = \sum_k \Psi_k \mathbf{Z}_{n-k},$$

where $\Phi^{-1}(z)\Theta(z) = \Psi(z) = \sum_0^\infty \Psi_k z^k$ $(|z| \le 1)$. *Therefore,* $\{\mathbf{Y}_n\}_{n\in\mathbb{Z}}$ *has spectral density matrix*

$$f_\mathbf{Y}(\alpha) = \Psi(e^{-i\alpha})B_\tau \Psi^{-1}(e^{-i\alpha}).$$

8.9 APPLICATION IN CLIMATIC AND HYDROLOGICAL RESEARCH

The theory and algorithms of the ARMA models stated in Sections 8.1–8.8 have been directly applied in various climatic and hydrological predictions. Here, we describe some representative case studies.

Since ARMA models can deal with stationary time series well, while wavelet and empirical mode decomposition (EMD) can deal with non-stationary time series well, Karthikeyan and Nagesh Kumar (2013) combined ARMA models with wavelet/EMD analyses for flood modeling. In detail, they used wavelet/EMD analyses to decompose a hydrological time series into independent components with both time and frequency localizations. Then each component series were fit with specific ARMA models to obtain forecasts. Finally, these forecasts were combined to obtain the actual predictions.

Accurate forecasting of the inflow reservoir has a significant importance in water resource management. Valipour et al. (2013) used ARMA models and ARIMA models to forecast the inflow of the Dez dam reservoir in Iran. Inflow of the dam reservoir in the preceding 12 months showed that ARIMA model has less error than the ARMA model.

Gámiz-Fortis et al. (2010) studied the predictability of the Douro river in Spain by using the combination of a time series approach (ARMA) and previous seasonal sea surface temperature anomalies. Their combined sea surface temperature and ARMA(4,3) model explains 76% of the total variance for spring Douro streamflow series.

Drought prediction also plays an important role in water resource management. Durdu (2010) chose the Standardized Precipitation Index as an indicator of drought severity and used ARIMA models to detect the drought severity in the Buyuk Menderes river basin in western Turkey. This ARIMA model can be applied to forecast drought impacts for the Buyuk Menderes river basin and gives reasonably good results up to 2 months ahead.

The runoff coefficient reflects the rainfall-runoff relationship. It is defined as the ratio of the total runoff to the total rainfall in a specific time period. In hydrological modeling, runoff coefficients represent the lumped effects of many processes, including antecedent soil moisture, evaporation, rainfall, and snowmelt. Pektas and Cigizoglu (2013) used univariate ARIMA and multivariate ARIMA to model and predict runoff coefficients

Premonsoon rainfall over India is highly variable. Narayanan et al. (2013) used a univariate ARIMA model for premonsoon rainfall and discovered that there is a significant rise in the premonsoon rainfall in northwestern India.

Kumar et al. (2009) fit sensible heat fluxes between 1 and 2 m and between 1 and 4 m in height into the ARIMA models. These models can provide a reasonably good prediction of fluxes for 1 h in advance.

PROBLEMS

8.1 Let $\{Y_n\}_{n\in\mathbb{Z}}$ be an ARMA(p, q) process satisfying

$$Y_n - \sum_{1}^{p} \varphi_j Y_{n-j} = Z_n + \sum_{1}^{q} \theta_j Z_{n-j} \quad (n \in \mathbb{Z}),$$

$$\{Z_n\}_{n\in\mathbb{Z}} \sim \text{WN}(0, \sigma^2),$$

where φ_j, θ_j are constants. Let

$$\varphi(z) = 1 - \sum_{j} \varphi_j z^j,$$

$$\theta(z) = 1 + \sum_{j} \theta_j z^j$$

have no common zeros in $|z| < 1$. Try to prove that this process is invertible if and only if $\theta(z) \neq 0$ ($|z| \leq 1$).

8.2 Suppose that $\{Y_n\}_{n\in\mathbb{Z}}$ is a ARMA(p, q) process satisfying

$$\varphi(T)Y_n = \theta(T)Z_n \quad (n \in \mathbb{Z}),$$

$$\{Z_n\}_{n\in\mathbb{Z}} \sim \text{WN}(0, \sigma^2),$$

where $\varphi(z) \neq 0$, $\theta(z) \neq 0$ ($|z| \geq 1$) and the two functions have only real zeros. Try to prove that $\{Y_n\}_{n\in\mathbb{Z}}$ must be a causal invertible ARMA(p, q) process satisfying

$$\widetilde{\varphi}(T)Y_n = \widetilde{\theta}(T)Z_n^* \quad (n \in \mathbb{Z}),$$

$$\{Z_n^*\}_{n\in\mathbb{Z}} \sim \text{WN}(0, \sigma^2),$$

where

$$\widetilde{\varphi}(z) = \varphi(z) \prod_{1}^{p} \frac{1 - a_j z}{z - a_j},$$

$$\widetilde{\theta}(z) = \theta(z) \prod_{1}^{q} \frac{1 - b_j z}{z - b_j},$$

and a_j $(j = 1, \ldots, p)$ and b_j $(j = 1, \ldots, q)$ are zeros of $\varphi(z)$ and $\theta(z)$ in $|z| < 1$, respectively.

8.3 Let an ARMA process $\{Y_n\}_{n \in \mathbb{Z}}$ satisfy

$$Y_n - \frac{5}{2} Y_{n-1} + Y_{n-2} = Z_n + Z_{n-1} \quad (n \in \mathbb{Z}),$$

$$\{Z_n\}_{n \in \mathbb{Z}} \sim \text{WN}(0, \sigma^2).$$

Try to find the solution of the ARMA equation, the covariance matrix, and the spectral density.

8.4 Try to prove that the sample covariance matrix $\hat{\Sigma}_n$

$$\hat{\Sigma}_n = (\hat{\gamma}_{|l-j|})_{l,j=1,\ldots,n} = \begin{pmatrix} \hat{\gamma}_0 & \hat{\gamma}_1 & \cdots & \hat{\gamma}_{n-1} \\ \hat{\gamma}_1 & \hat{\gamma}_0 & \cdots & \hat{\gamma}_{n-2} \\ \vdots & \vdots & \ddots & \vdots \\ \hat{\gamma}_{n-1} & \hat{\gamma}_{n-2} & \cdots & \hat{\gamma}_0 \end{pmatrix}$$

is non-negative definite.

BIBLIOGRAPHY

Box, G.E.P., Jenkins, G.M., 1970. Time Series Analysis: Forecasting and Control. Holden-Day, San Francisco, CA.

Bras, R.L., Rodriguez-Iturbe, I., 1985. Random Functions and Hydrology. Addison-Wesley, Reading, MA.

Durdu, O.F., 2010. Application of linear stochastic models for drought forecasting in the Buyuk Menderes river basin, western Turkey. Stoch. Environ. Res. Risk Assess. 24, 1145-1162.

Fatimah, M.H., Ghaffar, R.A., 1986. Univariate approach towards cocoa price forecasting. Malay. J. Agric. Econ. 3, 1-11.

Gámiz-Fortis, S.R., Esteban-Parra, M.J., Trigo, R.M., Castro-Dez, Y., 2010. Potential predictability of an Iberian river flow based on its relationship with previous winter global SST. J. Hydrol. 385, 143-149.

Han, P., Wang, P., Tian, M., Zhang, S., Liu, J., Zhu, D., 2013. Application of the ARIMA models in drought forecasting using the standardized precipitation index. Comput. Comput. Technol. Agric. VI 392, 352-358.

Karthikeyan, L., Nagesh Kumar, D., 2013. Predictability of nonstationary time series using wavelet and EMD based ARMA models. J. Hydrol. 502, 103-119.

Keskin, M.E., Taylan, D., Terzi, O., 2006. Adaptive neural-based fuzzy inference system (ANFIS) approach for modelling hydrological time series. Hydrol. Sci. J. 51, 588-598.

Kumar, M., Kumar, A., Mahanti, N.C., Mallik, C., Shukla, R.K., 2009. Surface flux modelling using ARIMA technique in humid subtropical monsoon area. J. Atmos. Sol. Terr. Phys. 71, 1293-1298.

Mohammadi, K., Eslami, H.R., Kahawita, R., 2006. Parameter estimation of an ARMA model for river flow forecasting using goal programming. J. Hydrol. 331, 293-299.

Narayanan, P., Basistha, A., Sarkar, S., Sachdevaa, K., 2013. Trend analysis and ARIMA modelling of pre-monsoon rainfall data for western India. C. R. Geosci. 345, 22-27.

Pektas, A.O., Cigizoglu, H.K., 2013. ANN hybrid model versus ARIMA and ARIMAX models of runoff coefficient. J. Hydrol. 500, 21-36.

Salas, J.D., Delleur, J.W., Yevejevich, V., Lane, W.L., 1980. Applied Modeling of Hydrologic Timeseries. Water Resources, Littleton.

Stedinger, J.R., Lettenmaier, D.P., Vogel, R.M., 1985. Multisite ARMA (1,1) and disaggregation models for annual streamflow generation. Water Resour. Res. 21, 497-509.

Valipour, M., Banihabib, M.E., Behbahani, S.M.R., 2013. Comparison of the ARMA, ARIMA, and the autoregressive artificial neural network models in forecasting the monthly inflow of Dez dam reservoir. J. Hydrol. 476, 433-441.

Weeks, W.D., Boughton, W.C., 1987. Tests of ARMA model forms for rainfall-runoff modelling. J. Hydrol. 91, 29-47.

Chapter 9

Data Assimilation

Data assimilation is a powerful technique which has been widely applied in the investigations of the atmosphere, ocean, and land surface. It combines observation data and the underlying dynamical principles governing the system to provide an estimate of the state of the system which is better than could be obtained using just the data or the model alone. In this chapter, we introduce various data assimilation methods, including the Cressman analysis method, the optimal interpolation method, three-/four-dimensional variational analysis, and the Kalman filter. All these methods are based on least-squares methods, with the final estimate being chosen to minimize the uncertainty of the final estimate. The difference lies in the choice of the metric used to measure the uncertainty and the corresponding weight given to the observations and the prior estimate.

9.1 CONCEPT OF DATA ASSIMILATION

Effective climatic and environmental prediction require two sources of information. One source is well-distributed observation data of Earth. The other source is the models that embody the physical and chemical laws governing the behavior of Earth's land surface, oceans, and atmosphere. Both observations and models have errors. The errors of observations are of three kinds: random errors, systematic errors, and representativeness errors. The errors of models are produced owing to processes being omitted to make the problem tractable. The science of data assimilation is used to combine these two sources of information into successful prediction systems for weather, oceans, climate, and ecosystems. Because data assimilation techniques can improve forecasting or modeling and increase physical understanding of the systems considered, data assimilation now plays a very important role in studies of climate change.

The development of data assimilation techniques has experienced three stages: simple analysis, optimal interpolation, and variational analysis. The simple analysis method was the earliest basis of data assimilation and was used widely in the 1950s. In the 1960s and 1970s, the optimal interpolation method was used to assimilate observations into forecast models. In the 1980s and 1990s, data assimilation switched to variational methods, mainly including three- and four-dimensional variational data assimilation. These approaches attempt to combine observations and model information in an optimal way to produce the best possible estimate of the model initial state.

Mathematical and Physical Fundamentals of Climate Change

We use a simple example to explain the concept, idea, and method of data assimilation.

We estimate the temperature x_t. By using a thermometer, we measure the temperature of the room, and we get the observation information x_o. Suppose that this observation is unbiased and the thermometer possesses accuracy σ_o, i.e.,

$$E[x_o] = x_t,$$
$$\text{Var } x_o = \sigma_o^2.$$

The temperature estimated from a physical model can be treated as background information x_b. Suppose further that

$$E[x_b] = x_t,$$
$$\text{Var } x_b = \sigma_b^2.$$

The observation x_o and the background x_b can be combined to provide a better estimate x_a of the truth x_t. We want to look for a weighted average

$$x_a = kx_o + (1-k)x_b \quad \text{or} \quad x_a = x_b + k(x_o - x_b). \tag{9.1}$$

Denote the observation error, background error, and analysis error, respectively, by

$$\epsilon_o = x_o - x_t,$$
$$\epsilon_b = x_b - x_t,$$
$$\epsilon_a = x_a - x_t.$$

So

$$\epsilon_a = k\epsilon_o + (1-k)\epsilon_b,$$

and so

$$E[\epsilon_a] = kE[\epsilon_o] + (1-k)E[\epsilon_b] = 0.$$

The variance of analysis error ϵ_a is

$$\sigma_a^2 = \text{Var } \epsilon_a = E[(k\epsilon_o + (1-k)\epsilon_b)^2] = k^2 E[\epsilon_o^2] + 2k(1-k)E[\epsilon_o\epsilon_b] + (1-k)^2 E[\epsilon_b^2]$$

where

$$E[\epsilon_o^2] = \text{Var } \epsilon_o = \sigma_o^2,$$
$$E[\epsilon_b^2] = \text{Var } \epsilon_b = \sigma_b^2.$$

Since the observation and background are uncorrelated,

$$E[\epsilon_o\epsilon_b] = E[\epsilon_o]E[\epsilon_b] = 0,$$

and so

$$\sigma_a^2 = k^2\sigma_o^2 + (1-k)^2\sigma_b^2.$$

When $k = \frac{\sigma_b^2}{\sigma_o^2 + \sigma_b^2}$, the variance σ_a^2 attains the minimal value

$$\sigma_a^2 = \frac{\sigma_o^2 \sigma_b^2}{\sigma_o^2 + \sigma_b^2},$$

i.e.,

$$\frac{1}{\sigma_a^2} = \frac{1}{\sigma_o^2} + \frac{1}{\sigma_b^2}.$$

From this, we also see that the analysis error variance is less than or equal to the observation error variance and the background error variance, i.e., $\sigma_a^2 \leq \min\{\sigma_o^2, \sigma_b^2\}$. From this and (9.1), when $k = \frac{\sigma_b^2}{\sigma_o^2 + \sigma_b^2}$, the analysis

$$x_a = x_a^* = \frac{\dfrac{x_o}{\sigma_o^2} + \dfrac{x_b}{\sigma_b^2}}{\dfrac{1}{\sigma_o^2} + \dfrac{1}{\sigma_b^2}} \tag{9.2}$$

is the optimal analysis, i.e., x_a is the best estimate of the true temperature x_t derived by combining the observation and the background.

Define a cost function of the analysis:

$$J(x) = \frac{(x - x_b)^2}{\sigma_b^2} + \frac{(x - x_o)^2}{\sigma_o^2} = J_b(x) + J_o(x), \tag{9.3}$$

where $J_b(x)$ is the background cost and $J_o(x)$ is the observation cost. It is easily checked that the cost function attains the minimal value at $x = x_a$ (see (9.2)). This simple example explains how the data assimilation problem can be reduced to an ordinary extreme problem if some statistics are known. Notice that

$$J''(x) = 2 \left(\frac{1}{\sigma_b^2} + \frac{1}{\sigma_o^2} \right).$$

Then

$$\sigma_a^2 = \left(\frac{1}{2} J'' \right)^{-1}.$$

Therefore, we can determine the analysis error variance by the second-order derivative of the cost function J. If the observation error ϵ_o and the background error ϵ_b are both Gaussian random variables with

$$E[\epsilon_o] = E[\epsilon_b] = 0,$$

$$\text{Var}\, \epsilon_o = \sigma_o^2,$$

$$\text{Var}\, \epsilon_b = \sigma_b^2,$$

then the observation error and the background error probability density functions are, respectively,

$$\rho_o(x) = \frac{1}{\sqrt{2\pi}\sigma_o} e^{-\frac{(y-x)^2}{2\sigma_o^2}},$$

$$\rho_b(x) = \frac{1}{\sqrt{2\pi}\sigma_b} e^{-\frac{(x_b-x)^2}{2\sigma_b^2}}.$$

The analysis error probability density function is defined as the Bayesian product of $\rho_o(x)$ and $\rho_b(x)$, i.e.,

$$\rho_a(x) = \rho_o(x)\rho_b(x) = Ce^{-\frac{1}{2}\left(\frac{(y-x)^2}{\sigma_o^2} + \frac{(x-x_b)^2}{\sigma_b^2}\right)} = Ce^{-\frac{1}{2}J(x)},$$

where C is a constant and $J(x)$ is the cost function which is stated in (9.3). $J(x)$ attains the minimal value at $x = x_a$, so $x = x_a$ is the maximal likelihood estimator of the real temperature x_t.

9.2 CRESSMAN METHOD

The model state is assumed to be univariate and represented as grid-point values. Denote a previous estimate of the model state (background) by a n-dimensional vector $\mathbf{x}_b = (x_b(1), \ldots, x_b(n))^T$ and an observed vector by an n-dimensional vector $\mathbf{y}_b = (y_b(1), \ldots, y_b(n))^T$. Cressman analysis gives an analysis model

$$\mathbf{x}_a = (x_a(1), \ldots, x_a(n))^T$$

by the following update equation

$$\mathbf{x}_a(j) = \mathbf{x}_b(j) + \frac{\sum_{i=1}^n \omega(i,j)(\mathbf{y}(i) - \mathbf{x}_b(i))}{\sum_{i=1}^n \omega(i,j)},$$

where

$$\omega(i,j) = \max\left\{0, \frac{R^2 - d_{ij}^2}{R^2 + d_{ij}^2}\right\},$$

$$d_{ij} = |i - j|,$$

and R is a user-defined control parameter. Since

$$\omega(i,j) = 1 \quad \text{if } i = j,$$
$$\omega(i,j) = 0 \quad \text{if } d_{ij} > R,$$

the parameter R is called the *influence radius*. There are several improvements of the Cressman method, for example, the Barnes weight $\omega(i,j) = e^{-\frac{d_{ij}^2}{2R^2}}$ is used to replace the Cressman weight. In addition, the updates can be performed several times in order to enhance the smoothness of corrections.

The Cressman method has many limitations. All observations are assumed to have a similar error variance since the weighting is based only on distance. On the other hand, we do not know how to decide the shape of the function ω. Because of its simplicity, the Cressman method can be a useful starting tool, but it is impossible to give a good-quality analysis.

9.3 OPTIMAL INTERPOLATION ANALYSIS

In a forecast scheme, the set of numbers representing the state is called the *state vector* **x**. One must distinguish between reality itself $\mathbf{x}_t$ and the best possible representation $\mathbf{x}_b$, which often is given by physical models. $\mathbf{x}_b$ is also called the *background state vector*. In analysis, we use an observation vector **y** and compare it with the state vector. In practice there are fewer observations than variables in the background model such that the only correct way to compare observations with the state vector is to use an observation operator h from model state space to observation space.

Suppose that we know an observation vector $\mathbf{y} = (y^1, \dots, y^p)^T$ and a state vector $\mathbf{x}_b = (x_b^1, \dots, x_b^n)^T$ ($n \geq p$) from background models. In order to combine the observation vector **y** with the state vector $\mathbf{x}_b$, one needs to introduce a linear operator h from an n-dimensional space to a p-dimensional space. This operator corresponds to a $p \times n$ matrix H and $h(\mathbf{x}_b) = H\mathbf{x}_b$. Let

$$H = (H_{ij})_{i=1,\dots,p;j=1,\dots,n},$$

$$h(\mathbf{x}_b) = (h_1, \dots, h_p)^T.$$

Then

$$h_i = \sum_{j=1}^{n} H_{ij} x_b^j \quad (i = 1, \dots, p).$$

Denote the $n \times n$ covariance matrix of the background error $\mathbf{x}_b - \mathbf{x}_t$ by B and denote the $p \times p$ covariance matrix of the observation error $\mathbf{y} - h(\mathbf{x}_b)$ by R. Moreover, the background error $x_b - x_t$ and the observation error $\mathbf{y} - h(\mathbf{x}_b)$ are uncorrelated and

$$E[\mathbf{x}_b - \mathbf{x}_t] = \mathbf{0},$$

$$E[\mathbf{y} - h(\mathbf{x}_b)] = \mathbf{0}. \tag{9.4}$$

Define an n-dimensional analysis vector $\mathbf{x}_a$:

$$\mathbf{x}_a = \mathbf{x}_b + K(\mathbf{y} - h(\mathbf{x}_b)),$$

where K is an $n \times p$ weight matrix. We choose K such that the variance of the analysis error $\mathbf{x}_a - \mathbf{x}_t$ attains the minimal value. Suppose that

$$\mathbf{x}_a - \mathbf{x}_t = (x_a^1 - x_t^1, \dots, x_a^n - x_t^n)^T.$$

For convenience, assume that $E[x_a^i - x_t^i] = 0 (i = 1, \ldots, n)$. Hence, the variance of the analysis error is

$$\epsilon = \text{Var}\,(\mathbf{x}_a - \mathbf{x}_t) = \sum_1^n \text{Var}(x_a^i - x_t^i).$$

Theorem 9.1 (Optimal Interpolation Analysis). *Let $\mathbf{x}_b$, $\mathbf{y}$, and h be the background vector, the observation vector, and the observation operator, respectively, which are stated as above. If the analysis vector $\mathbf{x}_a$ is defined as*

$$\mathbf{x}_a = \mathbf{x}_b + K(\mathbf{y} - h(\mathbf{x}_b)), \tag{9.5}$$

where K is an $n \times p$ matrix, then

(i) *the analysis vector $\mathbf{x}_a$ is such that $\text{Var}(\mathbf{x}_a - \mathbf{x}_t)$ attains the minimal value if and only if*

$$K = K^* = BH^T (HBH^T + R)^{-1}, \tag{9.6}$$

where matrices B, H and R are stated as above.

(ii) *If (9.6) holds, the covariance matrix of analysis error $\mathbf{x}_a - \mathbf{x}_t$ is*

$$A = (I - K^*H)B$$

and the analysis error $\epsilon = \text{Var}(\mathbf{x}_a - \mathbf{x}_t) = \text{tr}((I - K^*H)B)$, where $\text{tr}(S)$ is the trace of the square matrix S, i.e., $\text{tr}(S)$ is the sum of diagonal elements of the matrix S.

Proof. We first find the covariance matrix of $\mathbf{x}_a - \mathbf{x}_t$.

Let $\epsilon_b = \mathbf{x}_b - \mathbf{x}_t$, $\epsilon_a = \mathbf{x}_a - \mathbf{x}_t$, and $\epsilon_o = \mathbf{y} - h(\mathbf{x}_t)$. Then, by (9.5),

$$\epsilon_a - \epsilon_b = \mathbf{x}_a - \mathbf{x}_b = K(\mathbf{y} - h(\mathbf{x}_b)).$$

Since $h(\mathbf{x}) - h(\mathbf{x}_1) = H(\mathbf{x} - \mathbf{x}_1)$,

$$\mathbf{y} - h(\mathbf{x}_b) = (\mathbf{y} - h(\mathbf{x}_t)) - (h(\mathbf{x}_b) - h(\mathbf{x}_t)) = \epsilon_o - H(\mathbf{x}_b - \mathbf{x}_t) = \epsilon_o - H(\epsilon_b).$$

So $\epsilon_a - \epsilon_b = K(\epsilon_o - H(\epsilon_b))$, and so

$$\epsilon_a = \epsilon_b + K\epsilon_o - KH(\epsilon_b) = (I - KH)\epsilon_b + K\epsilon_o.$$

Notice that

$$\begin{aligned} \epsilon_a \epsilon_a^T &= ((I - KH)\epsilon_b + K\epsilon_o)(\epsilon_b^T(I - K^TH^T) + \epsilon_o^T K^T) \\ &= (I - KH)\epsilon_b \epsilon_b^T(I - K^TH^T) + K\epsilon_o \epsilon_b^T(I - K^TH^T) \\ &\quad + K\epsilon_o \epsilon_o^T K^T + (I - KH)\epsilon_b \epsilon_o^T K^T. \end{aligned}$$

Since B and R are covariance matrices of the background error and the observation error, respectively, by (9.4),

$$E[\epsilon_b \epsilon_b^T] = B,$$
$$E[\epsilon_o \epsilon_o^T] = R.$$

This implies that the covariance matrix of the analysis error is

$$A(K) = E[\epsilon_a \epsilon_a^T]$$
$$= (I - KH)B(I - KH)^T + KE[\epsilon_0 \epsilon_b^T](I - K^T H^T)$$
$$+ KRK^T + (I - KH)E[\epsilon_b \epsilon_0^T]K^T.$$

Since observation error and background error are uncorrelated,

$$E[\epsilon_0 \epsilon_b^T] = 0,$$
$$E[\epsilon_b \epsilon_0^T] = 0,$$

the covariance matrix of the analysis error is represented as follows:

$$A(K) = E[\epsilon_a \epsilon_a^T] = (I - KH)B(I - KH)^T + KRK^T. \qquad (9.7)$$

Now we consider the trace of matrix A. Since the trace of a matrix is linear,

$$\text{tr}(A(K)) = \text{tr}((I - KH)B(I - H^T K^T)) + \text{tr}(KRK^T)$$
$$= \text{tr}(B) - \text{tr}(KHB) - \text{tr}(BH^T K^T)$$
$$+ \text{tr}(KHBH^T K^T) + \text{tr}(KRK^T).$$

Since covariance matrix B is a symmetric matrix,

$$\text{tr}(KHB) = \text{tr}((KHB)^T) = \text{tr}(BH^T K^T),$$

and so

$$\text{tr}(A(K)) = \text{tr}(B) - 2\text{tr}(BH^T K^T) + \text{tr}(KHBH^T K^T) + \text{tr}(KRK^T). \qquad (9.8)$$

Let L be an arbitrary $n \times p$ test matrix. Then

$$\text{tr}(A)(K + L) = \text{tr}(B) - 2\text{tr}(BH^T (K + L)^T)$$
$$+ \text{tr}((K + L)HBH^T (K + L)^T)$$
$$+ \text{tr}((K + L)R(K + L)^T).$$

From this and (9.8), we have

$$\text{tr}(A)(K + L) - \text{tr}(A)(K) = I_1' - 2I_2' + I_3',$$

where

$$I_1' = \text{tr}((K + L)HBH^T (K^T + L^T) - KHBH^T K^T)$$
$$= 2\text{tr}(KHBH^T L^T) + \text{tr}(LHBH^T L^T),$$
$$I_2' = \text{tr}(BH^T (K^T + L^T)) - \text{tr}(BH^T K^T) = \text{tr}(BH^T L^T),$$
$$I_3' = \text{tr}((K + L)R(K + L)^T) - \text{tr}(KRK^T) = 2\text{tr}(KRL^T) + \text{tr}(LRL^T).$$

In these computations, the formula

$$\text{tr}(V) = \text{tr}(V^T)$$

is used for any square matrix V. Therefore,

$$\text{tr}(A(K + L)) - \text{tr}(A(K)) = 2\text{tr}((K(HBH^T + R) - BH^T)L^T)$$
$$+ \text{tr}(L(HBH^T + R)L^T)$$
$$= I_1 + I_2.$$

Let

$$K(HBH^T + R) - BH^T = (\alpha_{ij})_{i=1,\ldots,n;j=1,\ldots,p}, \tag{9.9}$$

$$L = (l_{ij})_{i=1,\ldots,n;j=1,\ldots,p},$$

$$HBH^T + R = (\beta_{ij})_{i,j=1,\ldots,p}.$$

Then

$$(K(HBH^T + R) - BH^T)L^T = \left(\sum_{k=1}^{p} \alpha_{ik} l_{jk}\right)_{i,j=1,\ldots,n},$$

$$L(HBH^T + R)L^T = \left(\sum_{k=1}^{p}\sum_{s=1}^{p} \beta_{sk} l_{is} l_{jk}\right)_{i,j=1,\ldots,n}.$$

By the definition of traces of matrices,

$$I_1 = 2\sum_{i=1}^{n}\sum_{k=1}^{p} \alpha_{ik} l_{ik},$$

$$I_2 = \sum_{i=1}^{n}\sum_{k,s=1}^{p} \beta_{sk} l_{is} l_{ik}.$$

From this, it follows that

$$|I_2| \leq \frac{1}{2} \max_{k,s=1,\ldots,n} |\beta_{sk}| \left(\sum_{i=1}^{n}\sum_{k=1}^{p}\sum_{s=1}^{p} (l_{is}^2 + l_{ik}^2)\right)$$

$$= p \max_{k,s=1,\ldots,n} |\beta_{sk}| \left(\sum_{i=1}^{n}\sum_{k=1}^{p} l_{ik}^2\right)$$

$$= O\left(\| L \|^2\right),$$

where

$$\| L \| = \left(\sum_{i=1}^{n}\sum_{j=1}^{p} l_{ij}^2\right)^{1/2}$$

is the norm of the $n \times p$ matrix L or the norm of np-dimensional vector $l_{ij}(i = 1, \ldots, n; j = 1, \ldots, p)$, and the term $S := O(\| L \|^2)$ means that $|S| \leq M \| L \|^2$, where M is a constant independent of $\| L \|^2$. Therefore, for an arbitrary fixed $n \times p$ matrix A and any $n \times p$ test matrix $L = (l_{ij})_{i=1,\ldots,n; j=1,\ldots,p}$ with small norm,

$$\text{tr}(A(K + L)) - \text{tr}(A(K)) = 2 \sum_{i=1}^{n} \sum_{j=1}^{p} \alpha_{ij} l_{ij} + O(\| L \|^2).$$

Noticing that $2 \sum_{i=1}^{n} \sum_{j=1}^{p} \alpha_{ij} l_{ij}$ is the linear principal part of this increment, by multivariate calculus, we deduce that $\text{tr}(A(K))$ attains the minimal value if and only if the coefficients $\alpha_{ij} = 0 (i = 1, \ldots, n; j = 1, \ldots, p)$. From (9.9), this implies that

$$K = K^* = BH^{\text{T}}(HBH^{\text{T}} + R)^{-1},$$

i.e., the analysis error $\epsilon = \text{Var}(\mathbf{x}_a - \mathbf{x}_t)$ is minimal if and only if (9.6) holds.

If (9.6) holds, by (9.7), when $K = K^*$, the covariance matrix of $\mathbf{x}_a - \mathbf{x}_t$ is

$$\begin{aligned} A(K^*) &= (I - K^*H)B - (I - K^*H)BH^{\text{T}}K^{*\text{T}} + K^*RK^{*\text{T}} \\ &= (I - K^*H)B - BH^{\text{T}}K^{*\text{T}} + K^*(HBH^{\text{T}} + R)K^{*\text{T}} \\ &= (I - K^*H)B. \end{aligned}$$

From this and $\epsilon = \text{Var}(\mathbf{x}_a - \mathbf{x}_t) = \text{tr}(A)$, the analysis error $\epsilon = \text{tr}((I - K^*H)B)$.

$\square$

9.4 COST FUNCTION AND THREE-DIMENSIONAL VARIATIONAL ANALYSIS

In this section, we will define a cost function such that the computation of the weight function K^* (see (9.6)) in the optimal interpolation analysis is reduced to the problem of minimizing the cost function of the analysis.

Definition 9.1. Let $\mathbf{x}_b - \mathbf{x}_t$ be the background error with the covariance matrix B, and let $\mathbf{y} - h(\mathbf{x}_t)$ be the observation with the covariance matrix R. The *cost function* of the analysis is defined as

$$\begin{aligned} J(\mathbf{x}) &= (\mathbf{x} - \mathbf{x}_b)^{\text{T}} B^{-1} (\mathbf{x} - \mathbf{x}_b) + (\mathbf{y} - h(\mathbf{x}))^{\text{T}} R^{-1} (\mathbf{y} - h(\mathbf{x})) \\ &=: J_b(\mathbf{x}) + J_o(\mathbf{x}), \end{aligned} \tag{9.10}$$

where J_b is the background cost and J_o is the observation cost.

We will prove that when $\mathbf{x} = \mathbf{x}_a = \mathbf{x}_b + K^*(\mathbf{y} - h(\mathbf{x}_b))$, the cost function $J(\mathbf{x})$ attains the minimal value. Therefore, the optimal interpolation analysis is reduced to the extreme value problem of the cost function.

Let the background error probability density function $p_b(\mathbf{x})$ and the observation error probability density function $p_o(\mathbf{x})$ both be Gaussian, i.e.,

$$p_b(\mathbf{x}) = C_b e^{-\frac{1}{2}(\mathbf{x}-\mathbf{x}_b)^T B^{-1}(\mathbf{x}-\mathbf{x}_b)},$$

$$p_o(\mathbf{x}) = C_o e^{-\frac{1}{2}(\mathbf{y}-H\mathbf{x})^T R^{-1}(\mathbf{y}-H\mathbf{x})},$$

where

$$C_b = (2\pi)^{-\frac{n}{2}}(\det B)^{-\frac{1}{2}},$$

$$C_o = (2\pi)^{-\frac{n}{2}}(\det R)^{-\frac{1}{2}}.$$

The aim of the analysis is to find the maximal value of the conditional probability of the model state given by the observations and the backgrounds. By Bayes's theorem, the analysis error probability density function is

$$p_a(\mathbf{x}) = p_b(\mathbf{x})p_o(\mathbf{x}) = C_b C_o e^{-\frac{1}{2}J(\mathbf{x})},$$

where $J(\mathbf{x})$ is the cost function. When $\mathbf{x} = \mathbf{x}_a$, $J(\mathbf{x})$ attains the minimal value. So $\mathbf{x}_a$ is the maximal likelihood estimator of $\mathbf{x}_t$.

We have explained that in the above Gaussian case, when $\mathbf{x} = \mathbf{x}_a$, the cost function $J(\mathbf{x})$ attains the minimal value. To prove this result in the general case, we need the following propositions. Let $f(\mathbf{x}) = f(x_1, \ldots, x_n)$ be an n-dimensional function. Define

$$\nabla f(\mathbf{x}) = \left(\frac{\partial f}{\partial x_1}, \ldots, \frac{\partial f}{\partial x_n} \right)^T,$$

where ∇ is called a *gradient operator*.

Proposition 9.1. *Let $\mathbf{x}$ be an n-dimensional vector and C be an $n \times n$ positive definite matrix, and let $f(\mathbf{x}) = \mathbf{x}^T C \mathbf{x}$. Then the gradient of f satisfies $\nabla f(\mathbf{x}) = 2C\mathbf{x}$.*

Proof. Let

$$\mathbf{x} = (x_1, \ldots, x_n)^T,$$

$$C = (c_{ij})_{i,j=1,\ldots,n}.$$

Then

$$f(\mathbf{x}) = \mathbf{x}^T C \mathbf{x} = \sum_{i=1}^{n} \sum_{j=1}^{n} c_{i,j} x_i x_j.$$

In order to compute the gradient of f, we rewrite it as follows. For any $l = 1, \ldots, n$,

$$f(\mathbf{x}) = \sum_{i \neq l} x_i \left(\sum_{j=1}^{n} c_{ij} x_j \right) + x_l \sum_{j=1}^{n} c_{lj} x_j.$$

Therefore,

$$\frac{\partial f}{\partial x_l} = \sum_{i=1}^{n} c_{il} x_i + \sum_{j=1}^{n} c_{lj} x_j.$$

Since the matrix C is symmetric, $c_{il} = c_{li}$. This implies that for $l = 1, \ldots, n$,

$$\frac{\partial f}{\partial x_l} = 2 \sum_{i=1}^{n} c_{il} x_i.$$

and so $\nabla f(\mathbf{x}) = 2C\mathbf{x}$.
□

Proposition 9.2. *Let* $W = f(z_1, \ldots, z_p)$ *and*

$$z_1 = z_1(x_1, \ldots, x_n),$$
$$z_2 = z_2(x_1, \ldots, x_n),$$
$$\vdots$$
$$z_p = z_p(x_1, \ldots, x_n).$$

Then the gradient of the compound function $W = W(x_1, \ldots, x_n)$ *is*
$$\nabla W = D^{\mathrm{T}} \nabla f(z),$$
where D is the Jacobian matrix and

$$D = \frac{\partial(z_1, \ldots, z_p)}{\partial(x_1, \ldots, x_n)} = \begin{pmatrix} \frac{\partial z_1}{\partial x_1} & \cdots & \frac{\partial z_1}{\partial x_n} \\ \vdots & \cdots & \vdots \\ \frac{\partial z_p}{\partial x_1} & \cdots & \frac{\partial z_p}{\partial x_n} \end{pmatrix}. \tag{9.11}$$

Proof. From

$$\frac{\partial W}{\partial x_l} = \frac{\partial f}{\partial z_1} \frac{\partial z_1}{\partial x_l} + \frac{\partial f}{\partial z_2} \frac{\partial z_2}{\partial x_l} + \cdots + \frac{\partial f}{\partial z_p} \frac{\partial z_p}{\partial x_l} \quad (l = 1, \ldots, n),$$

it follows that

$$\nabla W = \begin{pmatrix} \frac{\partial W}{\partial x_1} \\ \vdots \\ \frac{\partial W}{\partial x_n} \end{pmatrix} = \frac{\partial f}{\partial z_1} \begin{pmatrix} \frac{\partial z_1}{\partial x_1} \\ \vdots \\ \frac{\partial z_1}{\partial x_n} \end{pmatrix} + \cdots + \frac{\partial f}{\partial z_p} \begin{pmatrix} \frac{\partial z_p}{\partial x_1} \\ \vdots \\ \frac{\partial z_p}{\partial x_n} \end{pmatrix} = D^{\mathrm{T}} \nabla f(z).$$

□

Theorem 9.2. *Let*
$$\mathbf{x}_a = \mathbf{x}_b + K^*(\mathbf{y} - h(\mathbf{x}_b)), \quad \text{where } K^* = BH^{\mathrm{T}}(HBH^{\mathrm{T}} + R)^{-1}.$$
Then the cost function $J(\mathbf{x})$ of the analysis takes the minimal value if and only if $\mathbf{x} = \mathbf{x}_a$. Moreover, $\nabla^2 J(\mathbf{x}_a) = (2A)^{-1}$, where A is the covariance matrix of $\mathbf{x}_a - \mathbf{x}_t$ and $A = (B^{-1} + H^{\mathrm{T}} R^{-1} H)^{-1}$.
Proof. Let $\mathbf{z} = h(\mathbf{x})$, i.e.,

$$z_1 = h_1(x_1, \ldots, x_n),$$
$$z_2 = h_2(x_1, \ldots, x_n),$$
$$\vdots$$
$$z_p = h_p(x_1, \ldots, x_n).$$

Then, by (9.10), the observation term in the cost function $J_0(\mathbf{x})$ can be regarded as a compound function

$$W = J_0(\mathbf{x}) = f(h(\mathbf{x})),$$

where $f(\mathbf{z}) = (\mathbf{y} - \mathbf{z})^{\mathrm{T}} R^{-1}(\mathbf{y} - \mathbf{z})$ and $\mathbf{z} = h(\mathbf{x})$. By Proposition 9.2, the gradient is

$$\nabla J_0(\mathbf{x}) = \left(\frac{\partial(h_1, \ldots, h_p)}{\partial(x_1, \ldots, x_n)} \right)^{\mathrm{T}} \nabla f(\mathbf{z}).$$

By Proposition 9.1,

$$\nabla f(\mathbf{z}) = -2R^{-1}(\mathbf{y} - h(\mathbf{x})).$$

Since $h(\mathbf{x})$ is a linear operator from an n-dimensional space to a p-dimensional space, $h(\mathbf{x})$ can be written in the matrix form $h(\mathbf{x}) = H\mathbf{x}$, where

$$H = \begin{pmatrix} H_{11} & \cdots & H_{1n} \\ \vdots & \ddots & \vdots \\ H_{p1} & \cdots & H_{pn} \end{pmatrix}, \quad \mathbf{x} = \begin{pmatrix} x_1 \\ \vdots \\ x_n \end{pmatrix},$$

i.e., $h_i(x_1, \ldots, x_n) = \sum_1^n H_{il}x_l$, and so

$$\frac{\partial h_i}{\partial x_j} = H_{ij} \quad (i = 1, \ldots, p; j = 1, \ldots, n).$$

Therefore,

$$\nabla J_0(\mathbf{x}) = -2H^{\mathrm{T}} R^{-1}(\mathbf{y} - h(\mathbf{x})).$$

On the other hand, by Proposition 9.1,

$$\nabla J_b(\mathbf{x}) = 2B^{-1}(\mathbf{x} - \mathbf{x}_b),$$

and so

$$\nabla J(\mathbf{x}) = \nabla J_0(\mathbf{x}) + \nabla J_b(\mathbf{x}) = 2B^{-1}(\mathbf{x} - \mathbf{x}_b) - 2H^{\mathrm{T}} R^{-1}(\mathbf{y} - h(\mathbf{x})). \quad (9.12)$$

From this and $h(\mathbf{x}_a) - h(\mathbf{x}_b) = H(\mathbf{x}_a - \mathbf{x}_b)$, it follows that $\nabla J(\mathbf{x}) = 0$ if and only if

$$B^{-1}(\mathbf{x} - \mathbf{x}_b) - H^{\mathrm{T}} R^{-1}(\mathbf{y} - h(\mathbf{x})) = 0,$$

which is equivalent to $(B^{-1} + H^{\mathrm{T}} R^{-1} H)(\mathbf{x} - \mathbf{x}_b) = H^{\mathrm{T}} R^{-1}(\mathbf{y} - h(\mathbf{x}_b))$ or

$$\mathbf{x} - \mathbf{x}_b = (B^{-1} + H^{\mathrm{T}} R^{-1} H)^{-1} H^{\mathrm{T}} R^{-1}(\mathbf{y} - h(\mathbf{x}_b)). \quad (9.13)$$

From

$$(B^{-1} + H^{\mathrm{T}} R^{-1} H)BH^{\mathrm{T}} = H^{\mathrm{T}} R^{-1}(HBH^{\mathrm{T}} + R),$$

it follows that

$$(B^{-1} + H^{\mathrm{T}} R^{-1} H)^{-1} H^{\mathrm{T}} R^{-1} = BH^{\mathrm{T}}(HBH^{\mathrm{T}} + R)^{-1}.$$

Therefore, (9.13) is equivalent to

$$\mathbf{x} - \mathbf{x}_b = BH^T(HBH^T + R)^{-1}(\mathbf{y} - h(\mathbf{x}_b)) = K^*(\mathbf{y} - h(\mathbf{x}_b)),$$

i.e., $\nabla J(\mathbf{x}) = 0$ if and only if $\mathbf{x} = \mathbf{x}_a$.

By (9.12) and $h(\mathbf{x}) = H\mathbf{x}$, it follows that

$$\nabla J(\mathbf{x}) = 2B^{-1}\mathbf{x} - 2B^{-1}\mathbf{x}_b - 2H^T R^{-1}\mathbf{y} + 2H^T R^{-1}H\mathbf{x}$$
$$= 2(B^{-1} + H^T R^{-1}H)\mathbf{x} - 2B^{-1}\mathbf{x}_b - 2H^T R^{-1}\mathbf{y}. \tag{9.14}$$

Clearly, ∇J is a n-dimensional vector, and $\nabla^2 J = \nabla(\nabla J)$ means to find the gradient of each component of ∇J. Therefore, $\nabla^2 J$ is a $n \times n$ matrix. Since the second term and the third term on the right-hand side of (9.14) are independent of $\mathbf{x}$, we have

$$\nabla^2 J(\mathbf{x}) = \nabla(\nabla J(\mathbf{x})) = 2\nabla((B^{-1} + H^T R^{-1}H)\mathbf{x}),$$

where $B^{-1} + H^T R^{-1}H$ is a $n \times n$ matrix and $\mathbf{x}$ is an n-dimensional vector. Denote

$$M = B^{-1} + H^T R^{-1}H = (\lambda_{ij})_{i,j=1,\dots,n},$$
$$\mathbf{x} = (x_1, \dots, x_n)^T.$$

Then the ith component of $M\mathbf{x}$ is

$$(M\mathbf{x})_i = \sum_{j=1}^{n} \lambda_{ij} x_j,$$

and so

$$\nabla^2 J(\mathbf{x}) = \nabla(M\mathbf{x}) = 2 \begin{pmatrix} \lambda_{11} & \cdots & \lambda_{1n} \\ \vdots & \ddots & \vdots \\ \lambda_{n1} & \cdots & \lambda_{nn} \end{pmatrix} = 2M,$$

i.e.,

$$\nabla^2 J(\mathbf{x}) = 2(B^{-1} + H^T R^{-1}H). \tag{9.15}$$

Now we prove that when

$$\mathbf{x} = \mathbf{x}_a = \mathbf{x}_b + K^*(\mathbf{y} - h(\mathbf{x}_b)) \quad (K^* = BH^T(HBH^T + R)^{-1}),$$

the covariance matrix of analysis error $\mathbf{x}_a - \mathbf{x}_t$

$$A = (B^{-1} + H^T R^{-1}H)^{-1}.$$

By $\nabla J(\mathbf{x}_a) = 0$ and (9.12),

$$2B^{-1}(\mathbf{x}_a - \mathbf{x}_b) - 2H^T R^{-1}(\mathbf{y} - h(\mathbf{x}_a)) = 0.$$

Replacing $\mathbf{x}_a - \mathbf{x}_b$ by $\mathbf{x}_a - \mathbf{x}_t + \mathbf{x}_t - \mathbf{x}_b$ and replacing $\mathbf{y} - h(\mathbf{x}_a)$ by $\mathbf{y} - h(\mathbf{x}_t) + H(\mathbf{x}_t - \mathbf{x}_a)$ in this equality, we get

$$(B^{-1} + H^{\mathrm{T}} R^{-1} H)(\mathbf{x}_a - \mathbf{x}_t) = B^{-1}(\mathbf{x}_b - \mathbf{x}_t) + H^{\mathrm{T}} R^{-1}(\mathbf{y} - h(\mathbf{x}_t)).$$

Its transpose is

$$(\mathbf{x}_a - \mathbf{x}_t)^{\mathrm{T}}(B^{-1} + H^{\mathrm{T}} R^{-1} H) = (\mathbf{x}_b - \mathbf{x}_t)^{\mathrm{T}} B^{-1} + (\mathbf{y} - h(\mathbf{x}_t))^{\mathrm{T}} R^{-1} H.$$

Multiplying these two equalities, we get

$$(B^{-1} + H^{\mathrm{T}} R^{-1} H)(\mathbf{x}_a - \mathbf{x}_t)(\mathbf{x}_a - \mathbf{x}_t)^{\mathrm{T}}(B^{-1} + H^{\mathrm{T}} R^{-1} H) =$$

$$(B^{-1}(\mathbf{x}_b - \mathbf{x}_t) + H^{\mathrm{T}} R^{-1}(\mathbf{y} - h(\mathbf{x}_t)))((\mathbf{x}_b - \mathbf{x}_t)^{\mathrm{T}} B^{-1} + (\mathbf{y} - h(\mathbf{x}_t))^{\mathrm{T}} R^{-1} H).$$

$$(9.16)$$

Since $\mathbf{x}_b - \mathbf{x}_t$ and $\mathbf{y} - h(\mathbf{x}_t)$ are independent, from $E[\mathbf{x}_b - \mathbf{x}_t] = 0$ and $E[\mathbf{y} - h(\mathbf{x}_t)] = 0$, this implies that

$$E[(\mathbf{x}_b - \mathbf{x}_t)(\mathbf{y} - h(\mathbf{x}_t))^{\mathrm{T}}] = E[(\mathbf{x}_b - \mathbf{x}_t)]E\left[\mathbf{y} - h(\mathbf{x}_t)^{\mathrm{T}}\right] = 0.$$

Since

$$E[(\mathbf{x}_a - \mathbf{x}_t)(\mathbf{x}_a - \mathbf{x}_t)^{\mathrm{T}}] = A,$$

$$E[(\mathbf{x}_b - \mathbf{x}_t)(\mathbf{x}_b - \mathbf{x}_t)^{\mathrm{T}}] = B,$$

$$E[(\mathbf{y} - h(\mathbf{x}_t))(\mathbf{y} - h(\mathbf{x}_t))^{\mathrm{T}}] = R,$$

taking the expectation in (9.16), we deduce that

$$(B^{-1} + H^{\mathrm{T}} R^{-1} H)A(B^{-1} + H^{\mathrm{T}} R^{-1} H) = B^{-1} B B^{-1} + H^{\mathrm{T}} R^{-1} R R^{-1} H$$

$$= B^{-1} + H^{\mathrm{T}} R^{-1} H.$$

So $A = (B^{-1} + H^{\mathrm{T}} R^{-1} H)^{-1}$.

From this and (9.15), $\nabla^2 J(\mathbf{x}) = (2A)^{-1}$. Since the covariance matrix A is positive definite, $(2A)^{-1}$ is also positive definite. Finally, since $\nabla^2 J(\mathbf{x}_a)$ is positive definite and $\nabla J(\mathbf{x}_a) = 0$, we know that the cost function $J(\mathbf{x})$ attains the minimal value at $\mathbf{x} = \mathbf{x}_a$. $\qquad\square$

From the definition of the cost function, we know that $J(\mathbf{x})$ is a quadric form and $\nabla^2 J(\mathbf{x})$ does not depend on $\mathbf{x}$. In the one-dimensional case, $J(\mathbf{x})$ is a parabola and $J''(\mathbf{x})$ is a positive constant. If $J''(\mathbf{x})$ is large, then the cost function has a strong convexity, so the quality of the analysis is high. If $J''(\mathbf{x})$ is small, then the cost function has a weak convexity, so the quality of the analysis is low. This explains the relationship between the two-order derivative and the quality of the analysis.

9.5 DUAL OF THE OPTIMAL INTERPOLATION

By Theorem 9.1, the optimal analysis is

$$\mathbf{x}_a = \mathbf{x}_b + B H^{\mathrm{T}} (H B H^{\mathrm{T}} + R)^{-1}(\mathbf{y} - H \mathbf{x}_b).$$

It can be decomposed into two equations:

$$\mathbf{x}_a = \mathbf{x}_b + BH^T\mathbf{w}_a,$$

$$\mathbf{w}_a = (HBH^T + R)^{-1}(\mathbf{y} - H\mathbf{x}_b).$$

From this, $\mathbf{w}_a$ is the solution of the system of linear equations,

$$(HBH^T + R)\mathbf{w} = \mathbf{y} - H\mathbf{x}_b,$$

which can be regarded as the dual of the optimal interpolation algorithm. The cost function

$$F(\mathbf{w}) = \mathbf{w}^T(HBH^T + R)\mathbf{w} - 2\mathbf{w}^T(\mathbf{y} - H\mathbf{x}_b) = I_b(\mathbf{w}) - I_0(\mathbf{w})$$

attains the minimal value at $\mathbf{w} = \mathbf{w}_a$.

$\nabla F(\mathbf{w}) = \nabla I_b(\mathbf{w}) - \nabla I_0(\mathbf{w})$. By Proposition 9.1,

$$\nabla I_b(\mathbf{w}) = 2(HBH^T + R)\mathbf{w}.$$

Denote

$$\mathbf{w} = (w_1, \ldots, w_p)^T,$$

$$\mathbf{y} - H\mathbf{x}_b = (c_1, \ldots, c_p)^T.$$

Then $I_0(\mathbf{w}) = 2(c_1 w_1 + \cdots + c_p w_p)$, and so

$$\nabla I_0(\mathbf{w}) = 2(c_1, \ldots, c_p)^T = 2(\mathbf{y} - H\mathbf{x}_b).$$

This implies that

$$\nabla F(\mathbf{w}) = 2(HBH^T + R)\mathbf{w} - 2(\mathbf{y} - H\mathbf{x}_0).$$

Therefore, $F(\mathbf{w})$ attains the minimal value at w^* if and only if $\nabla F(W) = 0$, i.e.,

$$w^* = (HBH^T + R)^{-1}(\mathbf{y} - H\mathbf{x}_0) = w_a.$$

9.6 FOUR-DIMENSIONAL VARIATIONAL ANALYSIS

The four-dimensional variational assimilation is a simple generalization of the three-dimensional variational assimilation. The observations are distributed among $N + 1$ times in the interval. The cost function can be generalized as

$$\widetilde{J}(\mathbf{x}) = (\mathbf{x} - \mathbf{x}_b)^T B^{-1}(\mathbf{x} - \mathbf{x}_b) + \sum_0^N (\mathbf{y}_i - H_i(\mathbf{x}_i))^T R_i^{-1}(\mathbf{y}_i - H_i(\mathbf{x}_i)). \quad (9.17)$$

The assimilation problem with minimal variance is reduced to looking for the analysis vector $\mathbf{x}_a$ such that $J(\mathbf{x})$ attains the minimal value at $\mathbf{x} = \mathbf{x}_a$.

Let $\mathbf{x}_i$ be a n-dimensional vector, $\mathbf{y}_i$ be a p-dimensional vector, and H_i be a linear operator which maps n-dimensional space to p-dimensional space.

Suppose that m_{oi} is a predefined model forecast operator from the initial time to i, i.e.,

$$\mathbf{x}_i = m_{oi}(\mathbf{x}).$$

It is an operator from n-dimensional space to n-dimensional space which is determined by understanding for the system observed such as the equation of state

$$H_i(\mathbf{x}_i) = H_i(m_{oi}(\mathbf{x})).$$

If the operator m_{oi} is continuously differentiable at $\mathbf{x}_b$, then we have the following first-order Taylor formula in the neighborhood of the background state $\mathbf{x}_b$:

$$\mathbf{x}_i = m_{oi}(\mathbf{x}) = \mathbf{x}_{ib} - M_{oi}(\mathbf{x}_b)(\mathbf{x} - \mathbf{x}_b) + O_i(\| \mathbf{x} - \mathbf{x}_b \|^2), \qquad (9.18)$$

where $\mathbf{x}_{ib} = m_{oi}(\mathbf{x}_b)$ and $M_{oi}(\mathbf{x}_b)$ is the Jacobian matrix of nonlinear operator m_{oi} at $\mathbf{x}_b$ (the definition of the Jacobian matrix is given in (9.11)). From this and (9.18), we get

$$H_i(\mathbf{x}_i) = H_i(\mathbf{x}_{ib}) + H_i M_{oi}(\mathbf{x}_b)(\mathbf{x} - \mathbf{x}_b) + \gamma_i,$$

where $\gamma_i = O_i(\| \mathbf{x} - \mathbf{x}_b \|^2)$. Again, by (9.17), the linear principal part of $\tilde{J}(\mathbf{x})$ is

$$\begin{aligned}
J(\mathbf{x}) &= (\mathbf{x} - \mathbf{x}_b)^T B^{-1}(\mathbf{x} - \mathbf{x}_b) \\
&\quad + \sum_0^N (H_i M_{oi}(\mathbf{x}_b)(\mathbf{x} - \mathbf{x}_b) - d_i)^T R_i^{-1}(H_i M_{oi}(\mathbf{x}_b)(\mathbf{x} - \mathbf{x}_b) - d_i) \\
&= J_1(\mathbf{x}) + J_2(\mathbf{x}), \qquad (9.19)
\end{aligned}$$

where $d_i = \mathbf{y}_i - H_i(\mathbf{x}_{ib})$ is a p-dimensional vector. By Proposition 9.1,

$$\nabla J_1(\mathbf{x}) = 2B^{-1}(\mathbf{x} - \mathbf{x}_b),$$

$$J_2(\mathbf{x}) = \sum_0^n F_i(g_i(\mathbf{x})),$$

where

$$F_i(\mathbf{z}_i) = \mathbf{z}_i^T R_i^{-1} \mathbf{z}_i,$$
$$\mathbf{z}_i = g_i(\mathbf{x}) = H_i M_{oi}(\mathbf{x}_b)(\mathbf{x} - \mathbf{x}_b) - d_i,$$

and $g_i(\mathbf{x})$ is a p-dimensional vector. By Propositions 9.1 and 9.2, we get

$$\begin{aligned}
\nabla J_2(\mathbf{x}) &= \sum_0^N \nabla F_i(g_i(\mathbf{x})) = \sum_0^N (H_i M_{oi}(\mathbf{x}_b))^T \nabla F_i(\mathbf{z}) \\
&= 2 \sum_0^N (H_i M_{oi}(\mathbf{x}_b))^T R_i^{-1}(H_i M_{oi}(\mathbf{x}_b)(\mathbf{x} - \mathbf{x}_b) - d_i).
\end{aligned}$$

Denote $G_i = H_i M_{0i}(\mathbf{x}_b)$ is a $p \times n$ matrix. Then

$$\nabla J_2(\mathbf{x}) = 2 \left(\sum_0^N G_i^T R_i^{-1} G_i \right) (\mathbf{x} - \mathbf{x}_b) - 2 \sum_0^N G_i^T R_i^{-1} d_i.$$

From this and (9.19), we have

$$\nabla J(\mathbf{x}) = 2 \left(B^{-1} + \sum_0^N G_i^T R_i^{-1} G_i \right) (\mathbf{x} - \mathbf{x}_b) - 2 \sum_0^N G_i^T R_i^{-1} d_i. \qquad (9.20)$$

Therefore, $\nabla J(\mathbf{x}) = \mathbf{0}$ if and only if

$$\mathbf{x} = \mathbf{x}_b + \left(B^{-1} + \sum_0^N G_i^T R_i^{-1} G_i \right)^{-1} \left(\sum_0^N R_i^{-1} G_i^T d_i \right). \qquad (9.21)$$

Since R_i is a $p \times p$ symmetric matrix,

$$S = \sum_0^N G_i^T R_i^{-1} G_i$$

is a $p \times p$ symmetric matrix. By a known result in the theory of linear algebra, there exists an orthogonal matrix $\widetilde{H}$ of order p such that $\widetilde{H} S \widetilde{H}^T = \widetilde{R}^{-1}$. So $\widetilde{R}$ is a diagonal matrix of order p and

$$S = \widetilde{H}^T \widetilde{R}^{-1} \widetilde{H}.$$

So

$$\sum_0^N G_i^T R_i^{-1} G_i = \widetilde{H}^T \widetilde{R}^{-1} \widetilde{H}. \qquad (9.22)$$

Notice that $\sum_1^N G_i^T d_i$ is a p-dimensional vector and $\widetilde{H}^T \widetilde{R}^{-1}$ is a $p \times p$ matrix. There exists a p-dimensional vector $\widetilde{\mathbf{d}}$ such that

$$\sum_1^N R_i^{-1} G_i^T d_i = \widetilde{H}^T \widetilde{R}^{-1} \widetilde{\mathbf{d}}. \qquad (9.23)$$

From this and (9.21), it follows that $\nabla J(\mathbf{x}) = 0$ if and only if $\mathbf{x} = \mathbf{x}_b + \widetilde{K}\widetilde{\mathbf{d}}$, where

$$\widetilde{K} = (B^{-1} + \widetilde{H}^T \widetilde{R}^{-1} \widetilde{H})^{-1} \widetilde{H}^T \widetilde{R}^{-1}.$$

Notice that $\widetilde{H}^T \widetilde{R}^{-1} (\widetilde{H} B \widetilde{H}^T + \widetilde{R}) = (B^{-1} + \widetilde{H}^T \widetilde{R}^{-1} \widetilde{H}) B \widetilde{H}^T$. This implies that

$$(B^{-1} + \widetilde{H}^T \widetilde{R}^{-1} \widetilde{H})^{-1} \widetilde{H}^T \widetilde{R}^{-1} = B \widetilde{H}^T (\widetilde{H} B \widetilde{H}^T + \widetilde{R})^{-1}.$$

So

$$\widetilde{K} = B \widetilde{H}^T (\widetilde{H} B \widetilde{H}^T + \widetilde{R})^{-1}. \qquad (9.24)$$

Combining (9.20), (9.22), and (9.23), we get

$$\nabla J(\mathbf{x}) = 2(B^{-1} + \widetilde{H}^{\mathrm{T}} \widetilde{R}^{-1} \widetilde{H})(\mathbf{x} - \mathbf{x}_b) - 2\widetilde{H}^{\mathrm{T}} \widetilde{R}^{-1} \widetilde{\mathbf{d}}.$$

Similarly to the argument of Theorem 9.2, we have

$$\nabla^2 J = 2(B^{-1} + \widetilde{H}^{\mathrm{T}} \widetilde{R}^{-1} \widetilde{H}).$$

Since the matrix on the right-hand side is symmetric and positive definite, the cost function $J(\mathbf{x})$ attains the minimal value at $\mathbf{x} = \mathbf{x}_b + \widetilde{K}\widetilde{\mathbf{d}}$, where $\widetilde{K}$ is stated in (9.24) in a region of the state space near the background, i.e.,

$$\mathbf{x}_a = \mathbf{x}_b + \widetilde{K}\widetilde{\mathbf{d}}.$$

9.7 KALMAN FILTER

The Kalman filter is widely applied in data assimilation in which each background is provided by a forecast that starts from the previous analysis. It is adapted to the real-time assimilation of observations distributed in time into a forecast model.

The model forecast operator from dates i to $i + 1$ is denoted by $M_{i \to i+1}$. The deviation of the prediction from the true evolution

$$M_{i \to i+1}(\mathbf{x}_t(i)) - \mathbf{x}_t(i + 1)$$

is called *the model error*, and the model error covariance matrix $Q(i)$ is known. The background vector and analysis vector are denoted by $\mathbf{x}_f(i)$ and $\mathbf{x}_a(i)$, respectively. The background vector $\mathbf{x}_f(i)$ is defined as

$$\mathbf{x}_f(i + 1) = M_{i \to i+1}\mathbf{x}_a(i). \tag{9.25}$$

$\mathbf{y}_i$ and h_i are the observation vector and the observation operator at time i, respectively, and $R(i)$ is the observation error covariance matrix.

Assume that the analysis error $\mathbf{x}_a(i) - \mathbf{x}_t(i)$ and model errors are uncorrelated; then the difference of the model prediction is a linear function

$$M_{i \to i+1}(\mathbf{x}(i)) - M_{i \to i+1}(\mathbf{x}_a(i)) = \widetilde{M}_{i \to i+1}(\mathbf{x}(i) - \mathbf{x}_a(i + 1)), \tag{9.26}$$

where $\widetilde{M}_{i \to i+1}$ is a matrix.

The Kalman filter algorithm is the following recurrence formula over the observation time i.

Theorem 9.3. *Denote the background and analysis error covariance matrices by $P_f(i)$ and $P_a(i)$. Then*

(i) *the background error covariance matrix for analysis at time $i + 1$*

$$P_f(i + 1) = \widetilde{M}_{i \to i+1} P_a(i) \widetilde{M}_{i \to i+1}^{\mathrm{T}} + Q(i),$$

where $Q(i)$ is the model error covariance matrix;

(ii) *the least-squares analysis is*

$$\mathbf{x}_a(i) = \mathbf{x}_f(i) + K(i)[\mathbf{y}(i) - h(i)\mathbf{x}_f(i)],$$

where $K(i) = P_f(i)H^T(i)(H(i)P_f(i)H^T(i) + R(i))^{-1}$.

Proof. The proof of (ii) is similar to that of Theorem 9.1. Now we prove (i). By (9.25) and (9.26), we have

$$\mathbf{x}_f(i+1) - \mathbf{x}_t(i+1) = \widetilde{M}_{i\to i+1}(\mathbf{x}_a(i) - \mathbf{x}_t(i)) + (\widetilde{M}_{i\to i+1}(\mathbf{x}_t(i)) - \mathbf{x}_t(i+1)).$$

From this, the covariance matrix of the background vector is

$$P_f(i+1) = E\left[(\mathbf{x}_f(i+1) - \mathbf{x}_t(i+1))(\mathbf{x}_f(i+1) - \mathbf{x}_t(i+1))^T\right]$$
$$= I_1 + I_2 + I_3 + I_4,$$

where

$$I_1 = E\left[\widetilde{M}_{i\to i+1}(\mathbf{x}_a(i) - \mathbf{x}_t(i))(\widetilde{M}_{i\to i+1}(\mathbf{x}_a(i) - \mathbf{x}_t(i)))^T\right],$$

$$I_2 = E\left[\widetilde{M}_{i\to i+1}(\mathbf{x}_a(i) - \mathbf{x}_t(i))(\widetilde{M}_{i\to i+1}\mathbf{x}_t(i) - \mathbf{x}_t(i+1))^T\right],$$

$$I_3 = E\left[(\widetilde{M}_{i\to i+1}(\mathbf{x}_a(i) - \mathbf{x}_t(i)))^T(\widetilde{M}_{i\to i+1}\mathbf{x}_t(i) - \mathbf{x}_t(i+1))\right],$$

$$I_4 = E\left[(\widetilde{M}_{i\to i+1}\mathbf{x}_t(i) - \mathbf{x}_t(i+1))(\widetilde{M}_{i\to i+1}\mathbf{x}_t(i) - \mathbf{x}_t(i+1))^T\right].$$

Since $M_{i\to i+1}$ is a deterministic matrix and we denote the analysis error in step i by $P_a(i)$, we have

$$I_1 = \widetilde{M}_{i\to i+1}P_a(i)\widetilde{M}^T_{i\to i+1}.$$

Since the model error of the subsequent forecast

$$M_{i\to i+1}\mathbf{x}_t(i) - \mathbf{x}_t(i+1)$$

and the analysis errors $\mathbf{x}_a(i) - \mathbf{x}_t(i)$ are uncorrelated, $I_2 = I_3 = 0$. By the definition, I_4 is the model error, i.e., $I_4 = Q(i)$. Finally, we have

$$P_f(i+1) = M_{i\to i+1}P_a(i)M^T_{i\to i+1} + Q(i),$$

i.e., (i). $\qquad\square$

PROBLEMS

9.1 Try to prove that
- **(i)** in the optimal interpolation algorithm, the analysis error probability density function is equal to the product of the background error probability density function and the observation error probability density function;
- **(ii)** if the background and observation error probability density functions are both Gaussian, then the likelihood function of the analysis error

is $Ce^{-\frac{J(x)}{2}}$, where $J(x)$ is the cost function of the optimal interpolation algorithm and C is a constant;

(iii) $\mathbf{x} = \mathbf{x}_a = \mathbf{x}_b + K^*(\mathbf{y} - H\mathbf{x}_b)$, where $K^* = BH^T(HBH^T + R)^{-1}$ is the maximum likelihood estimator of $\mathbf{x}_t$.

9.2 In the case $n = 2$ and $p = 1$, discuss the dual of the optimal interpolation algorithm. Let the background vector and the true state be $\mathbf{x}_b$ and $\mathbf{x}_t$, respectively. The covariance matrix of background error $\epsilon_b = \mathbf{x}_b - \mathbf{x}_t$ is $B = (b_{ij})_{i,j=1,2}$. The observation y is one-dimensional with the observation operator $h(\mathbf{x}) = H\mathbf{x}$, where $H = (h_1, h_2)$. Find the optimal analysis $\mathbf{x}_a$ and the cost function $F(w)$.

9.3 Compare three-dimensional variational analysis with four-dimensional variational analysis.

BIBLIOGRAPHY

Andreadis, K.M., Lettenmaier, D.P., 2006. Assimilating remotely sensed snow observation into a macroscale hydrology model. Adv. Water Resour. 29, 872-886.

Aubert, D., Loumagne, C., Oudin, L., 2003. Sequential assimilation of soil moisture and streamflow data into a conceptual rainfall-runoff model. J. Hydrol. 280, 145-161.

Barker, D.M., Huang, W., Guo, Y.R., Xiao, Q.N., 2004. A three-dimensional (3DVAR) data assimilation system for use with MM5: implementation and initial results. Mon. Weather Rev. 132, 897-914.

Barker, D., Huang, X.-Y., Liu, Z., Auligné, T., Zhang, X., Rugg, S., Ajjaji, R., Bourgeois, A., Bray, J., Chen, Y., Demirtas, M., Guo, Y.-R., Henderson, T., Huang, W., Lin, H.-C., Michalakes, J., Rizvi, S., Zhang, X., 2012. The weather research and forecasting model's community variational/ensemble data assimilation system: WRFDA. Bull. Am. Meteorol. Soc. 93, 831-843.

Cosgrove, B.A., Houser, P.R., 2002. The effect of errors in snow assimilation on land surface modeling. Preprints, 16th Conference on Hydrology. American Meteorological Society, Orlando.

Courtier, P., Andersson, E., Heckley, W., Pailleux, J., Vasiljevic, D., Hamrud, M., Hollingsworth, A., Rabier, F., Fisher, M., 1998. The ECMWF implementation of three-dimensional variational assimilation (3D-Var) Part 1: Formulation. Q. J. R. Meteorol. Soc. 124, 1783-1807.

Crow, W., 2003. Correcting land surface model predictions for the impact of temporally sparse rainfall rate measurements using an ensemble Kalman filter and surface brightness temperature observations. J. Hydrometeorol. 4, 960-973.

Daley, R., 1991. Atmospheric Data Analysis. Cambridge University Press, Cambridge.

Durand, M., Margulis, S.A., 2007. Correcting first-order errors in snow water equivalent estimates using a multifrequency, multiscale radiometric data assimilation scheme. J. Geophys. Res. 112, D13121.1-D13121.15.

Evensen, G., 2003. The ensemble Kalman filter: theoretical formulation and practical implementation. Ocean Dyn. 53, 343-367.

Ghil, M., 1989. Meteorological data assimilation for oceanographers. Part I: Description and theoretical framework. Dyn. Atmos. Oceans 13, 171-218.

Hebson, C., Wood, E., 1985. Partitioned state and parameter estimation for real-time flood forecasting. Appl. Math. Comput. 17, 357-374.

Huang, X.Y., Xiao, Q., Barker, D.M., Zhang, X., Michalakes, J., Huang, W., Henderson, T., Bray, J., Chen, Y., Ma, Z., Dudhia, J., Guo, Y., Zhang, X., Won, D.J., Lin, H.C., Kuo, Y.H., 2009.

Four-dimensional variational data assimilation for WRF: formulation and preliminary results. Mon. Weather Rev. 137, 299-314.

Hurkmans, R., Paniconi, C., Troch, P.A., 2006. Numerical assessment of a dynamical relaxation data assimilation scheme for a catchment hydrological model. Hydrol. Process. 20, 549-563.

Pauwels, V.R.N., De Lannoy, G.J.M., 2006. Improvement of modeled soil wetness conditions and turbulent fluxes through the assimilation of observed discharge. J. Hydrometeorol. 7, 458-477.

Rabier, F., Courtier, P., 1992. Four-dimensional assimilation in the presence of baroclinic instability. Q. J. R. Meteorol. Soc. 118, 649-672.

Thepaut, J.-N., Courtier, P., 1991. Four-dimensional data assimilation using the adjoint of a multi-level primitive-equation model. Q. J. R. Meteorol. Soc. 117, 1225-1254.

Wang, B., Zou, X., Zhu, J., 2000. Data assimilation and its applications. Proc. Natl. Acad. Sci. USA 97, 11143-11144.

Chapter 10

Fluid Dynamics

Earth's atmosphere and oceans exhibit complex patterns of fluid motion over a vast range of space and time scales. These patterns combine to establish the climate in response to solar radiation that is inhomogeneously absorbed by the materials composing air, water, and land. Therefore, fluid dynamics is fundamental for understanding, modeling, and prediction of climate change. In this chapter, we will introduce principles of fluid dynamics, including the continuity equation, Euler's equation, Bernoulli's equation, and the Kelvin law.

10.1 GRADIENT, DIVERGENCE, AND CURL

Gradient, divergence, and curl are three fundamental concepts. To give their definitions, we first introduce two symbols. The symbols Δ and ∇ are defined as follows:

$$\Delta = \frac{\partial^2}{\partial x^2} + \frac{\partial^2}{\partial y^2} + \frac{\partial^2}{\partial z^2},$$

$$\nabla = \frac{\partial}{\partial x}\mathbf{i} + \frac{\partial}{\partial y}\mathbf{j} + \frac{\partial}{\partial z}\mathbf{k},$$

where $\mathbf{i}, \mathbf{j}$, and $\mathbf{k}$ are unit vectors in the x-, y-, and z-directions, respectively. The symbol Δ is called the *Laplace operator*.

A field of the three-dimensional vectors in space has a formula like

$$\mathbf{F}(x, y, z) = M(x, y, z)\mathbf{i} + N(x, y, z)\mathbf{j} + P(x, y, z)\mathbf{k}.$$

If three component functions M, N, P are continuous, we say that the field is continuous. If three component functions M, N, P are differentiable, we say that the field is differentiable, and so on.

The gradient field consists of the gradient vectors. Let $u(x, y, z)$ be a differentiable function on a region Ω in space. The *gradient* (or gradient vector) of u at a point $(x, y, z) \in \Omega$ is defined as

$$\operatorname{grad} u = \nabla u = \frac{\partial u}{\partial x}\mathbf{i} + \frac{\partial u}{\partial y}\mathbf{j} + \frac{\partial u}{\partial z}\mathbf{k}.$$

Sometimes, write $\operatorname{grad} = \frac{\partial}{\partial x}\mathbf{i} + \frac{\partial}{\partial y}\mathbf{j} + \frac{\partial}{\partial z}\mathbf{k}$. The magnitude of the gradient is

$$|\operatorname{grad} u| = \sqrt{\left(\frac{\partial u}{\partial x}\right)^2 + \left(\frac{\partial u}{\partial y}\right)^2 + \left(\frac{\partial u}{\partial z}\right)^2},$$

and the direction of the gradient is $(\cos\alpha, \cos\beta, \cos\gamma)$, where

$$\cos\alpha = \frac{\partial u/\partial x}{|\operatorname{grad} u|},$$

$$\cos\beta = \frac{\partial u/\partial y}{|\operatorname{grad} u|},$$

$$\cos\gamma = \frac{\partial u/\partial z}{|\operatorname{grad} u|}.$$

Consider a field of the three-dimensional vectors in space:

$$\mathbf{F}(x, y, z) = M(x, y, z)\mathbf{i} + N(x, y, z)\mathbf{j} + P(x, y, z)\mathbf{k}.$$

Suppose that three components M, N, P have the continuous first partial derivatives on an open region Ω. The *divergence* of the vector $\mathbf{F}$ at a point $(x, y, z) \in \Omega$ is defined as

$$\operatorname{div} \mathbf{F} = \nabla \cdot \mathbf{F} = \frac{\partial M}{\partial x} + \frac{\partial N}{\partial y} + \frac{\partial P}{\partial z}.$$

The *curl* of the vector $\mathbf{F}$ at a point $(x, y, z) \in \Omega$ is defined as

$$\operatorname{curl} \mathbf{F} = \nabla \times \mathbf{F} = \left(\frac{\partial P}{\partial y} - \frac{\partial N}{\partial z}\right)\mathbf{i} + \left(\frac{\partial M}{\partial z} - \frac{\partial P}{\partial x}\right)\mathbf{j} + \left(\frac{\partial N}{\partial x} - \frac{\partial M}{\partial y}\right)\mathbf{k}.$$

The gradient, divergence, and curl have the following properties.

Property 10.1. Let u, v be scalar functions and c, d be constants. Then

$$\operatorname{grad}(cu + dv) = c\operatorname{grad} u + d\operatorname{grad} v,$$

$$\operatorname{grad}(uv) = u\operatorname{grad} v + v\operatorname{grad} u.$$

Property 10.2. Let $\mathbf{F}$ and $\mathbf{G}$ be two vectors. Then

$$\operatorname{div}(\mathbf{F} + \mathbf{G}) = \operatorname{div} \mathbf{F} + \operatorname{div} \mathbf{G},$$

$$\operatorname{curl}(\mathbf{F} + \mathbf{G}) = \operatorname{curl} \mathbf{F} + \operatorname{curl} \mathbf{G}.$$

Property 10.3. Let $\mathbf{F}$ be a vector and u be a scalar function. Then $\operatorname{div}(u\mathbf{F}) = u\operatorname{div} \mathbf{F} + \mathbf{F} \cdot \operatorname{grad} u$.

In fact, let $\mathbf{F} = f_1\mathbf{i} + f_2\mathbf{j} + f_3\mathbf{k}$. Then

$$u\mathbf{F} = (uf_1)\mathbf{i} + (uf_2)\mathbf{j} + (uf_3)\mathbf{k}.$$

By the definition of the divergence,

$$\operatorname{div}(u\mathbf{F}) = \frac{\partial(uf_1)}{\partial x} + \frac{\partial(uf_2)}{\partial y} + \frac{\partial(uf_3)}{\partial z}$$

$$= u \left(\frac{\partial f_1}{\partial x} + \frac{\partial f_2}{\partial y} + \frac{\partial f_3}{\partial z} \right) + \left(f_1 \frac{\partial u}{\partial x} + f_2 \frac{\partial u}{\partial y} + f_3 \frac{\partial u}{\partial z} \right)$$

$$= u \operatorname{div} \mathbf{F} + \left(f_1 \frac{\partial u}{\partial x} + f_2 \frac{\partial u}{\partial y} + f_3 \frac{\partial u}{\partial z} \right).$$

By the definition of the gradient, the last term is

$$f_1 \frac{\partial u}{\partial x} + f_2 \frac{\partial u}{\partial y} + f_3 \frac{\partial u}{\partial z} = \mathbf{F} \cdot \operatorname{grad} u.$$

Therefore, $\operatorname{div}(u\mathbf{F}) = u \operatorname{div}\mathbf{F} + \mathbf{F} \cdot \operatorname{grad} u$.

Property 10.4. Let $\mathbf{F}$ be a vector. Then $\operatorname{curl}(\frac{\partial \mathbf{F}}{\partial t}) = \frac{\partial}{\partial t}(\operatorname{curl} \mathbf{F})$.
In fact, let $\mathbf{F} = f_1 \mathbf{i} + f_2 \mathbf{j} + f_3 \mathbf{k}$. By the definition of the curl,

$$\operatorname{curl} \mathbf{F} = \left(\frac{\partial f_3}{\partial y} - \frac{\partial f_2}{\partial z} \right) \mathbf{i} + \left(\frac{\partial f_1}{\partial z} - \frac{\partial f_3}{\partial x} \right) \mathbf{j} + \left(\frac{\partial f_2}{\partial x} - \frac{\partial f_1}{\partial y} \right) \mathbf{k},$$

and so

$$\frac{\partial}{\partial t}(\operatorname{curl} \mathbf{F}) = \left(\frac{\partial^2 f_3}{\partial y \partial t} - \frac{\partial^2 f_2}{\partial z \partial t} \right) \mathbf{i} + \left(\frac{\partial^2 f_1}{\partial z \partial t} - \frac{\partial^2 f_3}{\partial x \partial t} \right) \mathbf{j} + \left(\frac{\partial^2 f_2}{\partial x \partial t} - \frac{\partial^2 f_1}{\partial y \partial t} \right) \mathbf{k}.$$

On the other hand, since

$$\frac{\partial \mathbf{F}}{\partial t} = \frac{\partial f_1}{\partial t} \mathbf{i} + \frac{\partial f_y}{\partial t} \mathbf{j} + \frac{\partial f_z}{\partial t} \mathbf{k},$$

by the definition of the curl,

$$\operatorname{curl} \left(\frac{\partial \mathbf{F}}{\partial t} \right) = \left(\frac{\partial^2 f_3}{\partial y \partial t} - \frac{\partial^2 f_2}{\partial z \partial t} \right) \mathbf{i} + \left(\frac{\partial^2 f_1}{\partial z \partial t} - \frac{\partial^2 f_3}{\partial x \partial t} \right) \mathbf{j} + \left(\frac{\partial^2 f_2}{\partial x \partial t} - \frac{\partial^2 f_1}{\partial y \partial t} \right) \mathbf{k}.$$

Therefore, $\operatorname{curl}(\frac{\partial \mathbf{F}}{\partial t}) = \frac{\partial}{\partial t}(\operatorname{curl} \mathbf{F})$.

Property 10.5. Let u be a scalar function. Then $\operatorname{curl}(\operatorname{grad} u) = \mathbf{0}$.
In fact, by the definition of the gradient and the curl,

$$\operatorname{curl}(\operatorname{grad} u) = \operatorname{curl} \left(\frac{\partial u}{\partial x} \mathbf{i} + \frac{\partial u}{\partial y} \mathbf{j} + \frac{\partial u}{\partial z} \mathbf{k} \right)$$

$$= \left(\frac{\partial^2 u}{\partial y \partial z} - \frac{\partial^2 u}{\partial z \partial y} \right) \mathbf{i} + \left(\frac{\partial^2 u}{\partial z \partial x} - \frac{\partial^2 u}{\partial x \partial z} \right) \mathbf{j}$$

$$+ \left(\frac{\partial^2 u}{\partial x \partial y} - \frac{\partial^2 u}{\partial y \partial x} \right) \mathbf{k}$$

$$= \mathbf{0}.$$

Property 10.6. Let $\mathbf{F}$ be a vector field. Then

$$\mathbf{F} \times \operatorname{curl} \mathbf{F} + (\mathbf{F} \cdot \operatorname{grad})\mathbf{F} = \frac{1}{2} \operatorname{grad}(\mathbf{F} \cdot \mathbf{F}).$$

Proof. Let $\mathbf{F} = f_1\mathbf{i} + f_2\mathbf{j} + f_3\mathbf{k}$. By the definition of the curl,

$$\operatorname{curl}\mathbf{F} = \mathbf{i}\left(\frac{\partial f_3}{\partial y} - \frac{\partial f_2}{\partial z}\right) + \mathbf{j}\left(\frac{\partial f_1}{\partial z} - \frac{\partial f_3}{\partial x}\right) + \mathbf{k}\left(\frac{\partial f_2}{\partial x} - \frac{\partial f_1}{\partial y}\right),$$

and so the vector product of $\mathbf{F}$ and $\operatorname{curl}\mathbf{F}$ is

$$\mathbf{F} \times \operatorname{curl}\mathbf{F} = \mathbf{i}\left(f_2\left(\frac{\partial f_2}{\partial x} - \frac{\partial f_1}{\partial y}\right) - f_3\left(\frac{\partial f_1}{\partial z} - \frac{\partial f_3}{\partial x}\right)\right)$$

$$+ \mathbf{j}\left(f_3\left(\frac{\partial f_3}{\partial y} - \frac{\partial f_2}{\partial z}\right) - f_1\left(\frac{\partial f_2}{\partial x} - \frac{\partial f_1}{\partial y}\right)\right)$$

$$+ \mathbf{k}\left(f_1\left(\frac{\partial f_1}{\partial z} - \frac{\partial f_3}{\partial x}\right) - f_2\left(\frac{\partial f_3}{\partial y} - \frac{\partial f_2}{\partial z}\right)\right).$$

Notice that $\mathbf{F} \cdot \operatorname{grad} = f_1\frac{\partial}{\partial x} + f_2\frac{\partial}{\partial y} + f_3\frac{\partial}{\partial z}$. Then

$$(\mathbf{F} \cdot \operatorname{grad})\mathbf{F} = \left(f_1\frac{\partial}{\partial x} + f_2\frac{\partial}{\partial y} + f_3\frac{\partial}{\partial z}\right)(f_1\mathbf{i} + f_2\mathbf{j} + f_3\mathbf{k})$$

$$= \mathbf{i}\left(f_1\frac{\partial f_1}{\partial x} + f_2\frac{\partial f_1}{\partial y} + f_3\frac{\partial f_1}{\partial z}\right)$$

$$+ \mathbf{j}\left(f_1\frac{\partial f_2}{\partial x} + f_2\frac{\partial f_2}{\partial y} + f_3\frac{\partial f_2}{\partial z}\right)$$

$$+ \mathbf{k}\left(f_1\frac{\partial f_3}{\partial x} + f_2\frac{\partial f_3}{\partial y} + f_3\frac{\partial f_3}{\partial z}\right).$$

Adding these two equalities together, we get

$$\mathbf{F} \times \operatorname{curl}\mathbf{F} + (\mathbf{F} \cdot \operatorname{grad})\mathbf{F} = \frac{1}{2}\left(\frac{\partial(f_1^2 + f_2^2 + f_3^2)}{\partial x}\mathbf{i} + \frac{\partial(f_1^2 + f_3^2 + f_3^2)}{\partial y}\mathbf{j}\right.$$

$$\left. + \frac{\partial(f_1^2 + f_2^2 + f_3^2)}{\partial z}\mathbf{k}\right).$$

So

$$\mathbf{F} \times \operatorname{curl}\mathbf{F} + (\mathbf{F} \cdot \operatorname{grad})\mathbf{F} = \frac{1}{2}\left(\frac{\partial(\mathbf{F} \cdot \mathbf{F})}{\partial x}\mathbf{i} + \frac{\partial(\mathbf{F} \cdot \mathbf{F})}{\partial y}\mathbf{j} + \frac{\partial(\mathbf{F} \cdot \mathbf{F})}{\partial z}\mathbf{k}\right)$$

$$= \frac{1}{2}\operatorname{grad}(\mathbf{F} \cdot \mathbf{F}).$$

$\square$

Property 10.7. Let $\mathbf{F}$ and $\mathbf{G}$ be two vectors. Then

$$\operatorname{curl}(\mathbf{F} \times \mathbf{G}) = (\mathbf{G} \cdot \operatorname{grad})\mathbf{F} - (\mathbf{F} \cdot \operatorname{grad})\mathbf{G} + \mathbf{F}(\operatorname{div}\mathbf{G}) - \mathbf{G}(\operatorname{div}\mathbf{F})$$

or

$$\nabla \times (\mathbf{F} \times \mathbf{G}) = (\mathbf{G} \cdot \nabla)\mathbf{F} - (\mathbf{F} \cdot \nabla)\mathbf{G} + \mathbf{F}(\nabla \cdot \mathbf{G}) - \mathbf{G}(\nabla \cdot \mathbf{F}).$$

Proof. Let $\mathbf{F} = f_1\mathbf{i} + f_2\mathbf{j} + f_3\mathbf{k}$ and $\mathbf{G} = g_1\mathbf{i} + g_2\mathbf{j} + g_3\mathbf{k}$. Then

$$\mathbf{F} \times \mathbf{G} = (f_2g_3 - f_3g_2)\mathbf{i} + (f_3g_1 - f_1g_3)\mathbf{j} + (f_1g_2 - f_2g_1)\mathbf{k}.$$

By the definition of the curl,

$$\text{curl}(\mathbf{F} \times \mathbf{G}) = \mathbf{i}\left(\frac{\partial(f_1g_2 - f_2g_1)}{\partial y} - \frac{\partial(f_3g_1 - f_1g_3)}{\partial z}\right)$$

$$+ \mathbf{j}\left(\frac{\partial(f_2g_3 - f_3g_2)}{\partial z} - \frac{\partial(f_1g_2 - f_2g_1)}{\partial x}\right)$$

$$+ \mathbf{k}\left(\frac{\partial(f_3g_1 - f_1g_3)}{\partial x} - \frac{\partial(f_2g_3 - f_3g_2)}{\partial y}\right).$$

On the other hand, notice that

$$\mathbf{G} \cdot \text{grad} = g_1\frac{\partial}{\partial x} + g_2\frac{\partial}{\partial y} + g_3\frac{\partial}{\partial z},$$

$$\mathbf{F} \cdot \text{grad} = f_1\frac{\partial}{\partial x} + f_2\frac{\partial}{\partial y} + f_3\frac{\partial}{\partial z}.$$

Then

$$(\mathbf{G} \cdot \text{grad})\mathbf{F} = \mathbf{i}\left(g_1\frac{\partial f_1}{\partial x} + g_2\frac{\partial f_1}{\partial y} + g_3\frac{\partial f_1}{\partial z}\right)$$

$$+ \mathbf{j}\left(g_1\frac{\partial f_2}{\partial x} + g_2\frac{\partial f_2}{\partial y} + g_3\frac{\partial f_2}{\partial z}\right)$$

$$+ \mathbf{k}\left(g_1\frac{\partial f_3}{\partial x} + g_2\frac{\partial f_3}{\partial y} + g_3\frac{\partial f_3}{\partial z}\right),$$

$$-(\mathbf{F} \cdot \text{grad})\mathbf{G} = \mathbf{i}\left(-f_1\frac{\partial g_1}{\partial x} - f_2\frac{\partial g_1}{\partial y} - f_3\frac{\partial g_1}{\partial z}\right)$$

$$+ \mathbf{j}\left(-f_1\frac{\partial g_2}{\partial x} - f_2\frac{\partial g_2}{\partial y} - f_3\frac{\partial g_2}{\partial z}\right)$$

$$+ \mathbf{k}\left(-f_1\frac{\partial g_3}{\partial x} - f_2\frac{\partial g_3}{\partial y} - f_3\frac{\partial g_3}{\partial z}\right).$$

Notice that

$$\text{div}\,\mathbf{F} = \frac{\partial f_1}{\partial x} + \frac{\partial f_2}{\partial y} + \frac{\partial f_3}{\partial z},$$

$$\text{div}\,\mathbf{G} = \frac{\partial g_1}{\partial x} + \frac{\partial g_2}{\partial y} + \frac{\partial g_3}{\partial z}.$$

Then

$$F(\text{div } G) = \mathbf{i}\left(f_1\frac{\partial g_1}{\partial x} + f_1\frac{\partial g_2}{\partial y} + f_1\frac{\partial g_3}{\partial z}\right)$$

$$+ \mathbf{j}\left(f_2\frac{\partial g_1}{\partial x} + f_2\frac{\partial g_2}{\partial y} + f_2\frac{\partial g_3}{\partial z}\right)$$

$$+ \mathbf{k}\left(f_3\frac{\partial g_1}{\partial x} + f_3\frac{\partial g_2}{\partial y} + f_3\frac{\partial g_3}{\partial z}\right).$$

$$-G(\text{div } F) = \mathbf{i}\left(-g_1\frac{\partial f_1}{\partial x} - g_1\frac{\partial f_2}{\partial y} - g_1\frac{\partial f_3}{\partial z}\right)$$

$$+ \mathbf{j}\left(-g_2\frac{\partial f_1}{\partial x} - g_2\frac{\partial f_2}{\partial y} - g_2\frac{\partial f_3}{\partial z}\right)$$

$$+ \mathbf{k}\left(-g_3\frac{\partial f_1}{\partial x} - g_3\frac{\partial f_2}{\partial y} - g_3\frac{\partial f_3}{\partial z}\right).$$

Adding these four equalities together, we get

$$(\mathbf{G} \cdot \text{grad})\mathbf{F} - (\mathbf{F} \cdot \text{grad})\mathbf{G} + \mathbf{F}(\text{div } \mathbf{G}) - \mathbf{G}(\text{div } \mathbf{F})$$

$$= \mathbf{i}\left(\frac{\partial(f_1g_2 - f_2g_1)}{\partial y} - \frac{\partial(f_3g_1 - f_1g_3)}{\partial z}\right)$$

$$+ \mathbf{j}\left(\frac{\partial(f_2g_3 - f_3g_2)}{\partial z} - \frac{\partial(f_1g_2 - f_2g_1)}{\partial x}\right)$$

$$+ \mathbf{k}\left(\frac{\partial(f_3g_1 - f_1g_3)}{\partial x} - \frac{\partial(f_2g_3 - f_3g_2)}{\partial y}\right).$$

So Property 10.7 follows. $\qquad\square$

In a similar way, we can consider the two-dimensional case.

The *gradient* (or gradient vector) of a scalar function $u(x, y)$ at a point (x, y) in the plane is defined as

$$\text{grad } u = \nabla u = \frac{\partial u}{\partial x}\mathbf{i} + \frac{\partial u}{\partial y}\mathbf{j}.$$

Its magnitude $|\text{grad } u| = \sqrt{(\frac{\partial u}{\partial x})^2 + (\frac{\partial u}{\partial y})^2}$. Its direction is $(\cos\alpha, \sin\alpha)$, where

$$\cos\alpha = \frac{\partial u/\partial x}{|\text{grad } u|},$$

$$\sin\alpha = \frac{\partial u/\partial y}{|\text{grad } u|}.$$

A field of the two-dimensional vectors in the plane has a formula like

$$\mathbf{F}(\mathbf{x}, \mathbf{y}) = M(x, y)\mathbf{i} + N(x, y)\mathbf{j}.$$

Suppose that $M(x, y)$ and $N(x, y)$ have the continuous first partial derivatives on an open region Ω. The *divergence* of the vector $\mathbf{F}$ at $(x, y) \in \Omega$ is defined as

$$\operatorname{div} \mathbf{F} = \nabla \cdot \mathbf{F} = \frac{\partial M}{\partial x} + \frac{\partial N}{\partial y},$$

and the *curl* of the vector $\mathbf{F}$ at $(x, y) \in \Omega$ is defined as

$$\operatorname{curl} \mathbf{F} = \nabla \times \mathbf{F} = \left(\frac{\partial N}{\partial x} - \frac{\partial M}{\partial y} \right) \mathbf{k}.$$

10.2 CIRCULATION AND FLUX

Suppose that a two-dimensional vector

$$\mathbf{F}(x, y) = M(x, y)\mathbf{i} + N(x, y)\mathbf{j} \quad ((x, y) \in \Omega)$$

is continuous and that a curve $C_{AB} \subset \Omega$ is a smooth curve joining two points $A, B \in \Omega$ and

$$C_{AB} : \mathbf{r}(t) = g(t)\mathbf{i} + h(t)\mathbf{j} \quad (a \le t \le b),$$

where $A = \mathbf{r}(a)$ and $B = \mathbf{r}(b)$. The *flow integral* of the vector $\mathbf{F}$ around the smooth curve C_{AB} is defined as

$$\int_{C_{AB}} \mathbf{F} \cdot \mathbf{T} \, dl,$$

where dl is the arc element and $\mathbf{T}$ is the unit tangent vector of a smooth curve C_{AB}. This is a curvilinear integral with respect to the arc length. The integrand of the integral is a scalar product of the vector and the unit tangent vector of the smooth curve. It represents a flow of the vector $\mathbf{F}$ along the curve C_{AB}. Hence, it is often written in the form

$$\text{Flow} = \int_{C_{AB}} \mathbf{F} \cdot \mathbf{T} \, dl.$$

If the smooth curve is a closed curve, then the integral is called a *circulation* of the vector $\mathbf{F}$ around the smooth closed curve C. It is often written in the form

$$\text{Circulation} = \oint_C \mathbf{F} \cdot \mathbf{T} \, dl.$$

Notice that $\mathbf{T} \, dl = d\mathbf{r}$ where $\mathbf{r}(t) = g(t)\mathbf{i} + h(t)\mathbf{j}$ $(a \le t \le b)$. Let $x = g(t)$ and $y = h(t)$. Then

$$\text{Flow} = \int_a^b \mathbf{F}(g(t), h(t)) \cdot \mathbf{r}'(t) \, dt.$$

From

$$\mathbf{F}(g(t), h(t)) = (M(g(t), h(t)), N(g(t), h(t))),$$
$$\mathbf{r}'(t) = (g'(t), h'(t)),$$

it follows that the integrand is

$$\mathbf{F}(g(t), h(t)) \cdot \mathbf{r}'(t) = M(g(t), h(t))g'(t) + N(g(t), h(t))h'(t),$$

and so

$$\text{Flow} = \int_a^b (M(g(t), h(t))g'(t) + N(g(t), h(t))h'(t)) \, dt.$$

Since $x = g(t)$ and $y = h(t)$, the formula computing the flow of $\mathbf{F}$ around C_{AB} is given by

$$\text{Flow} = \int_{C_{AB}} M dx + N dy.$$

Similarly, the circulation of the vector $\mathbf{F}$ around a smooth closed curve C is computed by the formula

$$\text{Circulation} = \oint_C M dx + N dy. \tag{10.1}$$

The *outward flux* of the vector $\mathbf{F}$ across a smooth curve C_{AB} or a smooth closed curve C is defined as

$$\text{Flux} = \int_{C_{AB}} \mathbf{F} \cdot \mathbf{n} \, dl,$$

$$\text{Flux} = \oint_C \mathbf{F} \cdot \mathbf{n} \, dl,$$

where dl is the arc element and $\mathbf{n}$ is the outward unit normal vector of C_{AB} or C. These two integrals are both curvilinear integrals with respect to the arc length. The integrand of the integral is the scalar product of the vector and the unit normal vector.

Notice the difference between flux and circulation. The outward flux of a vector $\mathbf{F}$ across a closed curve C is a curvilinear integral of the scalar product of the vector $\mathbf{F}$ and the outward unit normal vector $\mathbf{n}$ of the curve C, whereas the circulation of a vector $\mathbf{F}$ around a closed curve C is a curvilinear integral of the scalar product of the vector $\mathbf{F}$ and the unit tangent vector of the curve C.

Suppose that the closed smooth curve is in the xy-plane and its direction is the counterclockwise direction. We may take the outward unit normal vector $\mathbf{n} = \mathbf{T} \times \mathbf{k}$, where $\mathbf{k}$ is the unit vector in the z-direction. So

$$\mathbf{n} = \mathbf{T} \times \mathbf{k} = \left(\frac{dx}{dl} \mathbf{i} + \frac{dy}{dl} \mathbf{j} \right) \times \mathbf{k} = \frac{dy}{dl} \mathbf{i} - \frac{dx}{dl} \mathbf{j},$$

and so

$$\mathbf{F} \cdot \mathbf{n} \, dl = (M, N) \cdot (dy, -dx) = M dy - N dx.$$

From this, the outward flux of the vector $\mathbf{F}$ across the smooth curve C_{AB} or across the smooth closed curve C has the following computation formulas:

$$\text{Flux} = \int_{C_{AB}} \mathbf{F} \cdot \mathbf{n} \, dl = \int_{C_{AB}} M \, dy - N \, dx,$$

$$\text{Flux} = \oint_{C} \mathbf{F} \cdot \mathbf{n} \, dl = \oint_{C} M \, dy - N \, dx. \tag{10.2}$$

In a similar way, the flux across an oriented surface in space is defined as follows.

Suppose that a three-dimensional vector

$$\mathbf{F} = M(x, y, z)\mathbf{i} + N(x, y, z)\mathbf{j} + P(x, y, z)\mathbf{k} \quad ((x, y, z) \in S)$$

is continuous, S is an oriented surface, and $\mathbf{n}$ is the chosen unit normal vector of the surface S. The *flux* of $\mathbf{F}$ across S in the direction of $\mathbf{n}$ is defined as

$$\text{Flux} = \int\int_{S} \mathbf{F} \cdot \mathbf{n} \, ds,$$

where ds is the surface element.

10.3 GREEN'S THEOREM, DIVERGENCE THEOREM, AND STOKES'S THEOREM

Green's theorem transforms a curvilinear integral to a surface integral.

Green's Theorem. *The outward flux of a two-dimensional vector $\mathbf{F}$ across a closed smooth curve C is equal to the double integral of* div $\mathbf{F}$ *over the region S enclosed by the curve C,*

$$\oint_{C} \mathbf{F} \cdot \mathbf{n} \, dl = \int\int_{S} \text{div} \, \mathbf{F} \, dS,$$

where $\mathbf{n}$ is the outward unit normal vector of the curve C, dl is the arc element, and dS is the surface element.

The double integral on the right-hand side is called a *divergence integral*. If the curve C is in the xy-plane and $\mathbf{F} = M(x, y)\mathbf{i} + N(x, y)\mathbf{j}$, then this equality is written in the form:

$$\oint_{C} M \, dy - N \, dx = \int\int_{S} \left(\frac{\partial M}{\partial x} + \frac{\partial N}{\partial y} \right) dS.$$

The theorem corresponding to Green's theorem in three dimensions is called the divergence theorem. The divergence theorem transforms a surface integral to a volume integral.

Divergence Theorem. *The outward flux of a three-dimensional vector $\mathbf{F}$ across a closed surface S is equal to the triple integral of* div $\mathbf{F}$ *over the volume V enclosed by the closed surface S,*

$$\int\int_{S} \mathbf{F} \cdot \mathbf{n} \, dS = \int\int\int_{V} \text{div} \, \mathbf{F} \, dV,$$

where **n** *is the outward unit normal vector of S,* d*S is the surface element, and* d*V is the volume element.*

The triple integral on the right-hand side is also a *divergence integral.*

Stokes's theorem transforms a curvilinear integral to a surface integral.

Stokes's Theorem. *Suppose that S is an oriented surface and* **n** *is the surface's unit normal vector. Suppose that C is the boundary of the surface S in the counterclockwise direction and* **T** *is the unit tangent vector of C. Then, for the three-dimensional differentiable vector* **F**, *its counterclockwise circulation around C is equal to the double integral of* (curl **F**) · **n** *over S,*

$$\oint_C \mathbf{F} \cdot \mathbf{T} dl = \int \int_S (\text{curl } \mathbf{F}) \cdot \mathbf{n} \, dS.$$

The double integral on the right-hand side is called a *curl integral.* It is the flux of curl **F** across the surface *S* spanning *C* in the direction of **n**. Therefore, Stokes's theorem states that the counterclockwise circulation of **F** around *C* is equal to the flux of curl **F** across the surface *S* spanning *C* in the direction of **n**.

10.4 EQUATIONS OF MOTION

The state of a moving fluid can be described by the fluid velocity $\mathbf{v}(x, y, z, t)$, the fluid density $\rho(x, y, z, t)$, and the fluid pressure $p(x, y, z, t)$, where x, y, z are the coordinates and t is the time. Hence, if the velocity **v** and two thermodynamic quantities ρ, p are given, the state of the moving fluid is completely determined. Using these three quantities, we will derive the fundamental equations of fluid dynamics. These equations include the continuity equation, Euler's equation, and Bernoulli's equation.

10.4.1 Continuity Equation

The conservation of mass is often expressed by the *continuity equation.*

Suppose that *S* is a closed oriented surface in space, **n** is the surface's outward unit normal vector, and *V* is the volume enclosed by the closed surface *S*. Let ρ be the fluid density and **v** be the fluid velocity. The double integral

$$\int \int_S (\rho \mathbf{v}) \cdot \mathbf{n} \, dS$$

represents the outward flux of the vector $\rho \mathbf{v}$ across the closed surface *S* in the direction of **n**.

The divergence theorem says that this outward flux is equal to the triple integral of div$(\rho \mathbf{v})$ over *V*:

$$\int \int_S (\rho \mathbf{v}) \cdot \mathbf{n} \, dS = \int \int \int_V \text{div}(\rho \mathbf{v}) \, dV.$$

On the other hand, the conservation of mass says that this outward flux is the decrease per unit time in the mass of fluid in V. Since the mass of fluid in V is

$$\int\int\int_V \rho \, dV,$$

the decrease per unit time in the mass of fluid in V is the negative value of the derivative of this triple integral with respect to t,

$$-\frac{\partial}{\partial t} \int\int\int_V \rho \, dV.$$

According to the conservation of mass, this outward flux satisfies

$$\int\int_S (\rho \mathbf{v}) \cdot \mathbf{n} \, dS = -\frac{\partial}{\partial t} \int\int\int_V \rho \, dV.$$

Therefore,

$$\int\int\int_V \text{div}(\rho \mathbf{v}) \, dV = -\frac{\partial}{\partial t} \int\int\int_V \rho \, dV$$

which is equivalent to

$$\int\int\int_V \left(\frac{\partial \rho}{\partial t} + \text{div}(\rho \mathbf{v}) \right) dV = 0.$$

Since this equation must hold for any volume, the integrand must vanish, i.e.,

$$\frac{\partial \rho}{\partial t} + \text{div}(\rho \mathbf{v}) = 0.$$

This equation is called the *continuity equation*. The vector $\rho \mathbf{v}$ is called the *mass flux density*, its direction is that of the motion of the fluid, and its magnitude is equal to the mass of fluid flowing in unit time through unit area perpendicular to the direction of the velocity.

By Property 10.3, $\text{div}(\rho \mathbf{v}) = \rho \, \text{div} \, \mathbf{v} + \mathbf{v} \cdot \text{grad} \, \rho$, and so the alternative form of the continuity equation is

$$\frac{\partial \rho}{\partial t} + \rho \, \text{div} \, \mathbf{v} + \mathbf{v} \cdot \text{grad} \, \rho = 0.$$

In adiabatic motion, the entropy of any particle of a fluid remains constant as that particle moves about in space. Denote by s the entropy per unit mass. The condition for adiabatic motion can be expressed as

$$\frac{ds}{dt} = 0,$$

where the derivative of s with respect to time is the rate of change of entropy for a given fluid particle as that particle moves about in space. Applying the chain rule gives

$$\frac{ds}{dt} = \frac{\partial s}{\partial t} + \frac{\partial s}{\partial x}\frac{dx}{dt} + \frac{\partial s}{\partial y}\frac{dy}{dt} + \frac{\partial s}{\partial z}\frac{dz}{dt} = \frac{\partial s}{\partial t} + \mathbf{v}\cdot\operatorname{grad} s.$$

From $\frac{ds}{dt} = 0$, it follows that

$$\frac{\partial s}{\partial t} + \mathbf{v}\cdot\operatorname{grad} s = 0.$$

This is the *general equation describing the adiabatic motion of an ideal fluid.*
Multiplying both sides of this equation by ρ,

$$\rho\frac{\partial s}{\partial t} + \rho\mathbf{v}\cdot\operatorname{grad} s = 0.$$

Multiplying both sides of the continuous equation by s,

$$s\frac{\partial \rho}{\partial t} + s\operatorname{div}(\rho\mathbf{v}) = 0.$$

Adding the two equations together,

$$\rho\frac{\partial s}{\partial t} + s\frac{\partial \rho}{\partial t} + s\operatorname{div}(\rho\mathbf{v}) + \rho\mathbf{v}\cdot\operatorname{grad} s = 0.$$

By Property 10.3, $\operatorname{div}(s\rho\mathbf{v}) = s\operatorname{div}(\rho\mathbf{v}) + \rho\mathbf{v}\cdot\operatorname{grad} s$. From this and $\frac{\partial(\rho s)}{\partial t} = \rho\frac{\partial s}{\partial t} + s\frac{\partial \rho}{\partial t}$, it follows that

$$\frac{\partial(\rho s)}{\partial t} + \operatorname{div}(s\rho\mathbf{v}) = 0.$$

This equation is the *entropy form of the continuity equation.* The product $s\rho\mathbf{v}$ is called the *entropy flux density.*

10.4.2 Euler's Equation

Suppose that S is a closed oriented surface in space, $\mathbf{n}$ is the surface's outward unit normal vector, and V is the volume enclosed by the closed surface S. Let p be the fluid pressure. Then the total force acting on V is equal to the negative value of the double integral of $p\mathbf{n}$ over S:

$$-\int\int_S p\mathbf{n}\, dS.$$

This double integral is transformed to a volume integral as follows.
Define three vectors by

$$\mathbf{p_1} = (p, 0, 0),$$
$$\mathbf{p_2} = (0, p, 0),$$
$$\mathbf{p_3} = (0, 0, p).$$

Let $\mathbf{n} = (n_1, n_2, n_3)$. Then

$$\mathbf{p}_1 \cdot \mathbf{n} = pn_1,$$
$$\mathbf{p}_2 \cdot \mathbf{n} = pn_2,$$
$$\mathbf{p}_3 \cdot \mathbf{n} = pn_3.$$

Therefore,

$$p\mathbf{n} = (pn_1, pn_2, pn_3) = (\mathbf{p}_1 \cdot \mathbf{n})\mathbf{i} + (\mathbf{p}_2 \cdot \mathbf{n})\mathbf{j} + (\mathbf{p}_3 \cdot \mathbf{n})\mathbf{k},$$

and so the double integral becomes

$$-\int\int_S p\mathbf{n}\, dS = -\mathbf{i}\int\int_S \mathbf{p}_1 \cdot \mathbf{n}\, dS - \mathbf{j}\int\int_S \mathbf{p}_2 \cdot \mathbf{n}\, dS - \mathbf{k}\int\int_S \mathbf{p}_3 \cdot \mathbf{n}\, dS.$$

Applying the divergence theorem to each integral gives

$$-\int\int_S p\mathbf{n}\, dS = -\mathbf{i}\int\int\int_V \operatorname{div}\mathbf{p}_1\, dV - \mathbf{j}\int\int\int_V \operatorname{div}\mathbf{p}_2\, dV$$
$$-\mathbf{k}\int\int\int_V \operatorname{div}\mathbf{p}_3\, dV.$$

Notice that

$$\operatorname{div}\mathbf{p}_1 = \frac{\partial p}{\partial x},$$
$$\operatorname{div}\mathbf{p}_2 = \frac{\partial p}{\partial y},$$
$$\operatorname{div}\mathbf{p}_3 = \frac{\partial p}{\partial z}.$$

Then

$$-\int\int_S p\mathbf{n}\, dS = -\int\int\int_V \left(\frac{\partial p}{\partial x}\mathbf{i} + \frac{\partial p}{\partial y}\mathbf{j} + \frac{\partial p}{\partial z}\mathbf{k}\right) dV.$$

By the definition of the gradient, it is transformed to a triple integral of $\operatorname{grad} p$ as follows:

$$-\int\int_S p\mathbf{n}\, dS = -\int\int\int_V \operatorname{grad} p\, dV.$$

This equality means that a force $(-\operatorname{grad} p)$ acts on the unit volume of the fluid. This force is equal to the product of the mass per unit volume and the acceleration, i.e.,

$$-\operatorname{grad} p = \rho\frac{d\mathbf{v}}{dt}$$

or

$$-\frac{\operatorname{grad} p}{\rho} = \frac{d\mathbf{v}}{dt}, \tag{10.3}$$

where $(d\mathbf{v}/dt)$ is the rate of change of the velocity of a given fluid particle as that particle moves about in space. Applying the chain rule gives

$$\frac{d\mathbf{v}}{dt} = \frac{\partial \mathbf{v}}{\partial t} + \frac{\partial \mathbf{v}}{\partial x}\frac{dx}{dt} + \frac{\partial \mathbf{v}}{\partial y}\frac{dy}{dt} + \frac{\partial \mathbf{v}}{\partial z}\frac{dz}{dt} = \frac{\partial \mathbf{v}}{\partial t} + (\mathbf{v} \cdot \text{grad})\mathbf{v}. \tag{10.4}$$

By (10.3) and (10.4), it follows that

$$\frac{\partial \mathbf{v}}{\partial t} + (\mathbf{v} \cdot \text{grad})\mathbf{v} = -\frac{\text{grad}\,p}{\rho}.$$

This equation is called *Euler's equation.*

If the fluid is at rest, the velocity $\mathbf{v} = \mathbf{0}$. In this case, Euler's equation takes the form

$$\text{grad}\,p = \mathbf{0} \quad \text{or} \quad p = \text{constant}.$$

This means that the pressure is the same at every point for the fluid at rest.

If the fluid is in a gravitational field, a force $\rho\mathbf{g}$ must be added except for the force $(-\text{grad}\,p)$ acting on the unit volume of the fluid, where $\mathbf{g}$ is the gravitational acceleration. The sum of these two forces is equal to the product of the mass per unit volume and the acceleration, i.e.,

$$\rho\frac{d\mathbf{v}}{dt} = -\text{grad}\,p + \rho\mathbf{g} \quad \text{or} \quad \frac{d\mathbf{v}}{dt} = -\frac{\text{grad}\,p}{\rho} + \mathbf{g}.$$

Combining this with (10.4), we find Euler's equation takes the form

$$\frac{\partial \mathbf{v}}{\partial t} + (\mathbf{v} \cdot \text{grad})\mathbf{v} = -\frac{\text{grad}\,p}{\rho} + \mathbf{g}.$$

If the fluid is at rest in a uniform gravitational field, the velocity $\mathbf{v} = \mathbf{0}$. In this case, Euler's equation takes the form

$$\text{grad}\,p = \rho\mathbf{g}.$$

Suppose further that the fluid density is constant throughout the volume. Then Euler's equation can be expanded as

$$\frac{\partial p}{\partial x}\mathbf{i} + \frac{\partial p}{\partial y}\mathbf{j} + \frac{\partial p}{\partial z}\mathbf{k} = -\rho g\mathbf{k},$$

where $g = |\mathbf{g}|$ and $\mathbf{k} = (0, 0, 1)$. Comparing both sides of this equality, we get

$$\frac{\partial p}{\partial x} = \frac{\partial p}{\partial y} = 0,$$

$$\frac{\partial p}{\partial z} = -\rho g.$$

Therefore, the fluid pressure

$$p = -\rho g z + C,$$

where C is a constant. If $p = p_0$ at every point on the horizontal plane $z = h$, where p_0 is an external pressure, then $C = p_0 + \rho gh$, and so the fluid pressure

$$p = p_0 + \rho g(h - z).$$

Now we turn to consider isentropic motions. If the entropy is constant throughout the volume of the fluid at some initial instant, it retains everywhere the same constant value at all times and for any subsequent motion of the fluid. Such a motion is called an *isentropic motion*. In this case, the adiabatic equation can be written simply as $s =$ constant, where s is the entropy.

For isentropic motions, since s is constant,

$$\operatorname{grad} s = \left(\frac{\partial s}{\partial x}, \frac{\partial s}{\partial y}, \frac{\partial s}{\partial z} \right) = 0.$$

Let T be the fluid temperature and w be the enthalpy (the heat function per unit mass of fluid). Then the thermodynamic relation is given by

$$\operatorname{grad} w = T \operatorname{grad} s + \frac{\operatorname{grad} p}{\rho}.$$

Therefore,

$$\operatorname{grad} w = \frac{\operatorname{grad} p}{\rho}. \tag{10.5}$$

Combining this with Euler's equation, we get

$$\frac{\partial \mathbf{v}}{\partial t} + (\mathbf{v} \cdot \operatorname{grad})\mathbf{v} = -\operatorname{grad} w.$$

This equation is *Euler's equation for isentropic motions*.

Similarly, in a gravitational field, Euler's equation for isentropic motions takes the form

$$\frac{\partial \mathbf{v}}{\partial t} + (\mathbf{v} \cdot \operatorname{grad})\mathbf{v} = -\operatorname{grad} w + \mathbf{g}.$$

Euler's equation for isentropic motions can be rewritten as follows. Let $\mathbf{F} = \mathbf{v}$ in Property 10.6 and notice that $\mathbf{v} \cdot \mathbf{v} = v^2$. Then

$$(\mathbf{v} \cdot \operatorname{grad})\mathbf{v} = \frac{1}{2} \operatorname{grad} v^2 - \mathbf{v} \times \operatorname{curl} \mathbf{v},$$

and so Euler's equation for isentropic motions becomes

$$\frac{\partial \mathbf{v}}{\partial t} + \frac{1}{2} \operatorname{grad} v^2 - \mathbf{v} \times \operatorname{curl} \mathbf{v} = -\operatorname{grad} w. \tag{10.6}$$

Taking the curl on both sides of this equation, we get

$$\operatorname{curl} \left(\frac{\partial \mathbf{v}}{\partial t} \right) + \frac{1}{2} \operatorname{curl}(\operatorname{grad} v^2) - \operatorname{curl}(\mathbf{v} \times \operatorname{curl} \mathbf{v}) = -\operatorname{curl}(\operatorname{grad} w).$$

However, by Properties 10.4 and 10.5,

$$\text{curl}\left(\frac{\partial \mathbf{v}}{\partial t}\right) = \frac{\partial(\text{curl }\mathbf{v})}{\partial t},$$

$$\text{curl}(\text{grad }v^2) = \mathbf{0},$$

$$\text{curl}(\text{grad }w) = \mathbf{0}.$$

Therefore, Euler's equation for isentropic motions takes the form

$$\frac{\partial(\text{curl }\mathbf{v})}{\partial t} = \text{curl}(\mathbf{v} \times \text{curl }\mathbf{v}).$$

This form involves only the fluid velocity.

Similarly, in a gravitational field, Euler's equation for isentropic motions takes the form

$$\frac{\partial(\text{curl }\mathbf{v})}{\partial t} = \text{curl}(\mathbf{v} \times \text{curl }\mathbf{v}) + \text{curl }\mathbf{g}.$$

In Chapter 11, we will introduce the generalization of Euler's equation, i.e., the Navier-Stokes equation.

10.4.3 Bernoulli's Equation

The derivation of Bernoulli's equation needs three concepts: the directional derivative, steady flow, and streamline.

Let u be a scalar function, and let $l = \cos\alpha\mathbf{i} + \cos\beta\mathbf{j} + \cos\gamma\mathbf{k}$ be a unit vector satisfying

$$\cos^2\alpha + \cos^2\beta + \cos^2\gamma = 1.$$

The *directional derivative* of u along the direction l is defined as

$$\frac{\partial u}{\partial l} = \frac{\partial u}{\partial x}\cos\alpha + \frac{\partial u}{\partial y}\cos\beta + \frac{\partial u}{\partial z}\cos\gamma.$$

Notice that

$$\frac{\partial u}{\partial x}\cos\alpha + \frac{\partial u}{\partial y}\cos\beta + \frac{\partial u}{\partial z}\cos\gamma = \left(\frac{\partial u}{\partial x}\mathbf{i} + \frac{\partial u}{\partial y}\mathbf{j} + \frac{\partial u}{\partial z}\mathbf{k}\right) \cdot$$
$$(\cos\alpha\mathbf{i} + \cos\beta\mathbf{j} + \cos\gamma\mathbf{k}).$$

Since $\text{grad }u = \frac{\partial u}{\partial x}\mathbf{i} + \frac{\partial u}{\partial y}\mathbf{j} + \frac{\partial u}{\partial z}\mathbf{k}$, the directional derivative can be expressed as follows:

$$\frac{\partial u}{\partial l} = (\text{grad }u) \cdot l.$$

This means that the directional derivative of u along the direction l is equal to the scalar product of $\text{grad }u$ and the unit vector l.

If the fluid velocity is constant in time at any point occupied by fluid, then such a flow is called a *steady flow*. Therefore, for a steady flow, the fluid velocity **v** is a function of only the coordinates, and so

$$\frac{\partial \mathbf{v}}{\partial t} = 0.$$

If the direction of the tangent line at any point on a curve is the direction of the fluid velocity at that point, such a curve is called a *streamline*. Thus, the streamline is determined by the system of differential equations

$$\frac{dx}{v_1} = \frac{dy}{v_2} = \frac{dz}{v_3},$$

where the fluid velocity $\mathbf{v} = (v_1, v_2, v_3)$. In a steady flow, the streamlines do not vary with time and they coincide with the paths of the fluid particles.

In a steady flow, since $\frac{\partial \mathbf{v}}{\partial t} = 0$, by (10.6), Euler's equation for isentropic motions is

$$\mathbf{v} \times \operatorname{curl} \mathbf{v} = \operatorname{grad} w + \frac{1}{2}\operatorname{grad} v^2,$$

where w is the enthalpy and $v^2 = \mathbf{v} \cdot \mathbf{v}$. Consider a streamline in the steady flow. Denote by l the unit vector tangent to the streamline at each point. Taking the scalar product with the unit vector l on both sides, we get

$$(\mathbf{v} \times \operatorname{curl} \mathbf{v}) \cdot l = (\operatorname{grad} w) \cdot l + \frac{1}{2}(\operatorname{grad} v^2) \cdot l.$$

Consider each term of this equation. It is clear that the vector $(\mathbf{v} \times \operatorname{curl} \mathbf{v})$ is perpendicular to the direction of the velocity **v**. Since the direction of l at any point of the streamline is the direction of **v** at that point of the streamline, the vector $(\mathbf{v} \times \operatorname{curl} \mathbf{v})$ is perpendicular to l, and so

$$(\mathbf{v} \times \operatorname{curl} \mathbf{v}) \cdot l = 0.$$

Since the directional derivative can be expressed as,

$$(\operatorname{grad} w) \cdot l = \frac{\partial w}{\partial l},$$

$$(\operatorname{grad} v^2) \cdot l = \frac{\partial v^2}{\partial l}.$$

Therefore,

$$\frac{\partial}{\partial l}\left(w + \frac{1}{2}v^2\right) = 0.$$

Since the vector l is the unit tangent vector at any point of the streamline,

$$w + \frac{1}{2}v^2 = \text{constant}$$

along a streamline in a steady flow for isentropic motions, where $v^2 = \mathbf{v} \cdot \mathbf{v}$ and w is the enthalpy. This equation is called *Bernoulli's equation for steady flow* or *Bernoulli's equation*. The constant in Bernoulli's equation, in general, takes different values for different streamlines.

In a gravitational field, Euler's equation for isentropic motions is

$$\frac{\partial \mathbf{v}}{\partial t} + \frac{1}{2} \text{grad } v^2 - \mathbf{v} \times \text{curl } \mathbf{v} = -\text{grad } w - g\,\mathbf{k},$$

where $g = |\mathbf{g}|$ and $\mathbf{k} = (0, 0, 1)$. In the steady flow, since $\frac{\partial \mathbf{v}}{\partial t} = 0$, Euler's equation for isentropic motions becomes

$$\mathbf{v} \times \text{curl } \mathbf{v} = \text{grad } w + \frac{1}{2} \text{grad } v^2 + g\,\mathbf{k}.$$

Consider a streamline in the steady flow. Denote by l the unit vector tangent to the streamline at each point. Taking the scalar product with the vector l on both sides, we get

$$(\mathbf{v} \times \text{curl } \mathbf{v}) \cdot l = (\text{grad } w) \cdot l + \frac{1}{2} (\text{grad } v^2) \cdot l + g(\mathbf{k} \cdot l).$$

Notice that

$$(\mathbf{v} \times \text{curl } \mathbf{v}) \cdot l = 0,$$

$$(\text{grad } w) \cdot l = \frac{\partial w}{\partial l},$$

$$(\text{grad } v^2) \cdot l = \frac{\partial v^2}{\partial l}.$$

Then

$$\frac{\partial w}{\partial l} + \frac{1}{2} \frac{\partial v^2}{\partial l} + g(\mathbf{k} \cdot l) = 0.$$

Let θ be the angle between $\mathbf{k}$ and l. Notice that $|\mathbf{k}| = 1$, $|l| = 1$ and $\cos \theta = \frac{\partial z}{\partial l}$. Then

$$\mathbf{k} \cdot l = |\mathbf{k}||l| \cos \theta = \frac{\partial z}{\partial l},$$

and so

$$\frac{\partial w}{\partial l} + \frac{1}{2} \frac{\partial v^2}{\partial l} + g \frac{\partial z}{\partial l} = 0$$

or

$$\frac{\partial}{\partial l} \left(w + \frac{1}{2} v^2 + gz \right) = 0.$$

This implies that in a gravitational field,

$$\frac{1}{2} v^2 + w + gz = \text{constant}$$

along a streamline in a steady flow for the isentropic motion, where $v^2 = \mathbf{v} \cdot \mathbf{v}, w$ is the enthalpy, and $g = |\mathbf{g}|$. This is *Bernoulli's equation* in a gravitational field.

10.5   ENERGY FLUX AND MOMENTUM FLUX

The studies of the energy flux and the momentum flux are based on the fundamental equations of fluid dynamics. Let ρ be the fluid density, $\mathbf{v}$ be the fluid velocity, and p be the fluid pressure.

First, we study the energy flux.

The *energy* of a unit volume of fluid consists of two parts. One is the *kinetic energy* of a unit volume of fluid, the other is the *internal energy* of a unit volume of fluid, i.e.,

$$E = \frac{1}{2}\rho v^2 + \rho \epsilon,$$

where $v^2 = (\mathbf{v} \cdot \mathbf{v})$ and ϵ is the internal energy per unit mass. The first term, $\frac{1}{2}\rho v^2$, is the kinetic energy of a unit volume of fluid. The second term, $\rho \epsilon$, is the internal energy of a unit volume of fluid. The rate of change of energy is given by the partial derivative

$$\frac{\partial E}{\partial t} = \frac{\partial}{\partial t}\left(\frac{1}{2}\rho v^2 + \rho\epsilon\right) = \frac{\partial}{\partial t}\left(\frac{1}{2}\rho v^2\right) + \frac{\partial(\rho\epsilon)}{\partial t},$$

where the first term is the rate of change of the kinetic energy and the second term is the rate of change of the internal energy.

We compute the rate of change of the kinetic energy.

It is clear that

$$\frac{\partial}{\partial t}\left(\frac{1}{2}\rho v^2\right) = \frac{1}{2}v^2\frac{\partial\rho}{\partial t} + \frac{1}{2}\rho\frac{\partial(v^2)}{\partial t}. \tag{10.7}$$

Let $\mathbf{v} = (v_1, v_2, v_3)$. Then

$$v^2 = \mathbf{v} \cdot \mathbf{v} = v_1^2 + v_2^2 + v_3^2,$$

$$\frac{\partial\mathbf{v}}{\partial t} = \left(\frac{\partial v_1}{\partial t}, \frac{\partial v_2}{\partial t}, \frac{\partial v_3}{\partial t}\right),$$

and so

$$\frac{\partial(v^2)}{\partial t} = \frac{\partial}{\partial t}\left(v_1^2 + v_2^2 + v_3^2\right) = 2\left(v_1\frac{\partial v_1}{\partial t} + v_2\frac{\partial v_2}{\partial t} + v_3\frac{\partial v_3}{\partial t}\right)$$

$$= 2\mathbf{v} \cdot \frac{\partial\mathbf{v}}{\partial t}.$$

From this and (10.7),

$$\frac{\partial}{\partial t}\left(\frac{1}{2}\rho v^2\right) = \frac{1}{2}v^2\frac{\partial\rho}{\partial t} + \rho\mathbf{v} \cdot \frac{\partial\mathbf{v}}{\partial t}.$$

Using the continuity equation and Euler's equation, we get

$$\frac{\partial \rho}{\partial t} = -\mathrm{div}(\rho \mathbf{v}),$$

$$\frac{\partial \mathbf{v}}{\partial t} = -(\mathbf{v} \cdot \mathrm{grad})\mathbf{v} - \frac{\mathrm{grad}\, p}{\rho},$$

and so

$$\frac{\partial}{\partial t}\left(\frac{1}{2}\rho\, v^2\right) = -\frac{1}{2}v^2 \,\mathrm{div}(\rho \mathbf{v}) - \rho \mathbf{v} \cdot (\mathbf{v} \cdot \mathrm{grad})\mathbf{v} - \mathbf{v} \cdot \mathrm{grad}\, p. \tag{10.8}$$

If we replace $\mathbf{F}$ by $\mathbf{v}$, Property 10.6 becomes:

$$\mathbf{v} \times \mathrm{curl}\, \mathbf{v} + (\mathbf{v} \cdot \mathrm{grad})\mathbf{v} = \frac{1}{2}\mathrm{grad}\, v^2.$$

Taking the scalar product with $\mathbf{v}$ on both sides, we get

$$\mathbf{v} \cdot (\mathbf{v} \times \mathrm{curl}\, \mathbf{v}) + \mathbf{v} \cdot (\mathbf{v} \cdot \mathrm{grad})\mathbf{v} = \frac{1}{2}\mathbf{v} \cdot \mathrm{grad}\, v^2.$$

Since the vector $(\mathbf{v} \times \mathrm{curl}\, \mathbf{v})$ is perpendicular to $\mathbf{v}$, clearly $\mathbf{v} \cdot (\mathbf{v} \times \mathrm{curl}\, \mathbf{v}) = 0$. So

$$\mathbf{v} \cdot (\mathbf{v} \cdot \mathrm{grad})\mathbf{v} = \frac{1}{2}\mathbf{v} \cdot \mathrm{grad}\, v^2,$$

and so the second term on the right-hand side of (10.8)

$$\rho \mathbf{v} \cdot (\mathbf{v} \cdot \mathrm{grad})\mathbf{v} = \frac{1}{2}\rho \mathbf{v} \cdot \mathrm{grad}\, v^2.$$

By the thermodynamic relation

$$\mathrm{grad}\, p = \rho \,\mathrm{grad}\, w - \rho\, T \,\mathrm{grad}\, s,$$

the last term on the right-hand side of (10.8)

$$\mathbf{v} \cdot \mathrm{grad}\, p = \rho \mathbf{v} \cdot \mathrm{grad}\, w - \rho\, T \mathbf{v} \cdot \mathrm{grad}\, s.$$

Therefore, by (10.8), the rate of change of the kinetic energy is

$$\frac{\partial}{\partial t}\left(\frac{1}{2}\rho\, v^2\right) = -\frac{1}{2}v^2 \,\mathrm{div}(\rho \mathbf{v}) - \frac{1}{2}\rho \mathbf{v} \cdot \mathrm{grad}\, v^2 - \rho \mathbf{v} \cdot \mathrm{grad}\, w + \rho\, T \mathbf{v} \cdot \mathrm{grad}\, s.$$

We compute the rate of change of the internal energy.

Since the enthalpy $w = \epsilon + \frac{p}{\rho}$, from the thermodynamic relation $\mathrm{d}\epsilon = T\mathrm{d}s + \frac{p}{\rho^2}\mathrm{d}\rho$, it follows that

$$\frac{\partial(\rho\epsilon)}{\partial t} = \epsilon\frac{\partial \rho}{\partial t} + \rho\frac{\partial \epsilon}{\partial t} = \left(w - \frac{p}{\rho}\right)\frac{\partial \rho}{\partial t} + \rho\left(T\frac{\partial s}{\partial t} + \frac{p}{\rho^2}\frac{\partial \rho}{\partial t}\right)$$

$$= w\frac{\partial \rho}{\partial t} + \rho\, T\frac{\partial s}{\partial t}.$$

The continuity equation and the general equation for the adiabatic motion of an ideal fluid have been given by

$$\frac{\partial \rho}{\partial t} = -\operatorname{div}(\rho \mathbf{v}),$$

$$\frac{\partial s}{\partial t} = -\mathbf{v} \cdot \operatorname{grad} s.$$

So the rate of change of the internal energy is

$$\frac{\partial (\rho \epsilon)}{\partial t} = -w \operatorname{div}(\rho \mathbf{v}) - \rho \, T \mathbf{v} \cdot \operatorname{grad} s.$$

Therefore, the sum of the rates of change of the kinetic energy and the internal energy is

$$\frac{\partial}{\partial t} \left(\frac{1}{2} \rho v^2 \right) + \frac{\partial \rho \epsilon}{\partial t} = - \left(\frac{1}{2} v^2 + w \right) \operatorname{div}(\rho \mathbf{v}) - \rho \mathbf{v} \cdot \operatorname{grad} \left(\frac{1}{2} v^2 + w \right).$$

By Property 10.3, the right-hand side is equal to the negative value of the divergence of $\left(\frac{1}{2} v^2 + w \right) \rho \mathbf{v}$. So the rate of change of the energy

$$\frac{\partial}{\partial t} \left(\frac{1}{2} \rho \, v^2 + \rho \epsilon \right) = -\operatorname{div} \left(\left(\frac{1}{2} v^2 + w \right) \rho \mathbf{v} \right). \tag{10.9}$$

The meaning of equality (10.9) is as follows.

Choose some volume V fixed in space and enclosed by a closed surface S. Integrating both sides of (10.9) over V, we get

$$\int \int \int_V \frac{\partial}{\partial t} \left(\frac{1}{2} \rho \, v^2 + \rho \epsilon \right) dV = - \int \int \int_V \operatorname{div} \left(\left(\frac{1}{2} v^2 + w \right) \rho \mathbf{v} \right) dV.$$

The divergence theorem says that the volume integral on the right-hand side can be converted into a surface integral as follows:

$$\int \int \int_V \operatorname{div} \left(\left(\frac{1}{2} v^2 + w \right) \rho \mathbf{v} \right) dV = \int \int_S \left(\frac{1}{2} v^2 + w \right) \rho \mathbf{v} \cdot \mathbf{n} \, dS,$$

where $\mathbf{n}$ is the outward unit normal vector of the surface S. Therefore,

$$\frac{\partial}{\partial t} \int \int \int_V \left(\frac{1}{2} \rho \, v^2 + \rho \epsilon \right) dV = - \int \int_S \left(\frac{1}{2} v^2 + w \right) \rho \mathbf{v} \cdot \mathbf{n} \, dS.$$

The left-hand side is the rate of change of the energy of the fluid in the volume V. The right-hand side is the *energy flux* flowing out of the volume V in unit time. Hence, the expression

$$\left(\frac{1}{2} v^2 + w \right) \rho \mathbf{v}$$

is called the *energy flux density vector*. Its magnitude

$$\left(\frac{1}{2}v^2 + w\right)\rho\,v$$

is the amount of energy passing in unit time through unit area perpendicular to the direction of the velocity. Therefore, the meaning of (10.9) is that the rate of change of the energy consisting of the kinetic energy and the internal energy is equal to the negative value of the divergence of the energy flux density vector.

Since the enthalpy $w = \epsilon + \frac{p}{\rho}$, the energy flux through a closed surface S is rewritten in the form

$$\int\int_S \left(\frac{1}{2}v^2 + w\right)\rho\mathbf{v}\cdot\mathbf{n}\,\mathrm{d}S = \int\int_S \left(\frac{1}{2}v^2 + \epsilon\right)\rho\mathbf{v}\cdot\mathbf{n}\,\mathrm{d}S + \int\int_S p\,\mathbf{v}\cdot\mathbf{n}\,\mathrm{d}S.$$

The first term on the right-hand side is the kinetic energy and the internal energy transported through the surface in unit time by the mass of fluid. The second term on the right-hand side is the work done by pressure force on the fluid within the surface.

Second, we study the momentum flux.

The *momentum* of a fluid of unit volume is $\rho\mathbf{v}$. We compute the rate of change of the momentum.

Let $\mathbf{v} = (v_1, v_2, v_3)$. Then

$$\frac{\partial(\rho\mathbf{v})}{\partial t} = \left(\frac{\partial(\rho v_1)}{\partial t}, \frac{\partial(\rho v_2)}{\partial t}, \frac{\partial(\rho v_3)}{\partial t}\right),$$

where

$$\frac{\partial(\rho\,v_i)}{\partial t} = \rho\frac{\partial v_i}{\partial t} + v_i\frac{\partial\rho}{\partial t} \quad (i = 1, 2, 3) \tag{10.10}$$

is the rate of change of the ith component of the momentum of a fluid of unit volume.

The partial derivatives $\frac{\partial v_i}{\partial t}$ $(i = 1, 2, 3)$ and $\frac{\partial\rho}{\partial t}$ are computed by using Euler's equation and continuity equation as follows.

Notice that

$$\frac{\partial\mathbf{v}}{\partial t} = \frac{\partial v_1}{\partial t}\mathbf{i} + \frac{\partial v_2}{\partial t}\mathbf{j} + \frac{\partial v_3}{\partial t}\mathbf{k},$$

$$(\mathbf{v}\cdot\mathrm{grad})\mathbf{v} = \left(v_1\frac{\partial}{\partial x_1} + v_2\frac{\partial}{\partial x_2} + v_3\frac{\partial}{\partial x_3}\right)(v_1\mathbf{i} + v_2\mathbf{j} + v_3\mathbf{k})$$

$$= v_1\left(\frac{\partial v_1}{\partial x_1}\mathbf{i} + \frac{\partial v_2}{\partial x_1}\mathbf{j} + \frac{\partial v_3}{\partial x_1}\mathbf{k}\right)$$

$$+ v_2\left(\frac{\partial v_1}{\partial x_2}\mathbf{i} + \frac{\partial v_2}{\partial x_2}\mathbf{j} + \frac{\partial v_3}{\partial x_2}\mathbf{k}\right)$$

$$+ v_3 \left(\frac{\partial v_1}{\partial x_3}\mathbf{i} + \frac{\partial v_2}{\partial x_3}\mathbf{j} + \frac{\partial v_3}{\partial x_3}\mathbf{k} \right),$$

$$\operatorname{grad} p = \frac{\partial p}{\partial x_1}\mathbf{i} + \frac{\partial p}{\partial x_2}\mathbf{j} + \frac{\partial p}{\partial x_3}\mathbf{k}.$$

Euler's equation $\frac{\partial \mathbf{v}}{\partial t} + (\mathbf{v} \cdot \operatorname{grad})\mathbf{v} = -\frac{\operatorname{grad} p}{\rho}$ is equivalent to

$$\frac{\partial v_1}{\partial t}\mathbf{i} + \frac{\partial v_2}{\partial t}\mathbf{j} + \frac{\partial v_3}{\partial t}\mathbf{k} + v_1 \left(\frac{\partial v_1}{\partial x_1}\mathbf{i} + \frac{\partial v_2}{\partial x_1}\mathbf{j} + \frac{\partial v_3}{\partial x_1}\mathbf{k} \right)$$

$$+ v_2 \left(\frac{\partial v_1}{\partial x_2}\mathbf{i} + \frac{\partial v_2}{\partial x_2}\mathbf{j} + \frac{\partial v_3}{\partial x_2}\mathbf{k} \right) + v_3 \left(\frac{\partial v_1}{\partial x_3}\mathbf{i} + \frac{\partial v_2}{\partial x_3}\mathbf{j} + \frac{\partial v_3}{\partial x_3}\mathbf{k} \right)$$

$$= -\frac{1}{\rho} \left(\frac{\partial p}{\partial x_1}\mathbf{i} + \frac{\partial p}{\partial x_2}\mathbf{j} + \frac{\partial p}{\partial x_3}\mathbf{k} \right).$$

Comparing both sides gives

$$\frac{\partial v_i}{\partial t} = -\sum_1^3 v_k \frac{\partial v_i}{\partial x_k} - \frac{1}{\rho}\frac{\partial p}{\partial x_i} \quad (i = 1, 2, 3).$$

By the definition of the divergence, the continuity equation $\frac{\partial \rho}{\partial t} = -\operatorname{div}(\rho \mathbf{v})$ is equivalent to

$$\frac{\partial \rho}{\partial t} = -\sum_1^3 \frac{\partial(\rho\, v_k)}{\partial x_k}.$$

Therefore, by (10.10), the rate of change of the ith component of the momentum is

$$\frac{\partial(\rho\, v_i)}{\partial t} = -\rho \left(\sum_{k=1}^3 v_k \frac{\partial v_i}{\partial x_k} + \frac{1}{\rho}\frac{\partial p}{\partial x_i} \right) - v_i \sum_{k=1}^3 \frac{\partial(\rho\, v_k)}{\partial x_k}$$

$$= -\sum_{k=1}^3 \left(\rho\, v_k \frac{\partial v_i}{\partial x_k} + v_i \frac{\partial(\rho\, v_k)}{\partial x_k} \right) - \frac{\partial p}{\partial x_i}$$

$$= -\sum_{k=1}^3 \frac{\partial(\rho\, v_k\, v_i)}{\partial x_k} - \frac{\partial p}{\partial x_i} \quad (i = 1, 2, 3).$$

Let

$$\delta_{k,i} = \begin{cases} 1, & k = i, \\ 0, & k \neq i. \end{cases}$$

Then

$$\frac{\partial(\rho\, v_i)}{\partial t} = -\sum_{k=1}^3 \frac{\partial(\rho\, v_k v_i + \delta_{k,i} p)}{\partial x_k} \quad (i = 1, 2, 3).$$

The right-hand side is

$$\sum_{k=1}^{3} \frac{\partial(\rho\, v_k v_i + \delta_{k,i}p)}{\partial x_k} = \frac{\partial(\rho\, v_1 v_i + \delta_{1,i}p)}{\partial x_1} + \frac{\partial(\rho\, v_2 v_i + \delta_{2,i}p)}{\partial x_2}$$

$$+ \frac{\partial(\rho\, v_3 v_i + \delta_{3,i}p)}{\partial x_3} = \operatorname{div} \mathbf{F},$$

where

$$\mathbf{F} = (\rho\, v_1 v_i + \delta_{1,i}p)\mathbf{i} + (\rho\, v_2 v_i + \delta_{2,i}p)\mathbf{j} + (\rho\, v_3 v_i + \delta_{3,i}p)\mathbf{k}.$$

Therefore, the rate of change of the ith component of the momentum

$$\frac{\partial(\rho\, v_i)}{\partial t} = -\operatorname{div} \mathbf{F} \quad (i = 1, 2, 3).$$

This means that the rate of change of the ith component of the momentum of a fluid of unit volume is the negative value of the divergence of the vector $\mathbf{F}$.

Choose some volume V fixed in space and enclosed by a closed surface S. Integrating both sides over V, we get

$$\iiint_{V} \frac{\partial(\rho\, v_i)}{\partial t}\, dV = - \iiint_{V} \operatorname{div} \mathbf{F}\, dV \quad (i = 1, 2, 3).$$

The divergence theorem says that the volume integral on the right-hand side can be converted into a surface integral:

$$\iiint_{V} \operatorname{div} \mathbf{F}\, dV = \iint_{S} \mathbf{F} \cdot \mathbf{n}\, dS,$$

where $\mathbf{n} = (n_1, n_2, n_3)$ is the outward unit normal vector on the surface S. Therefore,

$$\iiint_{V} \frac{\partial(\rho\, v_i)}{\partial t}\, dV = - \iint_{S} \mathbf{F} \cdot \mathbf{n}\, dS \quad (i = 1, 2, 3). \tag{10.11}$$

The volume integral on the left-hand side is the rate of change of the ith component of the momentum contained in the volume considered. The surface integral on the right-hand side is the *momentum flux* flowing out through the closed surface in unit time. Notice that

$$\mathbf{F} \cdot \mathbf{n} = \sum_{k=1}^{3} (\rho\, v_k v_i + \delta_{k,i}p)n_k \quad (i = 1, 2, 3).$$

The expression $(\rho\, v_k v_i + \delta_{k,i}p)n_k$ is the flux of the ith component of the momentum through the surface's unit area. Hence, the term $\rho\, v_k v_i + \delta_{k,i}p$ is called the *momentum flux density tensor* and the vector $\mathbf{F}$ is called the *momentum flux density tensor vector*. Since

$$\mathbf{F} \cdot \mathbf{n} = \rho\, v_i \sum_{k=1}^{3} v_k n_k + p \sum_{k=1}^{3} \delta_{k,i} n_k = \rho\, v_i (\mathbf{v} \cdot \mathbf{n}) + p n_i \quad (i = 1, 2, 3),$$

Equality (10.11) can be rewritten in the form

$$\frac{\partial}{\partial t} \int \int \int_V \rho v_i dV = - \int \int_S (\rho \, v_i (\mathbf{v} \cdot \mathbf{n}) + p n_i) \, dS \quad (i = 1, 2, 3).$$

Multiplying both sides of these three equations by $\mathbf{i}, \mathbf{j}, \mathbf{k}$, respectively, and then adding them together, we get

$$\mathbf{i} \left(\frac{\partial}{\partial t} \int \int \int_V \rho \, v_1 \, dV \right) + \mathbf{j} \left(\frac{\partial}{\partial t} \int \int \int_V \rho \, v_2 \, dV \right) + \mathbf{k} \left(\frac{\partial}{\partial t} \int \int \int_V \rho \, v_3 \, dV \right)$$

$$= - \mathbf{i} \int \int_S (\rho \, v_1 (\mathbf{v} \cdot \mathbf{n}) + p n_1) \, dS - \mathbf{j} \int \int_S (\rho \, v_2 (\mathbf{v} \cdot \mathbf{n}) + p n_2) \, dS$$

$$- \mathbf{k} \int \int_S (\rho \, v_3 (\mathbf{v} \cdot \mathbf{n}) + p n_3) \, dS$$

which is equivalent to

$$\frac{\partial}{\partial t} \int \int \int_V \rho \mathbf{v} \, dV = - \int \int_S (\rho \mathbf{v} (\mathbf{v} \cdot \mathbf{n}) + p \mathbf{n}) \, dS.$$

In this equality, the left-hand side is the rate of change of the momentum contained in the volume considered. The vector $\rho \mathbf{v} (\mathbf{v} \cdot \mathbf{n}) + p \mathbf{n}$ is the *momentum flux* in the direction of $\mathbf{n}$, so the right-hand side is the *momentum flux* flowing out through the closed surface perpendicular to $\mathbf{n}$ in unit time.

10.6 KELVIN LAW

The Kelvin law is the *conservation law of circulation*.

Let $\mathbf{v}$ be the fluid velocity and the curve C be a closed fluid contour with the unit tangent vector $\mathbf{T}$. The integral

$$\Gamma = \oint_C \mathbf{v} \cdot \mathbf{T} \, dl$$

is called the *velocity circulation* around the closed fluid contour C, where dl is the arc element.

Notice that $\mathbf{T} \, dl = d\mathbf{r}$, where $\mathbf{r}$ is the position vector. Thus, the velocity circulation can be written as

$$\Gamma = \oint_C \mathbf{v} \cdot d\mathbf{r}.$$

To avoid confusion, we denote differentiation with respect to the coordinates by the symbol δ and denote differentiation with respect to the time by the symbol d.

As the fluid counter moves, the time derivative of the velocity circulation is

$$\frac{d\Gamma}{dt} = \frac{d}{dt} \oint_C \mathbf{v} \cdot \delta \mathbf{r} = \oint_C \frac{d\mathbf{v}}{dt} \cdot \delta \mathbf{r} + \oint_C \mathbf{v} \cdot \frac{d\delta \mathbf{r}}{dt}. \tag{10.12}$$

According to Stokes's theorem, the first integral on the right-hand side of (10.12) is transformed to a surface integral:

$$\oint_C \frac{d\mathbf{v}}{dt} \cdot \delta\mathbf{r} = \int\int_S \text{curl}\left(\frac{d\mathbf{v}}{dt}\right) \cdot \mathbf{n}\,\delta S,$$

where $\mathbf{n}$ is the outward unit normal vector on S. For an isentropic motion, by (10.5) and Euler's equation,

$$\text{grad}\, w = \frac{\text{grad}\, p}{\rho},$$

$$\frac{d\mathbf{v}}{dt} = -\frac{\text{grad}\, p}{\rho},$$

it follows that

$$\frac{d\mathbf{v}}{dt} = -\text{grad}\, w,$$

where w is the enthalpy. Taking the curl on both sides, we get

$$\text{curl}\left(\frac{d\mathbf{v}}{dt}\right) = -\text{curl}(\text{grad}\, w).$$

By Property 10.5, $\text{curl}(\text{grad}\, w) = \mathbf{0}$, and so the first integral on the right-hand side of (10.12) is

$$\oint_C \frac{d\mathbf{v}}{dt} \cdot \delta\mathbf{r} = -\int\int_S \text{curl}(\text{grad}\, w) \cdot \mathbf{n}\,\delta S = -\int\int_S \mathbf{0} \cdot \mathbf{n}\,\delta S = 0.$$

Notice that

$$\frac{d\delta\mathbf{r}}{dt} = \delta\left(\frac{d\mathbf{r}}{dt}\right) = \delta\mathbf{v},$$

$$\delta v^2 = \delta(\mathbf{v} \cdot \mathbf{v}) = 2\mathbf{v} \cdot \delta\mathbf{v}.$$

The second integral on the right-hand side of (10.12) is

$$\oint_C \mathbf{v} \cdot \frac{d\delta\mathbf{r}}{dt} = \oint_C \mathbf{v} \cdot \delta\mathbf{v}. = \frac{1}{2}\oint_C \delta v^2 = 0.$$

Therefore, by (10.12),

$$\frac{d}{dt}\oint_C \mathbf{v} \cdot d\mathbf{r} = \oint_C \frac{d\mathbf{v}}{dt} \cdot \delta\mathbf{r} + \oint_C \mathbf{v} \cdot \frac{d\delta\mathbf{r}}{dt} = 0,$$

i.e., the time derivative of the velocity circulation around a closed fluid contour is equal to zero. So

$$\oint_C \mathbf{v} \cdot d\mathbf{r} = \text{constant},$$

i.e., the velocity circulation around a closed fluid contour is a constant.

Kelvin Law. *In an ideal fluid, for isentropic motions, the velocity circulation around a closed fluid contour is constant in time.*

Stokes's theorem says that the circulation of the velocity $\mathbf{v}$ around a closed fluid contour C can be transformed to a double integral of curl $\mathbf{v}$ over the oriented surface S enclosed by C:

$$\oint_C \mathbf{v} \cdot \mathbf{T}\delta l = \int\int_S (\text{curl } \mathbf{v}) \cdot \mathbf{n}\,\delta S.$$

The Kelvin law says that the circulation of the velocity $\mathbf{v}$ around a closed fluid contour C is constant in time:

$$\oint_C \mathbf{v} \cdot \mathbf{T}\delta l = \text{constant}.$$

Therefore,

$$\int\int_S (\text{curl } \mathbf{v}) \cdot \mathbf{n}\,\delta S = \text{constant},$$

where the vector curl $\mathbf{v}$ is called the *vorticity* of the fluid flow.

10.7 POTENTIAL FUNCTION AND POTENTIAL FLOW

Let $\mathbf{F}$ be a three-dimensional differentiable field and f be a scalar function, and let both $\mathbf{F}$ and f be defined on a region Ω in space. If $\mathbf{F} = \text{grad} f$, then f is called a *potential function* or *potential* of $\mathbf{F}$.

For example, if $\mathbf{F}$ is an electric field, then f is the electric potential; if $\mathbf{F}$ is a gravitational field, then f is a gravitational potential; if $\mathbf{F}$ is a velocity field, then f is a velocity potential.

Proposition 10.1. $\mathbf{F} = \text{grad} f$ *on* Ω *if and only if the integral* $\int_{C_{AB}} \mathbf{F} \cdot \mathbf{T}\, dl$ *is path independent on* Ω*, where* C_{AB} *is any smooth curve joining two points* A, B *on* Ω *and* $\mathbf{T}$ *is the unit tangent vector of* C_{AB}*.*

Proof. We only prove "if" part. Suppose that $\mathbf{F} = \text{grad} f$ on Ω. Then

$$\mathbf{F} = \frac{\partial f}{\partial x}\mathbf{i} + \frac{\partial f}{\partial y}\mathbf{j} + \frac{\partial f}{\partial z}\mathbf{k}.$$

Suppose that the equation of the smooth curve C_{AB} is

$$\mathbf{r}(t) = x(t)\mathbf{i} + y(t)\mathbf{j} + z(t)\mathbf{k} \quad (a \le t \le b)$$

and $A = \mathbf{r}(a), B = \mathbf{r}(b)$. Then

$$\mathbf{F} \cdot \frac{d\mathbf{r}}{dt} = \left(\frac{\partial f}{\partial x}\mathbf{i} + \frac{\partial f}{\partial y}\mathbf{j} + \frac{\partial f}{\partial z}\mathbf{k}\right) \cdot \left(\frac{dx}{dt}\mathbf{i} + \frac{dy}{dt}\mathbf{j} + \frac{dz}{dt}\mathbf{k}\right) = \frac{df}{dt}.$$

Since $\mathbf{T}$ is the unit tangent vector on C_{AB}, $\mathbf{T}dl = d\mathbf{r}$, and so

$$\int_{C_{AB}} \mathbf{F} \cdot \mathbf{T}\, dl = \int_{C_{AB}} \mathbf{F} \cdot d\mathbf{r} = \int_a^b \left(\mathbf{F} \cdot \frac{d\mathbf{r}}{dt}\right) dt$$

$$= \int_a^b \frac{df}{dt} \, dt = \int_A^B df = f(B) - f(A).$$

Thus, the integral $\int_{C_{AB}} \mathbf{F} \cdot \mathbf{T} \, dl$ is path independent in Ω. $\qquad\square$

Proposition 10.2. curl $\mathbf{F} = \mathbf{0}$ *on* Ω *if and only if* $\mathbf{F} = \operatorname{grad} f$ *on* Ω, *where* f *is a scalar function.*

Proof. Suppose that curl $\mathbf{F} = \mathbf{0}$. Then the surface integral

$$\int\int_S (\operatorname{curl} \mathbf{F}) \cdot \mathbf{n} \, dS = 0,$$

where the surface S is enclosed by a closed curve $C \subset \Omega$. Applying Stokes's theorem gives

$$\oint_C \mathbf{F} \cdot \mathbf{T} \, dl = \int\int_S (\operatorname{curl} \mathbf{F}) \cdot \mathbf{n} \, dS = 0.$$

This implies that the line integral $\int_{C_{AB}} \mathbf{F} \cdot \mathbf{T} \, dl$ is path independent on Ω, where C_{AB} is any curve joining any two points A, B on Ω. By Proposition 10.1, $\mathbf{F} = \operatorname{grad} f$, where f is a scalar function.

Conversely, suppose that $\mathbf{F} = \operatorname{grad} f$, where f is a scalar function. Let

$$\mathbf{F} = M(x,y,z)\mathbf{i} + N(x,y,z)\mathbf{j} + P(x,y,z)\mathbf{k}.$$

Notice that $\operatorname{grad} f = \frac{\partial f}{\partial x}\mathbf{i} + \frac{\partial f}{\partial y}\mathbf{j} + \frac{\partial f}{\partial z}\mathbf{k}$. Then

$$M(x,y,z) = \frac{\partial f}{\partial x},$$

$$N(x,y,z) = \frac{\partial f}{\partial y},$$

$$P(x,y,z) = \frac{\partial f}{\partial z},$$

and so

$$\frac{\partial M}{\partial y} = \frac{\partial N}{\partial x}, \quad \frac{\partial M}{\partial z} = \frac{\partial P}{\partial x}, \quad \frac{\partial N}{\partial z} = \frac{\partial P}{\partial y}.$$

This implies that

$$\operatorname{curl} \mathbf{F} = \left(\frac{\partial P}{\partial y} - \frac{\partial N}{\partial z} \right)\mathbf{i} + \left(\frac{\partial M}{\partial z} - \frac{\partial P}{\partial x} \right)\mathbf{j} + \left(\frac{\partial N}{\partial x} - \frac{\partial M}{\partial y} \right)\mathbf{k} = \mathbf{0}.$$
$\qquad\square$

A flow for which curl $\mathbf{v} = \mathbf{0}$ everywhere is called a *potential flow* or *irrotational flow*. A flow for which curl $\mathbf{v}$ is not zero everywhere is called a *rotational flow*.

In a potential flow, since curl $\mathbf{v} = \mathbf{0}$, according to Stokes's theorem, the velocity circulation round any smooth closed fluid contour C is equal to zero, i.e.,

$$\oint_C \mathbf{v} \cdot \mathbf{T} \delta l = \int\int_S \operatorname{curl} \mathbf{v} \cdot \mathbf{n}\, \delta S = 0,$$

where S is the surface enclosed by the closed curve C and $\mathbf{n}$ is the surface's unit normal vector. However, if the closed fluid contour C is a closed streamline, since the direction of $\mathbf{T}$ at any point in the streamline is the direction of $\mathbf{v}$ at that point, the circulation along such a closed streamline can never be zero. Therefore, closed streamlines cannot exist in the potential flow.

In a potential flow, since $\operatorname{curl} \mathbf{v} = 0$, according to Proposition 10.2, there is a φ such that $\mathbf{v} = \operatorname{grad} \varphi$, where φ is the *velocity potential*. Combining this with (10.6), we get

$$\operatorname{grad}\left(\frac{\partial \varphi}{\partial t} + \frac{1}{2}v^2 + w\right) = 0,$$

and so

$$\frac{\partial \varphi}{\partial t} + \frac{1}{2}v^2 + w = g(t),$$

where $g(t)$ is an arbitrary function of time. Let $h(t) = \int_0^t g(u)\mathrm{d}u$. Then $\operatorname{grad} h = 0$, and so

$$\operatorname{grad}(\varphi + h) = \operatorname{grad} \varphi + \operatorname{grad} h = \mathbf{v},$$

i.e., $\varphi + h$ is also a velocity potential of $\mathbf{v}$. Therefore, the potential is not uniquely defined. It is easy to show that there exists a velocity potential $\widetilde{\varphi}$ in the potential flow such that $\operatorname{grad} \widetilde{\varphi} = \mathbf{v}$ and

$$\frac{\partial \widetilde{\varphi}}{\partial t} + \frac{1}{2}v^2 + w = \text{constant}.$$

This equation is called *Bernoulli's equation for potential flow*.

10.8 INCOMPRESSIBLE FLUIDS

Let $\mathbf{v}$ be the fluid velocity. If $\operatorname{div} \mathbf{v} = 0$ everywhere, then the fluid is said to be *incompressible*.

Consider a three-dimensional incompressible potential flow. Since it is a potential flow, there is a velocity potential φ such that $\mathbf{v} = \operatorname{grad} \varphi$. So

$$\operatorname{div} \mathbf{v} = \operatorname{div}(\operatorname{grad} \varphi) = \operatorname{div}\left(\frac{\partial \varphi}{\partial x}\mathbf{i} + \frac{\partial \varphi}{\partial y}\mathbf{j} + \frac{\partial \varphi}{\partial z}\mathbf{k}\right) = \frac{\partial^2 \varphi}{\partial x^2} + \frac{\partial^2 \varphi}{\partial y^2} + \frac{\partial^2 \varphi}{\partial z^2} = \Delta\varphi,$$

where Δ is the Laplace operator. Since the flow is incompressible, $\operatorname{div} \mathbf{v} = 0$, and so

$$\Delta\varphi = 0.$$

This equation is called the *Laplace equation*, i.e., the velocity potential satisfies the Laplace equation.

If the velocity distribution in a moving fluid depends on only two coordinates x, y and the velocity is everywhere parallel to the xy-plane, such a flow is called a *plane flow*.

Consider an incompressible plane flow. Let the velocity $\mathbf{v} = (v_1, v_2)$. Since the flow is an incompressible flow, div $\mathbf{v} = 0$. So

$$\frac{\partial v_1}{\partial x} + \frac{\partial v_2}{\partial y} = 0.$$

This implies that there is some function $\psi(x, y)$ such that

$$v_1 = \frac{\partial \psi}{\partial y},$$

$$v_2 = -\frac{\partial \psi}{\partial x}. \tag{10.13}$$

The function $\psi(x, y)$ is called a *stream function*.

In Section 10.4.2, we have given Euler's equation for isentropic motions as follows:

$$\frac{\partial(\text{curl } \mathbf{v})}{\partial t} = \text{curl}(\mathbf{v} \times \text{curl } \mathbf{v}). \tag{10.14}$$

By (10.13), the curl of $\mathbf{v}$ is

$$\text{curl } \mathbf{v} = \left(\frac{\partial v_2}{\partial x} - \frac{\partial v_1}{\partial y} \right) \mathbf{k} = \left(-\frac{\partial^2 \psi}{\partial x^2} - \frac{\partial^2 \psi}{\partial y^2} \right) \mathbf{k} = -(\Delta \psi)\mathbf{k},$$

where Δ is the Laplace operator. So the left-hand side of (10.14) is

$$\frac{\partial}{\partial t}(\text{curl } \mathbf{v}) = -\frac{\partial(\Delta \psi)}{\partial t} \mathbf{k}.$$

Since the vector product of $\mathbf{v}$ and curl $\mathbf{v}$ is

$$\mathbf{v} \times \text{curl } \mathbf{v} = -v_2 \, \Delta \psi \, \mathbf{i} + v_1 \, \Delta \psi \, \mathbf{j},$$

by the definition of the curl, the right-hand side of (10.14) is

$$\text{curl}(\mathbf{v} \times \text{curl } \mathbf{v}) = \text{curl}(-v_2 \, \Delta \psi \, \mathbf{i} + v_1 \, \Delta \psi \, \mathbf{j})$$

$$= \left(\frac{\partial(v_1 \, \Delta \psi)}{\partial x} + \frac{\partial(v_2 \, \Delta \psi)}{\partial y} \right) \mathbf{k}$$

$$= \left(v_1 \frac{\partial(\Delta \psi)}{\partial x} + \Delta \psi \frac{\partial v_1}{\partial x} + v_2 \frac{\partial(\Delta \psi)}{\partial y} + \Delta \psi \frac{\partial v_2}{\partial y} \right) \mathbf{k}$$

$$= \left(\frac{\partial \psi}{\partial y} \frac{\partial(\Delta \psi)}{\partial x} + \Delta \psi \frac{\partial^2 \psi}{\partial x \partial y} - \frac{\partial \psi}{\partial x} \frac{\partial(\Delta \psi)}{\partial y} - \Delta \psi \frac{\partial^2 \psi}{\partial y \partial x} \right) \mathbf{k}$$

$$= \left(\frac{\partial \psi}{\partial y} \frac{\partial(\Delta \psi)}{\partial x} - \frac{\partial \psi}{\partial x} \frac{\partial(\Delta \psi)}{\partial y} \right) \mathbf{k}.$$

Therefore, by (10.14), we get

$$-\frac{\partial(\Delta\psi)}{\partial t}\mathbf{k} = \left(\frac{\partial\psi}{\partial y}\frac{\partial(\Delta\psi)}{\partial x} - \frac{\partial\psi}{\partial x}\frac{\partial(\Delta\psi)}{\partial y}\right)\mathbf{k}.$$

Comparing both sides, we get

$$\frac{\partial(\Delta\psi)}{\partial t} = \frac{\partial\psi}{\partial x}\frac{\partial(\Delta\psi)}{\partial y} + \frac{\partial\psi}{\partial y}\frac{\partial(\Delta\psi)}{\partial x}.$$

This is the equation that must be satisfied by the stream function.

We know that the direction of the tangent line at any point in a streamline is the direction of the velocity at that point. Therefore, the streamline in the plane satisfies the equation

$$\frac{dx}{v_1} = \frac{dy}{v_2} \quad \text{or} \quad -v_2\,dx + v_1\,dy = 0,$$

where the velocity $\mathbf{v} = (v_1, v_2)$. Notice that $v_1 = \frac{\partial\psi}{\partial y}$ and $v_2 = -\frac{\partial\psi}{\partial x}$. Then

$$\frac{\partial\psi}{\partial x}\,dx + \frac{\partial\psi}{\partial y}\,dy = 0 \quad \text{or} \quad d\psi = 0,$$

and so $\psi = $ constant. Thus, the streamlines are the family of curves obtained by putting the stream function $\psi(x, y)$ equal to an arbitrary constant.

The stream function is often used for computing the mass flux. If C_{AB} is a curve joining two points A, B in the xy-plane, noticing that $\mathbf{v} = (v_1, v_2)$ and (10.2), the mass flux Q of $\mathbf{v}$ across C_{AB} is

$$Q = \rho\int_{C_{AB}} \mathbf{v}\cdot\mathbf{n}\,dl = \rho\int_{C_{AB}} -v_2\,dx + v_1\,dy,$$

where dl is the arc element and $\mathbf{n}$ is the outward unit normal vector of C_{AB}. Notice that

$$v_1 = \frac{\partial\psi}{\partial y},$$

$$v_2 = -\frac{\partial\psi}{\partial x}.$$

The mass flux Q of $\mathbf{v}$ across C_{AB} is

$$Q = \rho\int_{C_{AB}} \frac{\partial\psi}{\partial x}\,dx + \frac{\partial\psi}{\partial y}\,dy = \rho\int_{C_{AB}} d\psi = \rho(\psi_B - \psi_A), \qquad (10.15)$$

where ψ_A, ψ_B are values of ψ at A, B, respectively. Therefore, the mass flux of the velocity across a curve equals the density times the difference between the values of the stream function at endpoints, regardless of the shape of the curve.

The velocity $\mathbf{v} = (v_1, v_2)$ and the potential function φ satisfy grad $\varphi = \mathbf{v}$, so

$$v_1 = \frac{\partial\varphi}{\partial x}, \quad v_2 = \frac{\partial\varphi}{\partial y}.$$

By (10.13), the velocity and the stream function satisfy

$$v_1 = \frac{\partial \psi}{\partial y}, \quad v_2 = -\frac{\partial \psi}{\partial x}.$$

Therefore, the potential function φ and the stream function ψ satisfy the Cauchy-Riemann equations:

$$\frac{\partial \varphi}{\partial x} = \frac{\partial \psi}{\partial y}, \quad \frac{\partial \varphi}{\partial y} = -\frac{\partial \psi}{\partial x}.$$

So the function $U = \varphi + i\psi$ is an analytic function of the variable $z = x + iy$, where

$$\frac{dU}{dz} = \frac{\partial \varphi}{\partial x} + i\frac{\partial \psi}{\partial x} = v_1 - iv_2.$$

The function U is called the *complex potential* and $\frac{dU}{dz}$ is called the *complex velocity*.

We compute the integral of the complex velocity around any closed fluid contour. Notice that

$$\oint_C \frac{dU}{dz}\, dz = \oint_C (v_1 - i\,v_2)(dx + i\,dy) = \oint_C v_1\, dx + v_2\, dy$$
$$+ i\oint_C v_1\, dy - v_2\, dx.$$

By (10.1), the real part of this expression is the velocity circulation around the contour. By (10.2), the imaginary part, multiplied by ρ, is the mass flux across the contour. If there are no sources of fluid within the contour, by (10.15), this flux is zero, i.e., the imaginary part

$$\oint_C v_1\, dy - v_2\, dx = 0.$$

Therefore,

$$\oint_C \frac{dU}{dz}\, dz = \oint_C v_1\, dx + v_2\, dy. \tag{10.16}$$

The residue theorem in complex analysis says that the integral of the complex velocity around any closed contour is equal to $2\pi i$ times the sum of the residues of the complex velocity at its simple poles inside the closed contour, i.e.,

$$\oint_C \frac{dU}{dz}\, dz = 2\pi i \sum_k R_k,$$

where R_k are the residues of the complex velocity at the kth simple pole inside the closed contour. From this and (10.16), the velocity circulation around the closed contour is given by

$$\oint_C v_1\, dx + v_2\, dy = 2\pi i \sum_k R_k.$$

PROBLEMS

10.1 Show that $\mathrm{div}(\mathbf{C}\,u) = \mathbf{C} \cdot \mathrm{grad}\,u$, where u is a scalar function and $\mathbf{C}$ is a constant vector.

10.2 Show that $\mathrm{div}(\mathrm{grad}\,u) = \Delta u$, where u is a scalar function.

10.3 Show that $\mathrm{div}(\mathrm{curl}\,\mathbf{F}) = 0$, where $\mathbf{F}$ is a vector.

10.4 If the density is constant, derive the continuity equation $\mathrm{div}\,\mathbf{v} = 0$ from first principles.

10.5 If the velocity is constant in a uniform gravitational field, find the pressure with the help of Euler's equations.

BIBLIOGRAPHY

Cushman-Roisin, B., Beckers, J.-M., 2011. Introduction to Geophysical Fluid Dynamics: Physical and Numerical Aspects, second ed. Academic Press, New York.

Durran, D.R., 1999. Numerical Methods for Wave Equations in Geophysical Fluid Dynamics, Texts in Applied Mathematics. Springer-Verlag, New York.

McWilliams, J.C., 2011. Fundamentals of Geophysical Fluid Dynamics. Cambridge University Press, Cambridge.

Pedlosky, J., 1987. Geophysical Fluid Dynamics, second ed. Springer-Verlag, New York.

Tritton, D.J., 1990. Physical Fluid Dynamics, second ed. Oxford University Press, Oxford.

Chapter 11

Atmospheric Dynamics

Earth's atmosphere is composed of a mixture of gases such as nitrogen, oxygen, carbon dioxide, water vapor, and ozone. A wide variety of fluid flows take place in the atmosphere. In this chapter, we show how the theory of fluid dynamics in Chapter 10 is applied to the atmosphere. For this purpose, we will introduce the Navier-Stokes equation, hydrostatic and geostrophic approximations, the Boussinesq approximation, potential temperature and equivalent potential temperature, quasi-geostrophic potential vorticity, buoyancy frequency, and so on.

According to the variation of temperature with height, the atmosphere is conventionally divided into layers in the vertical direction. The layer from the ground up to about 15 km altitude is called the *troposphere*. The troposphere is bounded above by the tropopause, and in the troposphere the temperature decreases with height. The layer from the tropopause to about 50 km altitude is called the *stratosphere*. The stratosphere is bounded above by the stratopause, and in the stratosphere the temperature rises with height. The layer from the stratopause to about 85-90 km altitude is called the *mesosphere*. The mesosphere is bounded above by the mesopause, and in the mesosphere the temperature again decreases with height. The layer above the mesopause is called the *thermosphere*. In the thermosphere, the temperature again rises with height. Sometimes, the troposphere is also called the *lower atmosphere*. Most weather phenomena, such as rain, snow, thunder, and lightning, occur in the lower atmosphere. The stratosphere and mesosphere together are called the *middle atmosphere*. The ozone molecules stay in the lower stratosphere and form the ozone layer. The layer above the mesosphere is called the *upper atmosphere*.

11.1 TWO SIMPLE ATMOSPHERIC MODELS

Energy transfer in the atmosphere involves short-wave radiation emitted by the Sun and long-wave radiation emitted by Earth's surface and atmosphere. These two wavelength ranges represent spectral regions of blackbody emission at temperatures of about 6000 and 288 K, respectively. Planck's law states that the blackbody spectral radiance $B_\lambda(T)$ at temperature T is

$$B_\lambda(T) = \frac{2hc^2}{\lambda^5(e^{\frac{hc}{\lambda k_B T}} - 1)}, \tag{11.1}$$

Mathematical and Physical Fundamentals of Climate Change

where h is Planck's constant (6.626×10^{-34} J s), c is the speed of light, λ is the wavelength, and k_B is Boltzmann's constant (1.38×10^{-23} J/K).

Since the blackbody radiation is isotropic, the blackbody spectral irradiance $F_\lambda(\mathbf{r}, \mathbf{n})$ at a point $\mathbf{r}$ through a surface of normal $\mathbf{n}$ is obtained by integration over a hemisphere on one side of the surface,

$$F_\lambda(\mathbf{r}, \mathbf{n}) = \int_S B_\lambda(T) \mathbf{n} \cdot \mathbf{s} \, d\Omega(\mathbf{s}),$$

where S is the hemisphere and $d\Omega(\mathbf{s})$ is the element of solid angle in the direction $\mathbf{s}$. Let ϕ be the angle between $\mathbf{s}$ and $\mathbf{n}$. Then

$$\mathbf{n} \cdot \mathbf{s} = \cos\phi,$$
$$d\Omega = 2\pi \sin\phi \, d\phi.$$

Notice that $B_\lambda(T)$ is independent of $\mathbf{s}$ and $\mathbf{r}$. Then

$$F_\lambda(\mathbf{r}, \mathbf{n}) = \pi B_\lambda(T) \int_0^{\pi/2} 2\cos\phi \sin\phi \, d\phi.$$

Since $\int_0^{\pi/2} 2\cos\phi \sin\phi \, d\phi = 1$, the blackbody spectral irradiance is given by

$$F_\lambda(\mathbf{r}, \mathbf{n}) = \pi B_\lambda(T).$$

Moreover, the flux density $F(\mathbf{r}, \mathbf{n})$ at the point $\mathbf{r}$ through the surface of normal $\mathbf{n}$

$$F(\mathbf{r}, \mathbf{n}) = \int_0^\infty F_\lambda(\mathbf{r}, \mathbf{n}) \, d\lambda = \pi \int_0^\infty B_\lambda(T) \, d\lambda. \tag{11.2}$$

By (11.1),

$$\pi \int_0^\infty B_\lambda(T) \, d\lambda = \int_0^\infty \frac{c_1}{\lambda^5 (e^{\frac{c_2}{\lambda T}} - 1)} \, d\lambda,$$

where $c_1 = 2h\pi c^2$ and $c_2 = \frac{hc}{k_B}$. The constant c_1 is called *the first radiance constant*, and the constant c_2 is called *the second radiance constant*. Take a change of variable $X = \frac{c_2}{\lambda T}$. This implies that

$$\lambda = \frac{c_2}{XT},$$
$$d\lambda = -\frac{c_2}{T} \frac{dX}{X^2}.$$

Since $\int_0^\infty \frac{X^3}{e^X - 1} \, dX = \frac{\pi^4}{15}$,

$$\pi \int_0^\infty B_\lambda(T) \, d\lambda = \frac{c_1}{c_2^4} T^4 \int_0^\infty \frac{X^3}{e^X - 1} \, dX = \frac{c_1 \pi^4}{15 c_2^4} T^4.$$

From this and (11.2), the *Stefan-Boltzmann law* for the blackbody irradiance is given by

$$F(\mathbf{r}, \mathbf{n}) = \sigma T^4,$$

where

$$\sigma = \frac{c_1 \pi^4}{15 c_2^4} \approx 5.6703 \times 10^{-8} \, \text{W}/(\text{m}^2 \, \text{K}^4).$$

The constant σ is called the *Stefan-Boltzmann constant*.

Earth's mean surface temperature is about 288 K. This observational fact can be explained by the Stefan-Boltzmann law. First we illustrate this point by the single-layer atmospheric model. Then we introduce the two-layer atmospheric model which is an extension of the single-layer atmospheric model.

11.1.1   The Single-Layer Model

In the single-layer atmospheric model, the atmosphere is taken to be a layer at a uniform temperature T_a, and the ground is assumed to emit as a blackbody at a uniform temperature T_g. We consider this simple model.

The solar power per unit area at Earth's mean distance from the Sun is called the *solar constant*. It is well known that the solar constant $F_s = 1370 \, \text{W/m}^2$.

Assume that the Earth-atmosphere system has a planetary albedo $\alpha = 0.3$. This means that an amount $\alpha F_s \pi a^2$, where a is Earth's radius, of the solar power is reflected back to space and the remainder $(1 - \alpha) F_s \pi a^2$ is the unreflected incoming solar irradiance at the top of the atmosphere. However, the total surface area of Earth is $4\pi a^2$. Therefore, the mean unreflected incoming solar irradiance at the top of the atmosphere is

$$F_0 = \frac{(1 - \alpha) F_s \pi a^2}{4 \pi a^2} = \frac{1}{4}(1 - \alpha) F_s.$$

Substituting $F_s = 1370$ and $\alpha = 0.3$ into this equality, we obtain the mean unreflected incoming solar irradiance at the top of the atmosphere is

$$F_0 = \frac{1}{4}(1 - 0.3) \times 1370 \approx 240 \, (\text{W/m}^2).$$

Denote by $\mathcal{T}_{sw}$ the *transmittance* of any incident solar (short-wave) radiation. Then an amount $\mathcal{T}_{sw} F_0$ is absorbed by the ground. Denote by F_g the upward irradiance from the ground and denote by $\mathcal{T}_{lw}$ the *transmittance* of any incident thermal (long-wave) radiation. Then a upward emission $\mathcal{T}_{lw} F_g$ reaches the top of the atmosphere. Since the atmosphere is not a blackbody, the atmosphere emission F_a is both upward and downward. Assume that the whole system is in radiative equilibrium. Then the balance of irradiances on the top of the atmosphere implies

$$F_0 = F_a + \mathcal{T}_{lw} F_g \tag{11.3}$$

and the balance of irradiances between the atmosphere and the ground is

$$F_g = F_a + \mathcal{T}_{sw} F_0.$$

Eliminating F_a from these two equations, we get

$$F_0 - F_g = T_{lw}F_g - T_{sw}F_0,$$

so

$$F_g = F_0 \frac{1 + T_{sw}}{1 + T_{lw}}. \tag{11.4}$$

Combining (11.3) with (11.4), the atmosphere emission is

$$F_a = F_0 - T_{lw}F_g = F_0 - T_{lw}F_0 \frac{1 + T_{sw}}{1 + T_{lw}} = F_0 \frac{1 - T_{lw}T_{sw}}{1 + T_{lw}}.$$

By the Stefan-Boltzmann law,

$$F_a = (1 - T_{lw})\sigma T_a^4,$$

where σ is the Stefan-Boltzmann constant, and so

$$(1 - T_{lw})\sigma T_a^4 = F_0 \frac{1 - T_{lw}T_{sw}}{1 + T_{lw}}.$$

Thus, the temperature of the model atmosphere is

$$T_a = \sqrt[4]{\frac{F_0(1 - T_{lw}T_{sw})}{\sigma(1 - T_{lw})(1 + T_{lw})}}.$$

Taking transmittances $T_{sw} = 0.9$ (strong transmittance and weak absorption of solar radiation) and $T_{lw} = 0.2$ (weak transmittance and strong absorption of thermal radiation), the temperature of the model atmosphere is

$$T_a = \sqrt[4]{\frac{(0.7 \times 1370)(1 - 0.2 \times 0.9)}{4 \times 5.6703 \times 10^{-8} \times (1 - 0.2)(1 + 0.2)}} \approx 245 \, (\text{K}).$$

Notice that $F_0 \approx 240 \, \text{W/m}^2$. By (11.4), the upward irradiance of the ground is

$$F_g = F_0 \frac{1 + T_{sw}}{1 + T_{lw}} \approx 240 \times \frac{1 + 0.9}{1 + 0.2} = 380 \, (\text{W/m}^2).$$

By the Stefan-Boltzmann law, $F_g = \sigma T_g^4$. From this and (11.4), the temperature of the ground is

$$T_g = \sqrt[4]{\frac{F_g}{\sigma}} = \sqrt[4]{\frac{F_0(1 + T_{sw})}{\sigma(1 + T_{lw})}} = \sqrt[4]{\frac{(0.7 \times 1370)(1 + 0.9)}{4 \times 5.6703 \times 10^{-8} \times (1 + 0.2)}} \approx 286 \, (\text{K}).$$

This temperature is close to the observed mean surface temperature of about 288 K.

11.1.2 The Two-Layer Model

The two-layer atmospheric model is an extension of the single-layer atmospheric model. This model includes two atmospheric layers, say, the upper atmosphere and the lower atmosphere. The upper layer mimics the stratosphere at temperature T_{strat}. The lower layer mimics the troposphere at temperature T_{trop}.

Assume that the upper layer is transparent to solar radiation and optically thin, and its thermal absorptance is taken as $\epsilon \ll 1$. Kirchhoff's law shows that the thermal emittance of the upper layer is also ϵ and its thermal transmittance is $1 - \epsilon$. Assume that the lower layer has transmittances T_{sw} of solar radiation and T_{lw} of thermal radiation and that the ground emits as a blackbody at temperature T_g, and that the mean unreflected incoming solar irradiance F_0 is defined as before.

Under these assumptions, the emissions from the upper and lower layers, F_{strat} and F_{trop}, are both upward and downward, and the emission from the ground, F_g is upward. Owing to the optically thin layer, the incoming solar irradiance on the top of the lower layer is also F_0. The amounts $T_{sw}F_0$ and $T_{lw}F_{strat}$ are absorbed by the ground. The amount $T_{lw}F_g$ is absorbed by the lower layer. The amounts $(1 - \epsilon)F_{trop}$ and $(1 - \epsilon)T_{lw}F_g$ are absorbed by the upper layer.

Assume further that the whole system is in radiative equilibrium. Then the balance of irradiances above the upper atmosphere implies

$$F_0 = F_{strat} + (1 - \epsilon)(F_{trop} + T_{lw}F_g), \tag{11.5}$$

the balance of irradiances between the upper atmosphere and the lower atmosphere implies

$$F_{trop} + T_{lw}F_g = F_0 + F_{strat}, \tag{11.6}$$

and the balance of irradiances between the lower atmosphere and the ground implies

$$F_g = T_{sw}F_0 + T_{lw}F_{strat} + F_{trop}. \tag{11.7}$$

Eliminating F_0 from (11.5) and (11.6), we get

$$2F_{strat} = \epsilon(F_{trop} + T_{lw}F_g),$$

and so

$$F_{trop} + T_{lw}F_g = \frac{2}{\epsilon}F_{strat}.$$

Substituting this into (11.5), the emission from the upper atmosphere is

$$F_{strat} = \frac{\epsilon}{2 - \epsilon}F_0. \tag{11.8}$$

The combination of (11.8) and (11.6) gives

$$F_{trop} = \frac{2}{2 - \epsilon}F_0 - T_{lw}F_g. \tag{11.9}$$

Substituting (11.8) and (11.9) into (11.7), we get

$$F_g = \frac{(2 - \epsilon)T_{sw} + \epsilon T_{lw} + 2}{2 - \epsilon}F_0 - T_{lw}F_g,$$

and so the emission from the ground is

$$F_g = \frac{(2 - \epsilon)\mathcal{T}_{sw} + \epsilon\mathcal{T}_{lw} + 2}{(2 - \epsilon)(1 + \mathcal{T}_{lw})} F_0. \tag{11.10}$$

From this and (11.9), the emission from the lower atmosphere is

$$F_{trop} = \frac{2 - (2 - \epsilon)\mathcal{T}_{lw}\mathcal{T}_{sw} - \epsilon\mathcal{T}_{lw}^2}{(2 - \epsilon)(1 + \mathcal{T}_{lw})} F_0. \tag{11.11}$$

However, according to the Stefan-Boltzmann law, the emissions from the upper and lower layers and the ground are, respectively,

$$F_{strat} = \sigma\epsilon T_{strat}^4,$$

$$F_{trop} = \sigma(1 - \mathcal{T}_{lw})T_{trop}^4,$$

$$F_g = \sigma T_g^4,$$

where σ is the Stefan-Boltzmann constant. Therefore, by (11.8), (11.10), and (11.11), the temperature of the upper atmosphere is

$$T_{strat} = \sqrt[4]{\frac{F_{strat}}{\sigma\epsilon}} = \sqrt[4]{\frac{F_0}{\sigma(2 - \epsilon)}},$$

the temperature of the lower atmosphere is

$$T_{trop} = \sqrt[4]{\frac{F_{trop}}{\sigma(1 - \mathcal{T}_{lw})}} = \sqrt[4]{\frac{F_0(2 - \epsilon\mathcal{T}_{lw}^2 - (2 - \epsilon)\mathcal{T}_{lw}\mathcal{T}_{sw})}{\sigma(2 - \epsilon)(1 - \mathcal{T}_{lw}^2)}},$$

and the temperature of the ground is

$$T_g = \sqrt[4]{\frac{F_g}{\sigma}} = \sqrt[4]{\frac{F_0((2 - \epsilon)\mathcal{T}_{sw} + \epsilon\mathcal{T}_{lw} + 2)}{\sigma(2 - \epsilon)(1 + \mathcal{T}_{lw})}}.$$

11.2 ATMOSPHERIC COMPOSITION

We introduce the ideal gas law and some basic concepts on gases.

If the mass of 1 mol is M_m and the volume of 1 mol is V_m, then the density

$$\rho = \frac{M_m}{V_m}.$$

For an ideal gas of pressure p and temperature T, each mole of gas obeys the law

$$pV_m = RT,$$

where R is the *universal gas constant*. So the *ideal gas law* is given by

$$p = R_a T\rho,$$

where $R_a = R/M_m$ is the gas constant per unit mass of air.

Consider a small sample of air with volume V, temperature T, and pressure p. It is composed of a mixture of gases $G_i (i \in \mathbb{Z}_+)$. If the number of molecules of gas G_i in the sample is n_i, then the *total number* of molecules in the sample is

$$n = \sum n_i.$$

If the molecular mass of gas G_i in the sample is m_i, then the *total mass* of the sample is

$$m = \sum m_i n_i.$$

If the molar mass of gas G_i in the sample is M_i, then the *total molar mass* is

$$M = \sum M_i n_i,$$

where all the sums are taken over all the gases in the sample. Now the ideal gas law for this small sample is

$$pV = RT,$$

where $R = n k_B$ and k_B is Boltzmann's constant. If the molecules of gas G_i in the sample alone are to occupy the volume V at temperature T, by the ideal gas law, the pressure exerted by the molecules of G_i from the sample is

$$p_i = n_i \frac{k_B T}{V}.$$

This pressure is called the *partial pressure* of gas G_i. The partial pressure is sometimes used to quantify chemical concentrations. If the molecules of gas G_i in the sample alone are to be held at temperature T and pressure p, then, by the ideal gas law, the volume occupied by the molecules of gas G_i from the sample is

$$V_i = n_i \frac{k_B T}{p}.$$

This volume is called the *partial volume* of gas G_i.

Dalton's Law. *Let p_i, p, V_i, and V be stated as above. Then $\sum p_i = p$ and $\sum V_i = V$.*

In fact, notice that $n = \sum n_i$. It follows from the ideal gas law immediately that

$$\sum p_i = \sum n_i \frac{k_B T}{V} = \left(\sum n_i \right) \frac{k_B T}{V} = n \frac{k_B T}{V} = p,$$

$$\sum V_i = \sum n_i \frac{k_B T}{p} = \left(\sum n_i \right) \frac{k_B T}{p} = n \frac{k_B T}{p} = V.$$

Define the volume mixing ratio v_i of gas G_i as $v_i = \frac{V_i}{V}$.

Proposition 11.1. *Let n_i, n, p_i, p, and v_i be stated as above. Then*
$$v_i = \frac{n_i}{n} = \frac{p_i}{p}.$$
In fact, by the definition of the partial volume,

$$v_i = \frac{V_i}{V} = \frac{n_i \frac{k_B T}{p}}{n \frac{k_B T}{p}} = \frac{n_i}{n}.$$

Similarly, by the definition of the partial pressure,

$$\frac{p_i}{p} = \frac{n_i \frac{k_B T}{V}}{n \frac{k_B T}{V}} = \frac{n_i}{n}.$$

Thus, $v_i = \frac{n_i}{n} = \frac{p_i}{p}$.

Define the mean molecular mass of the sample as $\overline{m} = \frac{m}{n}$. Similarly, define the mean molar mass of the sample as $\overline{M} = \frac{M}{n}$.

Proposition 11.2. *Let v_i, m_i, M_i, and $\overline{m}$, $\overline{M}$ be stated as above. Then*

$$\overline{m} = \sum m_i v_i,$$
$$\overline{M} = \sum M_i v_i.$$

In fact, by the definitions of $\overline{m}$, $\overline{M}$, and Proposition 11.1,

$$\overline{m} = \frac{m}{n} = \frac{\sum m_i n_i}{n} = \sum m_i \frac{n_i}{n} = \sum m_i v_i,$$

$$\overline{M} = \frac{M}{n} = \frac{\sum M_i n_i}{n} = \sum M_i \frac{n_i}{n} = \sum M_i v_i.$$

Define the mass mixing ratio μ_i of gas G_i as

$$\mu_i = \frac{m_i}{\overline{m}} \frac{p_i}{p}.$$

Mass and volume mixing ratios are more convenient measures of the concentration of an atmospheric gas when the transport of chemicals is being studied.

Proposition 11.3. *Let μ_i, v_i, p_i, p, m_i, and $\overline{m}$ be stated as above. Then*
$$\mu_i = \frac{m_i}{\overline{m}} v_i.$$

11.3 HYDROSTATIC BALANCE EQUATION

We will give several forms of the hydrostatic balance equation by using the ideal gas law and then derive the basic properties involving the atmospheric pressure and density.

Consider a small cylinder of air with height Δz and horizontal cross-sectional area ΔA. The mass of the cylinder of air is $\Delta m = \rho \Delta A \Delta z$, where ρ is the density. Let $g = |\mathbf{g}|$, where $\mathbf{g}$ be the gravitational acceleration. There are three vertical forces acting on the small cylinder of air: the downward gravitational force $g\Delta m$, the upward pressure force $p(z)\Delta A$ on the bottom of the cylinder, and the downward pressure force $p(z + \Delta z)\Delta A$ on the top of the cylinder. Since the small cylinder of air is in hydrostatic equilibrium, the balance of the three forces implies

$$g\rho\Delta A\Delta z = p(z)\Delta A - p(z + \Delta z)\Delta A.$$

Using Taylor's expansion,

$$p(z + \Delta z) \approx p(z) + \frac{\mathrm{d}p}{\mathrm{d}z}\Delta z,$$

and then canceling $\Delta A\Delta z$, we obtain the *hydrostatic balance equation*:

$$\frac{\mathrm{d}p}{\mathrm{d}z} = -g\rho.$$

Assume that the air is an ideal gas and is in hydrostatic balance. Then the air obeys the ideal gas law:

$$p = \rho R_a T,$$

where p, ρ, T, and R_a are stated as in Section 11.2, and so the hydrostatic balance equation takes the form

$$\frac{\mathrm{d}p}{\mathrm{d}z} = -\frac{gp}{R_a T}$$

or

$$\frac{\mathrm{d}p}{p} = -\frac{g}{R_a T}\mathrm{d}z.$$

If the temperature T is a function of height z, and the pressure at the ground is p_0, integrating both sides with respect to z from the ground $z = 0$ upward, we get

$$\ln p - \ln p_0 = -\frac{g}{R_a}\int_0^z \frac{\mathrm{d}z'}{T(z')}.$$

Taking exponentials on the both sides,

$$p = p_0 e^{-\frac{g}{R_a}\int_0^z \frac{\mathrm{d}z'}{T(z')}}.$$

Therefore, the pressure p is a function of the height z.

The simplest case is that of an isothermal temperature profile, i.e., $T = T_0$ =constant. Then

$$p = p_0 e^{-\frac{gz}{R_a T_0}} = p_0 e^{-\frac{z}{H}},$$

where $H = \frac{R_a T_0}{g}$ is called the *pressure scale height*. Thus, in the isothermal case, the pressure decays exponentially with height.

Similarly, eliminating the pressure p from the ideal gas law and the hydrostatic balance equation, the alternative form of the hydrostatic balance equation is

$$\frac{d\rho}{dz} = -\frac{g\rho}{R_a T}$$

or

$$\frac{d\rho}{\rho} = -\frac{g}{R_a T} dz.$$

Integrating both sides with respect to z from the ground $z = 0$ upward,

$$\ln \rho - \ln \rho_0 = -\frac{g}{R_a} \int_0^z \frac{dz'}{T(z')},$$

where ρ_0 is the density at the ground. Taking exponentials on both sides, we get

$$\rho = \rho_0 e^{-\frac{g}{R_a} \int_0^z \frac{dz'}{T(z')}}.$$

Therefore, the density ρ is also a function of the height z.

In the isothermal case, $T = T_0 = $ constant, and so

$$\rho = \rho_0 e^{-\frac{gz}{R_a T_0}} = \rho_0 e^{-\frac{z}{H}},$$

where $H = \frac{R_a T_0}{g}$. Therefore, in the isothermal case, the density also decays exponentially with height.

11.4 POTENTIAL TEMPERATURE

Consider a small air parcel with volume V, pressure p, and temperature T. Assume that the air in the parcel is of unit mass and is in hydrostatic balance. Let w be the enthalpy and ϵ be the internal energy per unit mass. Since the enthalpy $w = \epsilon + \frac{p}{\rho}$ and the thermodynamic relation $d\epsilon = T ds + \frac{p}{\rho^2} d\rho$, where s is the entropy, a small change of enthalpy

$$dw = d\epsilon + \frac{1}{\rho} dp - \frac{p}{\rho^2} d\rho = T ds + \frac{1}{\rho} dp.$$

According the ideal gas law $p = R_a T \rho$, the small change of enthalpy becomes

$$dw = T ds + \frac{R_a T}{p} dp.$$

On the other hand, for unit mass of ideal gas, $\epsilon = c_v T$, where c_v is the specific heat capacity at constant volume and is independent of T. Using the ideal gas law $p = R_a T \rho$, we find the enthalpy is

$$w = \epsilon + \frac{p}{\rho} = c_v T + R_a T = c_p T,$$

where $c_p = c_v + R_a$. So the small change of enthalpy

$$dw = c_p\, dT.$$

Therefore,

$$T\, ds + \frac{R_a T}{p}\, dp = c_p\, dT. \tag{11.12}$$

Dividing both sides of (11.12) by T,

$$ds + \frac{R_a}{p}\, dp = \frac{c_p}{T}\, dT,$$

which is equivalent to

$$ds = c_p\, d(\ln T) - R_a\, d(\ln p). \tag{11.13}$$

Integrating both sides and then letting $\kappa = \frac{R_a}{c_p}$, we find the entropy is

$$s = c_p \ln T - R_a \ln p + s_0 = c_p(\ln T - \kappa \ln p) + s_0 = c_p \ln(Tp^{-\kappa}) + s_0,$$

where s_0 is a constant.

When the air in the parcel is compressed adiabatically from pressure p and temperature T to pressure p_0 and temperature θ, where p_0 is usually taken to be 1000 hPa, we find the temperature θ.

Since the process is adiabatic, the entropy satisfies the condition $ds = 0$. From this and (11.13),

$$c_p\, d(\ln T) = R_a\, d(\ln p).$$

Integrating both sides and using the end conditions $T = \theta$ and $p = p_0$, we get

$$c_p \ln\left(\frac{\theta}{T}\right) = R_a \ln\left(\frac{p_0}{p}\right), \tag{11.14}$$

and so

$$\theta = T\left(\frac{p_0}{p}\right)^{\kappa}, \quad \text{where } \kappa = \frac{R_a}{c_p} = \frac{R_a}{c_v + R_a}.$$

The quantity θ is called the *potential temperature*. It depends on temperature T and pressure p. The expression $\theta = T\left(\frac{p_0}{p}\right)^{\kappa}$ is equivalent to

$$\ln \theta = \ln(Tp^{-\kappa}) + \ln p_0^{\kappa}.$$

So the specific entropy

$$s = c_p \ln(Tp^{-\kappa}) + s_0 = c_p \ln \theta + s_1, \tag{11.15}$$

where $s_1 = s_0 - c_p \ln p_0^{\kappa} = s_0 - c_p\kappa \ln p_0 = s_0 - R_a \ln p_0$. This means that the specific entropy relates to the potential temperature. When a mass of air is subject to an adiabatic change, since the entropy is constant in the adiabatic process, it follows from (11.15) that its potential temperature is constant. Conversely, when the mass is subject to a nonadiabatic change, its potential temperature will change.

Differentiating $\theta = T\left(\frac{p_0}{p}\right)^\kappa$ with respect to t,

$$\frac{d\theta}{dt} = \left(\frac{p_0}{p}\right)^\kappa \frac{dT}{dt} + T\frac{d}{dt}\left(\frac{p_0}{p}\right)^\kappa.$$

Notice that $\kappa = \frac{R_a}{c_p}$. Then

$$\frac{d}{dt}\left(\frac{p_0}{p}\right)^\kappa = \kappa\left(\frac{p_0}{p}\right)^{\kappa-1}\left(-\frac{p_0}{p^2}\right)\frac{dp}{dt} = -\frac{R_a}{c_p p}\left(\frac{p_0}{p}\right)^\kappa \frac{dp}{dt},$$

and so

$$\frac{d\theta}{dt} = \frac{1}{c_p}\left(\frac{p_0}{p}\right)^\kappa \left(c_p\frac{dT}{dt} - \frac{TR_a}{p}\frac{dp}{dt}\right).$$

Applying the ideal gas law $p = R_a T\rho$ gives

$$\frac{d\theta}{dt} = \frac{1}{c_p}\left(\frac{p_0}{p}\right)^\kappa \left(c_p\frac{dT}{dt} - \frac{1}{\rho}\frac{dp}{dt}\right).$$

By (11.12), the part in brackets

$$c_p\frac{dT}{dt} - \frac{1}{\rho}\frac{dp}{dt} = T\frac{ds}{dt}.$$

Therefore,

$$\frac{d\theta}{dt} = \frac{1}{c_p}\left(\frac{p_0}{p}\right)^\kappa T\frac{ds}{dt}.$$

Here the product of T and $\frac{ds}{dt}$ is the adiabatic heating rate per unit mass, denoted by Q, i.e., $T\frac{ds}{dt} = Q$. So

$$\frac{d\theta}{dt} = \frac{Q}{c_p}\left(\frac{p_0}{p}\right)^\kappa.$$

This equation is called the *thermodynamic energy equation*.

11.5 LAPSE RATE

Denote by T the air temperature. The quantity

$$\Gamma(z) = -\frac{dT}{dz}$$

is called the *lapse rate of temperature* with height. It represents the rate of decrease of temperature with height. It is clear that

(i) if the temperature decreases with height in some region, then $\Gamma > 0$ in that region;

(ii) if the temperature increases with height in some region, then $\Gamma < 0$ in that region.

In the troposphere $\Gamma > 0$, and in the stratosphere $\Gamma < 0$.

11.5.1   Adiabatic Lapse Rate

As an air parcel rises adiabatically, the rate of decrease of temperature with height, following the adiabatic parcel, is called the *adiabatic lapse rate*, denoted by Γ_a.

Now we find the adiabatic lapse rate.

Consider an adiabatically rising air parcel with pressure p and temperature T. By (11.12),

$$ds + \frac{R_a}{p}\, dp = \frac{c_p}{T}\, dT, \qquad (11.16)$$

where R_a and c_p are stated as above. Notice that $ds = 0$ in an adiabatic process. By (11.16), it follows that the vertical derivatives of the temperature and pressure of the parcel satisfy

$$\frac{R_a}{p}\frac{dp}{dz} = \frac{c_p}{T}\frac{dT}{dz}$$

or

$$\frac{dT}{dz} = \frac{R_a T}{p c_p}\frac{dp}{dz}.$$

From this and the hydrostatic balance equation $\frac{dp}{dz} = -\frac{gp}{R_a T}$, the adiabatic lapse rate is

$$\Gamma_a = -\frac{dT}{dz} = \frac{g}{c_p},$$

where $g = |\mathbf{g}|$ and $c_p = c_v + R_a$. Here $\mathbf{g}$ is the gravitational acceleration, R_a is the gas constant per unit mass of air, and c_v is the specific heat capacity at constant volume.

The adiabatic lapse rate for dry air is called the *dry adiabatic lapse rate*, denoted by Γ_d. It is approximately 9.8 K/km. The actual lapse rate in the atmosphere generally differs from the dry adiabatic lapse rate. To investigate this, consider a dry air parcel that is originally at an equilibrium position at height z_0 with temperature T_0, pressure p_0, and density ρ_0, all equal to the values for the surroundings. Suppose that an instantaneous upward force is applied to the parcel so that the parcel rises adiabatically through a small height Δz from its equilibrium position at height z_0 to the height $z = z_0 + \Delta z$ without influencing its surroundings. Then the parcel temperature has increased from T_0 to T_p and

$$T_p = T_0 - \Gamma_d \Delta z, \quad \text{where } \Gamma_d = -\frac{dT_p}{dz}.$$

On the other hand, the environment temperature has increased from T_0 to T_e, and

$$T_e = T_0 - \Gamma_e \Delta z, \quad \text{where } \Gamma_e = -\frac{dT_e}{dz}.$$

If $\Gamma_e \neq \Gamma_d$, then there is a difference between the temperature of the parcel and the temperature of its surrounding environment:

$$T_p - T_e = (\Gamma_e - \Gamma_d)\Delta z. \tag{11.17}$$

Since the pressures inside and outside the parcel at height $z + \Delta z$ are same, these pressures are both equal to

$$p = p_0 + \frac{dp_e}{dz}\Delta z.$$

According to the ideal gas law $p = R_a T \rho$, the densities inside and outside the parcel are, respectively,

$$\rho_p = \frac{p}{R_a T_p},$$

$$\rho_e = \frac{p}{R_a T_e}.$$

From this and (11.17),

$$\frac{\rho_e - \rho_p}{\rho_p} = \frac{T_p - T_e}{T_e} = \frac{(\Gamma_e - \Gamma_d)\Delta z}{T_e}. \tag{11.18}$$

If $\Gamma_e < \Gamma_d$, then $T_p < T_e$ and $\rho_e < \rho_p$. So the parcel temperature is less than environment temperature and the parcel is heavier than its environment. In this case the atmosphere is said to be *statically stable*. If $\Gamma_e > \Gamma_d$, then $T_p > T_e$ and $\rho_e > \rho_p$. So the parcel temperature is higher than the environment temperature, and the parcel is lighter than its environment. In this case the atmosphere is said to be *statically unstable*. If $\Gamma_e = \Gamma_d$, then $T_p = T_e$ and $\rho_e = \rho_p$. In this case the atmosphere is said to have *neutral stability*.

11.5.2 Buoyancy Frequency

Suppose that V is the volume of the air parcel at height z, and ρ_p and ρ_e are the densities inside and outside the parcel, respectively. Then the masses inside and outside the parcel are $\rho_p V$ and $\rho_e V$, respectively. So the upward buoyancy force on the parcel at height z is equal to g times the difference between these two masses:

$$g(\rho_e V - \rho_p V) = gV(\rho_e - \rho_p).$$

On the other hand, since the acceleration of the parcel is $\frac{d^2(\Delta z)}{dt^2}$, according to Newton's second law, the upward buoyancy force is equal to

$$\rho_p V \frac{d^2(\Delta z)}{dt^2},$$

i.e., the mass inside the parcel times the acceleration of the parcel. Therefore,

$$gV(\rho_e - \rho_p) = \rho_p V \frac{d^2(\Delta z)}{dt^2}.$$

Dividing both sides by $\rho_p V$,

$$\frac{d^2(\Delta z)}{dt^2} = g\left(\frac{\rho_e - \rho_p}{\rho_p}\right).$$

Combining this with (11.18), we get

$$\frac{d^2(\Delta z)}{dt^2} = \frac{g(\Gamma_e - \Gamma_a)}{T_e}\Delta z$$

which is equivalent to

$$\frac{d^2(\Delta z)}{dt^2} + N^2 \Delta z = 0 \quad \left(N^2 = \frac{g(\Gamma_a - \Gamma_e)}{T_e}\right).$$

The quantity N is called the *buoyancy frequency* or the *Brunt-Väisälä frequency*. The buoyancy frequency is a useful measure of atmosphere stratification. It relates to the potential temperature of the environment as follows.

Let θ_e be the potential temperature of the environment at height z. By (11.14), it follows that

$$\ln \theta_e = \ln T_e + \frac{R_a}{c_p}\ln p_0 - \frac{R_a}{c_p}\ln p.$$

Differentiating both sides with respect to z, and then using the hydrostatic balance equation $\frac{dp}{dz} = -\frac{gp}{R_a T_e}$, we get

$$\frac{1}{\theta_e}\frac{d\theta_e}{dz} = \frac{1}{T_e}\frac{dT_e}{dz} - \frac{R_a}{c_p p}\frac{dp}{dz} = \frac{1}{T_e}\frac{dT_e}{dz} + \frac{g}{T_e c_p}.$$

Notice that lapse rates $\Gamma_a = \frac{g}{c_p}$ and $\Gamma_e = -\frac{dT_e}{dz}$. Then

$$\frac{1}{\theta_e}\frac{d\theta_e}{dz} = \frac{\Gamma_a - \Gamma_e}{T_e}.$$

If the potential temperature θ_e increases with height, $\frac{d\theta_e}{dz} > 0$, and so $\Gamma_e < \Gamma_a$. In this case, the atmosphere is statically stable. If the potential temperature θ_e decreases with height, $\frac{d\theta_e}{dz} < 0$, and so $\Gamma_e > \Gamma_a$. In this case, the atmosphere is statically unstable. Notice that $N^2 = \frac{g(\Gamma_a - \Gamma_e)}{T_e}$. Then

$$N^2 = \frac{g}{\theta_e} \frac{d\theta_e}{dz}.$$

This shows that the buoyancy frequency closely relates to the potential temperature of the environment.

11.6 CLAUSIUS-CLAPEYRON EQUATION

The relationship between the temperature of a liquid and its vapor pressure is not a straight line in a temperature-pressure diagram. The vapor pressure of water, for example, increases significantly more rapidly than the temperature. This behavior can be explained with the Clausius-Clapeyron equation:

$$\frac{dp}{dT} = \frac{Lp}{R_v T^2},$$

where p and T are the pressure and temperature at the phase transition, L is the latent heat of vaporization per unit mass, and R_v is the specific gas constant for the vapor.

Consider a parcel of moist air. Denote by p_v and p_{sv} the partial pressures of water vapor and the saturation water vapor at the phase transition, respectively. Replacing p by p_v or p_{sv} in the Clausius-Clapeyron equation, we get

$$\frac{dp_v}{dT} = \frac{Lp_v}{R_v T^2},$$

$$\frac{dp_{sv}}{dT} = \frac{Lp_{sv}}{R_v T^2}$$

or

$$\frac{dp_v}{p_v} = \frac{L}{R_v} \frac{dT}{T^2},$$

$$\frac{dp_{sv}}{p_{sv}} = \frac{L}{R_v} \frac{dT}{T^2}. \tag{11.19}$$

Notice that if L is a constant, integrating both sides of (11.19), we get the following for the partial pressures of water vapor and saturation water vapor:

$$p_v(T) = p_v(T_0) e^{\frac{L}{R_v}\left(\frac{1}{T_0} - \frac{1}{T}\right)},$$

$$p_{sv}(T) = p_{sv}(T_0) e^{\frac{L}{R_v}\left(\frac{1}{T_0} - \frac{1}{T}\right)},$$

where T_0 is a constant reference temperature.

11.6.1 Saturation Mass Mixing Radio

By m_v denote the molecular mass of water vapor and by $\overline{m}$ denote the mean molecular mass of moist air. By Proposition 11.3, the mass mixing ratio of water vapor

$$\mu = \frac{m_v}{\overline{m}} \frac{p_v}{p},$$

where $\frac{m_v}{\overline{m}}$ is constant and is approximately 0.622. Define the *saturation mass mixing ratio* as

$$\mu_s = \frac{m_v}{\overline{m}} \frac{p_{sv}}{p},$$

where p_{sv} is the partial pressure of the saturation water vapor and p is the pressure at the phase transition.

If $\mu < \mu_s$, then the air is said to be *unsaturated*, and if $\mu = \mu_s$, then the air is said to be *saturated*. In this case, the corresponding adiabatic lapse rate is called the *saturation adiabatic lapse rate*, denoted by Γ_s,

$$\Gamma_s = -\frac{dT}{dz},$$

where T is the temperature at the phase transition. If $\mu > \mu_s$, then the air is said to be *supersaturated*.

11.6.2 Saturation Adiabatic Lapse Rate

Consider a saturated air parcel of unit mass. Assume that the parcel rises a small distance Δz and its temperature increases by a small amount ΔT. Then the saturation mass mixing ratio has an increase $-\Delta \mu_s$, and so the small amount of latent heat given to the parcel is

$$-L\Delta \mu_s,$$

where L is the latent heat of vaporization per unit mass and μ_s is the saturation mass mixing ratio.

On the other hand, a small heat input is given by $T\Delta s$, where ds is an increase of the entropy into the parcel. By (11.12) and the hydrostatic balance equation, the small heat input

$$T\Delta s = c_p \Delta T - \frac{R_a T}{p} \Delta p = c_p \Delta T + g\Delta z,$$

where $c_p = c_v + R_a$ and $g = |\mathbf{g}|$ and $\mathbf{g}$ is the gravitational acceleration.

The small latent heat is equal to the small heat input at saturation, so

$$-L\Delta \mu_s = c_p \Delta T + g\Delta z.$$

This is equivalent to

$$d\mu_s = -\frac{c_p}{L} dT - \frac{g}{L} dz. \tag{11.20}$$

The increase of the saturation mass mixing ratio is directly calculated as follows.

By the definition,

$$\mu_s = \frac{m_v}{m} \frac{p_{sv}}{p},$$

where p_{sv} depends only on T, taking logarithms and then differentiating both sides, since $\frac{m_v}{m}$ is constant, we get

$$\frac{d\mu_s}{\mu_s} = \frac{dp_{sv}}{p_{sv}} - \frac{dp}{p}.$$

From this, (11.19), and the hydrostatic balance equation, it follows that

$$\frac{d\mu_s}{\mu_s} = \frac{L}{R_v T^2} dT + \frac{g}{R_a T} dz,$$

and so the increase of the saturation mass mixing ratio is

$$d\mu_s = \frac{\mu_s L}{R_v T^2} dT + \frac{\mu_s g}{R_a T} dz. \tag{11.21}$$

From (11.20) and (11.21) for $d\mu_s$, it follows that

$$-\frac{c_p}{L} dT - \frac{g}{L} dz = \frac{\mu_s L}{R_v T^2} dT + \frac{\mu_s g}{R_a T} dz$$

or

$$-c_p \left(\frac{L^2 \mu_s}{c_p R_v T^2} + 1 \right) dT = g \left(\frac{L \mu_s}{R_a T} + 1 \right) dz.$$

This is equivalent to

$$-\frac{dT}{dz} = \frac{g}{c_p} \frac{\left(\frac{L \mu_s}{R_a T} + 1 \right)}{\left(\frac{L^2 \mu_s}{c_p R_v T^2} + 1 \right)},$$

i.e., the saturated adiabatic lapse rate

$$\Gamma_s = \frac{g}{c_p} \frac{\left(\frac{L \mu_s}{R_a T} + 1 \right)}{\left(\frac{L^2 \mu_s}{c_p R_v T^2} + 1 \right)}.$$

Notice that the adiabatic lapse rate $\Gamma_a = \frac{g}{c_p}$. Therefore,

$$\Gamma_s = \Gamma_a \frac{\left(\frac{L \mu_s}{R_a T} + 1 \right)}{\left(\frac{L^2 \mu_s}{c_p R_v T^2} + 1 \right)}.$$

This shows that the saturated adiabatic lapse rate and the adiabatic lapse rate have a close relation.

11.6.3   Equivalent Potential Temperature

Dividing both sides of (11.20) by T,

$$\frac{d\mu_s}{T} = -\frac{c_p \, dT}{LT} - \frac{g \, dz}{LT}.$$

From this and the hydrostatic balance equation $\frac{dp}{p} = -\frac{g}{R_a T} dz$, it follows that

$$\frac{L \, d\mu_s}{c_p T} = -\frac{dT}{T} + \kappa \frac{dp}{p}, \text{ where } \kappa = \frac{R_a}{c_p},$$

which is equivalent to

$$d \left(\frac{L \mu_s}{c_p T} \right) = -d(\ln T) + \kappa \, d(\ln p).$$

Hence,

$$\frac{L \mu_s}{c_p T} + \ln T - \kappa \ln p = c,$$

where c is a constant. Taking exponentials on both sides,

$$e^{\frac{L \mu_s}{c_p T}} \frac{T}{p^\kappa} = e^c.$$

This is equivalent to

$$T \left(\frac{p_0}{p} \right)^\kappa e^{\frac{L \mu_s}{c_p T}} = e^c p_0^\kappa.$$

Let

$$\theta_e = T \left(\frac{p_0}{p} \right)^\kappa e^{\frac{L \mu_s}{c_p T}}.$$

The quantity θ_e is called the *equivalent potential temperature*. It depends on temperature T and pressure p. The potential temperature given in Section 11.4 is

$$\theta = T \left(\frac{p_0}{p} \right)^\kappa.$$

Therefore,

$$\theta_e = \theta e^{\frac{L \mu_s}{c_p T}},$$

where μ_s is the saturation mass mixing ratio, T is the temperature, L is the latent heat of vaporization per unit mass, and $c_p = c_v + R_a$. This equality shows a close relation between the equivalent potential temperature and the potential temperature.

When the air in a rising parcel remains unsaturated, the latent heat is not released. So $L = 0$, and so $\theta_e = \theta$, i.e., the equivalent potential temperature is equal to the potential temperature. Once saturation takes place, the latent heat is released. So $L \neq 0$, and so $\theta_e \neq \theta$, i.e., the equivalent potential temperature is not equal to the potential temperature.

11.7 MATERIAL DERIVATIVES

Consider a blob of the atmosphere fluid and denote by $\mathbf{r}(t)$ the position of the blob at time t. The Eulerian velocity of the blob at time t is equal to the current time rate of change of the blob's position:

$$\mathbf{v}(\mathbf{r}(t), t) = \frac{d\mathbf{r}}{dt}, \tag{11.22}$$

where

$$\mathbf{v}(\mathbf{r}(t), t) = (v_1(x(t), y(t), z(t), t),\ v_2(x(t), y(t), z(t), t),$$
$$v_3(x(t), y(t), z(t), t)),$$
$$\mathbf{r} = (x(t), y(t), z(t)).$$

Notice that

$$\frac{d\mathbf{r}}{dt} = \left(\frac{dx(t)}{dt}, \frac{dy(t)}{dt}, \frac{dz(t)}{dt} \right).$$

Then

$$(v_1(x(t), y(t), z(t), t),\ v_2(x(t), y(t), z(t), t),\ v_3(x(t), y(t), z(t), t))$$
$$= \left(\frac{dx(t)}{dt}, \frac{dy(t)}{dt}, \frac{dz(t)}{dt} \right).$$

Comparing both sides of this equality, we get

$$\frac{dx}{dt} = v_1(x(t), y(t), z(t), t),$$

$$\frac{dy}{dt} = v_2(x(t), y(t), z(t), t),$$

$$\frac{dz}{dt} = v_3(x(t), y(t), z(t), t). \tag{11.23}$$

Similarly, the Eulerian acceleration of the blob at time t is equal to the second derivative of the position vector:

$$\mathbf{a}(\mathbf{r}(t), t) = \frac{d^2\mathbf{r}}{dt^2}.$$

Notice that

$$\frac{d^2\mathbf{r}}{dt^2} = \left(\frac{d^2x}{dt^2}, \frac{d^2y}{dt^2}, \frac{d^2z}{dt^2} \right).$$

Then

$$\mathbf{a}(\mathbf{r}(t), t) = \left(\frac{d^2 x}{dt^2}, \frac{d^2 y}{dt^2}, \frac{d^2 z}{dt^2} \right).$$

Differentiating (11.23) once again, applying the chain rule, we find the three components on the right-hand side are

$$\frac{d^2 x}{dt^2} = \frac{\partial v_1}{\partial x} \frac{dx}{dt} + \frac{\partial v_1}{\partial y} \frac{dy}{dt} + \frac{\partial v_1}{\partial z} \frac{dz}{dt} + \frac{\partial v_1}{\partial t} = \mathbf{v} \cdot \text{grad } v_1 + \frac{\partial v_1}{\partial t},$$

$$\frac{d^2 y}{dt^2} = \frac{\partial v_2}{\partial x} \frac{dx}{dt} + \frac{\partial v_2}{\partial y} \frac{dy}{dt} + \frac{\partial v_2}{\partial z} \frac{dz}{dt} + \frac{\partial v_2}{\partial t} = \mathbf{v} \cdot \text{grad } v_2 + \frac{\partial v_2}{\partial t},$$

$$\frac{d^2 z}{dt^2} = \frac{\partial v_3}{\partial x} \frac{dx}{dt} + \frac{\partial v_3}{\partial y} \frac{dy}{dt} + \frac{\partial v_3}{\partial z} \frac{dz}{dt} + \frac{\partial v_3}{\partial t} = \mathbf{v} \cdot \text{grad } v_3 + \frac{\partial v_3}{\partial t}.$$

Therefore, the Eulerian acceleration

$$\mathbf{a} = (\mathbf{v} \cdot \nabla)\mathbf{v} + \frac{\partial \mathbf{v}}{\partial t}.$$

Define the *material derivative* as

$$\frac{D}{Dt} = \frac{\partial}{\partial t} + (\mathbf{v} \cdot \nabla)$$

or

$$\frac{D}{Dt} = \frac{\partial}{\partial t} + v_1 \frac{\partial}{\partial x} + v_2 \frac{\partial}{\partial y} + v_3 \frac{\partial}{\partial z}.$$

Especially, define the *material derivatives of the position vector* $\mathbf{r}$ as

$$\frac{D\mathbf{r}}{Dt} = \frac{d\mathbf{r}}{dt},$$

$$\frac{Dr}{Dt} = \frac{dr}{dt} \quad (r = |\mathbf{r}|).$$

So the Eulerian acceleration is written simply as

$$\mathbf{a} = \frac{D\mathbf{v}}{Dt}. \tag{11.24}$$

The material derivatives differ from the partial derivatives. The material derivative $\frac{D}{Dt}$ represents the rate of change with respect to time following the moving fluid blob, while the partial derivative $\frac{\partial}{\partial t}$ represents the rate of change with respect to time at a fixed point.

Property. The material derivative has the following simple properties.

(i) Let c be a constant. Then $\frac{Dc}{Dt} = 0$.

(ii) Let f and g be functions of time t. Then

$$\frac{D(f \pm g)}{Dt} = \frac{Df}{Dt} \pm \frac{Dg}{Dt},$$

$$\frac{D(fg)}{Dt} = \frac{Df}{Dt}g + f\frac{Dg}{Dt},$$

$$\frac{D\left(\frac{f}{g}\right)}{Dt} = \frac{1}{g^2}\left(\frac{Df}{Dt}g - f\frac{Dg}{Dt}\right).$$

(iii) Let **F** and **G** be vectors with respect to t. Then

$$\frac{D(\mathbf{F}\cdot\mathbf{G})}{Dt} = \frac{D\mathbf{F}}{Dt}\cdot\mathbf{G} + \mathbf{F}\cdot\frac{D\mathbf{G}}{Dt},$$

$$\frac{D(\mathbf{F}\times\mathbf{G})}{Dt} = \frac{D\mathbf{F}}{Dt}\times\mathbf{G} + \mathbf{F}\times\frac{D\mathbf{G}}{Dt}.$$

Proof. We only prove (iii) here.

Assume that **F** and **G** are both the three-dimensional vectors and $\mathbf{F} = (f_1, f_2, f_3)$ and $\mathbf{G} = (g_1, g_2, g_3)$. Then the scalar product of **F** and **G** is

$$\mathbf{F}\cdot\mathbf{G} = f_1 g_1 + f_2 g_2 + f_3 g_3.$$

So

$$\frac{\partial(\mathbf{F}\cdot\mathbf{G})}{\partial t} = \frac{\partial f_1}{\partial t}g_1 + f_1\frac{\partial g_1}{\partial t} + \frac{\partial f_2}{\partial t}g_2 + f_2\frac{\partial g_2}{\partial t} + \frac{\partial f_3}{\partial t}g_3 + f_3\frac{\partial g_3}{\partial t}. \quad (11.25)$$

On the other hand, by

$$\frac{\partial\mathbf{F}}{\partial t} = \left(\frac{\partial f_1}{\partial t}, \frac{\partial f_2}{\partial t}, \frac{\partial f_3}{\partial t}\right),$$

$$\frac{\partial\mathbf{G}}{\partial t} = \left(\frac{\partial g_1}{\partial t}, \frac{\partial g_2}{\partial t}, \frac{\partial g_3}{\partial t}\right),$$

it follows that the scalar product of $\frac{\partial\mathbf{F}}{\partial t}$ and **G** is

$$\frac{\partial\mathbf{F}}{\partial t}\cdot\mathbf{G} = \frac{\partial f_1}{\partial t}g_1 + \frac{\partial f_2}{\partial t}g_2 + \frac{\partial f_3}{\partial t}g_3,$$

and the scalar product of **F** and $\frac{\partial\mathbf{G}}{\partial t}$ is

$$\mathbf{F}\cdot\frac{\partial\mathbf{G}}{\partial t} = f_1\frac{\partial g_1}{\partial t} + f_2\frac{\partial g_2}{\partial t} + f_3\frac{\partial g_3}{\partial t}.$$

Adding these two equations together gives

$$\frac{\partial\mathbf{F}}{\partial t}\cdot\mathbf{G} + \mathbf{F}\cdot\frac{\partial\mathbf{G}}{\partial t} = \frac{\partial f_1}{\partial t}g_1 + \frac{\partial f_2}{\partial t}g_2 + \frac{\partial f_3}{\partial t}g_3 + f_1\frac{\partial g_1}{\partial t} + f_2\frac{\partial g_2}{\partial t} + f_3\frac{\partial g_3}{\partial t}.$$

From this and (11.25), it follows that

$$\frac{\partial(\mathbf{F}\cdot\mathbf{G})}{\partial t} = \frac{\partial\mathbf{F}}{\partial t}\cdot\mathbf{G} + \mathbf{F}\cdot\frac{\partial\mathbf{G}}{\partial t}.$$

Similarly,

$$v_1 \frac{\partial (\mathbf{F} \cdot \mathbf{G})}{\partial x} = v_1 \frac{\partial \mathbf{F}}{\partial x} \cdot \mathbf{G} + v_1 \mathbf{F} \cdot \frac{\partial \mathbf{G}}{\partial x},$$

$$v_2 \frac{\partial (\mathbf{F} \cdot \mathbf{G})}{\partial y} = v_2 \frac{\partial \mathbf{F}}{\partial y} \cdot \mathbf{G} + v_2 \mathbf{F} \cdot \frac{\partial \mathbf{G}}{\partial y},$$

$$v_3 \frac{\partial (\mathbf{F} \cdot \mathbf{G})}{\partial z} = v_3 \frac{\partial \mathbf{F}}{\partial z} \cdot \mathbf{G} + v_3 \mathbf{F} \cdot \frac{\partial \mathbf{G}}{\partial z}.$$

Adding these four equations together, by the definition of material derivative, we get

$$\frac{D(\mathbf{F} \cdot \mathbf{G})}{Dt} = \frac{D\mathbf{F}}{Dt} \cdot \mathbf{G} + \mathbf{F} \cdot \frac{D\mathbf{G}}{Dt},$$

i.e., the first equality of (iii). Similarly, the second equality of (iii) can be derived.
□

In Section 10.4, the Eulerian form of the continuity equation was given. By the definition of the material derivative, the continuity equation takes the form

$$\frac{D\rho}{Dt} + \rho \operatorname{div} \mathbf{v} = 0.$$

In fact, by the definition of the material derivative,

$$\frac{D\rho}{Dt} + \rho \operatorname{div} \mathbf{v} = \frac{\partial \rho}{\partial t} + (\mathbf{v} \cdot \nabla)\rho + \rho \operatorname{div} \mathbf{v}.$$

Let $\mathbf{v} = (v_1, v_2, v_3)$. Then the sum of the last two terms on the right-hand side is

$$(\mathbf{v} \cdot \nabla)\rho + \rho \operatorname{div} \mathbf{v} = \left(v_1 \frac{\partial \rho}{\partial x} + v_2 \frac{\partial \rho}{\partial y} + v_3 \frac{\partial \rho}{\partial z} \right) + \left(\rho \frac{\partial v_1}{\partial x} + \rho \frac{\partial v_2}{\partial y} + \rho \frac{\partial v_3}{\partial z} \right)$$

$$= \frac{\partial (\rho v_1)}{\partial x} + \frac{\partial \rho v_2}{\partial y} + \frac{\partial \rho v_3}{\partial z} = \operatorname{div} (\rho \mathbf{v}),$$

and so

$$\frac{D\rho}{Dt} + \rho \operatorname{div} \mathbf{v} = \frac{\partial \rho}{\partial t} + \operatorname{div} (\rho \mathbf{v}).$$

From this and the Eulerian form of the continuity equation: $\frac{\partial \rho}{\partial t} + \operatorname{div} (\rho \mathbf{v}) = 0$, we get

$$\frac{D\rho}{Dt} + \rho \operatorname{div} \mathbf{v} = 0.$$

This is called the *Lagrangian form* of the continuity equation.

Euler's equation in a gravitational field given in Section 10.4 is

$$\frac{\partial \mathbf{v}}{\partial t} + (\mathbf{v} \cdot \nabla)\mathbf{v} = -\frac{\nabla p}{\rho} - g\mathbf{k}.$$

By the definition of the material derivative, it is rewritten in the form

$$\frac{D\mathbf{v}}{Dt} = -\frac{\nabla p}{\rho} - g\mathbf{k}.$$

11.8 VORTICITY AND POTENTIAL VORTICITY

Vorticity and potential vorticity are two important concepts in the atmosphere.

The curl of the atmospheric fluid velocity $\mathbf{v}$ is called the *vorticity*, denoted by ϖ:

$$\varpi = \nabla \times \mathbf{v} = \text{curl } \mathbf{v}.$$

For a three-dimensional flow, by the definition of the curl, the vorticity is given by

$$\varpi = \text{curl } \mathbf{v} = \left(\frac{\partial v_3}{\partial y} - \frac{\partial v_2}{\partial z}\right)\mathbf{i} + \left(\frac{\partial v_1}{\partial z} - \frac{\partial v_3}{\partial x}\right)\mathbf{j} + \left(\frac{\partial v_2}{\partial x} - \frac{\partial v_1}{\partial y}\right)\mathbf{k},$$

where $\mathbf{v} = (v_1, v_2, v_3)$ is the fluid velocity.

For a two-dimensional flow in which the direction of the fluid velocity is parallel to the xy-plane and the fluid velocity is independent of z, the vorticity is given by

$$\varpi = \text{curl } \mathbf{v} = \left(\frac{\partial v_2}{\partial x} - \frac{\partial v_1}{\partial y}\right)\mathbf{k}.$$

where $\mathbf{v} = (v_1, v_2)$.

In order to investigate the relationship between vorticities in an inertial frame and in a rotating frame, the definition of vorticity shows that we need to know only the relationship between the velocities in these two frames.

Suppose that the rotating frame R rotates at a constant angular velocity $\mathbf{\Omega}$ with respect to the inertial frame I and that the z-axes of both frames are in the direction of $\mathbf{\Omega}$. If the position vector $\mathbf{r}(t)$ is viewed in the rotating frame, then the change of the position vector between t and $t + \Delta t$ is

$$(\Delta\mathbf{r})_R = \mathbf{r}(t + \Delta t) - \mathbf{r}(t).$$

If the position vector is viewed in the inertial frame, then the rotation gives an extra change $\mathbf{\Omega} \times \mathbf{r}\Delta t$ in $\mathbf{r}$. So the change of the position vector at times t and $t + \Delta t$ is

$$(\Delta\mathbf{r})_I = (\Delta\mathbf{r})_R + \mathbf{\Omega} \times \mathbf{r}\Delta t.$$

This is equivalent to

$$\left(\frac{\Delta\mathbf{r}}{\Delta t}\right)_I = \left(\frac{\Delta\mathbf{r}}{\Delta t}\right)_R + \mathbf{\Omega} \times \mathbf{r}.$$

Let $\Delta t \to 0$. Then

$$\left(\frac{d\mathbf{r}}{dt}\right)_I = \left(\frac{d\mathbf{r}}{dt}\right)_R + \mathbf{\Omega} \times \mathbf{r}.$$

By (11.22), $\left(\frac{d\mathbf{r}}{dt}\right)_I = \mathbf{v}_I$ and $\left(\frac{d\mathbf{r}}{dt}\right)_R = \mathbf{v}_R$. So the velocities in these two frames satisfy

$$\mathbf{v}_I = \mathbf{v}_R + \mathbf{\Omega} \times \mathbf{r},$$

i.e., the velocity in the inertial frame is equal to the velocity in the rotating frame plus the vector product of the angular velocity and the position vector.

Taking the curl on both sides, and then using Property 10.1, we get

$$\text{curl } \mathbf{v}_I = \text{curl } \mathbf{v}_R + \text{curl } (\mathbf{\Omega} \times \mathbf{r}),$$

i.e.,

$$\boldsymbol{\varpi}_I = \boldsymbol{\varpi}_R + \text{curl } (\mathbf{\Omega} \times \mathbf{r}). \tag{11.26}$$

This shows that the vorticity in the inertial frame is equal to the vorticity in the rotating frame plus the curl of the vector product of the angular vector and the position vector.

We compute the term $\text{curl } (\mathbf{\Omega} \times \mathbf{r})$. By the definition of the curl,

$$\text{curl } (\mathbf{\Omega} \times \mathbf{r}) = \nabla \times (\mathbf{\Omega} \times \mathbf{r}).$$

Replacing $\mathbf{F}$ and $\mathbf{G}$ by $\mathbf{\Omega}$ and $\mathbf{r}$, respectively, in Property 10.7, we get

$$\nabla \times (\mathbf{\Omega} \times \mathbf{r}) = (\mathbf{r} \cdot \nabla)\mathbf{\Omega} - (\mathbf{\Omega} \cdot \nabla)\mathbf{r} + \mathbf{\Omega}(\nabla \cdot \mathbf{r}) - \mathbf{r}(\nabla \cdot \mathbf{\Omega}).$$

Therefore,

$$\text{curl } (\mathbf{\Omega} \times \mathbf{r}) = (\mathbf{r} \cdot \nabla)\mathbf{\Omega} - (\mathbf{\Omega} \cdot \nabla)\mathbf{r} + \mathbf{\Omega}(\nabla \cdot \mathbf{r}) - \mathbf{r}(\nabla \cdot \mathbf{\Omega}). \tag{11.27}$$

We compute each term on the right-hand side of (11.27). Let

$$\mathbf{\Omega} = (\Omega_1, \Omega_2, \Omega_3),$$

$$\mathbf{r} = (x, y, z).$$

Notice that the angular velocity $\mathbf{\Omega}$ is a constant. Then

$$(\mathbf{r} \cdot \nabla)\mathbf{\Omega} = x\frac{\partial \mathbf{\Omega}}{\partial x} + y\frac{\partial \mathbf{\Omega}}{\partial y} + z\frac{\partial \mathbf{\Omega}}{\partial z} = 0,$$

$$(\mathbf{\Omega} \cdot \nabla)\mathbf{r} = \left(\Omega_1\frac{\partial}{\partial x} + \Omega_2\frac{\partial}{\partial y} + \Omega_3\frac{\partial}{\partial z}\right)(x, y, z)$$

$$= (\Omega_1, \Omega_2, \Omega_3) = \mathbf{\Omega}.$$

Notice that

$$\nabla \cdot \mathbf{r} = \frac{\partial x}{\partial x} + \frac{\partial y}{\partial y} + \frac{\partial z}{\partial z} = 3,$$

$$\nabla \cdot \mathbf{\Omega} = \frac{\partial \Omega_1}{\partial x} + \frac{\partial \Omega_2}{\partial y} + \frac{\partial \Omega_3}{\partial z} = 0.$$

Then

$$\mathbf{\Omega}(\nabla \cdot \mathbf{r}) = 3\mathbf{\Omega},$$

$$\mathbf{r}(\nabla \cdot \mathbf{\Omega}) = 0.$$

Combining these results with (11.27), we get

$$\text{curl}\,(\boldsymbol{\Omega} \times \mathbf{r}) = 2\boldsymbol{\Omega}.$$

From this and (11.26),

$$\varpi_I = \varpi_R + 2\boldsymbol{\Omega},$$

where $\boldsymbol{\Omega}$ is the constant rotation vector. Sometimes, ϖ_I is called the *absolute vorticity* and ϖ_R is called *the relative vorticity*. Thus, the absolute vorticity is equal to the relative vorticity plus twice the rotation vector.

Let ϖ_I be the absolute vorticity, θ the potential temperature, and ρ the fluid density. The quantity

$$P = \frac{\varpi_I \cdot \text{grad}\,\theta}{\rho}$$

is called the *potential vorticity*.

11.9 NAVIER-STOKES EQUATION

The Navier-Stokes equation is an equation of motion involving viscous fluids. Here Newton's second law is applied to a small moving blob of a viscous fluid, and then the Navier-Stokes equation is derived.

11.9.1 Navier-Stokes Equation in an Inertial Frame

Consider a blob of cuboidal shape instantaneously with sides Δx, Δy, and Δz. Its volume $\Delta V = \Delta x \Delta y \Delta z$ and its mass $\Delta m = \rho \Delta V$, where ρ is the density. Applying Newton's second law gives

$$\Delta \mathbf{F} = (\rho \Delta V)\mathbf{a}, \tag{11.28}$$

where $\mathbf{a}$ is the acceleration of the blob and $\Delta \mathbf{F}$ is the vector sum of the pressure force, the gravitational force, and the frictional force acting on the blob, i.e.,

$$\Delta \mathbf{F} = \Delta \mathbf{F}_p + \Delta \mathbf{F}_g + \Delta \mathbf{F}_v.$$

First, we compute the pressure force acting on the blob.

The pressure force at position x is $p(x, y, z)\Delta y \Delta z$ in the positive x-direction and the pressure force at position $x + \Delta x$ is $p(x + \Delta x, y, z)\Delta y \Delta z$ in the negative x-direction, where $\Delta y \Delta z$ is the area of the relevant wall of the blob. Using Taylor's theorem gives

$$p(x + \Delta x, y, z)\Delta y \Delta z = \left(p(x, y, z) + \frac{\partial p}{\partial x}(x, y, z)\Delta x + \cdots \right)\Delta y \Delta z.$$

Therefore, the net pressure force in the positive x-direction is

$$(p(x, y, z) - p(x + \Delta x, y, z))\Delta y \Delta z \approx -\frac{\partial p}{\partial x}\Delta V.$$

Similarly, the net pressure forces in the positive y-direction and the positive z-direction are, respectively,

$$(p(x, y, z) - p(x, y + \Delta y, z))\Delta x \Delta z \approx -\frac{\partial p}{\partial y}\Delta V,$$

$$(p(x, y, z) - p(x, y, z + \Delta z))\Delta x \Delta y \approx -\frac{\partial p}{\partial z}\Delta V.$$

Summarizing, the net pressure force in three directions is

$$\Delta \mathbf{F}_p = -\left(\frac{\partial p}{\partial x}\mathbf{i} + \frac{\partial p}{\partial y}\mathbf{j} + \frac{\partial p}{\partial z}\mathbf{k}\right)\Delta V = -(\nabla p)\Delta V,$$

where $\mathbf{i} = (1, 0, 0)$, $\mathbf{j} = (0, 1, 0)$, and $\mathbf{k} = (0, 0, 1)$.

Next, we compute the gravitational force acting on the blob.

The gravitational force is the mass Δm of the blob times g acting downward. Since $\Delta m = \rho \Delta V$, the gravitational force

$$\Delta \mathbf{F}_g = -\Delta m g \mathbf{k} = -(\rho \Delta V)g \mathbf{k},$$

where $g = |\mathbf{g}|$ and $\mathbf{k} = (0, 0, 1)$.

Finally, we compute the frictional force acting on the blob.

For simplicity, we consider a special case where the frictional force acting on the blob is provided by a horizontal force $\tau = (\tau_x, \tau_y, 0)$ alone. The net frictional force acting on the blob is the difference between the stresses on the top and the bottom,

$$(\tau_x(z + \Delta z) - \tau_x(z))\Delta x \Delta y \approx \frac{\partial \tau_x}{\partial z}\Delta V,$$

$$(\tau_y(z + \Delta z) - \tau_y(z))\Delta x \Delta y \approx \frac{\partial \tau_y}{\partial z}\Delta V,$$

where $\Delta x \Delta y$ is the area of the relevant wall of the blob. Therefore, the net frictional force is

$$\Delta \mathbf{F}_v = \left(\frac{\partial \tau_x}{\partial z}\mathbf{i} + \frac{\partial \tau_y}{\partial z}\mathbf{j}\right)\Delta V, \tag{11.29}$$

where $\mathbf{i} = (1, 0, 0)$ and $\mathbf{j} = (0, 1, 0)$. In general, the net frictional force acting on the blob is

$$\Delta \mathbf{F}_v = \begin{cases} \Delta V \eta \left(\nabla^2 \mathbf{v} + \frac{1}{3}\nabla(\nabla \cdot \mathbf{v})\right) & \text{if the fluid is compressible,} \\ \Delta V \eta \nabla^2 \mathbf{v} & \text{if the fluid is incompressible,} \end{cases}$$

where η is the dynamic viscosity and $\nabla^2 = \nabla \cdot \nabla$.

Combining the above results with (11.28), we get

$$(\rho \Delta V)\mathbf{a} = -(\nabla p)\Delta V - (\rho \Delta V)g \mathbf{k} + \Delta \mathbf{F}_v.$$

Canceling ΔV, and then combining the result with (11.24), we get

$$\frac{D\mathbf{v}}{Dt} = -\frac{\nabla p}{\rho} - g\mathbf{k} + \mathbf{F}_v, \tag{11.30}$$

where

$$\mathbf{F}_v = \frac{\Delta \mathbf{F}_v}{\rho \Delta V}.$$

If the frictional force acting on blob is provided by the horizontal force $\tau = (\tau_x, \tau_y, 0)$ alone, then, by (11.29),

$$\mathbf{F}_v = \frac{1}{\rho}\frac{\partial \tau_x}{\partial z}\mathbf{i} + \frac{1}{\rho}\frac{\partial \tau_y}{\partial z}\mathbf{j}. \tag{11.31}$$

In general,

$$\mathbf{F}_v = \begin{cases} \frac{\eta}{\rho}\left(\nabla^2 \mathbf{v} + \frac{1}{3}\nabla(\nabla \cdot \mathbf{v})\right) & \text{if the fluid is compressible,} \\ \frac{\eta}{\rho}\nabla^2 \mathbf{v} & \text{if the fluid is incompressible.} \end{cases}$$

Equation (11.30) is called the *Navier-Stokes equation* or the *momentum equation* in an inertial frame. Comparing the Navier-Stokes equation with Euler's equation given in the end of Section 11.7, we see that the frictional term $\mathbf{F}_v$ is added to Euler's equation. Therefore, the Navier-Stokes equation is a generalization of Euler's equation.

11.9.2 Navier-Stokes Equation in a Rotating Frame

For the large-scale atmospheric flows, the rotation of Earth cannot be ignored. So Earth's rotation must be incorporated into the Navier-Stokes equation, and this will modify the Navier-Stokes equation.

Consider an inertial frame I and a rotating frame R. Suppose that the frame R rotates at a constant angular velocity $\boldsymbol{\Omega}$ with respect to the frame I, and that the z-axes of the two frames are both in the direction of $\boldsymbol{\Omega}$. In Section 11.8 we obtained the following relationship between the velocities in both frames,

$$\left(\frac{d\mathbf{r}}{dt}\right)_I = \left(\frac{d\mathbf{r}}{dt}\right)_R + \boldsymbol{\Omega} \times \mathbf{r}.$$

A double application of this equation gives

$$\left(\frac{d^2\mathbf{r}}{dt^2}\right)_I = \left(\frac{d}{dt}\left(\frac{d\mathbf{r}}{dt}\right)_I\right)_I = \left(\frac{d}{dt}\left(\frac{d\mathbf{r}}{dt}\right)_I\right)_R + \boldsymbol{\Omega} \times \left(\frac{d\mathbf{r}}{dt}\right)_I$$

$$= \left(\frac{d}{dt}\left(\left(\frac{d\mathbf{r}}{dt}\right)_R + \boldsymbol{\Omega} \times \mathbf{r}\right)\right)_R + \boldsymbol{\Omega} \times \left(\left(\frac{d\mathbf{r}}{dt}\right)_R + \boldsymbol{\Omega} \times \mathbf{r}\right).$$

This is equivalent to

$$\left(\frac{d^2\mathbf{r}}{dt^2}\right)_I = \left(\frac{d}{dt}\left(\frac{d\mathbf{r}}{dt}\right)_R\right)_R + \left(\frac{d}{dt}(\boldsymbol{\Omega} \times \mathbf{r})\right)_R + \boldsymbol{\Omega} \times \left(\frac{d\mathbf{r}}{dt}\right)_R$$

$$+ \boldsymbol{\Omega} \times (\boldsymbol{\Omega} \times \mathbf{r}).$$

Clearly, the first term on the right-hand side

$$\left(\frac{d}{dt}\left(\frac{d\mathbf{r}}{dt}\right)_R\right)_R = \left(\frac{d^2\mathbf{r}}{d^2t}\right)_R.$$

Let $\mathbf{\Omega} = (\Omega_1, \Omega_2, \Omega_3)$ and $\mathbf{r} = (x, y, z)$. Notice that

$$\frac{d\mathbf{r}}{dt} = \left(\frac{dx}{dt}, \frac{dy}{dt}, \frac{dz}{dt}\right).$$

Then

$$\frac{d}{dt}(\mathbf{\Omega} \times \mathbf{r}) = \frac{d}{dt}\begin{vmatrix} \mathbf{i} & \mathbf{j} & \mathbf{k} \\ \Omega_1 & \Omega_2 & \Omega_3 \\ x & y & z \end{vmatrix} = \begin{vmatrix} \mathbf{i} & \mathbf{j} & \mathbf{k} \\ \Omega_1 & \Omega_2 & \Omega_3 \\ \frac{dx}{dt} & \frac{dy}{dt} & \frac{dz}{dt} \end{vmatrix} = \mathbf{\Omega} \times \frac{d\mathbf{r}}{dt},$$

i.e., the second term is equal to the third term on the right-hand side. Therefore,

$$\left(\frac{d^2\mathbf{r}}{dt^2}\right)_I = \left(\frac{d^2\mathbf{r}}{dt^2}\right)_R + 2\mathbf{\Omega} \times \left(\frac{d\mathbf{r}}{dt}\right)_R + \mathbf{\Omega} \times (\mathbf{\Omega} \times \mathbf{r}).$$

However, by (11.24), the accelerations in the inertial frame and the rotating frame are, respectively,

$$\mathbf{a}_I = \left(\frac{d^2\mathbf{r}}{dt^2}\right)_I,$$

$$\mathbf{a}_R = \left(\frac{d^2\mathbf{r}}{dt^2}\right)_R,$$

and by (11.22), the velocity in the rotating frame is

$$\mathbf{v}_R = \left(\frac{d\mathbf{r}}{dt}\right)_R.$$

Therefore,

$$\mathbf{a}_I = \mathbf{a}_R + 2\mathbf{\Omega} \times \mathbf{v}_R + \mathbf{\Omega} \times (\mathbf{\Omega} \times \mathbf{r}),$$

where the term $2\mathbf{\Omega} \times \mathbf{v}_R$ is the *Coriolis acceleration* and the term $\mathbf{\Omega} \times (\mathbf{\Omega} \times \mathbf{r})$ is the *centripetal acceleration*. On the other hand, by (11.30), the Navier-Stokes equation in an inertial frame takes the form

$$\mathbf{a}_I = -\frac{\nabla p}{\rho} - g\mathbf{k} + \mathbf{F}_v,$$

where $\mathbf{k} = (0, 0, 1)$. Therefore,

$$\mathbf{a}_R + 2\mathbf{\Omega} \times \mathbf{v}_R + \mathbf{\Omega} \times (\mathbf{\Omega} \times \mathbf{r}) = -\frac{\nabla p}{\rho} - g\mathbf{k} + \mathbf{F}_v.$$

By (11.24), $\mathbf{a}_R = \left(\frac{D\mathbf{v}}{Dt}\right)_R$. If we drop the subscript R, the Navier-Stokes equation in the rotating frame is

$$\frac{D\mathbf{v}}{Dt} = -\frac{\nabla p}{\rho} - 2\boldsymbol{\Omega} \times \mathbf{v} - \boldsymbol{\Omega} \times (\boldsymbol{\Omega} \times \mathbf{r}) - g\mathbf{k} + \mathbf{F_v}, \tag{11.32}$$

where $\mathbf{k} = (0, 0, 1)$.

11.9.3 Component Form of the Navier-Stokes Equation

By using Cartesian coordinates, we take unit vectors $\mathbf{i}$ pointing eastward and $\mathbf{j}$ pointing northward, and $\mathbf{k}$ pointing upward at a point on Earth's surface.

Consider small incremental distances:

$$dx = r\cos\phi \, d\lambda \quad \text{in the eastward (zonal) direction,}$$

$$dy = r \, d\phi \quad \text{in the northward (meridional) direction,}$$

$$dz = dr \quad \text{in the vertical direction,}$$

where ϕ, λ, and z are latitude, longitude, and the vertical distance from Earth's surface, respectively, and $r = a + z$, where a is Earth's radius.

Let the velocity $\mathbf{v} = v_1\mathbf{i} + v_2\mathbf{j} + v_3\mathbf{k}$. Notice that $\mathbf{i}, \mathbf{j}$, and $\mathbf{k}$ change with time. Then

$$\frac{D\mathbf{v}}{Dt} = \frac{Dv_1}{Dt}\mathbf{i} + v_1\frac{D\mathbf{i}}{Dt} + \frac{Dv_2}{Dt}\mathbf{j} + v_2\frac{D\mathbf{j}}{Dt} + \frac{Dv_3}{Dt}\mathbf{k} + v_3\frac{D\mathbf{k}}{Dt}.$$

First, we compute the material derivatives of the unit vector $\mathbf{k}$.

Notice that $\mathbf{k} = \frac{\mathbf{r}}{r}$, where r is the magnitude of position vector $\mathbf{r}$. Then

$$\frac{D\mathbf{k}}{Dt} = \frac{D}{Dt}\left(\frac{\mathbf{r}}{r}\right) = \frac{1}{r}\frac{D\mathbf{r}}{Dt} - \frac{\mathbf{r}}{r^2}\frac{Dr}{Dt} = \frac{1}{r}\left(\frac{D\mathbf{r}}{Dt} - \mathbf{k}\frac{Dr}{Dt}\right).$$

By the definition of the material derivative of the position vector given in Section 11.7,

$$\frac{D\mathbf{k}}{Dt} = \frac{1}{r}\left(\frac{d\mathbf{r}}{dt} - \mathbf{k}\frac{dr}{dt}\right).$$

By (11.22), (11.23) and $dz = dr$,

$$\frac{D\mathbf{k}}{Dt} = \frac{1}{r}(\mathbf{v} - v_3\mathbf{k}) = \frac{v_1\mathbf{i} + v_2\mathbf{j}}{r}.$$

Next, we compute the material derivative of the unit vector $\mathbf{j}$.

Let the constant angular velocity $\boldsymbol{\Omega} = \Omega(\mathbf{j}\cos\phi + \mathbf{k}\sin\phi)$. It follows from $\frac{D\boldsymbol{\Omega}}{Dt} = 0$ that

$$\cos\phi\frac{D\mathbf{j}}{Dt} + \mathbf{j}\frac{D\cos\phi}{Dt} + \sin\phi\frac{D\mathbf{k}}{Dt} + \mathbf{k}\frac{D\sin\phi}{Dt} = 0.$$

This is equivalent to

$$\frac{D\mathbf{j}}{Dt} = \mathbf{j}\tan\phi\frac{D\phi}{Dt} - \tan\phi\frac{D\mathbf{k}}{Dt} - \mathbf{k}\frac{D\phi}{Dt} = (\mathbf{j}\tan\phi - \mathbf{k})\frac{D\phi}{Dt} - \tan\phi\frac{v_1\mathbf{i} + v_2\mathbf{j}}{r}.$$

By the definition of the material derivative, (11.23), and $dy = r d\phi$, we get

$$\frac{D\phi}{Dt} = \frac{\partial\phi}{\partial t} + v_1 \frac{\partial\phi}{\partial x} + v_2 \frac{\partial\phi}{\partial y} + v_3 \frac{\partial\phi}{\partial z}$$

$$= \frac{\partial\phi}{\partial t} + \frac{\partial\phi}{\partial x}\frac{dx}{dt} + \frac{\partial\phi}{\partial y}\frac{dy}{dt} + \frac{\partial\phi}{\partial z}\frac{dz}{dt} = \frac{d\phi}{dt} = \frac{1}{r}\frac{dy}{dt} = \frac{v_2}{r}.$$

Therefore,

$$\frac{D\mathbf{j}}{Dt} = (\mathbf{j}\tan\phi - \mathbf{k})\frac{v_2}{r} - \tan\phi\frac{v_1\mathbf{i} + v_2\mathbf{j}}{r} = -\frac{v_1\tan\phi}{r}\mathbf{i} - \frac{v_2}{r}\mathbf{k}.$$

Finally, we compute the material derivative of the unit vector $\mathbf{i}$.

Notice that $\mathbf{i} = \mathbf{j} \times \mathbf{k}$. Applying property (iii) of the material derivative given in Section 11.7, we get

$$\frac{D\mathbf{i}}{Dt} = \frac{D(\mathbf{j} \times \mathbf{k})}{Dt} = \frac{D\mathbf{j}}{Dt} \times \mathbf{k} + \mathbf{j} \times \frac{D\mathbf{k}}{Dt} = \frac{-v_1\tan\phi\mathbf{i} - v_2\mathbf{k}}{r} \times \mathbf{k} + \mathbf{j} \times \frac{v_1\mathbf{i} + v_2\mathbf{j}}{r}.$$

By the orthogonality of the unit vectors,

$$\mathbf{i} \times \mathbf{k} = -\mathbf{j}, \quad \mathbf{k} \times \mathbf{k} = 0, \quad \mathbf{j} \times \mathbf{i} = -\mathbf{k}, \quad \mathbf{j} \times \mathbf{j} = 0,$$

and so

$$\frac{D\mathbf{i}}{Dt} = \frac{v_1\tan\phi\mathbf{j} - v_1\mathbf{k}}{r}.$$

Summarizing these results, we get

$$\frac{D\mathbf{v}}{Dt} = \frac{Dv_1}{Dt}\mathbf{i} + v_1\frac{v_1\tan\phi\mathbf{j} - v_1\mathbf{k}}{r} + \frac{Dv_2}{Dt}\mathbf{j} + v_2\frac{-v_1\tan\phi\mathbf{i} - v_2\mathbf{k}}{r} + \frac{Dv_3}{Dt}\mathbf{k}$$

$$+ v_3\frac{v_1\mathbf{i} + v_2\mathbf{j}}{r}.$$

The Coriolis term, the vector product of $2\mathbf{\Omega}$ and $\mathbf{v}$, is given by

$$2\mathbf{\Omega} \times \mathbf{v} = 2\Omega(v_3\cos\phi - v_2\sin\phi)\mathbf{i} + 2\Omega v_1\sin\phi\mathbf{j} - 2\Omega v_1\cos\phi\mathbf{k},$$

and the pressure gradient term is given by

$$-\frac{\nabla p}{\rho} = -\frac{1}{\rho}\left(\frac{\partial p}{\partial x}\mathbf{i} + \frac{\partial p}{\partial y}\mathbf{j} + \frac{\partial p}{\partial z}\mathbf{k}\right).$$

Disregard the centripetal acceleration $\mathbf{\Omega} \times (\mathbf{\Omega} \times \mathbf{r})$. Let

$$\mathbf{F}_v = F_x\mathbf{i} + F_y\mathbf{j} + F_z\mathbf{k}.$$

Then the Navier-Stokes equation (11.32) is written in the form

$$\frac{Dv_1}{Dt}\mathbf{i} + v_1\frac{v_1\tan\phi\mathbf{j} - v_1\mathbf{k}}{r} + \frac{Dv_2}{Dt}\mathbf{j} + v_2\frac{-v_1\tan\phi\mathbf{i} - v_2\mathbf{k}}{r} + \frac{Dv_3}{Dt}\mathbf{k}$$

$$+ v_3\frac{v_1\mathbf{i} + v_2\mathbf{j}}{r} = -\frac{1}{\rho}\left(\frac{\partial p}{\partial x}\mathbf{i} + \frac{\partial p}{\partial y}\mathbf{j} + \frac{\partial p}{\partial z}\mathbf{k}\right) - 2\Omega(v_3\cos\phi - v_2\sin\phi)\mathbf{i}$$

$$- 2\Omega v_1\sin\phi\mathbf{j} + 2\Omega v_1\cos\phi\mathbf{k} - g\mathbf{k} + F_x\mathbf{i} + F_y\mathbf{j} + F_z\mathbf{k}.$$

Collecting the terms in $\mathbf{i}, \mathbf{j}$, and $\mathbf{k}$, we can write the Navier-Stokes equation (11.32) in the component form:

$$\begin{cases} \dfrac{Dv_1}{Dt} = -\dfrac{1}{\rho}\dfrac{\partial p}{\partial x} + \left(2\Omega + \dfrac{v_1}{r\cos\phi}\right)(v_2\sin\phi - v_3\cos\phi) + F_x, \\[3mm] \dfrac{Dv_2}{Dt} = -\dfrac{1}{\rho}\dfrac{\partial p}{\partial y} - \dfrac{v_2 v_3}{r} - \left(2\Omega + \dfrac{v_1}{r\cos\phi}\right)v_1\sin\phi + F_y, \\[3mm] \dfrac{Dv_3}{Dt} = -\dfrac{1}{\rho}\dfrac{\partial p}{\partial z} + \dfrac{v_1^2 + v_2^2}{r} + 2\Omega v_1\cos\phi - g + F_z, \end{cases}$$

where the first two equations are called the *horizontal momentum equations* and the third equation is called the *vertical momentum equation*.

11.10 GEOSTROPHIC BALANCE EQUATIONS

The component form of the Navier-Stokes equation is complicated. We need to simplify it for motions associated with large-scale weather systems. Since the depth of the atmosphere is much less than Earth's radius, in the component form of the Navier-Stokes equation, we first can replace the distance r by Earth's radius a with negligible error,

$$\begin{cases} \dfrac{Dv_1}{Dt} = -\dfrac{1}{\rho}\dfrac{\partial p}{\partial x} + \left(2\Omega + \dfrac{v_1}{a\cos\phi}\right)(v_2\sin\phi - v_3\cos\phi) + F_x, \\[3mm] \dfrac{Dv_2}{Dt} = -\dfrac{1}{\rho}\dfrac{\partial p}{\partial y} - \dfrac{v_2 v_3}{a} - \left(2\Omega + \dfrac{v_1}{a\cos\phi}\right)v_1\sin\phi + F_y, \quad (11.33) \\[3mm] \dfrac{Dv_3}{Dt} = -\dfrac{1}{\rho}\dfrac{\partial p}{\partial z} + \dfrac{v_1^2 + v_2^2}{a} + 2\Omega v_1\cos\phi - g + F_z. \end{cases}$$

Assume that $|v_3\cos\phi| \ll |v_2\sin\phi|$ and $\frac{|v_1|}{a\cos\phi} \ll 2\Omega$, and the Coriolis parameter $f = 2\Omega\sin\phi$. This is a common assumption for weather systems. Then the first equation in (11.33) reduces to

$$\frac{Dv_1}{Dt} = -\frac{1}{\rho}\frac{\partial p}{\partial x} + fv_2 + F_x.$$

Similarly, for weather systems, we usually assume that $\frac{|v_2 v_3|}{a} \ll 2\Omega|v_1\sin\phi|$ and $\frac{|v_1|}{a\cos\phi} \ll 2\Omega$ and the Coriolis parameter $f = 2\Omega\sin\phi$. Then the second equation reduces to

$$\frac{Dv_2}{Dt} = -\frac{1}{\rho}\frac{\partial p}{\partial y} - fv_1 + F_y.$$

Also, we usually assume that $\frac{v_1^2 + v_2^2}{a} \ll g$ and $2\Omega v_1\cos\phi \ll g$. Then the third equation reduces to

$$\frac{Dv_3}{Dt} = -\frac{1}{\rho}\frac{\partial p}{\partial z} - g + F_z.$$

Combining these with (11.33), the component form of the Navier-Stokes equation is simplified further to

$$\begin{cases} \dfrac{Dv_1}{Dt} = -\dfrac{1}{\rho}\dfrac{\partial p}{\partial x} + fv_2 + F_x, \\[2mm] \dfrac{Dv_2}{Dt} = -\dfrac{1}{\rho}\dfrac{\partial p}{\partial y} - fv_1 + F_y, \\[2mm] \dfrac{Dv_3}{Dt} = -\dfrac{1}{\rho}\dfrac{\partial p}{\partial z} - g + F_z, \end{cases} \qquad (11.34)$$

where the Coriolis parameter $f = 2\Omega \sin \phi$.

The f-plane and the β-plane are two Cartesian coordinate systems. In the f-plane the Coriolis parameter is assumed to have constant value $f_0 = 2\Omega \sin \phi_0$, where ϕ_0 is a constant latitude. In the β-plane the Coriolis parameter is assumed to vary linearly with latitude $f = f_0 + \beta y$, where $\beta = \frac{2\Omega \cos \phi_0}{a}$ is the north-south variation of the Coriolis force and a is Earth's radius.

Assume further that $\frac{Dv_1}{Dt} \ll fv_2$ and that $\frac{\partial p}{\partial x}$ is the only term that can balance the large term fv_2. This is also a common assumption for weather systems. If we disregard frictional force, the first equation in (11.34) reduces to

$$fv_2 = \frac{1}{\rho}\frac{\partial p}{\partial x}.$$

Similarly, assume further that $\frac{Dv_2}{Dt} \ll fv_1$ and that $\frac{\partial p}{\partial y}$ is the only term that can balance the large term fv_1. If we disregard frictional force, the second equation in (11.34) reduces to

$$fv_1 = -\frac{1}{\rho}\frac{\partial p}{\partial y}.$$

These two equations are called the *geostrophic balance equations.*

The geostrophic balance equations are a good approximation for the horizontal momentum equations.

Usually, we assume further that $\frac{Dv_3}{Dt} \ll g$ and that the vertical pressure gradient $\frac{\partial p}{\partial z}$ is the only term that can balance the large g term. If we disregard frictional force, the third equation in (11.34) reduces to

$$\frac{\partial p}{\partial z} = -g\rho.$$

This equation is just the *hydrostatic balance equation* in Section 11.3. The hydrostatic balance equation is a good approximation for the vertical momentum equation.

11.11 BOUSSINESQ APPROXIMATION AND ENERGY EQUATION

In the Boussinesq approximation, the atmospheric flows are always assumed to satisfy div $\mathbf{v} = 0$, where $\mathbf{v}$ is the fluid velocity. This assumption implies that the continuity equation decouples into two equations:

$$\text{div } \mathbf{v} = 0,$$
$$\frac{D\rho}{Dt} = 0.$$

The first equation is equivalent to

$$\frac{\partial v_1}{\partial x} + \frac{\partial v_2}{\partial y} + \frac{\partial v_3}{\partial z} = 0,$$

where $\mathbf{v} = (v_1, v_2, v_3)$. The second equation is called the *density equation*. It states that for atmospheric flow, the density is constant on following a moving fluid blob. But this does not imply that the density is uniform everywhere. Therefore, we must still allow for vertical density stratification.

We separate the density into a background $\overline{\rho}$ depending only on height z and a deviation ρ' as follows:

$$\rho(x, y, z, t) = \rho'(x, y, z, t) + \overline{\rho}(z).$$

Substituting this equation into the density equation leads to

$$\frac{D\rho'}{Dt} + \frac{D\overline{\rho}}{Dt} = 0.$$

Since $\overline{\rho}$ depends only on z,

$$\frac{\partial \overline{\rho}}{\partial t} = \frac{\partial \overline{\rho}}{\partial x} = \frac{\partial \overline{\rho}}{\partial y} = 0,$$

and so

$$\frac{D\overline{\rho}}{Dt} = \frac{\partial \overline{\rho}}{\partial t} + v_1 \frac{\partial \overline{\rho}}{\partial x} + v_2 \frac{\partial \overline{\rho}}{\partial y} + v_3 \frac{\partial \overline{\rho}}{\partial z} = v_3 \frac{d\overline{\rho}}{dz}.$$

Therefore,

$$\frac{D\rho'}{Dt} + v_3 \frac{d\overline{\rho}}{dz} = 0.$$

Denote $\rho_0 = \overline{\rho}(0)$ and $N_B^2 = -\frac{g}{\rho_0} \frac{d\overline{\rho}}{dz}$. This equation is equivalent to

$$\frac{g}{\rho_0} \frac{D\rho'}{Dt} - v_3 N_B^2 = 0.$$

The quantity N_B is called the *Boussinesq buoyancy frequency* for the stratified atmospheric fluid with div $\mathbf{v} = 0$.

Similarly, the pressure is separated into a background $\bar{p}$ depending only on height z and a deviation p':

$$p(x, y, z, t) = p'(x, y, z, t) + \bar{p}(z),$$

and so

$$\frac{\partial p}{\partial z} = \frac{\partial p'}{\partial z} + \frac{d\bar{p}}{dz}.$$

However, by the hydrostatic balance equation $\frac{\partial p}{\partial z} = -g\rho$ and $\rho = \rho' + \bar{\rho}$, it follows that

$$\frac{\partial p}{\partial z} = -g\rho' - g\bar{\rho}.$$

Therefore,

$$\frac{\partial p'}{\partial z} + \frac{d\bar{p}}{dz} = -g\rho' - g\bar{\rho}.$$

From this equation, we see that if the background pressure satisfies the hydrostatic balance equation:

$$\frac{d\bar{p}}{dz} = -g\bar{\rho},$$

then the deviation also satisfies the hydrostatic balance equation:

$$\frac{\partial p'}{\partial z} = -g\rho'.$$

Since $\bar{p}$ depends only on z,

$$\frac{\partial \bar{p}}{\partial x} = \frac{\partial \bar{p}}{\partial y} = 0.$$

Notice that $p = p' + \bar{p}$. Then

$$\frac{\partial p}{\partial x} = \frac{\partial p'}{\partial x} + \frac{\partial \bar{p}}{\partial x} = \frac{\partial p'}{\partial x},$$

$$\frac{\partial p}{\partial y} = \frac{\partial p'}{\partial y} + \frac{\partial \bar{p}}{\partial y} = \frac{\partial p'}{\partial y}.$$

Assume that $\rho = \rho_0$. Then the first two equations of (11.34) reduce to

$$\frac{Dv_1}{Dt} = -\frac{1}{\rho_0} \frac{\partial p'}{\partial x} + fv_2 + F_x,$$

$$\frac{Dv_2}{Dt} = -\frac{1}{\rho_0} \frac{\partial p'}{\partial y} - fv_1 + F_y.$$

Summarizing up all results, a system of five equations with respect to p' and ρ' is obtained as follows:

$$\begin{cases} \dfrac{Dv_1}{Dt} = -\dfrac{1}{\rho_0}\dfrac{\partial p'}{\partial x} + fv_2 + F_x, \\[2mm] \dfrac{Dv_2}{Dt} = -\dfrac{1}{\rho_0}\dfrac{\partial p'}{\partial y} - fv_1 + F_y, \\[2mm] \dfrac{\partial p'}{\partial z} = -g\rho', \\[2mm] \dfrac{g}{\rho_0}\dfrac{D\rho'}{Dt} - N_B^2 v_3 = 0, \\[2mm] \dfrac{\partial v_1}{\partial x} + \dfrac{\partial v_2}{\partial y} + \dfrac{\partial v_3}{\partial z} = 0, \end{cases} \tag{11.35}$$

where $\rho_0 = \overline{\rho}(0)$, $N_B^2 = -\dfrac{g}{\rho_0}\dfrac{d\overline{\rho}}{dz}$, $\mathbf{v} = (v_1, v_2, v_3)$, and the material derivative

$$\frac{D}{Dt} = \frac{\partial}{\partial t} + v_1\frac{\partial}{\partial x} + v_2\frac{\partial}{\partial y} + v_3\frac{\partial}{\partial z}.$$

This system of equations is called the *Boussinesq equation*. Because of the presence of the quadratic terms in the material derivative, this system of equations is sometimes called the *nonlinear Boussinesq equation*.

We make the further approximation linearizing these equations. Dropping the quadratic terms in the material derivatives and disregarding frictional force, the system of equations (11.35) can be approximated by

$$\begin{cases} \dfrac{\partial v_1}{\partial t} = fv_2 - \dfrac{1}{\rho_0}\dfrac{\partial p'}{\partial x} + F_x, \\[2mm] \dfrac{\partial v_2}{\partial t} = -fv_1 - \dfrac{1}{\rho_0}\dfrac{\partial p'}{\partial y} + F_y, \\[2mm] \dfrac{\partial p'}{\partial z} = -g\rho', \\[2mm] \dfrac{g}{\rho_0}\dfrac{\partial \rho'}{\partial t} - N_B^2 v_3 = 0, \\[2mm] \dfrac{\partial v_1}{\partial x} + \dfrac{\partial v_2}{\partial y} + \dfrac{\partial v_3}{\partial z} = 0. \end{cases} \tag{11.36}$$

This system of equations is called the *linearized Boussinesq equation*.

In (11.36), disregarding friction and then multiplying the first equation by $\rho_0 v_1$, the second equation by $\rho_0 v_2$, the third equation by v_3, the fourth equation by $\frac{g\rho'}{N_B^2}$, and the fifth equation by p', we get

$$\rho_0 v_1 \frac{\partial v_1}{\partial t} - \rho_0 v_1 f v_2 + v_1 \frac{\partial p'}{\partial x} = 0,$$

$$\rho_0 v_2 \frac{\partial v_2}{\partial t} + \rho_0 v_2 f v_1 + v_2 \frac{\partial p'}{\partial y} = 0,$$

$$v_3 \frac{\partial p'}{\partial z} + v_3 g \rho' = 0,$$

$$\frac{g^2 \rho'}{N_B^2 \rho_0} \frac{\partial \rho'}{\partial t} - g \rho' v_3 = 0,$$

$$p' \frac{\partial v_1}{\partial x} + p' \frac{\partial v_2}{\partial y} + p' \frac{\partial v_3}{\partial z} = 0.$$

Adding these five equations together gives

$$\rho_0 v_1 \frac{\partial v_1}{\partial t} + \rho_0 v_2 \frac{\partial v_2}{\partial t} + v_1 \frac{\partial p'}{\partial x} + v_2 \frac{\partial p'}{\partial y} + v_3 \frac{\partial p'}{\partial z} + p' \frac{\partial v_1}{\partial x} + p' \frac{\partial v_2}{\partial y}$$

$$+ p' \frac{\partial v_3}{\partial z} + \frac{g^2 \rho'}{N_B^2 \rho_0} \frac{\partial \rho'}{\partial t} = 0.$$

A short calculation shows that this equation is equivalent to

$$\frac{\rho_0}{2} \frac{\partial}{\partial t} \left[v_1^2 + v_2^2 + \left(\frac{g \rho'}{N_B \rho_0} \right)^2 \right] + \nabla \cdot (\mathbf{v} p') = 0,$$

which is called the *energy equation*, where $\mathbf{v} = (v_1, v_2, v_3)$.

In the energy equation, the term $\frac{\rho_0}{2}(v_1^2 + v_2^2)$ is clearly the kinetic energy per unit volume of the horizontal motion, the term involving $(\rho')^2$ can be interpreted as the available potential energy, and the term $\mathbf{v} p'$ can be interpreted as an *energy flux*. Overall, the equation states that the kinetic energy and available potential energy within a volume increases if there is an energy flux into the volume, and the kinetic energy and available potential energy within a volume decreases if there is an energy flux out of the volume.

11.12 QUASI-GEOSTROPHIC POTENTIAL VORTICITY

Consider a geostrophic flow $\mathbf{v_g} = (v_1^g, v_2^g, 0)$. Start from Boussinesq equation (11.35). Assume further that $\frac{Dv_1^g}{Dt} \ll f_0 v_2^g$ and $\frac{Dv_2^g}{Dt} \ll f_0 v_1^g$, where f_0 is the Coriolis parameter $(f_0 = 2\Omega \sin \phi_0)$ and ϕ_0 is a constant latitude. This is a common assumption for weather systems. Then the first two equations in (11.35) reduce to geostrophic balance equations, i.e., the geostrophic flow satisfies the following geostrophic balance equations:

$$v_2^g = \frac{1}{f_0 \rho_0} \frac{\partial p'}{\partial x},$$

$$v_1^g = -\frac{1}{f_0 \rho_0} \frac{\partial p'}{\partial y}.$$

Let $\psi = \frac{p'}{f_0 \rho_0}$. Then $v_2^g = \frac{\partial \psi}{\partial x}$ and $v_1^g = -\frac{\partial \psi}{\partial y}$. The function ψ is called the *geostrophic streamfunction*.

The geostrophic flow $(v_1^g, v_2^g, 0)$ differs from the true velocity. The differences of components between the true velocity (v_1, v_2, v_3) and the geostrophic flow $(v_1^g, v_2^g, 0)$ are denoted by

$$v_1^a = v_1 - v_1^g,$$
$$v_2^a = v_2 - v_2^g,$$
$$v_3^a = v_3.$$

The vector $\mathbf{v}_a = (v_1^a, v_2^a, v_3^a)$ is called the *ageostrophic velocity*.

Another approximation beyond geostrophic balance on a β-plane is given by the *quasi-geostrophic equations*:

$$\begin{cases} \dfrac{D_g v_1^g}{Dt} - f_0 v_2^a - \beta y v_2^g = 0, \\[2mm] \dfrac{D_g v_2^g}{Dt} + f_0 v_1^a + \beta y v_1^g = 0, \\[2mm] \dfrac{\partial v_1^a}{\partial x} + \dfrac{\partial v_2^a}{\partial y} + \dfrac{\partial v_3^a}{\partial z} = 0, \\[2mm] \dfrac{g}{\rho_0} \dfrac{D_g \rho'}{Dt} = N_B^2 v_3^a, \\[2mm] \dfrac{\partial p'}{\partial z} = -g\rho', \end{cases} \qquad (11.37)$$

where

$$\frac{D_g}{Dt} = \frac{\partial}{\partial t} + v_1^g \frac{\partial}{\partial x} + v_2^g \frac{\partial}{\partial y}$$

is the material derivative following the geostrophic flow. In the quasi-geostrophic approximation, frictional force is disregarded and the Coriolis parameter

$$f = f_0 + \beta y = 2\Omega \sin\phi_0 + \frac{2\Omega \cos\phi_0}{a} y,$$

where ϕ_0 is a constant latitude and a is Earth's radius. The quasi-geostrophic equations hold in general for large-scale, low-frequency motions.

The quasi-geostrophic equations can conveniently be combined as follows.

Differentiating the first equation in (11.37) with respect to y, and then multiplying both sides by -1, we get

$$\frac{D_g}{Dt}\left(-\frac{\partial v_1^g}{\partial y}\right) + f_0 \frac{\partial v_2^a}{\partial y} + \beta y \frac{\partial v_2^g}{\partial y} + \beta v_2^g = 0.$$

Differentiating the second equation in (11.37) with respect to x, we get

$$\frac{D_g}{Dt}\left(\frac{\partial v_2^g}{\partial x}\right) + f_0 \frac{\partial v_1^a}{\partial x} + \beta y \frac{\partial v_1^g}{\partial x} = 0.$$

Adding these two equations gives

$$\frac{D_g}{Dt}\left(-\frac{\partial v_1^g}{\partial y}+\frac{\partial v_2^g}{\partial x}\right)+f_0\left(\frac{\partial v_1^a}{\partial x}+\frac{\partial v_2^a}{\partial y}\right)+\beta y\left(\frac{\partial v_1^g}{\partial x}+\frac{\partial v_2^g}{\partial y}\right)$$

$$+\beta v_2^g=0. \tag{11.38}$$

We compute each term on the left-hand side of (11.38).
Since $v_1^g=-\frac{\partial\psi}{\partial y}$ and $v_2^g=\frac{\partial\psi}{\partial x}$, the first term is

$$\frac{D_g}{Dt}\left(-\frac{\partial v_1^g}{\partial y}+\frac{\partial v_2^g}{\partial x}\right)=\frac{D_g}{Dt}\left(\frac{\partial^2\psi}{\partial x^2}+\frac{\partial^2\psi}{\partial y^2}\right).$$

From $\psi=\frac{p'}{f_0\rho_0}$ and the fifth equation in (11.37), it follows that

$$\frac{\partial\psi}{\partial z}=-\frac{g\rho'}{f_0\rho_0}.$$

Combining this and the fourth equation in (11.37), it follows that

$$v_3^a=\frac{D_g}{Dt}\left(\frac{g\rho'}{\rho_0 N_B^2}\right)=\frac{D_g}{Dt}\left(-\frac{f_0}{N_B^2}\frac{\partial\psi}{\partial z}\right).$$

Again, by the third equation of (11.37), the second term is

$$f_0\left(\frac{\partial v_1^a}{\partial x}+\frac{\partial v_2^a}{\partial y}\right)=-f_0\frac{\partial v_3^a}{\partial z}=\frac{D_g}{Dt}\left(\frac{\partial}{\partial z}\left(\frac{f_0^2}{N_B^2}\frac{\partial\psi}{\partial z}\right)\right).$$

Since $v_2^g=\frac{\partial\psi}{\partial x}$ and $v_1^g=-\frac{\partial\psi}{\partial y}$, the third term is

$$\beta y\left(\frac{\partial v_1^g}{\partial x}+\frac{\partial v_2^g}{\partial y}\right)=\beta y\left(-\frac{\partial^2\psi}{\partial y\partial x}+\frac{\partial^2\psi}{\partial x\partial y}\right)=0.$$

Notice that

$$\frac{\partial(f_0+\beta y)}{\partial t}=0,$$

$$v_1^g\frac{\partial(f_0+\beta y)}{\partial x}=0,$$

$$v_2^g\frac{\partial(f_0+\beta y)}{\partial y}=\beta v_2^g.$$

Adding these three equations together gives

$$\beta v_2^g=\frac{\partial(f_0+\beta y)}{\partial t}+v_1^g\frac{\partial(f_0+\beta y)}{\partial x}+v_2^g\frac{\partial(f_0+\beta y)}{\partial y}.$$

From this and $\frac{D_g}{Dt} = \frac{\partial}{\partial t} + v_1^g \frac{\partial}{\partial x} + v_2^g \frac{\partial}{\partial y}$, the last term is

$$\beta v_2^g = \frac{D_g(f_0 + \beta y)}{Dt}.$$

Summarizing all the results, by (11.38), the quasi-geostrophic equations are combined as follows

$$\frac{D_g \zeta}{Dt} = 0,$$

where

$$\frac{D_g}{Dt} = \frac{\partial}{\partial t} - \frac{\partial \psi}{\partial y} \frac{\partial}{\partial x} + \frac{\partial \psi}{\partial x} \frac{\partial}{\partial y}$$

and

$$\zeta = f_0 + \beta y + \frac{\partial^2 \psi}{\partial x^2} + \frac{\partial^2 \psi}{\partial y^2} + \frac{\partial}{\partial z} \left(\frac{f_0^2}{N_B^2} \frac{\partial \psi}{\partial z} \right) =: f_0 + \beta y + \mathcal{L}\psi.$$

The equation $\frac{D_g \zeta}{Dt} = 0$ is called the *quasi-geostrophic potential vorticity equation*. The quantity ζ is called the *quasi-geostrophic potential vorticity* associated with the geostrophic flow. The operator

$$\mathcal{L} = \frac{\partial^2}{\partial x^2} + \frac{\partial^2}{\partial y^2} + \frac{\partial}{\partial z} \left(\frac{f_0^2}{N_B^2} \frac{\partial}{\partial z} \right)$$

is called the *elliptic operator*, and

$$-\frac{D_g}{Dt} \left(\frac{\partial}{\partial z} \left(\frac{f_0^2}{N_B^2} \frac{\partial \psi}{\partial z} \right) \right) = f_0 \frac{\partial v_3^a}{\partial z}.$$

The term on the left-hand side,

$$-\frac{D_g}{Dt} \left(\frac{\partial}{\partial z} \left(\frac{f_0^2}{N_B^2} \frac{\partial \psi}{\partial z} \right) \right),$$

is often called the *stretching term*. It can generate vorticity by differential vertical motions. If a cylindrical blob of air enters a region where the vertical velocity increases with height, it is stretched vertically. Conversely, if the cylindrical blob of air enters a region where the vertical velocity decreases with height, it is shrunken vertically.

11.13 GRAVITY WAVES

Gravity waves are a class of atmospheric waves and are frequently observed in the atmosphere. The famous lee waves manifested as parallel bands of cloud downstream of mountain ranges are gravity waves. The linearized Boussinesq equation is a good tool to use to develop models of gravity waves.

11.13.1   Internal Gravity Waves

We concentrate on small-scale gravity waves. These waves are called *internal gravity waves*.

The horizontal scale of internal gravity waves is so small that Earth's rotation can be negligible. Therefore, this can be used to simplify the linearized Boussinesq equation by disregarding the Coriolis terms. In this case, the linearized Boussinesq equation (11.36) becomes a system of linear partial differential equations:

$$
\begin{cases}
\dfrac{\partial v_1}{\partial t} + \dfrac{1}{\rho_0}\dfrac{\partial p'}{\partial x} = 0, \\[2mm]
\dfrac{\partial v_2}{\partial t} + \dfrac{1}{\rho_0}\dfrac{\partial p'}{\partial y} = 0, \\[2mm]
\dfrac{\partial p'}{\partial z} = -g\rho', \\[2mm]
\dfrac{g}{\rho_0}\dfrac{\partial \rho'}{\partial t} - N_B^2 v_3 = 0, \\[2mm]
\dfrac{\partial v_1}{\partial x} + \dfrac{\partial v_2}{\partial y} + \dfrac{\partial v_3}{\partial z} = 0,
\end{cases}
\tag{11.39}
$$

where the velocity $\mathbf{v} = (v_1, v_2, v_3)$, ρ_0 is the constant, and N_B is the Boussinesq buoyancy frequency.

We look for linear plane-wave solutions, propagating in the xz-plane and independent of y, of the form

$$
\{v_1, v_2, v_3, p', \rho'\} = \mathrm{Re}\left\{ (\tilde{v}_1, \tilde{v}_2, \tilde{v}_3, \tilde{p}, \tilde{\rho})e^{i(kx+mz-\omega t)} \right\},
$$

where $\tilde{v}_1, \tilde{v}_2, \tilde{v}_3$ are complex amplitudes.

Substituting this expression into (11.39), we obtain a system of algebraic equations as follows:

$$
\begin{cases}
-i\omega\tilde{v}_1 + ik\dfrac{\tilde{p}}{\rho_0} = 0, \\[2mm]
-i\omega\tilde{v}_2 = 0, \\[2mm]
im\tilde{p} + g\tilde{\rho} = 0, \\[2mm]
i\omega g\dfrac{\tilde{\rho}}{\rho_0} + N_B^2\tilde{v}_3 = 0, \\[2mm]
ik\tilde{v}_1 + im\tilde{v}_3 = 0.
\end{cases}
$$

This system of algebraic equations is equivalent to

$$
\begin{cases}
\omega\tilde{v}_1 = k\dfrac{\tilde{p}}{\rho_0}, \\[2mm]
\omega\tilde{v}_2 = 0, \\[2mm]
im\tilde{p} = -g\tilde{\rho}, \\[2mm]
N_B^2\tilde{v}_3 = -i\omega g\dfrac{\tilde{\rho}}{\rho_0}, \\[2mm]
k\tilde{v}_1 = -m\tilde{v}_3
\end{cases}
$$

or

$$
\begin{cases}
\dfrac{\tilde{v}_1}{\tilde{p}} = \dfrac{k}{\omega\rho_0}, \\[2mm]
\dfrac{\tilde{v}_3}{\tilde{v}_1} = -\dfrac{k}{m}, \\[2mm]
\dfrac{\tilde{\rho}}{\tilde{v}_3} = \dfrac{\mathrm{i}N_{\mathrm{B}}^2\rho_0}{\omega g}, \\[2mm]
\dfrac{\tilde{p}}{\tilde{\rho}} = \dfrac{\mathrm{i}g}{m}, \\[2mm]
\tilde{v}_2 = 0.
\end{cases}
\tag{11.40}
$$

Multiplying the first four equations in (11.40), we get

$$
1 = \frac{\tilde{v}_1}{\tilde{p}}\frac{\tilde{v}_3}{\tilde{v}_1}\frac{\tilde{\rho}}{\tilde{v}_3}\frac{\tilde{p}}{\tilde{\rho}} = \left(\frac{k}{\omega\rho_0}\right)\left(-\frac{k}{m}\right)\left(\frac{\mathrm{i}N_{\mathrm{B}}^2\rho_0}{\omega g}\right)\left(\frac{\mathrm{i}g}{m}\right),
$$

and so

$$
\omega^2 = \frac{k^2 N_{\mathrm{B}}^2}{m^2}.
$$

This is the *dispersion relation* for internal gravity waves. It relates the angular frequency ω to the components k, m of the wave vector $\mathbf{k} = (k, 0, m)$.

On the other hand, solving the system of algebraic equations (11.40), we obtain a nontrivial solution $\tilde{v}_1, \tilde{v}_2, \tilde{v}_3$, and $\tilde{\rho}$ in favor $\tilde{p}$:

$$
\begin{cases}
\tilde{v}_1 = \dfrac{k\tilde{p}}{\rho_0\omega}, \\[2mm]
\tilde{v}_2 = 0, \\[2mm]
\tilde{v}_3 = -\dfrac{k^2\tilde{p}}{m\rho_0\omega}, \\[2mm]
\tilde{\rho} = -\dfrac{\mathrm{i}m\tilde{p}}{g}.
\end{cases}
$$

Arbitrarily choosing $\tilde{p}$ to be real, we find the linear plane-wave solution of (11.39) is

$$
\begin{cases}
p' = \mathrm{Re}\left\{\tilde{p}\mathrm{e}^{\mathrm{i}(kx+mz-\omega t)}\right\} = \tilde{p}\cos(kx + mz - \omega t), \\[2mm]
v_1 = \mathrm{Re}\left\{\tilde{v}_1\mathrm{e}^{\mathrm{i}(kx+mz-\omega t)}\right\} = \dfrac{k\tilde{p}}{\rho_0\omega}\cos(kx + mz - \omega t), \\[2mm]
v_2 = \mathrm{Re}\left\{\tilde{v}_2\mathrm{e}^{\mathrm{i}(kx+mz-\omega t)}\right\} = 0, \\[2mm]
v_3 = \mathrm{Re}\left\{\tilde{v}_3\mathrm{e}^{\mathrm{i}(kx+mz-\omega t)}\right\} = -\dfrac{k^2\tilde{p}}{m\rho_0\omega}\cos(kx + mz - \omega t), \\[2mm]
\rho' = \mathrm{Re}\left\{\tilde{\rho}\mathrm{e}^{\mathrm{i}(kx+mz-\omega t)}\right\} = \dfrac{m\tilde{p}}{g}\sin(kx + mz - \omega t).
\end{cases}
$$

These are called the *polarization relations* for internal gravity waves. The polarization relations may verify the precise phase relations between the velocity, density, and pressure disturbances.

By the dispersion relation $\omega^2 = \frac{k^2 N_B^2}{m^2}$, the angular frequency has two possible solutions $\omega = \pm \frac{N_B k}{m}$. Define a group velocity vector

$$\mathbf{c}_g = (c_g^{(x)}, 0, c_g^{(z)}) = \left(\frac{\partial \omega}{\partial k}, 0, \frac{\partial \omega}{\partial m} \right).$$

Its vertical component is

$$c_g^{(z)} = \frac{\partial \omega}{\partial m} = \mp \frac{N_B k}{m^2}.$$

The signs are chosen as follows.

For an atmospheric internal gravity wave generated near the ground and propagating information upward, $c_g^{(z)} > 0$. By convention, $k > 0$. So the vertical velocity $c_g^{(z)} = \frac{N_B k}{m^2}$ and the angular frequency $\omega = -\frac{N_B k}{m}$.

For this choice of signs, in the xz-plane, the phase surfaces

$$kx + mz - \omega t = \text{constant}$$

move obliquely downward in the direction of the wave vector $\mathbf{k} = (k, 0, m)$. However, the propagation of information represented by the group velocity vector is obliquely upward. The velocity vector $(v_1, 0, v_3)$ is parallel to the slanting phase surfaces and the fluid blobs oscillate up and down these surfaces.

Secondly, we look for plane-wave solutions, now allowing variations in y as well, of the form

$$\{v_1, v_2, v_3, p', \rho'\} = \text{Re} \left\{ (\widetilde{v}_1, \widetilde{v}_2, \widetilde{v}_3, \widetilde{p}, \widetilde{\rho}) e^{i(kx + ly + mz - \omega t)} \right\},$$

where $\widetilde{v}_1$, $\widetilde{v}_2$, $\widetilde{v}_3$ are complex amplitudes. Substituting this into (11.39), we obtain a system of algebraic equations:

$$
\begin{cases}
-i\omega \widetilde{v}_1 + ik\dfrac{\widetilde{p}}{\rho_0} = 0, \\[2mm]
-i\omega \widetilde{v}_2 + il\dfrac{\widetilde{p}}{\rho_0} = 0, \\[2mm]
ik\widetilde{v}_1 + il\widetilde{v}_2 + im\widetilde{v}_3 = 0, \\[2mm]
i\omega g \dfrac{\widetilde{\rho}}{\rho_0} + N_B^2 \widetilde{v}_3 = 0, \\[2mm]
im\widetilde{p} + g\widetilde{\rho} = 0.
\end{cases}
$$

It follows from the first two equations that $\widetilde{v}_2 = \frac{l}{k}\widetilde{v}_1$. Combining this with the third equation, we get $\frac{\widetilde{v}_3}{\widetilde{v}_1} = -\frac{k^2 + l^2}{mk}$. Therefore, this system of algebraic equations is equivalent to

$$\begin{cases} \dfrac{\widetilde{v}_1}{\widetilde{p}} = \dfrac{k}{\omega\rho_0}, \\[2mm] \dfrac{\widetilde{v}_3}{\widetilde{v}_1} = -\dfrac{k^2 + l^2}{mk}, \\[2mm] \dfrac{\widetilde{\rho}}{\widetilde{v}_3} = \dfrac{iN_{\mathrm{B}}^2\rho_0}{\omega g}, \\[2mm] \dfrac{\widetilde{p}}{\widetilde{\rho}} = \dfrac{ig}{m}, \\[2mm] \dfrac{\widetilde{v}_2}{\widetilde{p}} = \dfrac{l}{\omega\rho_0}. \end{cases}$$

Multiplying the first four equations, we get

$$1 = \frac{\widetilde{v}_1}{\widetilde{p}}\frac{\widetilde{v}_3}{\widetilde{v}_1}\frac{\widetilde{\rho}}{\widetilde{v}_3}\frac{\widetilde{p}}{\widetilde{\rho}} = \left(\frac{k}{\omega\rho_0}\right)\left(-\frac{k^2+l^2}{mk}\right)\left(\frac{iN_{\mathrm{B}}^2\rho_0}{\omega g}\right)\left(\frac{ig}{m}\right),$$

and so

$$\omega^2 = \frac{(k^2 + l^2)N_{\mathrm{B}}^2}{m^2}.$$

This is the *dispersion relation* for linear internal gravity waves. It relates the angular frequency ω to the components of the wave vector $\mathbf{k} = (k, l, m)$.

On the other hand, solving the system of algebraic equations, we obtain a nontrivial solution $\widetilde{v}_1$, $\widetilde{v}_2$, $\widetilde{v}_3$, and $\widetilde{\rho}$ in favor $\widetilde{p}$:

$$\begin{cases} \widetilde{v}_1 = \dfrac{k\widetilde{p}}{\rho_0\omega}, \\[2mm] \widetilde{v}_2 = \dfrac{l\widetilde{p}}{\rho_0\omega}, \\[2mm] \widetilde{v}_3 = -\dfrac{(k^2+l^2)\widetilde{p}}{m\rho_0\omega}, \\[2mm] \widetilde{\rho} = -\dfrac{im\widetilde{p}}{g}. \end{cases}$$

Arbitrarily choose $\widetilde{p}$ to be real. The *polarization relations* for internal gravity waves are

$$\begin{cases} p' = \mathrm{Re}\left\{\widetilde{p}e^{i(kx+mz-\omega t)}\right\} = \widetilde{p}\cos(kx + mz - \omega t), \\[2mm] v_1 = \mathrm{Re}\left\{\widetilde{v}_1 e^{i(kx+mz-\omega t)}\right\} = \dfrac{k\widetilde{p}}{\rho_0\omega}\cos(kx + mz - \omega t), \\[2mm] v_2 = \mathrm{Re}\left\{\widetilde{v}_2 e^{i(kx+mz-\omega t)}\right\} = \dfrac{l\widetilde{p}}{\rho_0\omega}\cos(kx + mz - \omega t), \\[2mm] v_3 = \mathrm{Re}\left\{\widetilde{v}_3 e^{i(kx+mz-\omega t)}\right\} = -\dfrac{(k^2 + l^2)\widetilde{p}}{m\rho_0\omega}\cos(kx + mz - \omega t), \\[2mm] \rho' = \mathrm{Re}\left\{\widetilde{\rho}e^{i(kx+mz-\omega t)}\right\} = \dfrac{m\widetilde{p}}{g}\sin(kx + mz - \omega t). \end{cases}$$

11.13.2   Inertia Gravity Waves

We concentrate on large-scale gravity waves. These waves are called *inertia gravity waves*. They have a horizontal scale of hundreds of kilometers and periods of several hours. Inertia gravity waves are the generalization of internal gravity waves to the case that the Coriolis parameter is not equal to zero, i.e., $f_0 \neq 0$. In this case, the linearized Boussinesq equation (11.36) on an f-plane becomes

$$
\begin{cases}
\dfrac{\partial v_1}{\partial t} - f_0 v_2 + \dfrac{1}{\rho_0}\dfrac{\partial p'}{\partial x} = 0, \\[2mm]
\dfrac{\partial v_2}{\partial t} + f_0 v_1 + \dfrac{1}{\rho_0}\dfrac{\partial p'}{\partial y} = 0, \\[2mm]
\dfrac{\partial v_1}{\partial x} + \dfrac{\partial v_2}{\partial y} + \dfrac{\partial v_3}{\partial z} = 0, \\[2mm]
\dfrac{g}{\rho_0}\dfrac{\partial \rho'}{\partial t} - N_{\mathrm{B}}^2 v_3 = 0, \\[2mm]
\dfrac{\partial p'}{\partial z} = -g\rho'.
\end{cases}
\tag{11.41}
$$

We look for linear plane-wave solutions, propagating in the xz-plane and independent of y, of the form

$$
\{v_1, v_2, v_3, p', \rho'\} = \mathrm{Re}\left\{(\tilde{v}_1, \tilde{v}_2, \tilde{v}_3, \tilde{p}, \tilde{\rho})\mathrm{e}^{\mathrm{i}(kx+mz-\omega t)}\right\}.
$$

Substituting this expression into (11.41), we obtain a system of algebraic equations:

$$
\begin{cases}
-\mathrm{i}\omega\tilde{v}_1 - f_0\tilde{v}_2 + \dfrac{\mathrm{i}k}{\rho_0}\tilde{p} = 0, \\[2mm]
-\mathrm{i}\omega\tilde{v}_2 + f_0\tilde{v}_1 = 0, \\[2mm]
\mathrm{i}k\tilde{v}_1 + \mathrm{i}m\tilde{v}_3 = 0, \\[2mm]
-\dfrac{\mathrm{i}\omega g}{\rho_0}\tilde{\rho} + N_{\mathrm{B}}^2\tilde{v}_3 = 0, \\[2mm]
\mathrm{i}m\tilde{p} + g\tilde{\rho} = 0.
\end{cases}
$$

By the second equation, $\tilde{v}_1 = \frac{\mathrm{i}\omega}{f_0}\tilde{v}_2$. Combining this with the first equation, we get

$$
\frac{\tilde{p}}{\tilde{v}_2} = \frac{\mathrm{i}\rho_0(\omega^2 - f_0^2)}{kf_0}.
$$

Therefore, this system of algebraic equations is equivalent to

$$\begin{cases} \dfrac{\widetilde{p}}{\widetilde{v}_2} = \dfrac{\mathrm{i}\rho_0(\omega^2 - f_0^2)}{kf_0}, \\[2mm] \dfrac{\widetilde{v}_2}{\widetilde{v}_1} = \dfrac{f_0}{\mathrm{i}\omega}, \\[2mm] \dfrac{\widetilde{v}_1}{\widetilde{v}_3} = -\dfrac{m}{k}, \\[2mm] \dfrac{\widetilde{v}_3}{\widetilde{\rho}} = \dfrac{\omega g}{\mathrm{i}N_B^2\rho_0}, \\[2mm] \dfrac{\widetilde{\rho}}{\widetilde{p}} = -\dfrac{\mathrm{i}m}{g}. \end{cases}$$

Multiplying these equations, we get

$$\frac{\widetilde{p}}{\widetilde{v}_2}\frac{\widetilde{v}_2}{\widetilde{v}_1}\frac{\widetilde{v}_1}{\widetilde{v}_3}\frac{\widetilde{v}_3}{\widetilde{\rho}}\frac{\widetilde{\rho}}{\widetilde{p}} = \left(\frac{\mathrm{i}\rho_0(\omega^2 - f_0^2)}{kf_0}\right)\left(\frac{f_0}{\mathrm{i}\omega}\right)\left(-\frac{m}{k}\right)\left(\frac{\omega g}{\mathrm{i}N_B^2\rho_0}\right)\left(-\frac{\mathrm{i}m}{g}\right),$$

i.e., $1 = \frac{m^2(\omega^2 - f_0^2)}{k^2 N_B^2}$. So

$$\omega^2 = f_0^2 + \frac{k^2 N_B^2}{m^2}.$$

This is the *dispersion relation* for inertia gravity waves.

On the other hand, solving the system of algebraic equations, we obtain a nontrivial solution $\widetilde{v}_1$, $\widetilde{v}_2$, $\widetilde{v}_3$, and $\widetilde{\rho}$ in favor $\widetilde{p}$:

$$\begin{cases} \widetilde{v}_1 = \dfrac{k\omega\widetilde{p}}{\rho_0(\omega^2 - f_0^2)}, \\[3mm] \widetilde{v}_2 = -\mathrm{i}\dfrac{kf_0\widetilde{p}}{\rho_0(\omega^2 - f_0^2)}, \\[3mm] \widetilde{v}_3 = -\dfrac{k^2\omega\widetilde{p}}{m\rho_0(\omega^2 - f_0^2)}, \\[3mm] \widetilde{\rho} = -\mathrm{i}\dfrac{m\widetilde{p}}{g}. \end{cases}$$

Arbitrarily choosing $\widetilde{p}$ to be real, we find the plane-wave solution is

$$\begin{cases} p' = \mathrm{Re}\left\{\widetilde{p}\mathrm{e}^{\mathrm{i}(kx+mz-\omega t)}\right\} = \widetilde{p}\cos(kx + mz - \omega t), \\[3mm] v_1 = \mathrm{Re}\left\{\widetilde{v}_1\mathrm{e}^{\mathrm{i}(kx+mz-\omega t)}\right\} = \dfrac{k\omega\widetilde{p}}{\rho_0(\omega^2 - f_0^2)}\cos(kx + mz - \omega t), \\[3mm] v_2 = \mathrm{Re}\left\{\widetilde{v}_2\mathrm{e}^{\mathrm{i}(kx+mz-\omega t)}\right\} = \dfrac{kf_0\widetilde{p}}{\rho_0(\omega^2 - f_0^2)}\sin(kx + mz - \omega t), \\[3mm] v_3 = \mathrm{Re}\left\{\widetilde{v}_3\mathrm{e}^{\mathrm{i}(kx+mz-\omega t)}\right\} = -\dfrac{k^2\omega\widetilde{p}}{m\rho_0(\omega^2 - f_0^2)}\cos(kx + mz - \omega t), \\[3mm] \rho' = \mathrm{Re}\left\{\widetilde{\rho}\mathrm{e}^{\mathrm{i}(kx+mz-\omega t)}\right\} = \dfrac{m\widetilde{p}}{g}\sin(kx + mz - \omega t). \end{cases}$$

These are the *polarization relations* for inertia gravity waves.

11.14 ROSSBY WAVES

Rossby waves are another class of atmospheric waves. These waves are also called *planetary waves*. They have horizontal scales of thousands of kilometers and periods of several days. Rossby waves are associated with many observed large-scale disturbances in the troposphere and the stratosphere, so they are very important for understanding many large-scale atmospheric phenomena. The quasi-geostrophic potential vorticity equation allows us to set up a simple model for the large-scale Rossby waves.

Consider a uniform zonal background flow $(U, 0, 0)$, where U is a constant. Denote by ψ^* its geostrophic stream function. The geostrophic stream function satisfies

$$U = -\frac{\partial \psi^*}{\partial y},$$

$$0 = \frac{\partial \psi^*}{\partial x}.$$

The second equation implies that ψ^* is independent of x. Integrating the first equation with respect to y, we obtain a geostrophic stream function given by

$$\psi^* = -Uy.$$

If we have a zonal background flow plus a small disturbance, the geostrophic stream function of the total flow is

$$\psi = \psi^* + \psi' = -Uy + \psi',$$

where ψ' is due to a small disturbance. From Section 11.12, the quasi-geostrophic potential vorticity equation of the total flow is

$$\frac{D_g \zeta}{Dt} = 0,$$

where

$$\frac{D_g}{Dt} = \frac{\partial}{\partial t} - \frac{\partial \psi}{\partial y}\frac{\partial}{\partial x} + \frac{\partial \psi}{\partial x}\frac{\partial}{\partial y},$$

and the quasi-geostrophic potential vorticity of the total flow is

$$\zeta = f_0 + \beta y + \frac{\partial^2 \psi}{\partial x^2} + \frac{\partial^2 \psi}{\partial y^2} + \frac{\partial}{\partial z}\left(\frac{f_0^2}{N_B^2}\frac{\partial \psi}{\partial z}\right).$$

It is clear that

$$\frac{\partial^2 \psi}{\partial x^2} + \frac{\partial^2 \psi}{\partial y^2} + \frac{\partial}{\partial z}\left(\frac{f_0^2}{N_B^2}\frac{\partial \psi}{\partial z}\right) = \frac{\partial^2(-Uy + \psi')}{\partial x^2} + \frac{\partial^2(-Uy + \psi')}{\partial y^2}$$

$$+ \frac{\partial}{\partial z}\left(\frac{f_0^2}{N_B^2}\frac{\partial(-Uy + \psi')}{\partial z}\right).$$

Notice that U is a constant and

$$\frac{\partial^2(-Uy + \psi')}{\partial x^2} = \frac{\partial^2 \psi'}{\partial x^2},$$

$$\frac{\partial^2(-Uy + \psi')}{\partial y^2} = \frac{\partial^2 \psi'}{\partial y^2},$$

$$\frac{\partial}{\partial z}\left(\frac{f_0^2}{N_B^2}\frac{\partial(-Uy + \psi')}{\partial z}\right) = \frac{\partial}{\partial z}\left(\frac{f_0^2}{N_B^2}\frac{\partial \psi'}{\partial z}\right).$$

Then

$$\frac{\partial^2 \psi}{\partial x^2} + \frac{\partial^2 \psi}{\partial y^2} + \frac{\partial}{\partial z}\left(\frac{f_0^2}{N_B^2}\frac{\partial \psi}{\partial z}\right) = \frac{\partial^2 \psi'}{\partial x^2} + \frac{\partial^2 \psi'}{\partial y^2} + \frac{\partial}{\partial z}\left(\frac{f_0^2}{N_B^2}\frac{\partial \psi'}{\partial z}\right).$$

So the quasi-geostrophic potential vorticity for the total flow becomes

$$\zeta = f_0 + \beta y + \mathcal{L}\psi',$$

where

$$\mathcal{L} = \frac{\partial^2}{\partial x^2} + \frac{\partial^2}{\partial y^2} + \frac{\partial}{\partial z}\left(\frac{f_0^2}{N_B^2}\frac{\partial}{\partial z}\right),$$

and the quasi-geostrophic potential vorticity equation becomes

$$\frac{D_g(f_0 + \beta y)}{Dt} + \frac{D_g(\mathcal{L}\psi')}{Dt} = 0, \tag{11.42}$$

where

$$\frac{D_g}{Dt} = \frac{\partial}{\partial t} - \frac{\partial \psi}{\partial y}\frac{\partial}{\partial x} + \frac{\partial \psi}{\partial x}\frac{\partial}{\partial y}.$$

Notice that $\frac{\partial \psi^*}{\partial x} = 0$ and $\frac{\partial \psi^*}{\partial y} = -U$. It is clear from $\psi = \psi^* + \psi'$ that

$$\frac{\partial \psi}{\partial x} = \frac{\partial \psi'}{\partial x},$$

$$\frac{\partial \psi}{\partial y} = -U + \frac{\partial \psi'}{\partial y},$$

and so the two terms on the left-hand side of (11.42) are, respectively,

$$\frac{D_g(f_0 + \beta y)}{Dt} = \left(\frac{\partial}{\partial t} - \frac{\partial \psi}{\partial y}\frac{\partial}{\partial x} + \frac{\partial \psi}{\partial x}\frac{\partial}{\partial y}\right)(f_0 + \beta y) = \beta\frac{\partial \psi}{\partial x} = \beta\frac{\partial \psi'}{\partial x},$$

$$\frac{D_g(\mathcal{L}\psi')}{Dt} = \mathcal{L}\left(\frac{\partial \psi'}{\partial t} - \frac{\partial \psi}{\partial y}\frac{\partial \psi'}{\partial x} + \frac{\partial \psi}{\partial x}\frac{\partial \psi'}{\partial y}\right)$$

$$= \mathcal{L}\left(\frac{\partial \psi'}{\partial t} - \left(-U + \frac{\partial \psi'}{\partial y}\right)\frac{\partial \psi'}{\partial x} + \frac{\partial \psi'}{\partial x}\frac{\partial \psi'}{\partial y}\right)$$

$$= \mathcal{L}\left(\frac{\partial \psi'}{\partial t} + U\frac{\partial \psi'}{\partial x}\right) = \left(\frac{\partial}{\partial t} + U\frac{\partial}{\partial x}\right)\mathcal{L}\psi'.$$

From this and (11.42), the quasi-geostrophic potential vorticity equation for the total flow is

$$\left(\frac{\partial}{\partial t} + U\frac{\partial}{\partial x}\right)\mathcal{L}\psi' + \beta\frac{\partial \psi'}{\partial x} = 0.$$

Now we look for a plane-wave solution of the form

$$\psi' = \text{Re}\left\{\widetilde{\psi}'e^{i(kx+ly+mz-\omega t)}\right\},$$

where ω is the angular frequency, and k, l, and m are the components of the wave vector $\mathbf{k} = (k, l, m)$.

Substituting this into the quasi-geostrophic potential vorticity equation, we obtain an algebraic equation:

$$\left(\frac{\partial}{\partial t} + U\frac{\partial}{\partial x}\right)\mathcal{L}\left(\text{Re}\left\{\widetilde{\psi}'e^{i(kx+ly+mz-\omega t)}\right\}\right)$$

$$+ \beta\frac{\partial\left(\text{Re}\left\{\widetilde{\psi}'e^{i(kx+ly+mz-\omega t)}\right\}\right)}{\partial x} = 0$$

or

$$\text{Re}\left(\left(\frac{\partial}{\partial t} + U\frac{\partial}{\partial x}\right)\mathcal{L}\left\{\widetilde{\psi}'e^{i(kx+ly+mz-\omega t)}\right\} + \beta\frac{\partial\left(\widetilde{\psi}'e^{i(kx+ly+mz-\omega t)}\right\}}{\partial x}\right) = 0.$$

This implies that

$$\left(\frac{\partial}{\partial t} + U\frac{\partial}{\partial x}\right)\mathcal{L}\left(\widetilde{\psi}'e^{i(kx+ly+mz-\omega t)}\right\} + \beta\frac{\partial\left(\widetilde{\psi}'e^{i(kx+ly+mz-\omega t)}\right\}}{\partial x} = 0. \quad (11.43)$$

Assume that N_B is a constant. Then

$$\mathcal{L} = \frac{\partial^2}{\partial x^2} + \frac{\partial^2}{\partial y^2} + \frac{\partial}{\partial z}\left(\frac{f_0^2}{N_B^2}\frac{\partial}{\partial z}\right) = \frac{\partial^2}{\partial x^2} + \frac{\partial^2}{\partial y^2} + \frac{f_0^2}{N_B^2}\frac{\partial^2}{\partial z^2},$$

and so

$$\mathcal{L}\left(\widetilde{\psi}'e^{i(kx+ly+mz-\omega t)}\right)$$

$$= \left(\frac{\partial^2}{\partial x^2} + \frac{\partial^2}{\partial y^2} + \frac{f_0^2}{N_B^2}\frac{\partial^2}{\partial z^2}\right)\left(\widetilde{\psi}'e^{i(kx+ly+mz-\omega t)}\right)$$

$$= -\left(k^2 + l^2 + m^2\frac{f_0^2}{N_B^2}\right)\widetilde{\psi}'e^{i(kx+ly+mz-\omega t)}.$$

From this, the first term of (11.43) is

$$\left(\frac{\partial}{\partial t} + U\frac{\partial}{\partial x}\right)\mathcal{L}\left(\widetilde{\psi}\mathrm{e}^{\mathrm{i}(kx+ly+mz-\omega t)}\right)$$

$$= -\left(k^2 + l^2 + m^2\frac{f_0^2}{N_B^2}\right)\left(\frac{\partial}{\partial t} + U\frac{\partial}{\partial x}\right)\widetilde{\psi}\mathrm{e}^{\mathrm{i}(kx+ly+mz-\omega t)}$$

$$= \mathrm{i}(\omega - Uk)\left(k^2 + l^2 + m^2\frac{f_0^2}{N_B^2}\right)\widetilde{\psi}\mathrm{e}^{\mathrm{i}(kx+ly+mz-\omega t)},$$

and the second term of (11.43) is

$$\beta\frac{\partial\left(\widetilde{\psi}\mathrm{e}^{\mathrm{i}(kx+ly+mz-\omega t)}\right)}{\partial x} = \mathrm{i}k\beta\widetilde{\psi}\mathrm{e}^{\mathrm{i}(kx+ly+mz-\omega t)}.$$

By (11.43), we get

$$\mathrm{i}(\omega - Uk)\left(k^2 + l^2 + m^2\frac{f_0^2}{N_B^2}\right)\widetilde{\psi}\mathrm{e}^{\mathrm{i}(kx+ly+mz-\omega t)}$$

$$+ \mathrm{i}k\beta\widetilde{\psi}\mathrm{e}^{\mathrm{i}(kx+ly+mz-\omega t)} = 0.$$

This is simplified as

$$(\omega - Uk)\left(k^2 + l^2 + m^2\frac{f_0^2}{N_B^2}\right) + k\beta = 0,$$

and so

$$\omega = Uk - \frac{k\beta}{k^2 + l^2 + m^2\frac{f_0^2}{N_B^2}}.$$

This is the *dispersion relation* for Rossby waves.

Let $\beta = 0$. Then $\omega = kU$. The waves are merely carried along with the background flow. Therefore, β is crucial to the existence of Rossby waves.

The zonal phase speed of the waves is

$$c = \frac{\omega}{k} = U - \frac{\beta}{k^2 + l^2 + m^2\frac{f_0^2}{N_B^2}}.$$

Assume that k, l, and m are real and that m is nonzero. Notice that

$$\beta = \frac{2\Omega\cos\phi_0}{a} > 0.$$

Then

$$0 < U - c = \frac{\beta}{k^2 + l^2 + m^2\frac{f_0^2}{N_B^2}} < \frac{\beta}{k^2 + l^2}.$$

Since the crests and troughs of stationary waves do not move with respect to the ground, $\omega = 0$, and so $c = 0$. This implies that there will be a vertical propagation and

$$0 < U < \frac{\beta}{k^2 + l^2}.$$

This inequality is called the *Charney-Drazin criterion*.

Let

$$U_c = \frac{\beta}{k^2 + l^2}.$$

Then U_c increases with increasing horizontal wavelength. In the Northern Hemisphere stratosphere, it has been observed that the winter background winds are eastward and stationary Rossby waves have large horizontal scales, while the summer background winds are westward and stationary Rossby waves do not exist.

Define a group velocity vector

$$\mathbf{c}_g = (c_g^{(x)}, c_g^{(y)}, c_g^{(z)}) = \left(\frac{\partial \omega}{\partial k}, \frac{\partial \omega}{\partial l}, \frac{\partial \omega}{\partial m} \right).$$

Its vertical component is

$$c_g^{(z)} = \frac{\partial \omega}{\partial m} = \frac{2 f_0^2 \beta k m}{N_B^2 \left(k^2 + l^2 + m^2 \frac{f_0^2}{N_B^2} \right)^2}.$$

By convention, $k > 0$. If $m > 0$, then the vertical component of the group velocity is positive and the waves propagate information upward. For upward-propagating waves, the phase surfaces

$$kx + ly + mz - \omega t = \text{constant}$$

slope westward with height.

We use moving fluid blobs A, B, C, etc., lying along a line of latitude to describe the behavior of Rossby waves. Owing to the Coriolis term in the quasi-geostrophic potential vorticity, according to conservation of potential vorticity, a northward-moving blob loses some disturbance vorticity, and then a southward-moving blob must gain some disturbance vorticity. When blob A moves southward, the increase of the disturbance vorticity associated with blob A causes blob B to move southward, and then the increase of the disturbance vorticity associated with blob B will make blob C move southward and make blob A move northward again. After a short time, it is observed that the sinusoidal pattern of the blobs has moved westward, although each individual blob oscillates only north-south.

11.15 ATMOSPHERIC BOUNDARY LAYER

Since the frictional force has only a small impact on gravity waves and Rossby waves in the atmosphere, friction effects are usually ignored. However, friction effects are sometimes very important in the region near Earth's surface, especially in the lowest several kilometers of the atmosphere. The region near Earth's surface is called the *atmospheric boundary layer*.

By (11.36), noticing that

$$\frac{\partial p'}{\partial x} = \frac{\partial p}{\partial x},$$

$$\frac{\partial p'}{\partial y} = \frac{\partial p}{\partial y},$$

the first two equations of the linearized Boussinesq equations on an f-plane reduce to

$$\frac{\partial v_1}{\partial t} = -\frac{1}{\rho_0}\frac{\partial p}{\partial x} + f_0 v_2 + F_x,$$

$$\frac{\partial v_2}{\partial t} = -\frac{1}{\rho_0}\frac{\partial p}{\partial y} - f_0 v_1 + F_y,$$

where v_1, v_2 are the horizontal components of the velocity, f_0 is the Coriolis parameter, and $\rho_0 = \overline{\rho}(0)$. Assume that frictional stress is a horizontal force $\tau = (\tau_x, \tau_y, 0)$. Replacing ρ by ρ_0 in (11.31), we get

$$\mathbf{F}_v = \frac{1}{\rho_0}\frac{\partial \tau_x}{\partial z}\mathbf{i} + \frac{1}{\rho_0}\frac{\partial \tau_y}{\partial z}\mathbf{j}.$$

So

$$F_x = \frac{1}{\rho_0}\frac{\partial \tau_x}{\partial z},$$

$$F_y = \frac{1}{\rho_0}\frac{\partial \tau_y}{\partial z},$$

and so

$$\frac{\partial v_1}{\partial t} = -\frac{1}{\rho_0}\frac{\partial p}{\partial x} + f_0 v_2 + \frac{1}{\rho_0}\frac{\partial \tau_x}{\partial z},$$

$$\frac{\partial v_2}{\partial t} = -\frac{1}{\rho_0}\frac{\partial p}{\partial y} - f_0 v_1 + \frac{1}{\rho_0}\frac{\partial \tau_y}{\partial z}.$$

These two equations are both linear.

In order to separate these two equations, the atmospheric flow is separated into a pressure-driven flow and a frictional-stress-driven flow. Corresponding to this separation, the horizontal velocity components for the atmospheric flow are separated into sums of those of the velocities for the pressure-driven flow and the frictional-stress-driven flow, i.e.,

$$v_1 = v_1^p + v_1^\tau,$$
$$v_2 = v_2^p + v_2^\tau,$$

where the subscripts p and τ denote the pressure-driven flow and the frictional stress-driven flow, respectively. Then the above two linear equations are separated into two systems of equations, i.e., the pressure-driven flow satisfies the system of equations

$$\frac{\partial v_1^p}{\partial t} - f_0 v_2^p = -\frac{1}{\rho_0} \frac{\partial p}{\partial x},$$
$$\frac{\partial v_2^p}{\partial t} + f_0 v_1^p = -\frac{1}{\rho_0} \frac{\partial p}{\partial y}, \tag{11.44}$$

and the frictional-stress-driven flow satisfies the system of equations

$$\frac{\partial v_1^\tau}{\partial t} - f_0 v_2^\tau = \frac{1}{\rho_0} \frac{\partial \tau_x}{\partial z},$$
$$\frac{\partial v_2^\tau}{\partial t} + f_0 v_1^\tau = \frac{1}{\rho_0} \frac{\partial \tau_y}{\partial z}. \tag{11.45}$$

If the frictional-stress-driven flow is steady, then

$$\frac{\partial v_1^\tau}{\partial t} = \frac{\partial v_2^\tau}{\partial t} = 0.$$

From this and (11.45), it follows that

$$v_2^\tau = -\frac{1}{f_0 \rho_0} \frac{\partial \tau_x}{\partial z},$$
$$v_1^\tau = \frac{1}{f_0 \rho_0} \frac{\partial \tau_y}{\partial z}.$$

Assume that the frictional stress exists significantly only in a boundary layer of depth d above the flat ground at $z = 0$. This layer is called a *frictional boundary layer*. Then τ_x and τ_y are nonzero for $0 \le z < d$ but vanish for $z \ge d$. Denote by τ_x^0 and τ_y^0 the surface stresses exerted by the ground on the lowest layer of the atmosphere. Integrating these equations through the depth of this boundary layer, we get

$$\int_0^d v_2^\tau \, dz = \frac{\tau_x^0}{f_0 \rho_0},$$
$$\int_0^d v_1^\tau \, dz = -\frac{\tau_y^0}{f_0 \rho_0}.$$

Denote

$$V_1^\tau = \int_0^d v_1^\tau \, dz,$$

$$V_2^\tau = \int_0^d v_2^\tau \, dz.$$

The quantities V_1^τ and V_2^τ are called the *Ekman volume transports*, and represent the horizontal fluxes of volume within this boundary layer. Then

$$V_1^\tau = -\frac{\tau_y^0}{f_0 \rho_0},$$

$$V_2^\tau = \frac{\tau_x^0}{f_0 \rho_0},$$

and so

$$(V_1^\tau, V_2^\tau, 0) = \frac{1}{f_0 \rho_0} (-\tau_y^0, \tau_x^0, 0).$$

Let the surface stress $\tau^0 = (\tau_x^0, \tau_y^0, 0)$. Then the scalar product of these two vectors

$$(V_1^\tau, V_2^\tau, 0) \cdot \tau^0 = -\tau_y^0 \tau_x^0 + \tau_x^0 \tau_y^0 = 0,$$

i.e., the Ekman volume transport in the frictional boundary layer is perpendicular to the surface stress.

If the pressure-driven flow is steady, then

$$\frac{\partial v_1^p}{\partial t} = \frac{\partial v_2^p}{\partial t} = 0.$$

From this and (11.44), it follows that

$$v_2^p = \frac{1}{f_0 \rho_0} \frac{\partial p}{\partial x},$$

$$v_1^p = -\frac{1}{f_0 \rho_0} \frac{\partial p}{\partial y},$$

and so

$$\frac{\partial v_1^p}{\partial x} + \frac{\partial v_2^p}{\partial y} = -\frac{1}{f_0 \rho_0} \frac{\partial^2 p}{\partial y \partial x} + \frac{1}{f_0 \rho_0} \frac{\partial^2 p}{\partial x \partial y} = 0.$$

Assume that the atmospheric flow satisfies div $\mathbf{v} = 0$, i.e.,

$$\frac{\partial v_1}{\partial x} + \frac{\partial v_2}{\partial y} + \frac{\partial v_3}{\partial z} = 0.$$

Notice that $v_1 = v_1^p + v_1^\tau$ and $v_2 = v_2^p + v_2^\tau$. Then

$$\frac{\partial v_3}{\partial z} = -\left(\frac{\partial v_1}{\partial x} + \frac{\partial v_2}{\partial y} \right) = -\left(\frac{\partial v_1^p}{\partial x} + \frac{\partial v_2^p}{\partial y} \right) - \left(\frac{\partial v_1^\tau}{\partial x} + \frac{\partial v_2^\tau}{\partial y} \right)$$

$$= -\left(\frac{\partial v_1^\tau}{\partial x} + \frac{\partial v_2^\tau}{\partial y} \right).$$

Integrating both sides through the depth of the boundary layer, we get

$$v_3|_0^d = -\left(\frac{\partial}{\partial x}\int_0^d v_1^\tau dz + \frac{\partial}{\partial y}\int_0^d v_2^\tau dz\right) = -\left(\frac{\partial V_1^\tau}{\partial x} + \frac{\partial V_2^\tau}{\partial y}\right),$$

where V_1^τ and V_2^τ are the Ekman volume transports. Since the ground is flat, $v_3 = 0$ at $z = 0$. Denote by v_3^d the value of the vertical velocity at $z = d$. Then

$$v_3^d = -\left(\frac{\partial V_1^\tau}{\partial x} + \frac{\partial V_2^\tau}{\partial y}\right).$$

In this equality, the left-hand side is a upward flow out of the top of the boundary layer and the right-hand side is the horizontal convergence of the Ekman volume transports, i.e., the horizontal convergence of the Ekman volume transports must be balanced by a upward flow out of the top of the boundary layer. Conversely, horizontal divergence of the Ekman volume transports is balanced by a downward flow into the top of the boundary layer. The velocity v_3^d is called the *Ekman pumping velocity*. It depends only on the frictional-stress-driven flow. From

$$(V_1^\tau, V_2^\tau, 0) = \frac{1}{f_0\rho_0}(-\tau_y^0, \tau_x^0, 0),$$

it follows that

$$\frac{\partial V_1^\tau}{\partial x} = -\frac{1}{f_0\rho_0}\frac{\partial \tau_y^0}{\partial x},$$

$$\frac{\partial V_2^\tau}{\partial y} = \frac{1}{f_0\rho_0}\frac{\partial \tau_x^0}{\partial x},$$

and so the Ekman pumping velocity is

$$v_3^d = -\left(\frac{\partial V_1^\tau}{\partial x} + \frac{\partial V_2^\tau}{\partial y}\right) = \frac{1}{f_0\rho_0}\left(\frac{\partial \tau_y^0}{\partial x} - \frac{\partial \tau_x^0}{\partial y}\right). \tag{11.46}$$

Notice that the surface stress $\tau^0 = (\tau_x^0, \tau_y^0, 0)$. By the definition of the curl, the curl of the surface stress is

$$\mathrm{curl}\,\tau^0 = -\frac{\partial \tau_y^0}{\partial z}\mathbf{i} + \frac{\partial \tau_x^0}{\partial z}\mathbf{j} + \left(\frac{\partial \tau_y^0}{\partial x} - \frac{\partial \tau_x^0}{\partial y}\right)\mathbf{k}.$$

Therefore, the vertical component of the curl of the surface stress is

$$\mathrm{curl}_z\,\tau^0 = \frac{\partial \tau_y^0}{\partial x} - \frac{\partial \tau_x^0}{\partial y},$$

and so

$$v_3^d = \frac{1}{f_0\rho_0}\mathrm{curl}_z\,\tau^0.$$

This shows that the Ekman pumping velocity is proportional to the vertical component of the curl of the surface stress.

Imitating the kinetic theory in fluid dynamics, the stress components and the vertical derivatives of the horizontal velocity are assumed to satisfy

$$\tau_x = \rho_0 \nu \frac{\partial v_1^\tau}{\partial z},$$

$$\tau_y = \rho_0 \nu \frac{\partial v_2^\tau}{\partial z}, \tag{11.47}$$

where the quantity ν is called the *kinematic eddy viscosity*. Assume further that the flow is steady, ν is a constant, and the pressure-driven flow (v_1^p, v_2^p) is independent of z within the boundary layer and equal to the large-scale purely zonal flow satisfying

$$v_1^p = V(y),$$

$$v_2^p = 0.$$

Since there can be no flow at the ground with friction, the boundary conditions on the total flow are

$$(v_1, v_2) \to (0, 0) \quad \text{as } z \to 0,$$

$$(v_1, v_2) \to (V(y), 0) \quad (z \gg d).$$

In terms of the stress-driven flow, by $v_1 = v_1^p + v_1^\tau$ and $v_2 = v_2^p + v_2^\tau$, these become

$$(v_1^\tau, v_2^\tau) \to (-V(y), 0) \quad \text{as } z \to 0,$$

$$(v_1^\tau, v_2^\tau) \to (0, 0) \quad (z \gg d).$$

By (11.45) and (11.47), the steady stress-driven flow satisfies

$$f_0 v_2^\tau = -\nu \frac{\partial^2 v_1^\tau}{\partial z^2},$$

$$f_0 v_1^\tau = \nu \frac{\partial^2 v_2^\tau}{\partial z^2}.$$

Let $\lambda_\tau = v_1^\tau + i v_2^\tau$. Then it follows from the coupled differential equations that

$$\frac{\partial^2 \lambda_\tau}{\partial z^2} = \frac{\partial^2 v_1^\tau}{\partial z^2} + i \frac{\partial^2 v_2^\tau}{\partial z^2} = -\frac{f_0 v_2^\tau}{\nu} + i \frac{f_0 v_1^\tau}{\nu} = i \frac{f_0}{\nu}(v_1^\tau + i v_2^\tau) = i \frac{f_0}{\nu} \lambda_\tau,$$

i.e., λ_τ satisfies the second-order equation

$$\frac{\partial^2 \lambda_\tau}{\partial z^2} = i \frac{f_0}{\nu} \lambda_\tau.$$

A direct check shows that the second-order equation has two solutions:

$$\lambda_\tau = A e^{\pm(1+i)\sqrt{\frac{f_0}{2\nu}} z},$$

where A is a constant. Since
$$(v_1^\tau, v_2^\tau) \to (0, 0) \quad (z \gg d),$$
the minus sign must be chosen and the plus sign is deleted. So the solution of the second-order equation is
$$\lambda_\tau = A e^{-(1+i)\sqrt{\frac{f_0}{2\nu}}z}.$$
Taking the real and imaginary parts and noticing that $\lambda_\tau = v_1^\tau + i v_2^\tau$, we get
$$v_1^\tau = A e^{-\sqrt{\frac{f_0}{2\nu}}z} \cos\left(\sqrt{\frac{f_0}{2\nu}}z\right),$$

$$v_2^\tau = -A e^{-\sqrt{\frac{f_0}{2\nu}}z} \sin\left(\sqrt{\frac{f_0}{2\nu}}z\right).$$

Applying the boundary condition $v_1^\tau \to -V(y)$ as $z \to 0$ gives
$$A = -V(y),$$
and so
$$v_1^\tau = -V(y) e^{-\sqrt{\frac{f_0}{2\nu}}z} \cos\left(\sqrt{\frac{f_0}{2\nu}}z\right),$$

$$v_2^\tau = V(y) e^{-\sqrt{\frac{f_0}{2\nu}}z} \sin\left(\sqrt{\frac{f_0}{2\nu}}z\right).$$

Since the pressure-driven flow is
$$v_1^p = V(y),$$
$$v_2^p = 0,$$
the full solution is
$$v_1 = v_1^p + v_1^\tau = V(y)\left(1 - e^{-\frac{z}{h}}\cos\frac{z}{h}\right),$$
$$v_2 = v_2^p + v_2^\tau = V(y) e^{-\frac{z}{h}}\sin\frac{z}{h},$$

where $h = \sqrt{\frac{2\nu}{f_0}}$. This solution is called *Ekman's solution*, and the corresponding boundary layer is called the *Ekman layer*. From Ekman's solution, we see that for any fixed y, the horizontal velocity vector (v_1, v_2), as a function of $\frac{z}{h}$, represents a spiral which is called *Ekman's spiral*. Ekman's spiral shows that the deflection of the wind in the boundary layer is mostly to the low-pressure side of the geostrophic, large-scale flow.

By (11.46) and (11.47),
$$v_3^d = \frac{\nu}{f_0}\left(\frac{\partial^2 v_2}{\partial z \partial x} - \frac{\partial^2 v_1}{\partial z \partial y}\right)\bigg|_{z=0}.$$

From the full solution, it follows that

$$\frac{\partial v_1}{\partial y} = \left(1 - e^{-\frac{z}{h}} \cos \frac{z}{h}\right) \frac{dV}{dy},$$

$$\frac{\partial v_2}{\partial x} = 0,$$

and so

$$\left.\frac{\partial^2 v_1}{\partial z \partial y}\right|_{z=0} = \frac{1}{h} e^{-\frac{z}{h}} \left(\cos \frac{z}{h} + \sin \frac{z}{h}\right)\bigg|_{z=0} \frac{dV}{dy} = \frac{1}{h}\frac{dV}{dy},$$

$$\left.\frac{\partial^2 v_2}{\partial z \partial x}\right|_{z=0} = 0.$$

Notice that $h = \sqrt{\frac{2\nu}{f_0}}$. Then

$$v_3^d = -\frac{\nu}{f_0 h}\frac{dV}{dy} = \frac{1}{2}h\xi,$$

where $\xi = -\frac{dV}{dy}$ is the relative vorticity of the free-atmosphere flow. If the free-atmosphere flow is not a purely zonal flow, this relationship also holds. This relationship shows that the Ekman pumping velocity is upward under a cyclone in the free atmosphere for $\xi > 0$ and downward under an anticyclone for $\xi < 0$.

PROBLEMS

11.1 Show that

$$\int_0^\infty \frac{X^3}{e^X - 1}\, dX = \frac{\pi^4}{15}$$

which is used to derive the Stefan-Boltzmann law for the black-body irradiance.

11.2 Show that if the solar constant $F_s = 1370\,\text{W/m}^2$ and the planetary albedo $\alpha = 0.3$, then Earth's effective emitting temperature $T \approx 255\,\text{K}$.

11.3 Show that if the solar luminosity $F_s = 2619\,\text{W/m}^2$ and the planetary albedo $\alpha = 0.7$, then the effective temperature $T \approx 242\,\text{K}$. These are values appropriate to the planet Venus.

11.4 Let $\mathbf{v}$ be the atmospheric fluid velocity. Show that the material acceleration may be expressed by

$$\frac{D\mathbf{v}}{Dt} = \frac{\partial \mathbf{v}}{\partial t} + \text{grad}\left(\frac{|\mathbf{v}|^2}{2}\right) + (\text{curl } \mathbf{v}) \times \mathbf{v}.$$

11.5 Consider an air parcel with local speed $\mathbf{v}$ satisfying $\mathbf{v} = |\mathbf{v}|\mathbf{s}$, where $\mathbf{s}$ is a unit vector. Show that its material acceleration can be described by

$$\frac{D\mathbf{v}}{Dt} = \frac{D|\mathbf{v}|}{Dt}\mathbf{s} + |\mathbf{v}|\frac{D\mathbf{s}}{Dt}.$$

11.6 Let ρ_{sv} be the density of water vapor at saturation, L be the latent heat of vaporization per unit mass, and R_v be the specific gas constant for the vapor. Show that the density of water vapor at saturation is a function of temperature given by

$$\rho_{sv} = \frac{c}{T}e^{-\frac{d}{T}},$$

where c is a constant and $d = \frac{L}{R_v}$.

11.7 Suppose that the Rossby waves are independent of height and suppose also that there is no background flow. For these Rossby waves, show their quasi-geostrophic potential vorticity and their dispersion relation between the angular frequency and the wave-vector components.

BIBLIOGRAPHY

Boas, M.L., 1983. Mathematical Methods in the Physical Sciences. Wiley, New York.

Bolton, D., 1980. The computation of equivalent potential temperature. Mon. Weather Rev. 108, 1046-1053.

Durran, D.R., 1993. Is the Coriolis force really responsible for the inertial oscillation? Bull. Am. Meteorol. Soc. 74, 2179-2184.

Gill, A.E., 1982. Atmosphere-Ocean Dynamics. Academic Press, London.

Holton, J.R., 2004. An Introduction to Dynamic Meteorology, fourth ed. Academic Press, Burlington, MA.

Landau, L.D., Lifshitz, E.M., 1975. Statistical Physics, Part I, third ed. Pergamon, Oxford.

Lorenz, E.N., 1955. Available potential energy and the maintenance of the general circulation. Tellus 7, 157-167.

Marshall, J., Plumb, R., 2008. Atmosphere, Ocean and Climate Dynamics, An Introductory Text. Academic Press, New York.

Pedlosky, J., 1987. Geophysical Fluid Dynamics, second ed. Springer-Verlag, New York.

Thuburn, J., Craig, G.C., 2000. Stratospheric influence on tropopause height: the radiative constraint. J. Atmos. Sci. 57, 17-28.

Chapter 12

Oceanic Dynamics

The oceans are an important component of the climate system. They have a profound influence on global climate and ecosystems. Therefore, an understanding of oceanic dynamics is a prerequisite for understanding the present climate, including both the mean climate state and the superimposed natural climate variability. In this chapter, we will cover various aspects of oceanic dynamics in order to understand the circulation and dynamics of the ocean on small, regional, and global scales.

By international agreement there are four oceans on Earth's surface: the Atlantic Ocean, the Pacific Ocean, the Indian Ocean, and the Arctic Ocean. The oceans and adjacent seas cover 70.8% of the surface of Earth, which amounts to $361,254,000 \, \text{km}^2$.

12.1 SALINITY AND MASS

Conservation of mass and salt can be used to give very useful information about flows in the ocean.

Suppose that S_i is the salinity of the flow into a basin with volume V_i and density ρ_i of water flowing in. Suppose that S_o is the salinity of the flow out of the basin with volume V_o and density ρ_o of water flowing out. Then *conservation of mass* says that the mass flowing in equals the mass flowing out:

$$\rho_i V_i = \rho_o V_o.$$

Because salt is not deposited or removed from the sea, *conservation of salt* says that the salt flowing in equals the salt flowing out:

$$\rho_i V_i S_i = \rho_o V_o S_o.$$

We usually assume $\rho_i = \rho_o$ with little error. Then the conservation of salt becomes

$$v_i S_i = V_o S_o.$$

Using the estimated value of V_i and the measured salinity S_i, S_o, we can find the volume of water flowing out of the basin from conservation of salt:

$$V_o = \frac{V_i S_i}{S_o}.$$

Mathematical and Physical Fundamentals of Climate Change

If there is precipitation P and evaporation E at the surface of the basin and river inflow R, then the conservation of mass becomes

$$V_i + R + P = V_o + E.$$

If V_i and V_o are known, then

$$P + R - E = V_o - V_i.$$

This states that water flowing into the basin must balance precipitation plus river inflow minus evaporation.

12.2 INERTIAL MOTION

The Cartesian coordinate system is the coordinate system used most commonly in studies of oceanic dynamics. The standard convention is that x is to the east, y is to the north, and z is up. The f-plane is a Cartesian coordinate system in which the Coriolis parameter is assumed to be constant. The β-plane is also a Cartesian coordinate systems in which the Coriolis parameter is assumed to vary linearly with latitude.

If the water on the sea surface moves only under the influence of Coriolis force, no other force acts on the water, then such a motion is called to an *inertial motion*.

By (11.34),

$$
\begin{cases}
\dfrac{\partial v_1}{\partial t} + v_1 \dfrac{\partial v_1}{\partial x} + v_2 \dfrac{\partial v_1}{\partial y} + v_3 \dfrac{\partial v_1}{\partial z} = -\dfrac{1}{\rho}\dfrac{\partial p}{\partial x} + f v_2 + F_x, \\[2mm]
\dfrac{\partial v_2}{\partial t} + v_1 \dfrac{\partial v_2}{\partial x} + v_2 \dfrac{\partial v_2}{\partial y} + v_3 \dfrac{\partial v_2}{\partial z} = -\dfrac{1}{\rho}\dfrac{\partial p}{\partial y} - f v_1 + F_y, \\[2mm]
\dfrac{\partial v_3}{\partial t} + v_1 \dfrac{\partial v_3}{\partial x} + v_2 \dfrac{\partial v_3}{\partial y} + v_3 \dfrac{\partial v_3}{\partial z} = -\dfrac{1}{\rho}\dfrac{\partial p}{\partial z} - g + F_z,
\end{cases}
$$

where the velocity $\mathbf{v} = (v_1, v_2, v_3)$, the friction $\mathbf{F} = (F_x, F_y, F_y)$, and the Coriolis parameter $f = 2\Omega \sin\phi$. The first two equations are the horizontal momentum equations and the third equation is the vertical momentum equation.

Dropping the quadratic terms in the material derivatives, we obtain the horizontal momentum equations as

$$
\begin{cases}
\dfrac{\partial v_1}{\partial t} = -\dfrac{1}{\rho}\dfrac{\partial p}{\partial x} + f v_2 + F_x, \\[2mm]
\dfrac{\partial v_2}{\partial t} = -\dfrac{1}{\rho}\dfrac{\partial p}{\partial y} - f v_1 + F_y.
\end{cases}
$$

If only Coriolis force acts on the water, there must be no horizontal pressure gradient,

$$\frac{\partial p}{\partial x} = \frac{\partial p}{\partial y} = 0,$$

and so

$$
\begin{cases}
\dfrac{\partial v_1}{\partial t} = f v_2 + F_x, \\[2mm]
\dfrac{\partial v_2}{\partial t} = -f v_1 + F_y.
\end{cases}
\tag{12.1}
$$

For a frictionless ocean, $F_x = F_y = 0$. So (12.1) reduces to the two coupled, first-order, linear, differential equations

$$
\begin{cases}
\dfrac{\partial v_1}{\partial t} = f v_2, \\[2mm]
\dfrac{\partial v_2}{\partial t} = -f v_1.
\end{cases}
$$

This system of equations can be solved with standard techniques as follows.
Solving the second equation for v_1 gives

$$
v_1 = -\frac{1}{f} \frac{\partial v_2}{\partial t}.
$$

Inserting it into the first equation gives

$$
-\frac{1}{f} \frac{\partial^2 v_2}{\partial t^2} = f v_2.
$$

Therefore, the inertial motion satisfies the system of equations

$$
\begin{cases}
v_1 + \dfrac{1}{f} \dfrac{\partial v_2}{\partial t} = 0, \\[2mm]
\dfrac{\partial^2 v_2}{\partial t^2} + f^2 v_2 = 0.
\end{cases}
$$

This system of equations has the solution

$$
\begin{cases}
v_1 = V \sin ft, \\
v_2 = V \cos ft, \\
V^2 = v_1^2 + v_2^2.
\end{cases}
$$

The solution is a parameter equation for a circle with diameter $D_i = 2V$ and period $T_i = \frac{2\pi}{f} = \frac{T_{sd}}{2\sin\phi}$, where $T_{sd} = \frac{2\pi}{\Omega}$ is a sidereal day. The period T_i is called the *inertial period*. The current described by it is called an *inertial current* or an *inertial oscillation*.

Inertial currents are the commonest currents in the ocean. They have been observed at all depths in the ocean and at all latitudes. The motions are transient and decay in a few days. Oscillations at different depths or at different nearby sites are usually incoherent.

12.3 OCEANIC EKMAN LAYER

Steady winds blowing on the sea surface produce a horizontal boundary layer which is at most a few hundred meters thick at the top of the ocean; this layer is called the *Ekman layer*. A similar boundary layer existing at the bottom of the ocean is called the *bottom Ekman layer*.

12.3.1 Ekman Currents

Fridtjof Nansen asked Vilhelm Bjerknes to let one of Bjerknes's students perform a theoretical study of the influence of Earth's rotation on wind-driven currents. Walfrid Ekman was chosen. He presented the results in his thesis at Uppsala. Ekman later expanded the study to include the influence of continents and differences in density of water (Ekman, 1905).

Ekman assumed a steady, homogeneous, horizontal flow with friction on a rotating Earth. The steady flow means that

$$\frac{\partial v_1}{\partial t} = \frac{\partial v_2}{\partial t} = 0.$$

So (12.1) become

$$\begin{cases} fv_2 + F_x = 0, \\ -fv_1 + F_y = 0. \end{cases} \tag{12.2}$$

Ekman further assumed a constant vertical eddy viscosity. So the components of friction are, respectively,

$$F_x = A_z \frac{\partial^2 v_1}{\partial z^2},$$

$$F_y = A_z \frac{\partial^2 v_2}{\partial z^2},$$

where A_z is a constant *eddy viscosity*. So the system (12.2) of differential equations is now

$$\begin{cases} fv_2 + A_z \dfrac{\partial^2 v_1}{\partial z^2} = 0, \\ -fv_1 + A_z \dfrac{\partial^2 v_2}{\partial z^2} = 0, \end{cases} \tag{12.3}$$

where the Coriolis parameter $f = 2\Omega \sin \phi$. It is easy to verify that the system (12.3) of differential equations has the solution

$$\begin{cases} v_1(z) = V_0 e^{az} \sin \left(\dfrac{\pi}{4} - az \right), \\ v_2(z) = V_0 e^{az} \cos \left(\dfrac{\pi}{4} - az \right), \end{cases}$$

where V_0 is the velocity of the current at the sea surface $z = 0$. This current is called the *Ekman current*.

At the sea surface $z = 0$, and since $e^0 = 1$, the Ekman current is

$$\begin{cases} v_1(0) = V_0 \sin \dfrac{\pi}{4}, \\[2mm] v_2(0) = V_0 \cos \dfrac{\pi}{4}. \end{cases}$$

The Ekman current has a speed of V_0 to the northeast. In general, the surface current is 45° to the right of the wind in the Northern Hemisphere; the surface current is 45° to the left of the wind in the Southern Hemisphere. Below the sea surface,

$$\sqrt{v_1^2(z) + v_2^2(z)} = V_0 e^{az}.$$

So the velocity of the Ekman current decays exponentially with depth.

Now we find the constants a and V_0.

The second-order derivative of the solution $v_1 = V_0 e^{az} \sin\left(\frac{\pi}{4} - az\right)$ is evaluated:

$$\frac{\partial^2 v_1}{\partial z^2} = -2a^2 V_0 e^{az} \cos\left(\frac{\pi}{4} - az\right).$$

Substituting this derivative and the solution $v_2 = V_0 e^{az} \cos\left(\frac{\pi}{4} - az\right)$ into the first equation in (12.2), we get

$$f V_0 e^{az} \cos\left(\frac{\pi}{4} - az\right) - 2a^2 V_0 A_z e^{az} \cos\left(\frac{\pi}{4} - az\right) = 0$$

or

$$f = 2a^2 A_z.$$

Solve this equation to give the constant $a = \sqrt{\frac{f}{2A_z}}$.

The derivative of the solution $v_2 = V_0 e^{az} \cos\left(\frac{\pi}{4} - az\right)$ is evaluated:

$$\frac{\partial v_2}{\partial z} = V_0 a e^{az} \cos\left(\frac{\pi}{4} - az\right) + V_0 a e^{az} \sin\left(\frac{\pi}{4} - az\right) = V_0 \sqrt{2} a e^{az} \cos(az).$$

So

$$\left. \frac{\partial v_2}{\partial z} \right|_{z=0} = \sqrt{2} a V_0.$$

When the wind is blowing to the north, the wind stress

$$T = T_{yz} = A_z \frac{\partial v_2}{\partial z}.$$

Therefore, the wind stress at the sea surface $z = 0$ is

$$T = A_z \sqrt{2} a V_0.$$

This implies that

$$V_0 = \frac{T}{\sqrt{2} a A_z}.$$

Notice that $a = \sqrt{\frac{f}{2A_z}}$. Then the velocity of the current at the sea surface

$$V_0 = \frac{T}{\sqrt{fA_z}}.$$

The thickness of the Ekman layer is arbitrary because the Ekman currents decrease exponentially with depth. Ekman proposed that when the thickness is the depth $D_E = \frac{\pi}{a}$, the current velocity is opposite the velocity at the surface. Notice that $a = \sqrt{\frac{f}{2A_z}}$. The Ekman layer depth

$$D_E = \pi \sqrt{\frac{2A_z}{f}}.$$

12.3.2 Ekman Mass Transport

Flow in the ocean carries mass. The *mass transport* is a vector, denoted by **M**. Its horizontal components, denoted by M_x and M_y, are defined by integrals of the density of seawater times the fluid horizontal velocities v_1 and v_2 from the sea surface to a depth h:

$$M_x = \int_{-h}^{0} \rho_w v_1 \, dz,$$

$$M_y = \int_{-h}^{0} \rho_w v_2 \, dz,$$

where ρ_w is the density of seawater. Flow in the Ekman layer at the top of the ocean carries mass. The *Ekman mass transport* is denoted by $\mathbf{M_E}$. The two horizontal components of Ekman mass transport $\mathbf{M_E}$, denoted by M_{Ex} and M_{Ey}, are given by integrals of the product of the density of seawater and the Ekman velocities v_1^E, v_2^E from the sea surface to a depth h below the Ekman layer:

$$M_{Ex} = \int_{-h}^{0} \rho_w v_1^E \, dz,$$

$$M_{Ey} = \int_{-h}^{0} \rho_w v_2^E \, dz.$$

By (12.2),

$$f v_2^E = -F_x,$$

$$f v_1^E = F_y,$$

and so

$$f M_{Ex} = \int_{-h}^{0} f \rho_w v_1^E \, dz = \int_{-h}^{0} \rho_w F_y \, dz,$$

$$fM_{Ey} = \int_{-h}^{0} f\rho_w v_2^E \, dz = - \int_{-h}^{0} \rho_w F_x \, dz.$$

However, the horizontal components of friction from Ekman's assumption are, respectively,

$$F_x = \frac{1}{\rho_w} \frac{\partial}{\partial z} \left(\rho_w A_z \frac{\partial v_1^E}{\partial z} \right),$$

$$F_y = \frac{1}{\rho_w} \frac{\partial}{\partial z} \left(\rho_w A_z \frac{\partial v_2^E}{\partial z} \right),$$

where A_z is a constant eddy viscosity. Therefore,

$$fM_{Ex} = \int_{-h}^{0} \frac{\partial}{\partial z} \left(\rho_w A_z \frac{\partial v_2^E}{\partial z} \right) \, dz,$$

$$fM_{Ey} = - \int_{-h}^{0} \frac{\partial}{\partial z} \left(\rho_w A_z \frac{\partial v_1^E}{\partial z} \right) \, dz.$$

Notice that two components of the wind stress in the x- and y-directions are, respectively,

$$T_{xz} = \rho_w A_z \frac{\partial v_1^E}{\partial z},$$

$$T_{yz} = \rho_w A_z \frac{\partial v_2^E}{\partial z},$$

and so

$$fM_{Ex} = \int_{-h}^{0} \frac{\partial T_{yz}}{\partial z} \, dz = T_{yz}(0) - T_{yz}(-h).$$

$$fM_{Ey} = - \int_{-h}^{0} \frac{\partial T_{xz}}{\partial z} \, dz = -T_{xz}(0) + T_{xz}(-h).$$

Since the Ekman velocities approach zero at a few hundred meters below the sea surface,

$$T_{yz}(-h) = 0,$$
$$T_{xz}(-h) = 0.$$

Notice that the Coriolis parameter $f = 2\Omega \sin \varphi$. Thus, the two components of Ekman mass transport are, respectively,

$$M_{Ex} = \frac{T_{yz}(0)}{f} = \frac{T_{yz}(0)}{2\Omega \sin \phi},$$

$$M_{Ey} = -\frac{T_{xz}(0)}{f} = -\frac{T_{xz}(0)}{2\Omega \sin \phi},$$

where $T_{xz}(0)$ and $T_{yz}(0)$ are the two components of the wind stress at the sea surface.

12.3.3 Ekman Pumping

According to conservation of mass, the spatial variability of Ekman mass transport must lead to vertical velocities at the top of the Ekman layer. In order to calculate these velocities, consider an integral from the sea surface $z = 0$ to a depth h below the Ekman layer:

$$\int_{-h}^{0} \left(\frac{\partial(\rho_w v_1^E)}{\partial x} + \frac{\partial(\rho_w v_2^E)}{\partial y} + \frac{\partial(\rho_w v_3^E)}{\partial z} \right) dz,$$

where ρ_w is the density of seawater and the Ekman velocity $\mathbf{v}^E = (v_1^E, v_2^E, v_3^E)$. Since

$$\frac{\partial(\rho_w v_1^E)}{\partial x} + \frac{\partial(\rho_w v_2^E)}{\partial y} + \frac{\partial(\rho_w v_3^E)}{\partial z} = \text{div}\,(\rho_w \mathbf{v}^E),$$

the integral becomes

$$\int_{-h}^{0} \left(\frac{\partial(\rho_w v_1^E)}{\partial x} + \frac{\partial(\rho_w v_2^E)}{\partial y} + \frac{\partial(\rho_w v_3^E)}{\partial z} \right) dz = \int_{-h}^{0} \text{div}\,(\rho_w \mathbf{v}^E)\, dz.$$

By the continuity equation,

$$\text{div}\,(\rho_w \mathbf{v}^E) = -\frac{\partial \rho_w}{\partial t} = 0,$$

and so

$$\int_{-h}^{0} \left(\frac{\partial(\rho_w v_1^E)}{\partial x} + \frac{\partial(\rho_w v_2^E)}{\partial y} + \frac{\partial(\rho_w v_3^E)}{\partial z} \right) dz = 0$$

which is equivalent to

$$\frac{\partial}{\partial x} \int_{-h}^{0} \rho_w v_1^E \, dz + \frac{\partial}{\partial y} \int_{-h}^{0} \rho_w v_2^E \, dz = - \int_{-h}^{0} \frac{\partial(\rho_w v_3^E)}{\partial z} \, dz. \tag{12.4}$$

Notice that

$$\int_{-h}^{0} \rho_w v_1^E \, dz = M_{Ex},$$

$$\int_{-h}^{0} \rho_w v_2^E \, dz = M_{Ey},$$

$$\int_{-h}^{0} \frac{\partial(\rho_w v_3^E)}{\partial z} \, dz = \rho_w(0) v_3^E(0) - \rho_w(-h) v_3^E(-h).$$

Equality (12.4) becomes

$$\frac{\partial M_{E_x}}{\partial x} + \frac{\partial M_{E_y}}{\partial y} = -\rho_w(0)v_3^E(0) + \rho_w(-h)v_3^E(-h).$$

Since the vertical velocity at the base of the Ekman layer must be zero, $v_3(-h) = 0$, and so

$$\frac{\partial M_{E_x}}{\partial x} + \frac{\partial M_{E_y}}{\partial y} = -\rho_w(0)v_3^E(0).$$

This equation states that the horizontal divergence of the Ekman mass transport M_E leads to a vertical velocity in the upper boundary layer of the ocean. Such a process is called *Ekman pumping*. Combining this with

$$M_{Ex} = \frac{T_{yz}(0)}{2\Omega \sin\phi},$$

$$M_{Ey} = -\frac{T_{xz}(0)}{2\Omega \sin\phi},$$

we find the Ekman vertical velocity at the sea surface related to the wind stress is given by

$$v_3^E(0) = -\frac{1}{\rho_w(0)}\left(\frac{\partial}{\partial x}\left(\frac{T_{yz}(0)}{2\Omega \sin\phi}\right) - \frac{\partial}{\partial y}\left(\frac{T_{xz}(0)}{2\Omega \sin\phi}\right)\right),$$

where $T_{xz}(0)$ and $T_{yz}(0)$ are two components of the wind stress at the sea surface. Because the vertical velocity at the sea surface must be zero, the Ekman vertical velocity $v_3^E(0)$ must be balanced by the vertical velocity $v_3^G(0)$ of an ocean's interior geostrophic flow,

$$v_3^E(0) = -v_3^G(0).$$

Ekman pumping $v_3^E(0)$ drives a vertical geostrophic current $-v_3^G(0)$ in the ocean's interior.

12.4   GEOSTROPHIC CURRENTS

The horizontal pressure gradients in the ocean are balanced by the Coriolis force resulting from horizontal currents; this balance is called *geostrophic balance*. The vertical pressure gradient in the ocean is balanced by the weight of seawater; this balance is called *hydrostatic balance*.

12.4.1   Surface Geostrophic Currents

By (11.34), the momentum equations are

$$\begin{cases} \dfrac{Dv_1}{Dt} = -\dfrac{1}{\rho}\dfrac{\partial p}{\partial x} + fv_2 + F_x, \\[2mm] \dfrac{Dv_2}{Dt} = -\dfrac{1}{\rho}\dfrac{\partial p}{\partial y} - fv_1 + F_y, \\[2mm] \dfrac{Dv_3}{Dt} = -\dfrac{1}{\rho}\dfrac{\partial p}{\partial z} - g + F_z, \end{cases}$$

where the velocity $\mathbf{v} = (v_1, v_2, v_3)$, the friction $\mathbf{F} = (F_x, F_y, F_y)$, the Coriolis parameter $f = 2\Omega \sin\phi$, and

$$\frac{D}{Dt} = \frac{\partial}{\partial t} + v_1\frac{\partial}{\partial X} + v_2\frac{\partial}{\partial y} + v_3\frac{\partial}{\partial z}.$$

The first two equations are the horizontal momentum equations and the third equation is the vertical momentum equation.

Assume that horizontal velocities of the flow are much larger than the vertical velocity, $v_1, v_2 \gg v_3$ and the only external force is gravity. Assume further that the flow has no acceleration and friction can be disregarded, i.e., the acceleration $\mathbf{a} = \frac{D\mathbf{v}}{Dt} = 0$ (by (11.24)) and the friction $\mathbf{F} = \mathbf{0}$. So $F_x = F_y = F_z = 0$ and

$$\frac{Dv_1}{Dt} = \frac{Dv_2}{Dt} = \frac{Dv_3}{Dt} = 0.$$

With these assumptions, the momentum equations become

$$\begin{cases} \dfrac{\partial p}{\partial x} = \rho f v_2, \\[2mm] \dfrac{\partial p}{\partial y} = -\rho f v_1, \\[2mm] \dfrac{\partial p}{\partial z} = -\rho g, \end{cases} \tag{12.5}$$

where gravity g is a function of latitude ϕ and height z and the density ρ is a function of height z, i.e.,

$$g = g(\phi, z),$$

$$\rho = \rho(z).$$

The first two equations in (12.5) are called the *geostrophic balance equations* and the third equation in (12.5) is called the *hydrostatic balance equation*.

Assume that the sea surface is above or below the surface $z = 0$. Denote by ζ the height of the sea surface relative to the surface $z = 0$. Then atmospheric pressure at the sea surface is the same as that at the surface $z = 0$. Consider a level surface at $z = -h$ below the sea surface. The level surface is a constant gravitational potential surface and no work is required to move along a frictionless level surface. To obtain the pressure at the depth h, we integrate the third equation in (12.5) with respect to z from $-h$ to ζ:

$$p = p_0 + \int_{-h}^{\zeta} \rho(z)g(\phi, z)\, dz,$$

where p_0 is atmospheric pressure at $z = 0$.

If the ocean is homogeneous and density and gravity are constant, and the level surface considered is slightly below the sea surface, then the pressure on the level surface $z = -h$ is

$$p = p_0 + \rho g \int_{-h}^{\zeta} dz = p_0 + \rho g(\zeta + h),$$

where p_0 is atmospheric pressure at $z = 0$. Inserting this into the first two equations in (12.5), we find that the horizontal velocities of the surface geostrophic current, denoted by v_1^s and v_2^s, are

$$v_1^s = -\frac{1}{\rho f}\frac{\partial p}{\partial y} = -\frac{1}{\rho f}\frac{\partial}{\partial y}(p_0 + \rho g(\zeta + h)) = -\frac{g}{f}\frac{\partial \zeta}{\partial y},$$

$$v_2^s = \frac{1}{\rho f}\frac{\partial p}{\partial x} = \frac{1}{\rho f}\frac{\partial}{\partial x}(p_0 + \rho g(\zeta + h)) = \frac{g}{f}\frac{\partial \zeta}{\partial x}.$$

This shows that geostrophic currents concentrated in the upper ocean are independent of depth. Therefore, computation of horizontal velocities of geostrophic currents requires only slopes at the sea surface. Let ψ^s be the stream function on the level surface $z = -h$. It is clear that

$$\psi^s = -\frac{g}{f}\zeta.$$

Therefore, the sea surface is a stream function on the level surface scaled by the factor $\frac{g}{f}$.

If the ocean is stratified, then, for the depth h, the pressure

$$p = p_0 + \int_{-h}^{\zeta} \rho(z)g(\phi, z)\, dz = p_0 + \int_{-h}^{0} \rho(z)g(\phi, z)\, dz + \int_{0}^{\zeta} \rho(z)g(\phi, z)\, dz,$$

where p_0 is atmospheric pressure at $z = 0$. Inserting this into the first two equations in (12.5), we get

$$v_1^s = -\frac{1}{\rho f}\frac{\partial p}{\partial y} = -\frac{1}{\rho f}\frac{\partial}{\partial y}\int_{-h}^{0}\rho(z)g(\phi, z)\, dz - \frac{1}{\rho f}\frac{\partial}{\partial y}\int_{0}^{\zeta}\rho(z)g(\phi, z)\, dz,$$

$$v_2^s = \frac{1}{\rho f}\frac{\partial p}{\partial x} = \frac{1}{\rho f}\frac{\partial}{\partial x}\int_{-h}^{0}\rho(z)g(\phi, z)\, dz + \frac{1}{\rho f}\frac{\partial}{\partial x}\int_{0}^{\zeta}\rho(z)g(\phi, z)\, dz.$$

However, the second integrals on the right-hand side are

$$\frac{\partial}{\partial y}\int_{0}^{\zeta}\rho(z)g(\phi, z)\, dz = \rho g\frac{\partial \zeta}{\partial y},$$

$$\frac{\partial}{\partial x}\int_{0}^{\zeta}\rho(z)g(\phi, z)\, dz = \rho g\frac{\partial \zeta}{\partial x},$$

Therefore, the horizontal velocities of the surface geostrophic current are

$$
v_1^s = -\frac{1}{\rho f}\frac{\partial}{\partial y}\int_{-h}^{0}\rho(z)g(\phi,z)\,\mathrm{d}z - \frac{g}{f}\frac{\partial\zeta}{\partial y},
$$
$$
v_2^s = \frac{1}{\rho f}\frac{\partial}{\partial x}\int_{-h}^{0}\rho(z)g(\phi,z)\,\mathrm{d}z + \frac{g}{f}\frac{\partial\zeta}{\partial x}. \tag{12.6}
$$

The first terms on the right-hand side of (12.6), the *relative velocities*, are due to variations in density. The second terms on the right-hand side are due to the slopes at the sea surface. Therefore, computation of geostrophic currents requires both the relative velocities and the slopes at the sea surface.

12.4.2 Geostrophic Currents from Hydrography

To compute geostrophic currents, oceanographers need to compute the horizontal pressure gradient within the ocean. An approach to do this is to compute the slope of a constant-pressure surface relative to a constant-geopotential surface. The *geopotential* at the constant-pressure surface is defined as

$$
\Phi = -\int_{z}^{0} g\,\mathrm{d}z,
$$

where g is gravity. How are hydrographic data used to evaluate the horizontal velocities of surface geostrophic currents?

Oceanographers first modify the hydrostatic balance equation $\frac{1}{\rho}\frac{\partial p}{\partial z} = -g$ as follows.

The vertical pressure gradient is written in the form

$$
\frac{\delta p}{\rho} = \alpha\delta p = -g\delta z = \delta\Phi, \tag{12.7}
$$

where $\alpha = \alpha(S,t,p)$ is the specific volume of seawater with salinity S, temperature t, and pressure p.

Differentiating (12.7) with respect to x, we get

$$
\frac{1}{\rho}\frac{\partial p}{\partial x} = \frac{\partial\Phi}{\partial x}.
$$

Differentiating (12.7) with respect to y, we get

$$
\frac{1}{\rho}\frac{\partial p}{\partial y} = \frac{\partial\Phi}{\partial y}.
$$

Notice that the Coriolis parameter $f = 2\Omega\sin\phi$. So geostrophic balance equations $\frac{1}{\rho}\frac{\partial p}{\partial x} = fv_2$ and $\frac{1}{\rho}\frac{\partial p}{\partial y} = -fv_1$ are written in terms of the slope of the constant-pressure surface P:

$$
\frac{\partial\Phi(P)}{\partial x} = 2\Omega v_2\sin\phi,
$$

$$\frac{\partial \Phi(P)}{\partial y} = -2\Omega v_1 \sin \phi.$$

Consider two constant-pressure surfaces P_1 and P_2 in the ocean and two hydrographic stations A and B a distance L meters apart in the x-direction on the lower surface P_1. The geopotential differences Φ_A and Φ_B between constant-pressure surfaces P_1 and P_2 at hydrographic stations A and B are given, respectively, by

$$\Phi_A = \Phi(P_{2A}) - \Phi(P_{1A}) = \int_{P_{1A}}^{P_{2A}} \alpha(S,t,p)\, dp,$$

$$\Phi_B = \Phi(P_{2B}) - \Phi(P_{1B}) = \int_{P_{1B}}^{P_{2B}} \alpha(S,t,p)\, dp.$$

The *standard geopotential distance* between two constant pressure surfaces P_1 and P_2 is defined as

$$(\Phi_2 - \Phi_1)_{\text{std}} = \int_{P_{1A}}^{P_{2A}} \alpha(35,0,p)\, dp$$

where $\alpha(35,0,p)$ is the specific volume of seawater with salinity of 35 ppt, temperature of $0\,^\circ$C, and pressure p. The standard geopotential distance is the same at stations A and B, so the geopotential differences between two surfaces at stations A and B, respectively, are

$$\Phi_A = (\Phi_2 - \Phi_1)_{\text{std}} + \Delta\Phi_A,$$

$$\Phi_B = (\Phi_2 - \Phi_1)_{\text{std}} + \Delta\Phi_B,$$

where $\Delta\Phi_A$ and $\Delta\Phi_B$ are the *anomalies* of two geopotential distances between two surfaces at stations A and B, respectively, and

$$\Delta\Phi_A = \int_{P_{1A}}^{P_{2A}} (\alpha(S,t,p) - \alpha(35,0,p))\, dp,$$

$$\Delta\Phi_B = \int_{P_{1B}}^{P_{2B}} (\alpha(S,t,p) - \alpha(35,0,p))\, dp.$$

Therefore,

$$\Phi_B - \Phi_A = \Delta\Phi_B - \Delta\Phi_A.$$

The slope of the upper constant-pressure surface P_2 in the x-direction is

$$\frac{\partial \Phi(P_2)}{\partial x} \approx \frac{\Phi_B - \Phi_A}{L} = \frac{\Delta\Phi_B - \Delta\Phi_A}{L}.$$

Combining this with $\frac{\partial \Phi(P_2)}{\partial x} = 2\Omega v_2 \sin \phi$ gives

$$\frac{\Delta \Phi_B - \Delta \Phi_A}{L} = 2\Omega v_2 \sin \phi.$$

So the horizontal velocity in the y-direction of the surface geostrophic current is

$$v_2 = \frac{\Delta \Phi_B - \Delta \Phi_A}{2L\Omega \sin \phi}.$$

Similarly, consider another two hydrographic stations C and D a distance L meters apart in the y-direction on the lower constant-pressure surface P_1. The slope of the upper surface P_2 in the y-direction is

$$\frac{\partial \Phi(P_2)}{\partial y} = \frac{\Delta \Phi_D - \Delta \Phi_C}{L},$$

where $\Delta \Phi_C$ and $\Delta \Phi_D$ are anomalies of geopotential distances between the constant-pressure surfaces P_1 and P_2 at stations C and D, respectively. Combining this with $\frac{\partial \Phi(P_2)}{\partial y} = -2\Omega v_1 \sin \phi$ gives

$$\frac{\Delta \Phi_D - \Delta \Phi_C}{L} = -2\Omega v_1 \sin \phi.$$

So the horizontal velocity in the x-direction of the surface geostrophic current is

$$v_1 = -\frac{\Delta \Phi_D - \Delta \Phi_C}{2L\Omega \sin \phi}.$$

12.5 SVERDRUP'S THEOREM

What drives the ocean currents? Harald Sverdrup showed that the circulation in the upper kilometer or so of the ocean is directly related to the curl of the wind stress. Sverdrup's theorem laid the foundation for a modern theory of ocean circulation.

Sverdrup assumed that the flow is stationary, and then the material derivatives of the horizontal velocities are zero:

$$\frac{Dv_1}{Dt} = \frac{Dv_2}{Dt} = 0.$$

Sverdrup further assumed that the horizontal components of friction near the sea surface can be described by a vertical eddy friction,

$$F_x = \frac{1}{\rho} \frac{\partial}{\partial z} \left(\rho A_z \frac{\partial v_1}{\partial z} \right),$$

$$F_y = \frac{1}{\rho} \frac{\partial}{\partial z} \left(\rho A_z \frac{\partial v_2}{\partial z} \right),$$

where A_z is the eddy viscosity, and that the flow varies with depth and the wind-driven circulation vanishes at some depth of no motion. With these assumptions, the first two equations in (11.34) become

$$\frac{\partial p}{\partial x} = 2\Omega \rho v_2 \sin\phi + \frac{\partial}{\partial z}\left(\rho A_z \frac{\partial v_1}{\partial z}\right),$$

$$\frac{\partial p}{\partial y} = -2\Omega \rho v_1 \sin\phi + \frac{\partial}{\partial z}\left(\rho A_z \frac{\partial v_2}{\partial z}\right).$$

Integrating these two equations with respect to z from the sea surface to a depth h at which the currents go to zero, Sverdrup obtained

$$\int_{-h}^{0} \frac{\partial p}{\partial x}\,dz = 2\Omega \sin\phi \int_{-h}^{0} \rho v_2\,dz + \int_{-h}^{0} \frac{\partial}{\partial z}\left(\rho A_z \frac{\partial v_1}{\partial z}\right)\,dz,$$

$$\int_{-h}^{0} \frac{\partial p}{\partial y}\,dz = -2\Omega \sin\phi \int_{-h}^{0} \rho v_1\,dz + \int_{-h}^{0} \frac{\partial}{\partial z}\left(\rho A_z \frac{\partial v_2}{\partial z}\right)\,dz.$$

Sverdrup defined

$$\frac{\partial P}{\partial x} = \int_{-h}^{0} \frac{\partial p}{\partial x}\,dz,$$

$$\frac{\partial P}{\partial y} = \int_{-h}^{0} \frac{\partial p}{\partial y}\,dz,$$

where $P = \int_{-h}^{0} p\,dz$, and defined

$$M_x = \int_{-h}^{0} \rho v_1(z)\,dz,$$

$$M_y = \int_{-h}^{0} \rho v_2(z)\,dz,$$

where M_x and M_y are the components of the mass transport in the wind-driven layer extending down to an assumed depth of no motion. So

$$\frac{\partial P}{\partial x} = 2\Omega \sin\phi M_y + \left(\rho A_z \frac{\partial v_1}{\partial z}\right)\Big|_{-h}^{0},$$

$$\frac{\partial P}{\partial y} = -2\Omega \sin\phi M_x + \left(\rho A_z \frac{\partial v_2}{\partial z}\right)\Big|_{-h}^{0}.$$

However, the horizontal boundary condition at the sea surface is the wind stress, i.e., at $z = 0$,

$$\rho A_z \frac{\partial v_1}{\partial z} = T_x,$$

$$\rho A_z \frac{\partial v_2}{\partial z} = T_y,$$

and the horizontal boundary condition at depth h is zero stress because the currents go to zero, i.e., at $z = -h$,

$$\rho A_z \frac{\partial v_1}{\partial z} = 0,$$

$$\rho A_z \frac{\partial v_2}{\partial z} = 0,$$

where T_x and T_y are the components of the wind stress in the x- and y-directions. Therefore,

$$\frac{\partial P}{\partial x} = 2\Omega \sin \phi M_y + T_x,$$

$$\frac{\partial P}{\partial y} = -2\Omega \sin \phi M_x + T_y.$$

Notice that $\phi = \frac{y}{a}$, where a is Earth's radius. Differentiating the first equation with respect to y and the second equation with respect to x,

$$\frac{\partial^2 P}{\partial x \partial y} = \frac{2\Omega \cos \phi}{a} M_y + 2\Omega \sin \phi \frac{\partial M_y}{\partial y} + \frac{\partial T_x}{\partial y},$$

$$\frac{\partial^2 P}{\partial y \partial x} = -2\Omega \sin \phi \frac{\partial M_x}{\partial x} + \frac{\partial T_y}{\partial x}.$$

Subtracting the second equation from the first equation,

$$\frac{2\Omega \cos \phi}{a} M_y + 2\Omega \sin \phi \left(\frac{\partial M_y}{\partial y} + \frac{\partial M_x}{\partial x} \right) + \frac{\partial T_x}{\partial y} - \frac{\partial T_y}{\partial x} = 0. \qquad (12.8)$$

On the other hand, the density ρ is a function of depth, so the continuity equation is reduced to

$$\text{div} (\rho \mathbf{v}) = 0$$

or

$$\frac{\partial(\rho v_1)}{\partial x} + \frac{\partial(\rho v_2)}{\partial y} + \frac{\partial(\rho v_3)}{\partial z} = 0.$$

Integrating this equation with respect to z from the sea surface to a depth h, Sverdrup obtained

$$\frac{\partial}{\partial x} \int_{-h}^{0} \rho v_1 \, dz + \frac{\partial}{\partial y} \int_{-h}^{0} \rho v_2 \, dz + (\rho v_3)|_{-h}^{0} = 0$$

or

$$\frac{\partial M_x}{\partial x} + \frac{\partial M_y}{\partial y} + (\rho v_3)|^0_{-h} = 0.$$

He assumed that the vertical velocity at the sea surface and the vertical velocity at depth h are both zero, i.e., $v_3 = 0$ at $z = 0, -h$. Then

$$\frac{\partial M_x}{\partial x} + \frac{\partial M_y}{\partial y} = 0.$$

Inserting this to (12.8),

$$\frac{2\Omega \cos \phi}{a} M_y = \frac{\partial T_y}{\partial x} - \frac{\partial T_x}{\partial y}. \qquad (12.9)$$

Notice that the curl of the wind stress $\mathbf{T} = (T_x, T_y, T_z)$ is

$$\text{curl } \mathbf{T} = \left(\frac{\partial T_z}{\partial y} - \frac{\partial T_y}{\partial z} \right) \mathbf{i} + \left(\frac{\partial T_x}{\partial z} - \frac{\partial T_z}{\partial x} \right) \mathbf{j} + \left(\frac{\partial T_y}{\partial x} - \frac{\partial T_x}{\partial y} \right) \mathbf{k}.$$

So the vertical component of curl $\mathbf{T}$, denoted by $\text{curl}_z \mathbf{T}$, is

$$\text{curl}_z \mathbf{T} = \frac{\partial T_y}{\partial x} - \frac{\partial T_x}{\partial y}.$$

From this with (12.9), Sverdrup finally obtained

$$\frac{2\Omega \cos \phi}{a} M_y = \text{curl}_z(\mathbf{T}).$$

Sverdrup's Theorem. *Let $\beta = \frac{2\Omega \cos \phi}{a}$ be the north-south variation of the Coriolis force, where ϕ is latitude, a is Earth's radius, and Ω is the rotation rate of Earth. Denote by M_y the northward mass transport of the wind-driven current. Then*

$$\beta M_y = \text{curl}_z \mathbf{T},$$

where $\text{curl}_z \mathbf{T}$ is the vertical component of the curl of the wind stress $\mathbf{T}$.

Sverdrup's theorem states that the northward mass transport of the wind-driven current relates directly to the vertical component of the curl of the wind stress. This theorem is an important and fundamental theorem in oceanography.

Over much of the open ocean, especially in the tropics, the wind is zonal and $\frac{\partial T_y}{\partial x}$ is sufficiently small that

$$M_y \approx -\frac{a}{2\Omega \cos \phi} \frac{\partial T_x}{\partial y}.$$

Notice that $\phi = \frac{y}{a}$, where a is Earth's radius. Differentiating both sides with respect to y,

$$\frac{\partial M_y}{\partial y} = -\frac{a}{2\Omega} \left(\frac{\sin \phi}{a \cos^2 \phi} \frac{\partial T_x}{\partial y} + \frac{1}{\cos \phi} \frac{\partial^2 T_x}{\partial y^2} \right)$$

$$= -\frac{1}{2\Omega \cos \phi} \left(\tan \phi \frac{\partial T_x}{\partial y} + a \frac{\partial^2 T_x}{\partial y^2} \right).$$

Substituting this into $\frac{\partial M_x}{\partial x} + \frac{\partial M_y}{\partial y} = 0$, Sverdrup found the eastward mass transport M_x satisfies

$$\frac{\partial M_x}{\partial x} = \frac{1}{2\Omega \cos \phi} \left(\tan \phi \frac{\partial T_x}{\partial y} + a \frac{\partial^2 T_x}{\partial y^2} \right).$$

Sverdrup integrated this equation from a north-south eastern boundary at $x = 0$ and assumed $M_x = 0$ at $x = 0$. Then

$$M_x = \frac{1}{2\Omega \cos \phi} \left(\tan \phi \int_0^{\Delta x} \frac{\partial T_x}{\partial y} \, dx + a \int_0^{\Delta x} \frac{\partial^2 T_x}{\partial y^2} \, dx \right),$$

where Δx is the distance from the eastern boundary of the ocean basin and

$$\frac{1}{\Delta x} \int_0^{\Delta x} \frac{\partial T_x}{\partial y} \, dx,$$

$$\frac{1}{\Delta x} \int_0^{\Delta x} \frac{\partial^2 T_x}{\partial y^2} \, dx$$

are the zonal averages of the wind stress.

12.6 MUNK'S THEOREM

What drives the ocean currents? To solve this problem, Sverdrup used a vertical eddy friction. In order to provide a further answer to this problem, on the basis of Sverdrup's idea, Walter Munk added lateral eddy viscosity and calculated the circulation of the upper layers of the Pacific. Munk's theorem is an extension of Sverdrup's theorem. Munk's theorem also laid the foundation for a modern theory of ocean circulation.

The derivation of Munk's theorem is also based on the first two equations in (11.34):

$$\frac{Dv_1}{Dt} = -\frac{1}{\rho} \frac{\partial p}{\partial x} + f v_2 + F_x,$$

$$\frac{Dv_2}{Dt} = -\frac{1}{\rho} \frac{\partial p}{\partial y} - f v_1 + F_y,$$

where the velocity $\mathbf{v} = (v_1, v_2, v_3)$ the friction $\mathbf{F} = (F_x, F_y, F_z)$, the Coriolis parameter $f = 2\Omega \sin \phi$, and

$$\frac{D}{Dt} = \frac{\partial}{\partial t} + v_1 \frac{\partial}{\partial x} + v_2 \frac{\partial}{\partial y} + v_3 \frac{\partial}{\partial z}.$$

Sverdrup used a vertical eddy friction with eddy viscosity A_z. Munk added lateral eddy frictions with constant A_H as follows:

$$F_x = A_H \frac{\partial^2 v_1}{\partial x^2} + A_H \frac{\partial^2 v_1}{\partial y^2} + \frac{1}{\rho} \frac{\partial}{\partial z} \left(\rho A_z \frac{\partial v_1}{\partial z} \right),$$

$$F_y = A_H \frac{\partial^2 v_2}{\partial x^2} + A_H \frac{\partial^2 v_2}{\partial y^2} + \frac{1}{\rho} \frac{\partial}{\partial z} \left(\rho A_z \frac{\partial v_2}{\partial z} \right).$$

With these assumptions, the first two equations in (11.34) become

$$\frac{\partial p}{\partial x} = 2\Omega \rho v_2 \sin \phi + \rho A_H \frac{\partial^2 v_1}{\partial x^2} + \rho A_H \frac{\partial^2 v_1}{\partial y^2} + \frac{\partial}{\partial z} \left(\rho A_z \frac{\partial v_1}{\partial z} \right),$$

$$\frac{\partial p}{\partial y} = -2\rho v_1 \Omega \sin \phi + \rho A_H \frac{\partial^2 v_2}{\partial x^2} + \rho A_H \frac{\partial^2 v_2}{\partial y^2} + \frac{\partial}{\partial z} \left(\rho A_z \frac{\partial v_2}{\partial z} \right).$$

Integrating these two equations with respect to z from the surface at $z = z_0$ to a depth h at which the currents go to zero, Munk obtained

$$\int_{-h}^{z_0} \frac{\partial p}{\partial x} \, dz = 2\Omega \sin \phi \int_{-h}^{z_0} \rho v_2 \, dz + A_H \frac{\partial^2}{\partial x^2} \int_{-h}^{z_0} \rho v_1 \, dz$$

$$+ A_H \frac{\partial^2}{\partial y^2} \int_{-h}^{z_0} \rho v_1 \, dz + \left(\rho A_z \frac{\partial v_1}{\partial z} \right) \Big|_{-h}^{z_0},$$

$$\int_{-h}^{z_0} \frac{\partial p}{\partial y} \, dz = -2\Omega \sin \phi \int_{-h}^{z_0} \rho v_1 \, dz + A_H \frac{\partial^2}{\partial x^2} \int_{-h}^{z_0} \rho v_2 \, dz$$

$$+ A_H \frac{\partial^2}{\partial y^2} \int_{-h}^{z_0} \rho v_2 \, dz + \left(\rho A_z \frac{\partial v_2}{\partial z} \right) \Big|_{-h}^{z_0}. \tag{12.10}$$

Munk defined

$$\frac{\partial P}{\partial x} = \int_{-h}^{z_0} \frac{\partial p}{\partial x} \, dz,$$

$$\frac{\partial P}{\partial y} = \int_{-h}^{z_0} \frac{\partial p}{\partial y} \, dz,$$

where $P = \int_{-h}^{z_0} p \, dz$, and defined

$$M_x = \int_{-h}^{z_0} \rho v_1 \, dz,$$

$$M_y = \int_{-h}^{z_0} \rho v_2 \, dz,$$

where M_x and M_y are the components of the mass transport in the wind-driven layer extending down to an assumed depth of no motion. Then these two equalities (12.10) are written in the form

$$\frac{\partial P}{\partial x} = 2\Omega \sin \phi M_y + A_H \frac{\partial^2 M_x}{\partial x^2} + A_H \frac{\partial^2 M_x}{\partial y^2} + \left(\rho A_z \frac{\partial v_1}{\partial z} \right) \Big|_{-h}^{z_0},$$

$$\frac{\partial P}{\partial y} = -2\Omega \sin \phi M_x + A_H \frac{\partial^2 M_y}{\partial x^2} + A_H \frac{\partial^2 M_y}{\partial y^2} + \left(\rho A_z \frac{\partial v_2}{\partial z}\right)\Big|_{-h}^{z_0}.$$

However, the horizontal boundary condition at the surface $z = z_0$ is the wind stress, i.e., at $z = z_0$,

$$\rho A_z \frac{\partial v_1}{\partial z} = T_x,$$

$$\rho A_z \frac{\partial v_2}{\partial z} = T_y,$$

and the horizontal boundary condition at depth h is zero stress because the currents go to zero, i.e., at $z = -h$,

$$\rho A_z \frac{\partial v_1}{\partial z} = 0,$$

$$\rho A_z \frac{\partial v_2}{\partial z} = 0.$$

Therefore, these two equations can be written in the form

$$\frac{\partial P}{\partial x} = 2\Omega \sin \phi M_y + A_H \frac{\partial^2 M_x}{\partial x^2} + A_H \frac{\partial^2 M_x}{\partial y^2} + T_x,$$

$$\frac{\partial P}{\partial y} = -2\Omega \sin \phi M_x + A_H \frac{\partial^2 M_y}{\partial x^2} + A_H \frac{\partial^2 M_y}{\partial y^2} + T_y.$$

Notice that $\phi = \frac{y}{a}$, where a is Earth's radius. Differentiating the first equation with respect to y and differentiating the second equation with respect to x,

$$\frac{\partial^2 P}{\partial x \partial y} = \frac{2\Omega \cos \phi}{a} M_y + 2\Omega \sin \phi \frac{\partial M_y}{\partial y} + A_H \frac{\partial^3 M_x}{\partial x^2 \partial y} + A_H \frac{\partial^3 M_x}{\partial y^3} + \frac{\partial T_x}{\partial y},$$

$$\frac{\partial^2 P}{\partial y \partial x} = -2\Omega \sin \phi \frac{\partial M_x}{\partial x} + A_H \frac{\partial^3 M_y}{\partial x^3} + A_H \frac{\partial^3 M_y}{\partial y^2 \partial x} + \frac{\partial T_y}{\partial x}.$$

Subtracting the second equation from the first equation,

$$\frac{2\Omega \cos \phi}{a} M_y + 2\Omega \sin \phi \left(\frac{\partial M_x}{\partial x} + \frac{\partial M_y}{\partial y}\right)$$

$$+ A_H \left(\frac{\partial^3 M_x}{\partial x^2 \partial y} - \frac{\partial^3 M_y}{\partial x^3} + \frac{\partial^3 M_x}{\partial y^3} - \frac{\partial^3 M_y}{\partial y^2 \partial x}\right) \tag{12.11}$$

$$= \frac{\partial T_y}{\partial x} - \frac{\partial T_x}{\partial y}.$$

On the other hand, since the density is a function of depth, the continuity equation reduces to

$$\operatorname{div}(\rho \mathbf{v}) = 0$$

or

$$\frac{\partial(\rho v_1)}{\partial x} + \frac{\partial(\rho v_2)}{\partial y} + \frac{\partial(\rho v_3)}{\partial z} = 0.$$

Munk integrated both sides with respect to z from the surface $z = z_0$ to a depth h and obtained

$$\frac{\partial}{\partial x} \int_{-h}^{z_0} \rho v_1 \, dz + \frac{\partial}{\partial y} \int_{-h}^{z_0} \rho v_2 \, dz + (\rho v_3)|_{-h}^{z_0} = 0$$

or

$$\frac{\partial M_x}{\partial x} + \frac{\partial M_y}{\partial y} + (\rho v_3)|_{-h}^{z_0} = 0$$

He assumed that the vertical velocities at the surface $z = z_0$ and at depth h are both zero, i.e., $(v_3)|_{-h}^{z_0} = 0$. Then

$$\frac{\partial M_x}{\partial x} + \frac{\partial M_y}{\partial y} = 0.$$

From this and (12.11), it follows that

$$\frac{2\Omega \cos \phi}{a} M_y + A_H \left(\frac{\partial^3 M_x}{\partial x^2 \partial y} - \frac{\partial^3 M_y}{\partial x^3} + \frac{\partial^3 M_x}{\partial y^3} - \frac{\partial^3 M_y}{\partial y^2 \partial x} \right) = \frac{\partial T_y}{\partial x} - \frac{\partial T_x}{\partial y}.$$

However, since $\frac{\partial M_x}{\partial x} + \frac{\partial M_y}{\partial y} = 0$, there is a mass-transport stream function Ψ satisfying

$$M_x = \frac{\partial \Psi}{\partial y},$$

$$M_y = -\frac{\partial \Psi}{\partial x}.$$

Therefore, the equality is rewritten in the form

$$\frac{2\Omega \cos \phi}{a} M_y - A_H \left(2\frac{\partial^4 \Psi}{\partial x^2 \partial y^2} + \frac{\partial^4 \Psi}{\partial x^4} + \frac{\partial^4 \Psi}{\partial y^4} \right) = \frac{\partial T_y}{\partial x} - \frac{\partial T_x}{\partial y}.$$

This is equivalent to

$$\frac{2\Omega \cos \phi}{a} M_y - A_H \nabla^4 \Psi = \frac{\partial T_y}{\partial x} - \frac{\partial T_x}{\partial y},$$

where

$$\nabla^4 = \frac{\partial^4}{\partial x^4} + 2\frac{\partial^4}{\partial x^2 \partial y^2} + \frac{\partial^4}{\partial y^4}$$

is the biharmonic operator. Combining this with $\text{curl}_z \mathbf{T} = \frac{\partial T_y}{\partial x} - \frac{\partial T_x}{\partial y}$, Munk obtained

$$\frac{2\Omega \cos\phi}{a} M_y - A_H \nabla^4 \Psi = \text{curl}_z \mathbf{T},$$

which is a fourth-order partial differential equation.

Munk's Theorem. *Let* $\beta = \frac{2\Omega \cos\phi}{a}$ *be the north-south variation of the Coriolis force, where ϕ is latitude, a is Earth's radius, and Ω is the rotation rate of Earth. Denote by M_y the northward mass transport of wind-driven currents and denote by Ψ the mass-transport stream function. Then*

$$\beta M_y - A_H \nabla^4 \Psi = \text{curl}_z \mathbf{T}$$

where $\text{curl}_z \mathbf{T}$ *is the vertical component of the curl of wind stress* $\mathbf{T}$, A_H *is the constant lateral eddy viscosity, and*

$$\nabla^4 = \frac{\partial^4}{\partial x^4} + 2\frac{\partial^4}{\partial x^2 \partial y^2} + \frac{\partial^4}{\partial y^4}$$

is the biharmonic operator.

12.7 TAYLOR-PROUDMAN THEOREM

Vorticity is due to the rotation of a fluid. The rate of rotation can be determined in various ways. The influence of vorticity due to Earth's rotation is most striking for geostrophic flow of a fluid. Taylor and Proudman studied fluid dynamics on the f-plane. The derivation of the Taylor-Proudman theorem is based on equations (12.5).

Taylor and Proudman considered the geostrophic flow with constant density ρ_0 on an f-plane in which the Coriolis parameter $f_0 = 2\Omega \sin\phi_0$, where ϕ_0 is latitude. By (12.5),

$$\begin{cases} \dfrac{\partial p}{\partial x} = \rho_0 f_0 v_2, \\[2mm] \dfrac{\partial p}{\partial y} = -\rho_0 f_0 v_1, \\[2mm] \dfrac{\partial p}{\partial z} = -\rho_0 g, \end{cases} \qquad (12.12)$$

where the fluid velocity $\mathbf{v} = (v_1, v_2, v_3)$ and the gravity g is a function of depth z.

Differentiating the first equation in (12.12) with respect to z, they obtained

$$\frac{\partial}{\partial z}\left(\frac{\partial p}{\partial x}\right) = \rho_0 f_0 \frac{\partial v_2}{\partial z}.$$

Noticing that $\frac{\partial}{\partial z}\left(\frac{\partial p}{\partial x}\right) = \frac{\partial}{\partial x}\left(\frac{\partial p}{\partial z}\right)$ and using the third equation of (12.12), they obtained

$$\rho_0 f_0 \frac{\partial v_2}{\partial z} = \frac{\partial}{\partial z}\left(\frac{\partial p}{\partial x}\right) = \frac{\partial}{\partial x}\left(\frac{\partial p}{\partial z}\right) = -\rho_0 \frac{\partial g}{\partial x}.$$

Since the gravity g depends only on z, $\frac{\partial g}{\partial x} = 0$, and so

$$\frac{\partial v_2}{\partial z} = 0.$$

Differentiating the second equation in (12.12) with respect to z, they obtained

$$\frac{\partial}{\partial z}\left(\frac{\partial p}{\partial y}\right) = -\rho_0 f_0 \frac{\partial v_1}{\partial z}.$$

Notice that $\frac{\partial}{\partial z}\left(\frac{\partial p}{\partial y}\right) = \frac{\partial}{\partial y}\left(\frac{\partial p}{\partial z}\right)$. Using the third equation in (12.12), they obtained

$$-\rho_0 f_0 \frac{\partial v_1}{\partial z} = \frac{\partial}{\partial z}\left(\frac{\partial p}{\partial y}\right) = \frac{\partial}{\partial y}\left(\frac{\partial p}{\partial z}\right) = -\rho_0 \frac{\partial g}{\partial y}.$$

Since the gravity g depends only on z, $\frac{\partial g}{\partial y} = 0$, and so

$$\frac{\partial v_1}{\partial z} = 0.$$

Differentiating the first equation in (12.12) with respect to y and the second equation in (12.12) with respect to x, they obtained

$$\frac{\partial}{\partial y}\left(\frac{\partial p}{\partial x}\right) = \rho_0 f_0 \frac{\partial v_2}{\partial y},$$

$$-\frac{\partial}{\partial x}\left(\frac{\partial p}{\partial y}\right) = \rho_0 f_0 \frac{\partial v_1}{\partial x}.$$

Adding these two equations together gives

$$\frac{\partial v_1}{\partial x} + \frac{\partial v_2}{\partial y} = 0.$$

If the fluid is incompressible, div $\mathbf{v} = 0$ or $\frac{\partial v_1}{\partial x} + \frac{\partial v_2}{\partial y} + \frac{\partial v_3}{\partial z} = 0$, and so

$$\frac{\partial v_3}{\partial z} = 0.$$

Taylor-Proudman Theorem. *For a geostrophic flow of a fluid with constant density on an f-plane with Coriolis parameter $f_0 = 2\Omega \sin \phi_0$, the vertical derivative of the horizontal velocities must be zero:*

$$\frac{\partial v_1}{\partial z} = \frac{\partial v_2}{\partial z} = 0.$$

If the fluid is incompressible, the vertical derivative of the vertical velocity must be zero:

$$\frac{\partial v_3}{\partial z} = 0,$$

where the fluid velocity $\mathbf{v} = (v_1, v_2, v_3)$.

Sverdrup considered the geostrophic flow of a fluid with constant density ρ_0 on a β-plane. The Coriolis parameter in the β-plane is

$$f = f_0 + \beta y,$$

where $f_0 = 2\Omega \sin \phi_0$ and $\beta = \frac{2\Omega \cos \phi_0}{a}$, where a is Earth's radius and ϕ_0 is latitude. By (12.5),

$$\begin{cases} \dfrac{\partial p}{\partial x} = \rho_0 f v_2, \\[2mm] \dfrac{\partial p}{\partial y} = -\rho_0 f v_1, \\[2mm] \dfrac{\partial p}{\partial z} = -\rho_0 g. \end{cases}$$

Differentiating the first two equations with respect to z and using the third equation, Sverdrup obtained

$$\rho_0 f \frac{\partial v_2}{\partial z} = \frac{\partial}{\partial z}\left(\frac{\partial p}{\partial x}\right) = \frac{\partial}{\partial x}\left(\frac{\partial p}{\partial z}\right) = -\rho_0 \frac{\partial g}{\partial x},$$

$$-\rho_0 f \frac{\partial v_1}{\partial z} = \frac{\partial}{\partial z}\left(\frac{\partial p}{\partial y}\right) = \frac{\partial}{\partial y}\left(\frac{\partial p}{\partial z}\right) = -\rho_0 \frac{\partial g}{\partial y}.$$

Since g depends only on z,

$$\frac{\partial g}{\partial x} = \frac{\partial g}{\partial y} = 0,$$

and so the vertical derivatives of the horizontal velocities on the β-plane must be zero:

$$\frac{\partial v_1}{\partial z} = 0,$$

$$\frac{\partial v_2}{\partial z} = 0.$$

Differentiating the first equation with respect to y and differentiating the second equation with respect to x, Sverdrup obtained

$$\frac{\partial}{\partial y}\left(\frac{\partial p}{\partial x}\right) = \rho_0 \beta v_2 + \rho_0 f \frac{\partial v_2}{\partial y},$$

$$-\frac{\partial}{\partial x}\left(\frac{\partial p}{\partial y}\right) = \rho_0 f \frac{\partial v_1}{\partial x}.$$

Adding these two equations together gives

$$f\left(\frac{\partial v_1}{\partial x} + \frac{\partial v_2}{\partial y}\right) = -\beta v_2.$$

If the fluid is incompressible, div $\mathbf{v} = 0$ or $\frac{\partial v_1}{\partial x} + \frac{\partial v_2}{\partial y} + \frac{\partial v_3}{\partial z} = 0$, and so

$$f\frac{\partial v_3}{\partial z} = \beta v_2.$$

Notice that $\beta y \ll f_0$ and $f = f_0 + \beta y$. This equation is approximately

$$f_0\frac{\partial v_3}{\partial z} = \beta v_2,$$

where v_3 is the vertical velocity of the geostrophic flow. This equation is called the *Sverdrup equation*.

Sverdrup's Theorem. *For a geostrophic flow of a fluid with constant density on a β-plane, the vertical derivatives of the horizontal velocities must be zero,*

$$\frac{\partial v_1}{\partial z} = \frac{\partial v_2}{\partial z} = 0.$$

If the fluid is incompressible, the vertical derivative of the vertical velocity relates to the horizontal velocity in the y-direction:

$$f_0\frac{\partial v_3}{\partial z} = \beta v_2,$$

where $f_0 = 2\Omega \sin \phi_0$ and $\beta = \frac{2\Omega \cos \phi_0}{a}$, and a is Earth's radius and ϕ_0 is latitude.

12.8   OCEAN-WAVE SPECTRUM

Ocean waves are produced by the wind. The faster the wind, the longer the wind blows, and the bigger the area over which the wind blows, the bigger the waves. The spectrum is an important concept for describing ocean-surface waves quantitatively.

12.8.1   Spectrum

The Fourier series of any signal $\zeta(t)$ is a superposition of sine waves and cosine waves with harmonic wave frequencies:

$$\zeta(t) = \frac{a_0}{2} + \sum_{n \in \mathbb{Z}_+} \left(a_n \cos \frac{2\pi nt}{T} + b_n \sin \frac{2\pi nt}{T} \right)$$

with the Fourier coefficients

$$a_0 = \frac{2}{T} \int_{-T/2}^{T/2} \zeta(t)\, dt,$$

$$a_n = \frac{2}{T} \int_{-T/2}^{T/2} \zeta(t) \cos \frac{2\pi nt}{T}\, dt \quad (n \in \mathbb{Z}_+),$$

$$b_n = \frac{2}{T} \int_{-T/2}^{T/2} \zeta(t) \sin \frac{2\pi nt}{T} \, dt \quad (n \in \mathbb{Z}_+),$$

where $f = \frac{1}{T}$ is called the *fundamental frequency* and nf are called the *harmonics* of the fundamental frequency.

Using Euler's formula $e^{i\frac{2\pi nt}{T}} = \cos \frac{2\pi nt}{T} + i \sin \frac{2\pi nt}{T}$, the Fourier series can be rewritten as

$$\zeta(t) = \sum_n Z_n e^{i\frac{2\pi nt}{T}}$$

with the Fourier coefficients are

$$Z_n = \frac{1}{T} \int_{-T/2}^{T/2} \zeta(t) e^{-i\frac{2\pi nt}{T}} \, dt \quad (n \in \mathbb{Z}).$$

The *spectrum* of the signal $\zeta(t)$ is defined as

$$S_n = Z_n \overline{Z}_n = |Z_n|^2,$$

where $\overline{Z}_n$ is the complex conjugate of Z_n.

With use of similar techniques, any surface $\zeta(x, y)$ can be represented by a bivariate Fourier series. Thus, the sea surface can be represented as an infinite sum of sine and cosine functions of different frequencies.

12.8.2 Digital Spectrum

Calculating the Fourier series is very difficult since it requires one to measure the height of the sea surface everywhere in an area for a time interval. So oceanographers have to digitize the height of the sea surface:

$$\zeta_k = \zeta(t_k), \quad t_k = k\Delta \quad (k = 0, 1, \ldots, N - 1),$$

where Δ is the time interval between the taking of the samples and N is the total number of samples. The length of the record is $N\Delta$. This converts a continuous function into a digitized function.

The discrete Fourier transform Z_n of the wave record ζ_k is

$$Z_n = \frac{1}{N} \sum_{k=0}^{N-1} \zeta_k e^{-in\frac{2\pi k}{N}} \quad (n = 0, 1, \ldots, N - 1),$$

and the inverse discrete Fourier transform of Z_n is

$$\zeta_k = \sum_{n=0}^{N-1} Z_n e^{ik\frac{2\pi n}{N}} \quad (n = 0, 1, \ldots, N - 1).$$

These equations can be computed quickly using the fast Fourier transform.

The simple *spectrum* S_n of ζ_k, which is called a *periodogram*, is defined as

$$S_0 = \frac{1}{N^2}|Z_0|^2,$$

$$S_n = \frac{1}{N^2}\left[|Z_n|^2 + |Z_{N-n}|^2\right] \quad \left(n = 1, 2, \ldots, \frac{N}{2} - 1\right),$$

$$S_{N/2} = \frac{1}{N^2}\left|Z_{N/2}\right|^2.$$

12.8.3 Pierson-Moskowitz Spectrum

Ocean waves are generated by winds. Strong winds of long duration generate large waves. The speed of propagation of the wave crest is called the *phase velocity*. When the water depth is much greater than the wave length, the phase velocity $c = \frac{g}{\omega}$, where ω is the wave frequency and g is the acceleration due to gravity. The average of the highest one-third of the waves is called the *significant wave height*, denoted by $H_{1/3}$.

The Pierson-Moskowitz spectrum is defined as

$$S(\omega) = \frac{\alpha g^2}{\omega^5} e^{-\beta\left(\frac{\omega_0}{\omega}\right)^4},$$

where $\omega = 2\pi f$, f is the wave frequency in hertz, g is the acceleration due to gravity, and

$$\alpha = 8.1 \times 10^{-3},$$

$$\beta = 0.74,$$

$$\omega_0 = \frac{g}{U_{19.5}},$$

where $U_{19.5}$ is the wind speed at a height of 19.5 m above the sea surface.

(i) Denote by ω_p the frequency of the peak of the Pierson-Moskowitz spectrum. It can be obtained by solving the differential equation $\frac{dS}{d\omega} = 0$ for ω_p.

Notice that

$$\frac{dS}{d\omega} = \frac{\alpha g^2}{\omega^6} e^{-\beta\left(\frac{\omega_0}{\omega}\right)^4}\left(-5 + 4\beta\left(\frac{\omega_0}{\omega}\right)^4\right).$$

Let $\frac{dS}{d\omega} = 0$. Then

$$-5 + 4\beta\left(\frac{\omega_0}{\omega}\right)^4 = 0.$$

Solving this equation gives

$$\omega = \omega_0 \sqrt[4]{\frac{4\beta}{5}}.$$

By $\omega_0 = \frac{g}{U_{19.5}}$ and $\beta = 0.74$, the frequency of the peak of the Pierson-Moskowitz spectrum is

$$\omega_p = \frac{g}{U_{19.5}} \sqrt[4]{\frac{4 \times 0.74}{5}} \approx \frac{0.877g}{U_{19.5}}.$$

(ii) Denote by c_p the speed of the waves at the peak of the Pierson-Moskowitz spectrum. With use of the formula $c_p = \frac{g}{\omega_p}$, the speed of the waves at the peak is

$$c_p = \frac{g}{\omega_p} = \frac{U_{19.5}}{0.877} \approx 1.14 U_{19.5}.$$

Hence, waves with frequency ω_p travel 14% faster than the wind at a height of 19.5 m.

(iii) The significant wave height at the peak of the Pierson-Moskowitz spectrum is obtained by using the formula

$$H_{1/3} = 4 \left(\int_0^\infty S(\omega)\, d\omega \right)^{1/2},$$

where the integral $\int_0^\infty S(\omega)\, d\omega$ is called the *standard deviation of surface displacement*.

Integrating the spectrum $S(\omega)$ over all ω, the standard deviation of surface displacement is found as follows:

$$\int_0^\infty S(\omega)\, d\omega = \int_0^\infty \frac{\alpha g^2}{\omega^5} e^{-\beta \left(\frac{\omega_0}{\omega} \right)^4}\, d\omega = \left. \frac{\alpha g^2}{4\beta \omega_0^4} e^{-\beta \left(\frac{\omega_0}{\omega} \right)^4} \right|_0^\infty = \frac{\alpha g^2}{4\beta \omega_0^4}.$$

With $\alpha = 8.1 \times 10^{-3}$, $\beta = 0.74$, and $\omega_0 = \frac{g}{U_{19.5}}$, the standard deviation of surface displacement is

$$\int_0^\infty S(\omega)\, d\omega = \frac{8.1 \times 10^{-3}(U_{19.5})^4}{4 \times 0.74 g^2} \approx 2.74 \times 10^{-3} \frac{(U_{19.5})^4}{g^2},$$

and so the significant wave height is

$$H_{1/3} = 4 \left(\int_0^\infty S(\omega)\, d\omega \right)^{1/2} = 4 \sqrt{2.74 \times 10^{-3} \frac{(U_{19.5})^4}{g^2}}$$

$$= \frac{4 \times 0.52}{10} \frac{(U_{19.5})^2}{g} \approx 0.21 \frac{(U_{19.5})^2}{g}.$$

(iv) The wave period is the time it takes two successive wave crests to pass a fixed point. With use of the formula $T = \frac{2\pi}{\omega_p}$, the wave period is

$$T = \frac{2\pi}{\omega_p} = \frac{2\pi}{0.877} \frac{U_{19.5}}{g} \approx 7.16 \frac{U_{19.5}}{g}.$$

12.9 OCEANIC TIDAL FORCES

Tides produce currents in the ocean. Tidal currents generate internal waves over seamounts, continental slopes, and mid-ocean ridges. Since oceanic tides lag behind the tide-generating potential, this produces tidal forces that transfer angular momentum between Earth and the tide-producing body (e.g., the Moon or the Sun).

The tide-generating potential at Earth's surface is due to the Earth-Moon system rotating about a common center of mass. Assume that Earth is an ocean-covered planet with no land and the ocean is very deep, and that Earth's rotation is disregarded. Then the rotation of the Moon about Earth produces a potential at any point P on Earth's surface, denoted by V_M:

$$V_M = -\frac{\gamma M}{r_1},$$

where γ is the gravitational constant, M is the Moon's mass, and r_1 is the distance between point P and the Moon's center. Let r be Earth's radius and R_M be the distance between Earth's center and the Moon's center. According to the cosine law of a triangle, it is clear that

$$r_1^2 = R_M^2 - 2rR_M \cos\phi + r^2,$$

where ϕ is the angle between the Earth-Moon line and Earth's radius vector through P. The line joining the Earth's center and the Moon's center is called the *Earth-Moon line*. Therefore, the potential produced by the rotation of the Moon about Earth is

$$V_M = -\frac{\gamma M}{\sqrt{R_M^2 - 2rR_M \cos\phi + r^2}}$$

$$= -\frac{\gamma M}{R_M}\left(1 - 2\frac{r}{R_M}\cos\phi + \frac{r^2}{R_M^2}\right)^{-1/2}. \tag{12.13}$$

The part in brackets in (12.13) is expanded into a Legendre series as follows.

Denote by $P_n(x)$ the Legendre polynomials. It is well known that for $|x| \leq 1$ and $|t| < 1$

$$(1 - 2tx + t^2)^{-1/2} = \sum_0^\infty P_n(x)t^n.$$

Let $x = \cos\phi$ and $t = \frac{r}{R_M}$. Then the part in brackets in (12.13) is

$$\left(1 - 2\frac{r}{R_M}\cos\phi + \frac{r^2}{R_M^2}\right)^{-1/2} = \sum_0^\infty P_n(x)\left(\frac{r}{R_M}\right)^n.$$

From this and (12.13), it follows that

$$V_M = -\frac{\gamma M}{R_M} \sum_0^\infty P_n(\cos\phi) \left(\frac{r}{R_M}\right)^n$$

$$= -\frac{\gamma M}{R_M} \left(P_0(\cos\phi) + P_1(\cos\phi)\frac{r}{R_M} + P_2(\cos\phi)\left(\frac{r}{R_M}\right)^2 + \cdots \right).$$

From Chapter 1, the Legendre polynomials, respectively, are

$$P_0(x) = 1,$$
$$P_1(x) = x,$$
$$P_2(x) = \frac{1}{2}(3x^2 - 1),$$
$$P_3(x) = \frac{1}{2}(5x^3 - 3x),$$
$$\vdots$$

Therefore, the potential produced by the rotation of the Moon about Earth is

$$V_M = -\frac{\gamma M}{R_M} \left\{ 1 + \cos\phi \left(\frac{r}{R_M}\right) + \frac{1}{2}(3\cos^2\phi - 1)\left(\frac{r}{R_M}\right)^2 \right.$$

$$\left. + \frac{1}{2}(5\cos^3\phi - 3\cos\phi)\left(\frac{r}{R_M}\right)^3 + \cdots \right\}.$$

If we disregard the higher-order terms, the previous expression is reduced to

$$V_M \approx -\frac{\gamma M}{R_M} \left\{ 1 + \cos\phi \left(\frac{r}{R_M}\right) + \frac{1}{2}(3\cos^2\phi - 1)\left(\frac{r}{R_M}\right)^2 \right\}.$$

The first term $-\frac{\gamma M}{R_M}$ produces no force. The second term $-\frac{\gamma Mr}{R_M^2}\cos\phi$ produces a constant force parallel to the Earth-Moon line and which keeps Earth in orbit about the center of mass of the Earth-Moon system. The third term $-\frac{\gamma Mr^2}{2R_M^3}(3\cos^2\phi - 1)$ produces the tides and is called the *tide-generating potential*, denoted by V. Therefore, the tide-generating potential due to the Earth-Moon system is

$$V = -\frac{\gamma Mr^2}{2R_M^3}(3\cos^2\phi - 1).$$

It is easy to see that the tide-generating potential is symmetric about the Earth-Moon line.

The tide-generating force can be decomposed into two components. One is the vertical component perpendicular to the sea surface, denoted by P; the other is the horizontal component parallel to the sea surface, denoted by H. The vertical component P is balanced by pressure on the seabed. The horizontal component H produces tides. Since $H = \frac{1}{r}\frac{\partial V}{\partial \phi}$, the horizontal component is obtained:

$$H = \frac{2G_M}{r}\sin 2\phi,$$

where G_M is the Moon's tidal force and

$$G_M = \frac{3}{4}\gamma M \frac{r^2}{R_M^3}.$$

Similarly, the tide-generating potential due to the Earth-Sun system is

$$V = -\frac{\gamma S r^2}{2R_S^3}(3\cos^2\phi - 1),$$

and the Sun's tidal force

$$G_S = \frac{3}{4}\gamma S \frac{r^2}{R_S^3},$$

where γ is the gravitational constant, r is Earth's radius, S is the Sun's mass, R_S is the distance between Earth's center and the Sun's center, and ϕ is the angle between Earth's radius vector through P and the Earth-Sun line.

PROBLEMS

12.1 Let the vertical eddy viscosity A_z be a constant and f be the Coriolis parameter, and let v_1 and v_2 be the fluid velocities in the eastward and northward directions, respectively. Verify that the horizontal momentum equations

$$f v_2 + A_z \frac{\partial^2 v_1}{\partial z^2} = 0,$$

$$-f v_1 + A_z \frac{\partial^2 v_2}{\partial z^2} = 0$$

have the solutions

$$v_1 = V_0 e^{az}\sin\left(\frac{\pi}{4} - az\right),$$

$$v_2 = V_0 e^{az}\cos\left(\frac{\pi}{4} - az\right),$$

where $a = \sqrt{\frac{f}{2A_z}}$.

12.2 Let ψ^s be the stream function on a constant gravitational potential surface. Show that the sea surface ζ is a stream function on this level surface scaled by the factor $\frac{g}{f}$, i.e., $\psi^s = -\frac{g}{f}\zeta$.

12.3 Let $\zeta(t)$ be a rectangular wave,

$$\zeta(t) = \begin{cases} 1, 0 \leq t < 1, \\ 0, -1 \leq t < 0, \end{cases} \quad \text{and} \quad \zeta(t+2) = \zeta(t).$$

Find the spectrum of $\zeta(t)$?

12.4 Let γ be the gravitational constant, r be Earth's radius, S be the Sun's mass, R_S be the distance between Earth's center and the Sun's center, and ϕ be the angle between the Earth-Sun line and Earth's radius vector through any point on Earth's surface. Show that the tide-generating potential due to the Earth-Sun system is

$$V = -\frac{\gamma S r^2}{2R_S^3}(3\cos^2\phi - 1)$$

and the Sun's tidal force is

$$G_S = \frac{3}{4}\gamma S \frac{r^2}{R_S^3}.$$

BIBLIOGRAPHY

Bjekness, V., Sandström, J.W., 1910. Dynamic Meteorology and Hydrography, Part I. Statics, Publication 88. Carnegie Institution of Washington, Washington, DC.

Ekman, V.W., 1905. On the influence of the Earth's rotation on ocean currents. Ark. Mat. Astron. Fys. 2, 1–52.

Foster, E.L., Iliescu, T., Wang, Z., 2013. A finite element discretization of the streamfunction formulation of the stationary quasi-geostrophic equations of the ocean. Comput. Methods Appl. Mech. Eng. 261-262, 105-117.

Foster, E.L., Iliescu, T., Wells, D.R., 2013. A two-level finite element discretization of the streamfunction formulation of the stationary quasi-geostrophic equations of the ocean. Comput. Math. Appl. 66, 1261-1271.

Gill, A.E., 1982. Atmosphere-Ocean Dynamics. Academic Press, London.

Mana, P.P., Zanna, L., 2014. Toward a stochastic parameterization of ocean mesoscale eddies. Ocean Model. 79, 1-20.

Munk, W.H., 1950. On the wind-driven ocean circulation. J. Meteorol. 7, 79-93.

Munk, W.H., 1966. Abyssal recipes. Deep-Sea Res. 13, 707-730.

Munk, W.H., Cartwright, D.E., 1966. Tidal spectroscopy and prediction. Philos. Trans. R. Soc. Lond. A, 259, 533-581.

Munk, W.H., Wunsch, C., 1998. Abyssal recipes II. Deep-Sea Res. 45, 1976-2009.

Philander, S.G., Yamagata, H.T., Pacanowski, R.C., 1984. Unstable air-sea interactions in the tropics. J. Atmos. Res. 41, 604-613.

Pierson, W.J., Moskowitz, L., 1964. A proposed spectral form for fully developed wind seas based on the similarity theory of S.A. Kitaigordskii. J. Geophys. Res. 69, 5181-5190.

Proudman, J., 1916. On the motion of solids in a liquid possessing vorticity. Proc. R. Soc. Lond. A 92, 408-424.

Pugh, D.T., 1987. Tides, Surges, and Mean Sea-Level. John Wiley & Sons, Chichester.

Rebollo, T.C., Hecht, F., Marmol, M.G., Orzetti, G., Rubino, S., 2014. Numerical approximation of the Smagorinsky turbulence model applied to the primitive equations of the ocean. Math. Comput. Simul. 99, 54-70.

San, O., Staples, A.E., Iliescu, T., 2013. Approximate deconvolution large eddy simulation of a stratified two-layer quasigeostrophic ocean model. Ocean Model. 63, 1-20.

Sverdrup, H.U., 1947. Wind-driven currents in a baroclinic ocean: with application to the equatorial currents of the eastern Pacific. Proc. Natl. Acad. Sci. U. S. A., 33, 318-326.

Sverdrup, H.U., Johnson, M.W., Fleming, R.H., 1942. The Oceans: Their Physics, Chemistry, and General Biology. Prentice-Hall, Englewood Cliffs.

Taylor, G.I., 1921. Experiments with rotating fluids. Proc. R. Soc. Lond. A 100, 114-121.

Ueckermann, M.P., Lermusiaux, P.F.J., Sapsis, T.P., 2013. Numerical schemes for dynamically orthogonal equations of stochastic fluid and ocean flows. J. Comput. Phys. 233, 272-294.

Chapter 13

Glaciers and Sea Level Rise

Glaciers and ice sheets cover about 10% of Earth's land surface. Most mountain glaciers have been retreating since the end of the "Little Ice Age". The present volume of Earth's glacier ice, if totally melted, represents about 80 m in potential sea level rise. Sea level changes, especially in densely populated, low-lying coastal areas and on islands, have significant effects on human activities and facilities. Recent research on the current sea level rise budget indicates that the contribution from land ice has increased by 60% over the last decade. In this chapter, we will introduce glacier modeling with various degrees of complexity and estimate glacial contributions to sea level rise.

13.1 STRESS AND STRAIN

When glaciers flow downslope under their own weight, glacier stress occurs, and the strain rate is the ice response to this stress. So understanding how and why glaciers flow means that we must understand glacier stress and strain.

Given a point A in a body and a vector $\mathbf{v}$. Take a small disk $D(A, \delta)$ and $D(A, \delta) \perp \mathbf{v}$. The material on one side of $D(A, \delta)$ exerts a force $\Delta \mathbf{F}$ on the material on the other side. In general, the direction of $\Delta \mathbf{F}$ is different from that of $\mathbf{v}$. The limit vector

$$\lim_{\delta \to 0} \frac{\Delta \mathbf{F}}{\pi \delta^2} =: S_{\mathbf{v}}$$

is called the *stress* of the surface at point A along the vector $\mathbf{v}$. Here $\pi \delta^2$ is the area of the disk $D(A, \delta)$.

Define a $x_1 x_2 x_3$-*rectangular coordinate system* with the x_1-axis in direction $\mathbf{v}$. Since $D(A, \delta) \perp \mathbf{v}$, the disk $D(A, \delta)$ lies in the $x_2 x_3$-plane. The limit vector $S_{\mathbf{v}}$ has three components, denoted by σ_{11}, σ_{12}, and σ_{13}, respectively. The component σ_{11} is called a *normal stress*. The components σ_{12} and σ_{13} are called the *shear stresses*. Similarly, for the x_2-axis and the x_3-axis of the rectangular coordinate system, we can obtain six components $\sigma_{21}, \sigma_{22}, \sigma_{23}$ and $\sigma_{31}, \sigma_{32}, \sigma_{33}$, respectively. Write them as a 3×3 matrix:

$$\sigma = \begin{pmatrix} \sigma_{11} & \sigma_{12} & \sigma_{13} \\ \sigma_{21} & \sigma_{22} & \sigma_{23} \\ \sigma_{31} & \sigma_{32} & \sigma_{33} \end{pmatrix}.$$

Mathematical and Physical Fundamentals of Climate Change
© 2015 Elsevier Inc. All rights reserved.

The matrix σ is called the *full stress tensor*. It is a symmetric matrix, i.e., shear stresses satisfy

$$\sigma_{12} = \sigma_{21},$$
$$\sigma_{13} = \sigma_{31},$$
$$\sigma_{23} = \sigma_{32}.$$

The *mean value* of the normal stresses at a point is

$$\sigma_M = \frac{1}{3}(\sigma_{11} + \sigma_{22} + \sigma_{33}).$$

Here σ_M also gives the pressure for compression. When the axes are rotated, the mean value σ_M is invariant.

Define the *deviatoric stress tensor*:

$$\tau = \begin{pmatrix} \tau_{11} & \tau_{12} & \tau_{13} \\ \tau_{21} & \tau_{22} & \tau_{23} \\ \tau_{31} & \tau_{32} & \tau_{33} \end{pmatrix} = \begin{pmatrix} \sigma_{11} - \sigma_M & \sigma_{12} & \sigma_{13} \\ \sigma_{21} & \sigma_{22} - \sigma_M & \sigma_{23} \\ \sigma_{31} & \sigma_{32} & \sigma_{33} - \sigma_M \end{pmatrix},$$

i.e.,

$$\tau_{ii} = \sigma_{ii} - \sigma_M \quad (i = 1, 2, 3),$$
$$\tau_{ij} = \sigma_{ij} \quad (i \neq j, \ i, j = 1, 2, 3).$$

Its matrix form is

$$\tau = \sigma - \sigma_M I,$$

where I is the 3×3 unit matrix.

Define

$$\tau_E = \sqrt{\frac{1}{2} \mathrm{tr}\,(\tau \tau^T)},$$

where

$$\mathrm{tr}\,(\tau \tau^T) = \tau_{11}^2 + \tau_{22}^2 + \tau_{33}^2 + 2\tau_{12}^2 + 2\tau_{23}^2 + 2\tau_{13}^2 = \sum_{i=1}^{3} \sum_{j=1}^{3} \tau_{ij}^2.$$

It is also invariant for coordinate transform. So τ_E is usually called the *effective stress*.

Let the velocity $\mathbf{v} = (v_1, v_2, v_3)$. For a small deformation, the *shear strain rates* are

$$\dot{\epsilon}_{12} = \frac{1}{2}\left(\frac{\partial v_1}{\partial x_2} + \frac{\partial v_2}{\partial x_1}\right), \quad \dot{\epsilon}_{21} = \dot{\epsilon}_{12},$$

$$\dot{\epsilon}_{23} = \frac{1}{2}\left(\frac{\partial v_2}{\partial x_3} + \frac{\partial v_3}{\partial x_2}\right), \quad \dot{\epsilon}_{32} = \dot{\epsilon}_{23},$$

$$\dot{\epsilon}_{31} = \frac{1}{2}\left(\frac{\partial v_3}{\partial x_1} + \frac{\partial v_1}{\partial x_3}\right), \quad \dot{\epsilon}_{13} = \dot{\epsilon}_{31},$$

and the *normal strain rates* are

$$\dot{\epsilon}_{11} = \frac{\partial v_1}{\partial x_1},$$

$$\dot{\epsilon}_{22} = \frac{\partial v_2}{\partial x_2},$$

$$\dot{\epsilon}_{33} = \frac{\partial v_3}{\partial x_3}.$$

These equations can be written simply in the form

$$\dot{\epsilon}_{ij} = \frac{1}{2} \left(\frac{\partial v_i}{\partial x_j} + \frac{\partial v_j}{\partial x_i} \right) \quad (i, j = 1, 2, 3). \tag{13.1}$$

The *strain rate tensor* $\dot{\epsilon}$ is

$$\dot{\epsilon} = \begin{pmatrix} \dot{\epsilon}_{11} & \dot{\epsilon}_{12} & \dot{\epsilon}_{13} \\ \dot{\epsilon}_{21} & \dot{\epsilon}_{22} & \dot{\epsilon}_{23} \\ \dot{\epsilon}_{31} & \dot{\epsilon}_{32} & \dot{\epsilon}_{33} \end{pmatrix}.$$

When the axes are rotated, the shear strain rate

$$\dot{\epsilon}_E = \sqrt{\frac{1}{2} \mathrm{tr} \left(\dot{\epsilon} \dot{\epsilon}^T \right)}$$

is also invariant, where

$$\mathrm{tr} \left(\dot{\epsilon} \dot{\epsilon}^T \right) = \dot{\epsilon}_{11}^2 + \dot{\epsilon}_{22}^2 + \dot{\epsilon}_{33}^2 + 2\dot{\epsilon}_{12}^2 + 2\dot{\epsilon}_{13}^2 + 2\dot{\epsilon}_{23}^2 = \sum_{i=1}^{3} \sum_{j=1}^{3} \dot{\epsilon}_{ij}^2.$$

So $\dot{\epsilon}_E$ is usually called the *effective strain rate*.

13.2 GLEN'S LAW AND GENERALIZED GLEN'S LAW

For the range of stresses in ice sheets (50-150 kPa), Glen's law relating to the shear strain rate $\dot{\epsilon}$ and the shear stress τ is as follows:

$$\dot{\epsilon} = A\tau^n,$$

where n is a constant. The values of n range from 1.5 to 4.2, with a mean 3, so one often assumes $n = 3$. The coefficient A is called the *flow parameter*, and depends on the temperature and the material. The Arrhenius law shows that at a temperature below $-10\,^\circ$C, the flow parameter A and the absolute temperature T satisfy

$$A = A_0 e^{-\frac{Q}{R(T+\beta p)}},$$

where A_0 is a constant, Q is the low-temperature activation energy ($Q \approx 60\,\mathrm{kJ/mol}$), R is the gas constant ($R = 8.314\,\mathrm{J/(mol\,K)}$), p is pressure and $\beta = 7 \times 10^{-8}\,\mathrm{K/Pa}$.

Glen's law can be applied to only one component of the stress. In order to model glacier dynamics, this law needs to be generalized to the three-dimensional case. Nye assumed that

$$\dot{\epsilon}_{ij} = \lambda \tau_{ij} \quad (i,j = 1,2,3), \tag{13.2}$$

i.e., assumed that the strain rate is proportional to the corresponding deviatoric stress component, where the constant λ is determined by the physical properties of the deforming material and is independent of the choice of coordinate axes. Nye proposed that $\dot{\epsilon}_E$ and τ_E obey power-law behavior for ice, i.e.,

$$\dot{\epsilon}_E = A\tau_E^n.$$

By the definitions of $\dot{\epsilon}_E$ and τ_E, and (13.2), it follows that

$$\dot{\epsilon}_E = \lambda \tau_E.$$

Therefore,

$$\lambda \tau_E = A\tau_E^n,$$

and so $\lambda = A\tau_E^{n-1}$. From this and (13.2), $\dot{\epsilon}_{ij} = A\tau_E^{n-1}\tau_{ij}$, and so

$$\tau_{ij} = A^{-1}\tau_E^{1-n}\dot{\epsilon}_{ij} = A^{-1}(A^{-1}\dot{\epsilon}_E)^{\frac{1-n}{n}}\dot{\epsilon}_{ij} = A^{-\frac{1}{n}}(\dot{\epsilon}_E)^{\frac{1-n}{n}}\dot{\epsilon}_{ij} =: 2\eta\dot{\epsilon}_{ij}, \tag{13.3}$$

where $\eta = \frac{1}{2}A^{-\frac{1}{n}}(\dot{\epsilon}_E)^{\frac{1-n}{n}}$.

13.3 DENSITY OF GLACIER ICE

For the upper layers of ice sheets, when the temperature is near $0\,°C$, the *density of pure glacier ice* is

$$\rho_i = 917\,\text{kg/m}^3.$$

Glacier ice contains abundant bubbles (air or water). Denote by ρ_b the *density of fluid in the bubbles*. Then the density ρ of a glacier is

$$\rho = \nu\rho_b + (1 - \nu)\rho_i,$$

i.e., ρ is the mean of ρ_b and ρ_i with weight ν. When pressure p increases, the density of pure glacier ice also increases. For solid ice, the compressibility is

$$\gamma = \frac{1}{\rho_i}\frac{d\rho_i}{dp} \approx 1.2 \times 10^{-10}/\text{Pa}.$$

For the density of glacier ice at depth h, there is an empirical formula as follows:

$$\rho(h) = \rho_i - (\rho_i - \rho_s)e^{-h/c},$$

where ρ_s is the density of surface snow and c represents a different constant at different sites, e.g., $c = 68$ at the South Pole and $c = 43$ in Greenland. The accumulation rate $\dot{b}$ is defined as *meters of ice* added to the surface per year. If $\dot{b}$ is not zero, then the density ρ changes with time and

$$\frac{d\rho}{dt} = \frac{d\rho}{dx_3}\frac{dx_3}{dt} = \frac{d\rho}{dx_3}\frac{\dot{b}\rho_i}{\rho}. \tag{13.4}$$

The derivative of ρ with respect to t is called the *densification rate*. The driving force for densification of dry firn is the weight of the load. The grain-load stress satisfies

$$p_* = \frac{\rho_i p}{\rho},$$

where

$$p = g \int \rho \, dx_3.$$

The densification rate increases with p_* and satisfies the following formula:

$$\frac{1}{\rho}\frac{d\rho}{dt} = f_0 e^{-\frac{Q}{RT}}\left(\frac{\rho_i}{\rho} - 1\right)^3 p_*^3, \tag{13.5}$$

where f_0 is a constant, T is the absolute temperature, Q is the effective activation energy, and R is the gas constant. By (13.4) and (13.5),

$$\frac{d\rho}{dx_3} = \frac{f_0\rho^2}{\dot{b}\rho_i}e^{-\frac{Q}{RT}}\left(\frac{\rho_i}{\rho} - 1\right)^3 p_*^3.$$

From this, the density ρ can be solved out by numerical integration.

13.4 GLACIER MASS BALANCE

The flow of ice is governed by the conservation of mass. Ice is a nearly incompressible fluid with density $910 \, \text{kg/m}^3$. From this and the continuity equation in Chapter 10, it follows that

$$\frac{\partial v_1}{\partial x_1} + \frac{\partial v_2}{\partial x_2} + \frac{\partial v_3}{\partial x_3} = 0,$$

where $\mathbf{v} = (v_1, v_2, v_3)$ is the fluid velocity. Integrating both sides from h_0 to h_s, we get

$$\int_{h_0}^{h_s}\left(\frac{\partial v_1}{\partial x_1} + \frac{\partial v_2}{\partial x_2}\right) dx_3 + \int_{h_0}^{h_s}\frac{\partial v_3}{\partial x_3} dx_3 = 0,$$

where $h_0 = h_0(x_1, x_2, t)$ is the elevation of the bed and $h_s = h_s(x_1, x_2, t)$ is the elevation of the ice surface. So

$$v_3(h_s) - v_3(h_0) = -\int_{h_0}^{h_s} \left(\frac{\partial v_1}{\partial x_1} + \frac{\partial v_2}{\partial x_2} \right) dx_3 = -\left(\frac{\partial \bar{v}_1}{\partial x_1} + \frac{\partial \bar{v}_2}{\partial x_2} \right) H, \quad (13.6)$$

where $(\bar{v}_1, \bar{v}_2)$ is the mean value of horizontal velocity along the x_3-axis, where

$$\bar{v}_1(x_1, x_2) = \frac{1}{h_s - h_0} \int_{h_0}^{h_s} v_1(x_1, x_2, x_3) \, dx_3,$$

$$\bar{v}_2(x_1, x_2) = \frac{1}{h_s - h_0} \int_{h_0}^{h_s} v_2(x_1, x_2, x_3) \, dx_3,$$

and H is the ice thickness:

$$H = h_s - h_0.$$

Using the symbol $\nabla = (\frac{\partial}{\partial x_1}, \frac{\partial}{\partial x_2})$, we can rewrite (13.6) in the form

$$v_3(h_s) - v_3(h_0) = -(\nabla \cdot (\bar{v}_1, \bar{v}_2))H. \quad (13.7)$$

From $H = H(x_1, x_2, t) = h_s(x_1, x_2, t) - h_0(x_1, x_2, t)$, it follows that

$$\frac{\partial H}{\partial t} = \frac{\partial h_s(x_1, x_2, t)}{\partial t} - \frac{\partial h_0(x_1, x_2, t)}{\partial t} = v_3(h_s) - v_3(h_0).$$

If the surface mass balance rate b_s and the basal melt rate b_0 are also considered, then

$$\frac{\partial H}{\partial t} = v_3(h_s) - v_3(h_0) + b_s - b_0.$$

From this and (13.7),

$$\frac{\partial H}{\partial t} = -(\nabla \cdot (\bar{v}_1, \bar{v}_2))H + b_s - b_0.$$

13.5 GLACIER MOMENTUM BALANCE

The momentum conservation equation used in glacier modeling is another form of the Navier-Stokes equation in Section 11.9. It is stated as follows:

$$\rho \frac{D\mathbf{v}}{Dt} = \nabla \sigma + \rho \mathbf{g}$$

or

$$\rho \left(\frac{\partial \mathbf{v}}{\partial t} + (\mathbf{v} \cdot \nabla)\mathbf{v} \right) = \nabla \sigma + \rho \mathbf{g},$$

where the stress tensor σ is a 3×3 matrix, $\mathbf{g}$ is the gravitational acceleration, and

$$\nabla = \left(\frac{\partial}{\partial x_1}, \frac{\partial}{\partial x_2}, \frac{\partial}{\partial x_3} \right).$$

Notice that

$$(\mathbf{v} \cdot \nabla)\mathbf{v} = (\mathbf{v} \cdot \nabla)(v_1, v_2, v_3) = ((\mathbf{v} \cdot \nabla)v_1, (\mathbf{v} \cdot \nabla)v_2, (\mathbf{v} \cdot \nabla)v_3),$$

where the three components are, respectively,

$$(\mathbf{v} \cdot \nabla)v_i = \mathbf{v} \cdot \left(\frac{\partial v_i}{\partial x_1}, \frac{\partial v_i}{\partial x_2}, \frac{\partial v_i}{\partial x_3} \right) = v_1 \frac{\partial v_i}{\partial x_1} + v_2 \frac{\partial v_i}{\partial x_2} + v_3 \frac{\partial v_i}{\partial x_3} \quad (i = 1, 2, 3).$$

Denote $\sigma = (\sigma_{ij})_{i,j=1,2,3}$. Then

$$\nabla \sigma = \left(\frac{\partial \sigma_{11}}{\partial x_1} + \frac{\partial \sigma_{12}}{\partial x_2} + \frac{\partial \sigma_{13}}{\partial x_3}, \frac{\partial \sigma_{12}}{\partial x_1} + \frac{\partial \sigma_{22}}{\partial x_2} + \frac{\partial \sigma_{23}}{\partial x_3}, \frac{\partial \sigma_{13}}{\partial x_1} + \frac{\partial \sigma_{23}}{\partial x_2} + \frac{\partial \sigma_{33}}{\partial x_3} \right).$$

From this and $\rho \mathbf{g} = (0, 0, -\rho g)$, the momentum conservation equation can be written in the component forms as follows:

$$\rho \left(\frac{\partial v_1}{\partial t} + v_1 \frac{\partial v_1}{\partial x_1} + v_2 \frac{\partial v_1}{\partial x_2} + v_3 \frac{\partial v_1}{\partial x_3} \right) = \frac{\partial \sigma_{11}}{\partial x_1} + \frac{\partial \sigma_{12}}{\partial x_2} + \frac{\partial \sigma_{13}}{\partial x_3},$$

$$\rho \left(\frac{\partial v_2}{\partial t} + v_1 \frac{\partial v_2}{\partial x_1} + v_2 \frac{\partial v_2}{\partial x_2} + v_3 \frac{\partial v_2}{\partial x_3} \right) = \frac{\partial \sigma_{12}}{\partial x_1} + \frac{\partial \sigma_{22}}{\partial x_2} + \frac{\partial \sigma_{23}}{\partial x_3},$$

$$\rho \left(\frac{\partial v_3}{\partial t} + v_1 \frac{\partial v_3}{\partial x_1} + v_2 \frac{\partial v_3}{\partial x_2} + v_3 \frac{\partial v_3}{\partial x_3} \right) = \frac{\partial \sigma_{13}}{\partial x_1} + \frac{\partial \sigma_{23}}{\partial x_2} + \frac{\partial \sigma_{33}}{\partial x_3} - \rho g.$$

The acceleration terms on the left-hand side of these equations are set to zero, so the component forms of the momentum conservation equation are simplified as

$$\frac{\partial \sigma_{11}}{\partial x_1} + \frac{\partial \sigma_{12}}{\partial x_2} + \frac{\partial \sigma_{13}}{\partial x_3} = 0,$$

$$\frac{\partial \sigma_{12}}{\partial x_1} + \frac{\partial \sigma_{22}}{\partial x_2} + \frac{\partial \sigma_{23}}{\partial x_3} = 0,$$

$$\frac{\partial \sigma_{13}}{\partial x_1} + \frac{\partial \sigma_{23}}{\partial x_2} + \frac{\partial \sigma_{33}}{\partial x_3} = \rho g. \tag{13.8}$$

The deviatoric stresses and strain rates satisfy

$$\tau_{ii} = \sigma_{ii} - \sigma_M \ (i = 1, 2, 3), \qquad \tau_{ij} = \sigma_{ij} \ (i \neq j, i, j = 1, 2, 3),$$

$$\tau_{ij} = 2\eta \dot{\epsilon}_{ij},$$

$$\dot{\epsilon}_{ij} = \frac{1}{2} \left(\frac{\partial v_i}{\partial x_j} + \frac{\partial v_j}{\partial x_i} \right) \quad (i, j = 1, 2, 3), \tag{13.9}$$

where σ_M is the mean value of the normal stresses and η is stated in (13.3).

When the variational stress is disregarded, the last equation in (13.8) reduces to

$$\frac{\partial \sigma_{33}}{\partial x_3} \approx \rho g.$$

Disregarding atmospheric pressure, we obtain an expression for σ_{33} by integrating this equation from the surface s to a height h in the ice body:

$$\sigma_{33} = \int_s^h \frac{\partial \sigma_{33}}{\partial x_3} \, dx_3 = (h - s)\rho g.$$

By (13.8), it follows that

$$\frac{\partial(2\tau_{11} + \tau_{22})}{\partial x_1} = \frac{\partial(2\sigma_{11} + \sigma_{22})}{\partial x_1} - \frac{\partial(\sigma_{11} + \sigma_{22} + \sigma_{33})}{\partial x_1}$$

$$= -\frac{\partial \tau_{12}}{\partial x_2} - \frac{\partial \tau_{13}}{\partial x_3} - \frac{\partial \sigma_{33}}{\partial x_1},$$

and so

$$\frac{\partial(2\tau_{11} + \tau_{22})}{\partial x_1} + \frac{\partial \tau_{12}}{\partial x_2} + \frac{\partial \tau_{13}}{\partial x_3} = \rho g \frac{\partial s}{\partial x_1}. \tag{13.10}$$

Similarly,

$$\frac{\partial(2\tau_{22} + \tau_{11})}{\partial x_2} + \frac{\partial \tau_{12}}{\partial x_1} + \frac{\partial \tau_{23}}{\partial x_3} = \rho g \frac{\partial s}{\partial x_2}. \tag{13.11}$$

Equations (13.10) and (13.11) are often referred to as a "higher-order" approximation of the full Navier-Stokes equations.

Assume that horizontal derivatives of the vertical velocity are small compared with the vertical derivative of the horizontal velocity:

$$\frac{\partial v_3}{\partial x_1} \ll \frac{\partial v_1}{\partial x_3},$$

$$\frac{\partial v_3}{\partial x_2} \ll \frac{\partial v_2}{\partial x_3}.$$

By (13.9), a system of two equations with two unknowns v_1 and v_2 is derived from (13.10) and (13.11) as follows.

By (13.6), it follows that

$$\dot{\epsilon}_{33}^2 = \left(\frac{\partial v_3}{\partial x_3}\right)^2 = \left(\frac{\partial v_1}{\partial x_1} + \frac{\partial v_2}{\partial x_2}\right)^2 = (\dot{\epsilon}_{11} + \dot{\epsilon}_{22})^2,$$

and so

$$\dot{\epsilon}_E^2 = \dot{\epsilon}_{11}^2 + \dot{\epsilon}_{22}^2 + \dot{\epsilon}_{11}\dot{\epsilon}_{22} + \dot{\epsilon}_{12}^2 + \dot{\epsilon}_{23}^2 + \dot{\epsilon}_{13}^2. \tag{13.12}$$

By (13.3),

$$2\tau_{11} + \tau_{22} = 4\eta\dot{\epsilon}_{11} + 2\eta\dot{\epsilon}_{22} = 4\eta\frac{\partial v_1}{\partial x_1} + 2\eta\frac{\partial v_2}{\partial x_2},$$

$$2\tau_{22} + \tau_{11} = 4\eta\frac{\partial v_2}{\partial x_2} + 2\eta\frac{\partial v_1}{\partial x_1},$$

$$\tau_{12} = 2\eta\dot{\epsilon}_{12} = \eta\left(\frac{\partial v_1}{\partial x_2} + \frac{\partial v_2}{\partial x_1}\right),$$

$$\tau_{13} = 2\eta\dot{\epsilon}_{13} = \eta\frac{\partial v_1}{\partial x_3}.$$

$$\tau_{23} = \eta\frac{\partial v_2}{\partial x_3}. \tag{13.13}$$

This implies from (13.10) and (13.11) that

$$\frac{\partial}{\partial x_1}\left(4\eta\frac{\partial v_1}{\partial x_1} + 2\eta\frac{\partial v_2}{\partial x_2}\right) + \frac{\partial}{\partial x_2}\left(\eta\frac{\partial v_1}{\partial x_2} + \eta\frac{\partial v_2}{\partial x_1}\right) + \frac{\partial}{\partial x_3}\left(\eta\frac{\partial v_1}{\partial x_3}\right) = \rho g\frac{\partial s}{\partial x_1},$$

$$\frac{\partial}{\partial x_2}\left(4\eta\frac{\partial v_2}{\partial x_2} + 2\eta\frac{\partial v_1}{\partial x_1}\right) + \frac{\partial}{\partial x_1}\left(\eta\frac{\partial v_1}{\partial x_2} + \eta\frac{\partial v_2}{\partial x_1}\right) + \frac{\partial}{\partial x_3}\left(\eta\frac{\partial v_2}{\partial x_3}\right) = \rho g\frac{\partial s}{\partial x_2}.$$

So v_1 and v_2 are found by these two equalities and then, from (13.6), v_3 is found as follows:

$$v_3(h) - v_3(b) = -\int_b^h \left(\frac{\partial v_1}{\partial x_1} + \frac{\partial v_2}{\partial x_2}\right) dx_3.$$

By (13.3) and (13.12),

$$\eta = \frac{1}{2}A^{-\frac{1}{n}}\left\{\left(\frac{\partial v_1}{\partial x_1}\right)^2 + \left(\frac{\partial v_2}{\partial x_2}\right)^2 + \frac{\partial v_1}{\partial x_1}\frac{\partial v_2}{\partial x_2} + \frac{1}{4}\left(\frac{\partial v_1}{\partial x_2} + \frac{\partial v_2}{\partial x_1}\right)^2\right.$$

$$\left. + \frac{1}{4}\left(\frac{\partial v_1}{\partial x_3}\right)^2 + \frac{1}{4}\left(\frac{\partial v_2}{\partial x_3}\right)^2\right\}^{\frac{1-n}{2n}}.$$

This shows that η may be determined by v_1 and v_2.

13.6   GLACIER ENERGY BALANCE

The energy conservation equation used in glacier modeling is

$$\rho\frac{D(cT)}{Dt} = \nabla \cdot (K\nabla T) + \Phi$$

or

$$\rho\left(\frac{\partial(cT)}{\partial t} + (\mathbf{v} \cdot \nabla)(cT)\right) = \nabla \cdot (K\nabla T) + \Phi,$$

where c is the ice heat capacity and K is the heat conductivity.

Notice that

$$\nabla \cdot (K\nabla T) = K\nabla \cdot \nabla T = K\Delta T = K \left(\frac{\partial^2 T}{\partial x_1^2} + \frac{\partial^2 T}{\partial x_2^2} + \frac{\partial^2 T}{\partial x_3^2} \right),$$

$$(\mathbf{v} \cdot \nabla)(cT) = c\mathbf{v} \cdot \nabla T = c \left(v_1 \frac{\partial T}{\partial x_1} + v_2 \frac{\partial T}{\partial x_2} + v_3 \frac{\partial T}{\partial x_3} \right).$$

Then the energy conservation equation is written in the form

$$c\rho \left(\frac{\partial T}{\partial t} + v_1 \frac{\partial T}{\partial x_1} + v_2 \frac{\partial T}{\partial x_2} + v_3 \frac{\partial T}{\partial x_3} \right) = K \left(\frac{\partial^2 T}{\partial x_1^2} + \frac{\partial^2 T}{\partial x_2^2} + \frac{\partial^2 T}{\partial x_3^2} \right) + \Phi.$$

If we disregard horizontal diffusion $\frac{\partial^2 T}{\partial x_1^2}$ and $\frac{\partial^2 T}{\partial x_2^2}$, the energy conservation equation becomes

$$c\rho \frac{\partial T}{\partial t} = K\frac{\partial^2 T}{\partial x_3^2} - c\rho \left(v_1 \frac{\partial T}{\partial x_1} + v_2 \frac{\partial T}{\partial x_2} + v_3 \frac{\partial T}{\partial x_3} \right) + \Phi.$$

13.7 SHALLOW-ICE AND SHALLOW-SHELF APPROXIMATIONS

Two lower-order approximations—the shallow-ice approximation (SIA) and the shallow-shelf approximation (SSA)—are widely used.

The SIA is the commonest approximation. The first two terms on the left-hand side of (13.10) and (13.11) are disregarded, i.e.,

$$\frac{\partial(2\tau_{11} + \tau_{22})}{\partial x_1} + \frac{\partial \tau_{12}}{\partial x_2} = 0,$$

$$\frac{\partial(2\tau_{22} + \tau_{11})}{\partial x_2} + \frac{\partial \tau_{12}}{\partial x_1} = 0.$$

Then

$$\frac{\partial \tau_{13}}{\partial x_3} = \rho g \frac{\partial s}{\partial x_1},$$

$$\frac{\partial \tau_{23}}{\partial x_3} = \rho g \frac{\partial s}{\partial x_2},$$

where s is the surface elevation. The SIA is valid in the slow-moving interior of ice sheets. Integration from the surface s to height x_3 gives

$$\tau_{13} = (x_3 - s)\rho g \frac{\partial s}{\partial x_1},$$

$$\tau_{23} = (x_3 - s)\rho g \frac{\partial s}{\partial x_2}.$$

Notice that $\frac{\partial v_3}{\partial x_1} \ll \frac{\partial v_1}{\partial x_3}$ and $\frac{\partial v_3}{\partial x_2} \ll \frac{\partial v_2}{\partial x_3}$. From (13.1) and $\dot{\epsilon}_{ij} = A\tau_E^{n-1}\tau_{ij}$, it follows that

$$\frac{1}{2}\frac{\partial v_1}{\partial x_3} \approx \dot{\epsilon}_{13} = A(s, x_3)\tau_E^{n-1}\tau_{13},$$

$$\frac{1}{2}\frac{\partial v_2}{\partial x_3} \approx \dot{\epsilon}_{23} = A(s, x_3)\tau_E^{n-1}\tau_{23}.$$

Notice that

$$|\nabla s| = \sqrt{\left(\frac{\partial s}{\partial x_1}\right)^2 + \left(\frac{\partial s}{\partial x_2}\right)^2},$$

where $\nabla = (\frac{\partial}{\partial x_1}, \frac{\partial}{\partial x_2})$. Then for the SIA,

$$\tau_E \approx \sqrt{\tau_{13}^2 + \tau_{23}^2} = (x_3 - s)\rho g|\nabla s|.$$

This implies that

$$\frac{1}{2}\frac{\partial v_1}{\partial x_3} = A(s, x_3)(x_3 - s)^n(\rho g)^n|\nabla s|^{n-1}\frac{\partial s}{\partial x_1},$$

$$\frac{1}{2}\frac{\partial v_2}{\partial x_3} = A(s, x_3)(x_3 - s)^n(\rho g)^n|\nabla s|^{n-1}\frac{\partial s}{\partial x_2}.$$

Integration from h^* to h gives

$$v_1(h) - v_1(h^*) = 2(\rho g)^n|\nabla s|^{n-1}\frac{\partial s}{\partial x_1}\int_{h^*}^h A(s, x_3)(x_3 - s)^n \, dx_3,$$

$$v_2(h) - v_2(h^*) = 2(\rho g)^n|\nabla s|^{n-1}\frac{\partial s}{\partial x_2}\int_{h^*}^h A(s, x_3)(x_3 - s)^n \, dx_3.$$

Another approach is the SSA. The third terms on the left-hand side of (13.10) and (13.11) are disregarded, i.e.,

$$\frac{\partial(2\tau_{11} + \tau_{22})}{\partial x_1} + \frac{\partial \tau_{12}}{\partial x_2} = \rho g \frac{\partial s}{\partial x_1}.$$

$$\frac{\partial(2\tau_{22} + \tau_{11})}{\partial x_2} + \frac{\partial \tau_{12}}{\partial x_1} = \rho g \frac{\partial s}{\partial x_2}.$$

The SSA is valid for floating ice shelves where there is little or no vertical shear.

13.8 DYNAMIC ICE SHEET MODELS

Various ice sheet models have been developed to simulate the evolution, dynamics, and thermodynamics of glaciers and ice sheets.

The Community Ice Sheet Model (CISM) is the ice dynamics component of the Community Climate System Model (CCSM). Based on ideas in Sections 13.4–13.7, CISM uses a finite-difference method to numerically solve basic fluid equations in order to model glacier dynamics. CISM can be used for predicting ice sheet retreat and sea level rise in a warming climate. It is freely available and easy to use.

Potsdam Parallel Ice Sheet Model (PISM-PIK), developed at the Potsdam Institute for Climate Impact Research, is used for simulations of large-scale ice sheet-shelf systems. It is derived from the Parallel Ice Sheet Model. PISM-PIK is a three-dimensional thermodynamically coupled shallow model using a finite-difference discretization and SIA/SSA.

Elmer/Ice is a full-Stokes ice sheet model developed by CSC-IT Center for Science in Finland. Different from CISM and PISM-PIK, Elmer/Ice uses the finite-element method to numerically solve basic fluid equations in order to model glacier dynamics. Elmer/Ice builds on Elmer, an open-source, parallel, finite-element code.

Zhang and Moore (2014) are developing a new ice sheet model. Their basic idea is to use a wavelet method to numerically solve basic fluid equations in order to model glacier dynamics. Wavelets are a new tool in numerical solutions of basic fluid equations. Wavelets offer considerable advantages over the finite-difference method or the finite-element method. Its main advantages are as follows: (a) different resolutions can be used in different regions of space; (b) the coupling between different resolution levels is easy; (c) there are few topological constraints for increased-resolution regions compared with conventional numerical methods; (d) the numerical effort scales linearly with system size.

13.9 SEA LEVEL RISE

Sea level rise will be one of the most visible, costly, and globally widespread consequences of future climate change; an estimated 150 million people live within 1 m of the present-day sea level. Sea level varies considerably from year to year and from decade to decade, and the relatively short-term changes are due to redistribution of Earth's water budget, e.g., El Niño Southern Oscillation cycles and volcanic eruptions are known to impact the global balance of evaporation and precipitation, changing regional sea levels by up to 1 m and global sea levels by 10 cm or so for a year or two. For long-term changes, the Fourth Assessment Report of Intergovernmental Panel on Climate Change stated that the mean observational rate of sea level rise was 1.8 ± 0.5 mm/year from 1961 to 2003.

The total sea level budget ΔS as a sum of five terms is described as follows:

$$\Delta S = \Delta S_r + \Delta M_g + \Delta G_{is} + \Delta A_{is} + \Delta S_{nc},$$

where ΔS_r is the contribution from thermosteric expansion of ocean water, ΔM_g is from mountain glaciers and ice caps, ΔG_{is} is from the Greenland ice sheet, ΔA_{is} is from Antarctic ice sheet mass losses, and ΔS_{nc} is a nonclimate source of sea level rise, for example, from building dams and groundwater extraction. In all components of the sea level budget, the role of the ice sheets has the largest potential for unexpected contributions to sea level in the coming centuries. For a marine ice sheet, only the mass of ice above the flotation level contributes. Therefore, for a marine ice body with thickness H and depth-averaged density ρ resting on a bed at depth L below sea level, per unit area, the mass contributing to sea level rise is

$$\rho H - \rho_w L,$$

where ρ_w is the seawater density ($\rho_w = 1028\,\text{kg/m}^3$).

For the last two decades, different remote-sensing techniques have allowed systematic monitoring of the mass balance of the ice sheets. The Greenland and Antarctica contributions to sea level rise (Church and White, 2011) are

	1961-2008	1993-2010
Greenland	0.11 ± 0.17 mm/year	0.31 ± 0.17 mm/year
Antarctic	0.25 ± 0.20 mm/year	0.43 ± 0.20 mm/year

13.10 SEMIEMPIRICAL SEA LEVEL MODELS

Projections by semiempirical models are based on the assumption that the sea level in the future will respond as it has in the past to imposed climate forcing, e.g., Gornitz et al. (2001) assumed the following linear relationship between sea level and global temperature holds:

$$S = a(T - T_0) + b,$$

where S is the global mean sea level, T is the global temperature, T_0 is a reference temperature at a time when the sea level was in equilibrium, a is a sensitivity constant, and b is a constant. The latest improvement by Kemp et al. (2011) is

$$\frac{dS}{dt} = a_1(T(t) - T_{00}) + a_2(T(t) - T_0(t)) + b\frac{dT}{dt},$$

where

$$\frac{dT_0}{dt} = \frac{T(t) - T_0(t)}{\nu}$$

and a_1, a_2, b, ν, and T_{00} are constants.

Grinsted et al. (2010) introduced a different semiempirical model that make use of a response time to be determined by the data that represent the presumed centennial- or millennium-scale response of oceans and ice sheets. Jevrejeva et al. (2009) used radiative forcing as the forcing variable in a semiempirical formulation based on the model of Grinsted et al. (2010)—the respective publishing dates do not match the formulation history.

Various semiempirical models suggest that the prospects for keeping sea level rise below 1 m by 2100 rest on keeping the temperature rise below about 2 °C. The fundamental limits of semiempirical models are the assumption of linearity between the climate of the past and sea level response and that of the future. However, all models show that sea levels will continue to rise beyond 2100, perhaps by 2-3 m by 2300.

PROBLEMS

13.1 Download the Community Ice Sheet Model from http://oceans11.lanl. gov/trac/CISM/ and learn to use this model to simulate dynamics and thermodynamics of ice sheets.

13.2 Study various semiempirical sea level models from the Bibliography section and list the strengths and weaknesses of each model.

BIBLIOGRAPHY

Bard, E., et al., 1996. Sea level record from Tahiti corals and the timing of deglacial meltwater discharge. Nature 382, 241-244.

Bittermann, K., Rahmstorf, S., Perrette, M., Vermeer, M., 2013. Predictability of twentieth century sea-level rise from past data. Environ. Res. Lett. 8, 014013.

Bueler, E., Brown, J., 2009. Shallow shelf approximation as a sliding law in a thermomechanically coupled ice sheet model. J. Geophys. Res. 114, F03008.

Cazenave, A., Llovel, W., 2010. Contemporary sea level rise. Ann. Rev. Mar. Sci. 2, 145-173.

Church, J.A., White, N.J., 2011. Sea-level rise from the late 19th to the early 21st century. Surv. Geophys. 32, 585-602.

Gagliardini, O., Cohen, D., Raack, P., Zwinger, T., 2007. Finite element modeling of subglacial cavities and related friction law. J. Geophys. Res. 112, F02027.

Gornitz, V., 2001. Impoundment, groundwater mining, and other hydrologic transformations: impacts on global sea level rise. In: Douglas, B.C., Kearney, M.S., Leatherman, S.P. (Eds.), Sea Level Rise, History and Consequences. International Geophysics Series. Academic Press, New York.

Grinsted, A., Moore, J.C., Jevrejeva, S., 2010. Reconstructing sea level from paleo and projected temperatures 200 to 2100AD. Clim. Dyn. 34, 461-472.

Jevrejeva, S., Grinsted, A., Moore, J.C., 2009. Anthropogenic forcing dominates sea level rise since 1850, Geophys. Res. Lett., 36, L20706.

Jevrejeva, S., Moore, J.C., Grinsted, A., 2012a. Sea level projections with new generation of scenarios for climate change. Global Planet. Change 80, 14-20.

Jevrejeva, S., Moore, J.C., Grinsted, A., 2012b. Potential for bias in 21st century sea level projections from semiempirical models. J. Geophys. Res. 117, D20116.

Kemp, A., Horton, B., Donnelly, J., Mann, M., Vermeer, M., Rahmstorf, S., 2011. Climate related sea level variations over the past two millennia. Proc. Natl. Acad. Sci. U. S. A. 108, 11,017-11,022.

Marzeion, B., Jarosch, A.H., Hofer, M., 2012. Past and future sea-level change from the surface mass balance of glaciers. Cryosphere 6, 1295.

Moore, J.C., Jevrejeva, S., Grinsted, A., 2011. The historical sea level budget. Ann. Glaciol. 52, 8-14.

Moore, J.C., Grinsted, A., Zwinger, T., Jevrejeva, S., 2013. Semiempirical and process-based global sea level projections. Rev. Geophys. 51.

Pattyn, F., et al., 2012. Results of the marine ice sheet model intercomparison project, MISMIP. Cryosphere Discuss. 6, 267-308.

Rignot, E., Mouginot, J., Scheuchl, B., 2011. Ice flow of the Antarctic ice sheet. Science 333, 1427-1430.

Winkelmann, R., Martin, M.A., Haseloff, M., Albrecht, T., Bueler, E., Khroulev, C., Levermann, A., 2011. The Potsdam Parallel Ice Sheet Model (PISM-PIK) part 1: model description. Cryosphere 5, 715-726.

Zhang, Z., Moore, J.C., 2014. Wavelet-Based Ice Sheet Models. Technical report.

Chapter 14

Climate and Earth System Models

Various models are used to study the climate system and its natural variability, and to simulate the interaction between the physical climate and the biosphere, and the chemical constituents of the land, atmosphere, and ocean. Models are the best tools available to test hypotheses about the factors causing climate change and to assess future Earth system developments. In this chapter we will introduce basic physical principles in energy balance models (EBMs), radiative convective models (RCMs), statistical dynamical models (SDMs), and Earth system models (ESMs) consisting of atmospheric models, ocean models, land surface models, sea ice models, and couplers. Then we will introduce the Coupled Model Intercomparison Project (CMIP) and Geoengineering Model Intercomparison Project (GeoMIP).

14.1 ENERGY BALANCE MODELS

Global energy balance is used for the construction of the simplest climate models. Budyko (1969) and Sellers (1969) constructed two EBMs. Their works prompted much of the interest in simulation of climatic change.

14.1.1 Zero-Dimensional EBM

The zero-dimensional EBM considers Earth as a single point in space having a global mean effective temperature. It is called sometimes a *global EBM*. If the atmosphere of Earth absorbs thermal radiation, its surface temperature is greater than its effective temperature. Denote their difference by ΔT:

$$\Delta T = T_s - T_e,$$

where T_s is the surface temperature and T_e is the effective temperature. The difference ΔT is called the *greenhouse increment*. It is well known that Earth's effective temperature is 255 K. If Earth's greenhouse increment is about 33 K, then Earth's surface temperature is about 288 K.

In a simple zero-dimensional EBM of Earth, the rate of change of temperature with respect to time is caused by a difference between net incoming and net outgoing radiative fluxes per unit area at the top of the atmosphere:

Mathematical and Physical Fundamentals of Climate Change

$$mc\frac{\Delta T}{\Delta t} = (R_i - R_o)S_e, \tag{14.1}$$

where m is the mass of Earth, c is the specific heat capacity of Earth, R_i is the net incoming radiative flux, R_o is the net outgoing radiative flux, and S_e is the area of Earth. In the equilibrium climate state, the change in temperature has ceased, i.e., the rate of change of temperature is

$$\frac{\Delta T}{\Delta t} = 0,$$

and so it follows from (14.1) that

$$R_i = R_o,$$

i.e., the net incoming radiative flux is equal to the net outgoing radiative flux in the equilibrium climate state. The net incoming radiative flux is

$$R_i = (1 - \alpha)\frac{F_s}{4},$$

where F_s is the solar constant and α is the planetary albedo. This shows that the net incoming radiative flux is a function of the solar constant and the planetary albedo. So

$$(1 - \alpha)F_s = 4R_o. \tag{14.2}$$

According to the Stefan-Boltzmann law, the net outgoing radiative flux is

$$R_o = \tau_a \sigma T^4,$$

where T is the surface temperature, σ is the Stefan-Boltzmann constant, and τ_a is the infrared transmissivity of the atmosphere. This shows that the net outgoing radiative flux is a function of the surface temperature. If we combine this with (14.2), the equilibrium solution of (14.1) is

$$(1 - \alpha)F_s = 4\tau_a \sigma T^4,$$

and so the surface temperature is

$$T = \left(\frac{(1 - \alpha)F_s}{4\tau_a \sigma}\right)^{1/4}.$$

If the values of F_s, α, and τ_a are given, then the surface temperature of Earth can be obtained.

14.1.2 One-Dimensional EBM

The one-dimensional EBM considers the temperature as being latitudinally resolved. It is called sometimes a *zonal EBM*. The equilibrium state of each latitude zone is

$$(1 - \alpha_j)F_{s_j} = R_o^j + G_j, \tag{14.3}$$

where the additional term G_j is the loss of energy by a latitude zone to its colder neighbor or neighbors. Comparing (14.3) with (14.2), we see that the zero-dimensional EBM is a simplification of the one-dimensional EBM.

Each term in (14.3) is a function of the surface temperature T_j of zone j. The surface albedo of zone j is influenced by temperature in that it is increased when snow and ice form. The horizontal flux out of the zone depends not only on the zonal temperature but also on the global mean temperature. Therefore, we may denote them by

$$\alpha_j = \alpha(T_j),$$

$$R_o^j = R_0(T_j),$$

$$G_j = G(T_j).$$

The surface albedo is described by a step function:

$$\alpha_j = \alpha(T_j) = \begin{cases} 0.6 & T_j \le T_c, \\ 0.3 & T_j > T_c, \end{cases}$$

where T_c is the temperature at the snow line. Because of the relatively small range of temperatures involved, radiation leaving the top of the latitude zone can be approximated by a linear function:

$$R_o^j = R_0(T_j) = A + BT_j,$$

where A and B are two empirical constants. The additional term G_j is proportional to the difference between the zonal temperature and the global mean temperature, so it can be represented by

$$G_j = G(T_j) = k_j(T_j - \overline{T}),$$

where k_j is an empirical constant, T_j is the zonal temperature, and $\overline{T}$ is the global mean temperature. Substituting these two equalities into (14.3), we get

$$(1 - \alpha_j)F_{s_j} = A + BT_j + k_j(T_j - \overline{T}).$$

Solving this equation, we find the surface temperature of zone j is

$$T_j = \frac{(1 - \alpha_j)F_{s_j} + k_j\overline{T} - A}{B + k_j}. \tag{14.4}$$

Given a first-guess temperature distribution and by devising an appropriate weighting scheme to distribute the solar radiation over the globe, applying successively (14.4), we can obtain the equilibrium solution.

14.2 RADIATIVE CONVECTIVE MODELS

RCMs are used to model the temperature profile by considering radiative and convective energy transport up through the atmosphere. The one-dimensional RCMs divide the atmosphere into layers in order to derive a temperature profile for the atmosphere (see Section 11.1). Some RCMs include cloud prediction schemes. The two-dimensional RCMs further consider the temperature as being latitudinally resolved. In each latitude belt, the surface heat balance equation for land is given by

$$C_L D \frac{\partial T_L}{\partial t} = R_g - I - H_s - H_L,$$

where C_L is the heat capacity of the land-surface layer, D is the effective depth of the land-surface layer, T_L is the temperature of the land-surface layer, R_g is the solar radiation absorbed, I is the infrared heat flux, H_s is the sensible heat flux, and H_L is the latent heat flux. The atmospheric heat balance can be written as

$$\frac{\partial T}{\partial t} = Q_s + Q_I + Q_L + A,$$

where T is the atmospheric temperature, Q_s is the rate of solar heating, Q_I is the rate of long-wave heating, Q_L is the rate of latent heating, and A is the heating rate due to dynamical redistribution of heat.

The ocean parameterization includes the role of the ocean biomass in climate through its uptake of carbon. It includes downward transport of substances by phytoplankton and the subsequent settling of marine grazer feces. The simplified food web includes only phytoplankton and detritus. The phytoplankton is governed by

$$\frac{dB}{dt} = B \left(P_{\text{max}} f(I) \frac{N}{N+k} - r - m \right),$$

where B is the phytoplankton biomass, P_{max} is the maximum production rate, $f(I)$ is a light limitation function, N is the organic nitrogen, k is the half-saturation fraction for N, r is the respiration rate, and m is the mortality rate. The detritus is governed by

$$\frac{dD}{dt} = mB - sD,$$

where D is the detritus concentration and s is the setting rate for detritus.

14.3 STATISTICAL DYNAMICAL MODELS

The two-dimensional SDMs with usually one horizontal dimension and one vertical dimension are developed to simulate horizontal energy flows and processes that disrupt them. Statistical relationships are used to define the wind speed and wind direction within SDMs. The two-dimensional SDMs are based on numerical solution of the following basic equations.

Let v_1, v_2, and v_3 be the velocities in the eastward x, northward y, and vertical z directions. Let

$$v_i' = v_i - \langle v_i \rangle \quad (i = 1, 2, 3),$$

where $\langle v_i \rangle$ is the zonal average value of v_i.

The equation for the *zonal momentum* expresses that changes in zonal momentum with time are balanced by the Coriolis term and the eddy transport of momentum in the poleward direction as well as a frictional dissipation term, i.e.,

$$\frac{\partial \langle v_1 \rangle}{\partial t} = f \langle v_2 \rangle - \frac{\partial \langle v_1' v_2' \rangle}{\partial y} + \mathbf{F},$$

where f is the Coriolis parameter, $\langle v_1' v_2' \rangle$ is the zonal average value of $v_1' v_2'$, and $\mathbf{F}$ is the friction.

The equation for *geostrophic balance* expresses that the pressure gradient force in the poleward direction is balanced by Coriolis term, i.e.,

$$\frac{\partial \langle p \rangle}{\partial y} = -f \langle v_1 \rangle \langle \rho \rangle,$$

where $\langle p \rangle$ and $\langle \rho \rangle$ are the zonal average values of the pressure p and the density ρ, respectively. This equation is also called the equation for *meridional momentum*.

The equation for *hydrostatic balance* expresses that the pressure gradient force in the vertical direction is balanced by gravity, i.e.,

$$\frac{\partial \langle p \rangle}{\partial z} = -g \langle \rho \rangle,$$

where $g = |\mathbf{g}|$ and $\mathbf{g}$ is the acceleration due to gravity.

The equation for *thermodynamic balance* expresses that the temporal rate of change of zonally averaged temperature is balanced by two eddy transports of heat in the northward and vertical directions and the vertical transport of heat as well as the zonal diabatic heating, i.e.,

$$\frac{\partial \langle T \rangle}{\partial t} = -\frac{\partial \langle v_2' T' \rangle}{\partial y} - \frac{\partial \langle v_3' T' \rangle}{\partial z} - \langle v_3 \rangle \left(\frac{g}{\langle \rho \rangle c_p} + \frac{\partial \langle T \rangle}{\partial z} \right) + \frac{Q}{\langle \rho \rangle c_p},$$

where T is the temperature, $T' = T - \langle T \rangle$, c_p is the specific heat at constant pressure, and Q is the zonal diabatic heating.

For zonal averages, the change in the eastward direction has been averaged out and the changes in the northward and vertical directions remain. So the *continuity equation* becomes

$$\frac{\partial \langle \rho \rangle \langle v_2 \rangle}{\partial y} + \frac{\partial \langle \rho \rangle \langle v_3 \rangle}{\partial z} = 0.$$

For two-dimensional SDMs, the eddy momentum flux $\langle v_1' v_2' \rangle$ also can drive the meridional circulations. Its vertically and latitudinally varying distribution is used to estimate zonal wind, meridional wind, vertical wind, and temperature.

Eddy transport can be used to determine the equator-to-pole temperature gradient and the vertically zonal wind field. Since baroclinic waves are driven by the meridional temperature gradient, the eddy heat flux is given by

$$\langle v_2' T' \rangle = -K_T \frac{\partial \langle T \rangle}{\partial y}$$

and the eddy momentum flux is given by

$$\langle v_1' v_2' \rangle = -K_m \frac{\partial \langle v_1 \rangle}{\partial y},$$

where K_T and K_m are empirical coefficients for temperature and momentum.

14.4 EARTH SYSTEM MODELS

Various ESMs are used to study the climate system and its natural variability, and simulate the interaction between the physical climate and the biosphere, and the chemical constituents of the atmosphere and ocean. ESMs include processes, impacts, and complete feedback cycles. The most important use of ESMs is to study how Earth's climate might respond to increasing concentrations of CO_2 in the atmosphere. Other important uses of ESMs include studies of El Ninõ and the meridional overturning circulation.

ESMs are based on an atmospheric circulation model coupled with an oceanic circulation model, with representations of land, sea ice and glacier dynamics.

14.4.1 Atmospheric Models

Atmospheric models are built from fundamental conservation laws governing the physical behavior of the atmosphere (see Chapter 11). The simplest atmospheric models involve a minimum number of physical components and are described by mathematical equations that can be solved analytically. These models provide basic physical intuition. The simplest atmospheric models are also called *toy models*. The intermediate atmospheric models involve a small number of physical components, but the corresponding mathematical equations are solved numerically by computers. These models do not give accurate simulations of actual atmospheric behavior. The complex atmospheric models involve a large number of physical processes, and the corresponding mathematical equations are solved numerically by supercomputers. These models provide accurate simulations of actual atmospheric behavior. Such models are called *general circulation models*.

The Navier-Stokes equation describing the atmospheric dynamics processes is a partial differential equation involving partial time derivatives and partial spatial derivatives. All popular atmospheric models solve the Navier-Stokes equation numerically.

One numerical method is the finite-difference method. The partial time derivatives and the partial spatial derivatives in the Navier-Stokes equation can be approximated by finite differences. For example, the partial time derivative of the eastward velocity can be approximated by finite difference from the Taylor expansion with error of order $(\Delta t)^2$:

$$\frac{\partial v_1}{\partial t} \approx \frac{v_1(x, y, z, t + \Delta t) - v_1(x, y, z, t)}{\Delta t},$$

where Δt is the time interval. The partial spatial derivative of the eastward velocity can be approximated by the central difference:

$$\frac{\partial v_1}{\partial x} \approx \frac{v_1(x + \Delta x, y, z, t) - v_1(x - \Delta x, y, z, t)}{2\Delta x},$$

where Δx is the spacing in the eastward direction. In a similar way, other partial derivatives of the velocity and the pressure gradient can be represented.

Another numerical method is the spectral method. Fourier's theorem in Chapter 1 states that any periodic signal can be expanded as a Fourier series which is a summation of sine and cosine waves. Thus, the periodic signal can be approximated by partial sums of its Fourier series. For example, the eastward velocity v_1 can be approximated by partial sums of its Fourier series, i.e.,

$$v_1 \approx a_0 + \sum_{n=1}^{N} \left(a_n \cos \frac{2n\pi x}{L} + b_n \sin \frac{2n\pi x}{L} \right),$$

where L is the length of the latitude circle and the coefficients are as follows:

$$a_0 = \frac{2}{L} \int_{-L/2}^{L/2} v_1 \, dx,$$

$$a_n = \frac{2}{L} \int_{-L/2}^{L/2} v_1 \cos \frac{2n\pi x}{L} \, dx \quad (n = 1, 2, \ldots, N),$$

$$b_n = \frac{2}{L} \int_{-L/2}^{L/2} v_1 \sin \frac{2n\pi x}{L} \, dx \quad (n = 1, 2, \ldots, N).$$

The real number N is called the *truncation limit*. So the partial spatial derivative of the eastward velocity with respect to x is

$$\frac{\partial v_1}{\partial x} \approx \sum_{n=1}^{N} \frac{2n\pi}{L} \left(b_n \cos \frac{2n\pi x}{L} - a_n \sin \frac{2n\pi x}{L} \right).$$

14.4.2 Oceanic Models

Oceanic models are divided into two classes: mechanistic models and simulation models. Mechanistic models are simplified models used for studying processes, so their outputs are easier to interpret than those from more complex models. Many different types of simplified models have been developed, including models for describing planetary waves, the interaction of the flow with seafloor features, or the response of the upper ocean to the wind. Simulation models are used for calculating realistic circulation of oceanic regions. The models are often very complex because all important processes are included.

The *Bryan-Cox model* is the first simulation model that calculates realistic circulation of oceanic regions (Bryan, 1969; Cox, 1975). This model is the foundation for most current ocean global circulation models. The Bryan-Cox model calculates the three-dimensional flow in the ocean using the continuity equation and the Navier-Stokes equation with hydrostatic and Boussinesq approximations as well as a simplified equation of state. Such a model is also called a *primitive equation model* because it uses the most basic equations of motion. The Bryan-Cox model uses large horizontal and vertical viscosity and diffusion to eliminate turbulent eddies of diameters smaller than about 500 km. It also has complex coastlines, smoothed seafloor features, and a rigid lid. The Bryan-Cox model is used to predict how climate changes are determined by changes in the natural factors that control climate, such as ocean and atmospheric currents and temperature.

Currently, the Bryan-Cox model is evolving into many models providing impressive views of the global ocean circulation. They include the influence of heat and water fluxes, eddy dynamics, and the meridional overturning circulation (Semtner, 1995).

The Geophysical Fluid Dynamics Laboratory Modular Ocean Model (MOM) is the most widely used model growing out the original Bryan-Cox model. The model consists of a large set of modules that can be configured to run on many different computers to model many different aspects of the circulation. The source code is open and free. This model is widely used for studying the ocean's circulation over a wide range of space and time scales (Pacanowski and Griffies, 1999). It can also be coupled to atmospheric models. The latest version is MOM 5.1.0, released in March 2014.

The Semtner-Chervin global model is the first high-resolution global eddy-admitting model derived from the Bryan-Cox model (Semtner and Chervin, 1988). This model has a simple eddy viscosity which varies with scale. The Parallel Ocean Program (POP) is the latest version of the Semtner-Chervin global model (Barnier et al., 1995). Like the Bryan-Cox model, POP solves the primitive fluid equations on a sphere under the hydrostatic and Boussinesq approximations and uses depth as the vertical coordinate. In the horizontal, POP supports any generalized orthogonal grid. Because POP is a public code, many improvements to its physical parameterizations have resulted from external collaborations with other model groups, e.g., POP is the ocean component of the Community Earth System Model (CESM).

14.4.3   Land Surface Models

The land surface includes forests, grasslands, lakes, marshes, agricultural areas, and seasonal/perennial snow cover. Modeling and understanding the response of terrestrial ecosystems to changing environmental conditions and land use change is one of primary goals of climate mitigation policy. Land surface models compute the energy, water, and carbon balance at the land surface. They have become more credible owing to continuous improvements in the representation of land surface processes.

The Budyko model is the first land surface model and was introduced in 1969. It is also called the *bucket model*. This model has some maximum depth, usually termed by modelers *field capacity*. The bucket fills when precipitation exceeds evaporation and when it is full, excess water runs off. The bucket model has been demonstrated to be inadequate when the host model includes a diurnal cycle.

Currently, the Community Land Model (CLM) is the most popular land surface model and consists of biogeophysics, hydrologic cycle, biogeochemistry (carbon, nitrogen, dust, volatile organic compounds), and dynamic vegetation. CLM is a collaborative project between scientists in the Terrestrial Sciences Section and the Climate and Global Dynamics Division at the National Center for Atmospheric Research and the CESM Land Model and Biogeochemistry working groups. The latest CLM version is CLM4.5, which was released in June 2013. It runs with a half-hourly time step and has been extensively used to evaluate and predict the net carbon uptake and loss from terrestrial biomes, particularly forests.

14.4.4   Sea Ice Models

Sea ice controls the exchange of heat and freshwater between the atmosphere and the ocean in polar regions. As the ice cover is stretched, leads and polynyas form, allowing more energy transfer from the ocean. As the ice cover is compressed, ridges form, thickening the ice and changing the surface roughness and modifying heat transport. As the dominant component of the summertime surface energy balance in polar regions is solar radiation, it is essential that the large-scale surface albedo must be parameterized correctly in sea ice modeling.

The sea ice models lagged behind other components of ESMs in the twentieth century. Any sea ice model must consist of thermodynamics and dynamics. The earliest models of sea ice dealt with only thermodynamic processes. Thermodynamic models use forcing data from the atmosphere and ocean, such as ocean temperature, snowfall rate, and air temperature, to predict a growth rate for the ice. Semtner's three-layer sea ice model is one of the simplest sea ice models (Semtner, 1976). This model predicts two ice temperatures and a snow temperature. The dynamics of sea ice is influenced by the wind and ocean currents and internal stresses in the sea ice cover. The two most widely used techniques for solving the sea-ice momentum equations are the viscoplastic and elastic-viscoplastic approaches. The Flato-Hibler sea ice model simplifies the viscoplastic approach by treating sea ice as a cavitating fluid, while the Hunke-Dukowicz sea ice model treats sea ice as an elastic-viscoplastic material.

14.5 COUPLED MODEL INTERCOMPARISON PROJECT

In order to collect outputs from ESMs, CMIP was established in 1995 by a working group on coupled models of the world climate research program. In 2008, 20 Earth system modeling groups introduced a new set of model experiments which was called the fifth phase of CMIP (CMIP5). The aim of CMIP5 is to evaluate how realistic the models are in simulating the recent past, provide projections of future climate change in the near term (out to about 2035) and long term (out to 2100 and beyond), and understand some factors responsible for differences in model projections, including quantifying some key feedbacks involving clouds and the carbon cycle. All of the CMIP5 model outputs can now be freely downloaded through any one the following:

- http://pcmdi9.llnl.gov/;
- http://esgf-index1.ceda.ac.uk;
- http://esgf-data.dkrz.de;
- http://esg2.nci.org.au.

In the CMIP5 framework, the experiments are grouped into core set, tier 1 and tier 2 experiments. The tier 1 experiments examine specific aspects of climate model forcing, response, and processes, and the tier 2 experiments go deeper into those aspects. The main experiments in CMIP5 include

- a preindustrial control run;
- a historical run (1850-2005);
- a future projection (2006-2300) forced by representative concentration pathway RCP4.5, RCP8.5, or RCP2.6;
- a benchmark 1% per year increase in CO_2 level (to quadrupling);
- quadrupling of CO_2 level abruptly, then holding it fixed;
- climatological sea surface temperatures and sea ice imposed from the preindustrial control run, quadrupling of CO_2 level imposed, or aerosols specified from year 2000 of the historical run;
- zonally uniform sea surface temperatures imposed on an ocean-covered Earth;
- a historical simulation but with natural forcing only;
- a historical simulation but with greenhouse gas forcing only;
- a historical simulation but with other individual forcing agents or combinations of forcings:
- hindcasts but without volcanoes;
- natural forcing for 850-1850;
- Last Glacial Maximum conditions;
- a decadal forecast with a Pinatubo-like eruption in 2010.

Currently, a lot of research is being carried out using CMIP5 model outputs, e.g., Bellenger et al. (2014) examined the El Ninõ Southern Oscillation (ENSO) representation in CMIP3 and CMIP5. Compared with CMIP3, the CMIP5 multimodel ensemble displays an encouraging 30% reduction of the pervasive cold bias in the western Pacific, but no quantum leap in ENSO performance compared. The too large diversity in CMIP3 ENSO amplitude is, however, reduced by a factor of two in CMIP5, and the ENSO life cycle (location of surface temperature anomalies, seasonal phase locking) is modestly improved.

14.6 GEOENGINEERING MODEL INTERCOMPARISON PROJECT

Reducing fossil fuel burning by using energy-saving and emission-reduction technologies in industry and agriculture is clearly the most direct strategy to combat the ongoing change in the global climate. Negotiations on carbon dioxide emission reduction have largely failed because of the lack of international trust and the unwillingness of most governments to pursue anything except blind short-term self-interest. The Kyoto Protocol and subsequent emission negotiations have been obstructed repeatedly. In response, some scientists have proposed using geoengineering or climate engineering to artificially cool the planet (Royal Society, 2009). Geoengineering is the intentional large-scale manipulation of the environment to reduce undesired anthropogenic climate change (Keith, 2000). The main attraction of geoengineering lies in schemes that offer low-energy costs and short lead times for technical implementation. These geoengineering schemes would act rapidly to lower temperatures, with significant decreases occurring within 1-2 years (Bala, 2009), and would maybe produce side effects at the same time (Moriarty and Honnery, 2010). Prolonged geoengineering would curb sea level rise, which is arguably the greatest climate risk since 150 million people live within 1 m of high tide. Moderate geoengineering options could constrain sea level rise to about 50 cm above 2000 levels in the RCP3PD and RCP4.5 future climate scenarios, but only aggressive geoengineering similarly constrains the RCP8.5 future climate scenario (Moore et al., 2010). Importantly once started, geoengineering must be maintained for a very long period. Otherwise, when it is terminated, the climate will revert rapidly to maintain a global energy balance. If greenhouse gas concentrations continue to rise, then unprecedented and highly damaging rapid climate change will occur (the so-called termination shock; Jones et al., 2013a).

Various international treaties may limit some geoengineering experiments in the real world—although it is not clear how this would work in practice (Royal Society, 2011). The technical risks and uncertainties of geoengineering the climate are huge. The costs and benefits of geoengineering are likely to differ widely spatially over the planet, with some countries and regions gaining considerably, while others may be faced with a worse set of circumstances than would be the case without geoengineering. Although some features of geoengineering strategies may be testable on small scales or in the laboratory, since we have only one actual Earth, for the moment almost all tests of global geoengineering must be done using ESMs. A suite of standardized climate modeling experiments are being performed by 12 mainstream Earth system modeling groups—the GeoMIP (Kravitz et al., 2011).

The first two experiments in GeoMIP are related to an albedo geoengineering proposal which is to position sun shields in space to reflect the solar radiation:

• G1. The experiment is started from the preindustrial climate control run. An instantaneous quadrupling of CO_2 concentration from preindustrial levels is balanced by a reduction in the solar constant (this is equivalent to increasing albedo in the real world) and the experiment is run for 50 years to allow many medium-term feedbacks to occur.

• G2. The experiment is started from the preindustrial climate control run. The positive radiative forcing of an increase in CO_2 concentration of 1% per year is balanced by a decrease in the solar constant until year 50, then the geoengineering is switched off and the experiment run with just greenhouse gas forcing for a further 20 years.

Until now, 12 Earth system modeling groups, such as CESM, HadCM3, CanESM2, CSIRO Mk3L, GISS-E2-R, NorESM1-M, BNU-ESM, and MIROC-ESM, have participated in GeoMIP and have submitted the corresponding experiment results on G1/G2.

G1 is a completely artificial experiment and cannot be interpreted as a realistic geoengineering scheme, so the results from G1 are designed to discover the main impacts of balancing long-wave greenhouse radiative forcing with short-wave reductions and may help to interpret the results of more "realistic" geoengineering experiments. Under the G1 scenario, Kravitz et al. (2013a) showed that the global temperatures are well constrained to preindustrial levels, although the polar regions are relatively warmer by approximately 0.8 °C, while the tropics are relatively cooler by approximately 0.3 °C. Tilmes et al. (2013) showed that a global decrease in precipitation of 0.12 mm/day (4.9%) over land and 0.14 mm/day (4.5%) over the ocean can be expected. Moore et al. (2014) showed that for the Arctic region, G1 returns Arctic sea ice concentrations and extent to preindustrial levels, with the intermodel spread of seasonal ice extent being much greater than the difference in the ensemble means of preindustrial and G1 levels. Regional differences in concentration across the Arctic amount to 20% and the overall ice thickness and mass flux are greatly reduced.

Compared with G1, G2 is a relatively realistic geoengineering experiment. Jones et al. (2013a) focused on the impact of the sudden termination of geoengineering after 50 years of offsetting a 1% per year increase in CO_2 concentration and found that significant climate change would rapidly ensue on the termination of geoengineering, with temperature, precipitation, and sea-ice cover very likely changing considerably faster than would be experienced under the influence of rising greenhouse gas concentrations in the absence of geoengineering.

G3 and G4 experiments in GeoMIP are used to evaluate an aerosol geoengineering proposal which balances radiative forcing from greenhouse gases with reduced short-wave forcing by stratospheric aerosol injection. The details of the experiments are as follows:

- G3 assumes an RCP4.5 scenario. Sulfate aerosols are injected at the beginning in 2020 to balance the anthropogenic forcing and attempt to keep the net forcing constant (at 2020 levels) at the top of the atmosphere.
- G4 assumes an RCP4.5 scenario. Starting in 2020, stratospheric aerosols are injected at a rate of $5\,Tg\,SO_2$ per year to reduce global average temperature to about 1980 values.

Five ESMs—BNU-ESM, GISS-E2-R, HadGEM2-ES, MIROC-ESM, and MIROC-ESM-CHEM—have been used to run G3 and G4 experiments. By analyzing these ESM outputs, Berdahl et al. (2014) indicated that stratospheric geoengineering is successful at producing some global annual average temperature cooling. During the geoengineering period from 2020 to 2070, the global mean rate of warming in RCP4.5 from 2020 to 2070 is 0.03 K per year, while it is 0.02 K per year for G4 and 0.01 K per year for G3. In Arctic regions, summer temperature warming for RCP4.5 is 0.04 K per year, while it is 0.03 K per year and 0.01 K per year for G4 and G3, respectively. But neither G3 nor G4 is capable of retaining 2020 September sea ice extents throughout the entire geoengineering period (Berdahl et al., 2014).

Scientific discussion and research on geoengineering is today far more acceptable than it was just a few years ago. The Fourth Assessment report (2007) of the Intergovernmental Panel on Climate Change of the United Nations did not consider geoengineering worth more than a passing mention, while the Fifth Assessment Report (2013) included several sections on geoengineering and discussed it in the final paragraph of the summary by policymakers. The general public seems to be against geoengineering at present. Few of the population want to believe a future where the alternatives are between catastrophic climate change and the myriad risks associated with global geoengineering, and even fewer want to acknowledge that their lifestyle will lead them to this choice. But given the lack of political will for serious mitigation, it appears increasingly likely that actually those are the only choices available. Although geoengineering proposals can act rapidly to mitigate climate change, with significant global mean temperature decreases, unwanted side effects, such as diminished rainfall, would certainly also occur alongside the intended effect. Importantly, once started, geoengineering must be maintained for a very long period. Otherwise, when it is terminated, the climate will revert rapidly. The drawbacks of geoengineering remain formidable, and not easily overcome. The GeoMIP provides a framework of coordinated experiments for all earth system modelling groups, eventually allowing for robustness of results to be achieved. However these experiments used on a global scale have difficulty with accurate resolution of regional and local impacts, so future research on geoengineering is expect to be done by combining earth system models with regional climate models.

PROBLEMS

14.1 In a zero-dimensional EBM, if

$$F_s = 1370 \, \text{W/m}^2, \quad \alpha = 0.3, \quad \tau_a = 0.62, \quad \sigma = 5.67 \times 10^{-8} \, \text{W/(m}^2 \, \text{K}^4),$$

find the surface temperature of Earth.

14.2 Download the UVic model from http://climate.uvic.ca/model/ and learn to run the UVic model.

14.3 Download some CMIP5 model outputs.

14.4 Compare GeoMIP with CMIP.

BIBLIOGRAPHY

Bala, G., 2009. Problems with geoengineering schemes to combat climate change. Curr. Sci. 96, 41-48.

Barnier, B., Siefridt, L., Marchesiello, P., 1995. Thermal forcing for a global ocean circulation model using a three-year climatology of ECMWF analyses. J. Mar. Syst. 6, 380-393.

Bellenger, H., Guilyardi, E., Leloup, J., Lengaigne, M., Vialard, J., 2014. ENSO representation in climate models: from CMIP3 to CMIP5. Clim. Dyn. 42, 1999-2018.

Berdahl, M., Robock, A., Ji, D., Moore, J., Jones, A., Kravitz, B., Watanabe, S., 2014. Arctic cryosphere response in the geoengineering model intercomparison project (GeoMIP) G3 and G4 scenarios. J. Geophys. Res. 119, 1308-1321.

Bryan, K., 1969. A numerical method for the study of the world ocean. J. Comput. Phys. 17, 347-376.

Budyko, M.I., 1969. The effect of solar radiation variations on the climate of the Earth. Tellus 21, 611-619.

Cox, M.D., 1975. A baroclinic model of the world ocean: preliminary results. In: Numerical Models of Ocean Circulation. National Academy of Sciences, Washington, pp. 107-120.

de Haan, B.J., Jonas, M., Klepper, O., Krabec, J., Krol, M.S., Olendrzynski, K., 1994. An atmosphere-ocean model for integrated assessment of global change. Water Air Soil Pollut. 76, 283-318.

Flato, G.M., Hibler, W.D., 1992. Modelling sea ice as a cavitating fluid. J. Phys. Oceanogr. 22, 626-651.

Hunke, E.C., Dukowicz, J.K., 1997. An elastic-viscous-plastic model for sea ice dynamics. J. Phys. Oceanogr. 27, 1849-1867.

Jones, A., Haywood, J.M., Alterskjer, K., Boucher, O., Cole, J.N.S., Curry, C.L., Irvine, P.J., Ji, D., Kravitz, B., Kristjnsson, J.E., Moore, J.C., Niemeier, U., Robock, A., Schmidt, H., Singh, B., Tilmes, S., Watanabe, S., Yoon, J.-H., 2013a. The impact of abrupt suspension of solar radiation management (termination effect) in experiment G2 of the Geoengineering Model Intercomparison Project (GeoMIP). J. Geophys. Res. 118(17), 9743-9752.

Jones, C., Williamson, P., Haywood, J., Lowe, J., Wiltshire, A., Lenton, T., et al., 2013b. LWEC Geoengineering Report. A forward look for UK research on climate impacts of geoengineering. Living With Environmental Change (LWEC).

Keith, D.W., 2000. Geoengineering the climate: history and prospect. Ann. Rev. Energy Environ. 25, 245-284.

Kravitz, B., Robock, A., Boucher, O., Schmidt, H., Taylor, K.E., Stenchikov, G., Schulz, M., 2011. The Geoengineering Model Intercomparison Project (GeoMIP). Atmos. Sci. Lett. 12, 162-167.

Kravitz, B., Caldeira, K., Boucher, O., Robock, A., Rasch, P.J., Alterskjer, K., Karam, D.B., Cole, J.N.S., Curry, C.L., Haywood, J.M., Irvine, P.J., Ji, D., Jones, A., Lunt, D.J., Kristjnsson, J.E., Moore, J.C., Niemeier, U., Ridgwell, A., Schmidt, H., Schulz, M., Singh, B., Tilmes, S., Watanabe, S., Yoon, J.-H., 2013a. Climate model response from the Geoengineering Model Intercomparison Project (GeoMIP). J. Geophys. Res. 118, 8320-8332.

Kravitz, B.K., Rasch, P.J., Forster, P.M., Andrews, T., Cole, J.N.S., Irvine, P.J., Ji, D., Kristjánsson, J.E., Moore, J.C., Muri, H., Niemeier, U., Robock, A., Singh, B., Tilmes, S., Watanabe, S., Yoon, J.-H., 2013b. An energetic perspective on hydrologic cycle changes in the Geoengineering Model Intercomparison Project (GeoMIP). J. Geophys. Res. 118, 13087-13102.

Langehaug, H.R., Geyer, F., Smedsrud, L.H., Gao, Y., 2013. Arctic sea ice decline and ice export in the CMIP5 historical simulations. Ocean Model. 71, 114-126.

Manabe, S., Wetherald, R.T., 1967. Thermal equilibrium of the atmosphere with a given distribution of relative humidity. J. Atmos. Sci. 24, 241-259.

McGuffie, K., Sellers, A.H., 2005. A Climate Modelling Primer. John Wiley & Sons, New York.

Moore, J.C., Jevrejeva, S., Grinsted, A., 2010. Efficacy of geoengineering to limit 21st century sea-level rise. Proc. Natl. Acad. Sci. USA 107, 15699-15703.

Moore, J.C., Rinke, A., Yu, X., Ji, D., Cui, X., Li, Y., Alterskjær, K., Kristjánsson, J.E., Muri, H., Boucher, O., Huneeus, N., Kravitz, B., Robock, A., Niemeier, U., Schmidt, H., Schulz, M., Tilmes, S., Watanabe, S., 2014. Arctic sea ice and atmospheric circulation under the GeoMIP G1 scenario. J. Geophys. Res. 119, 567-583.

Moriarty, P., Honnery, D., 2010. A human needs approach to reducing atmospheric carbon. Energy Policy 38, 695-700.

Pacanowski, R.C., Griffies, S.M., 1999. MOM 3.0 Manual. NOAA/Geophysical Fluid Dynamics Laboratory, Princeton, USA.

Royal Society, 2009. Geoengineering the Climate: Science, Governance and Uncertainty.

Royal Society, 2011. Solar Radiation Management: The Governance of Research.

Sanap, S.D., Ayantika, D.C., Pandithurai, G., Niranjan, K., 2014. Assessment of the aerosol distribution over Indian subcontinent in CMIP5 models. Atmos. Environ. 87, 123-137.

Sellers, W.D., 1969. A global climatic model based on the energy balance of the Earth-atmosphere system. J. Appl. Meteorol. 8, 392-400.

Semtner, A.J., 1976. A model for the thermodynamic growth of sea-ice in numerical investigations of climate. J. Phys. Oceanogr. 6, 379-389.

Semtner, A.J., 1995. Modelling ocean circulation. Science 269, 1379-1385.

Semtner, A.J., Chervin, R.M., 1988. A simulation of the global ocean with resolved eddies. J. Geophys. Res. 93.

Semtner, A.J., Chervin, R.M., 1992. Ocean general circulation from a global eddy-resolving model. J. Geophys. Res. 97, 5493-5550.

Shukla, J., Mintz, Y., 1982. Influence of land-surface evapotranspiration on the earth's climate. Science 215, 1498-1501.

Taylor, K.E., Stouffer, R.J., Meehl, G.A., 2012. An overview of CMIP5 and the experiment design. Bull. Am. Meteorol. Soc. 93, 485-498.

Tilmes, S., Fasullo, J., Lamarque, J.-F., Marsh, D.R., Mills, M., Alterskjaeer, K., Boucher, O., Cole, J.N.S., Curry, C.L., Haywood, J.H., Irvine, P.J., Ji, D., Jones, A., Karam, D.B., Kravitz, B., Kristjánsson, J.E., Moore, J.C., Muri, H.O., Niemeier, U., Rasch, P.J., Robock, A., Schmidt, H., Schulz, M., Shuting, Y., Singh, B., Watanabe, S., Yoon, J.-H., 2013. The hydrological impact of geo-engineering in the Geoengineering Model Intercomparison Project (GeoMIP). J. Geophys. Res. 118, 11036-11058.

Wigley, T.M.L., Schlesinger, M.E., 1985. Analytical solution for the effect of increasing CO_2 on global mean temperature. Nature 315, 649-652.

VEMPA, 1995. Vegetation/ecosystem modelling and analysis project (VEMAP): comparing bio-geography and biogeochemistry models in a continental scale study of terrestrial ecosystem responses to climate change and CO_2 doubling. Global Biogeochem. Cycles 9, 407-437.

Zhang, Z., 2014. Tree-Rings, a Key Ecological Indicator of Environment and Climate Change, Ecol. Ind. in press.

Zhang, Z., Moore, J.C., Huisingh D., Zhao, Y., 2014. Review of geoengineering approaches to mitigating climate change, J. Clean. Prod. in press.

Index

Printed in the United States
By Bookmasters